Crystallisation

Crystallisation

J. W. Mullin
Ph.D., D.Sc., C.Eng., F.R.I.C., F.I.Chem.E

Professor of Chemical Engineering,
University College London

CRC PRESS

A DIVISION OF
THE **CHEMICAL RUBBER** CO.
CLEVELAND, OHIO

INTERNATIONAL SCIENTIFIC SERIES

English edition first published in 1972 by
Butterworth & Co (Publishers) Ltd
88 Kingsway, London WC2B 6AB

First published in the USA by
The Chemical Rubber Company
18901 Cranwood Parkway
Cleveland, Ohio 44128

Library of Congress Catalog Card Number 72-80105

Filmset by Photoprint Plates, Rayleigh, Essex

Printed in England by Hazell, Watson & Viney Ltd., Aylesbury, Bucks

Preface to the Second Edition

Ten years have elapsed since the publication of the first edition of this book, and in that time there has been a world-wide upsurge in the attention paid and research effort devoted to the science and technology of crystallisation. The publication explosion, which has affected most branches of science, has been particularly violent in the case of crystallisation; it has been estimated that more than 10 000 relevant papers on this subject were published in 1970, compared with fewer than 2000 in 1960.

This second edition has been completely rewritten and up-dated. New material has been included in every chapter, and two new chapters have been added. The alterations to Chapters 1–4 (The Crystalline State, Solutions and Solubility, Physical and Thermal Properties and Phase Equilibria) have been relatively minor. Chapter 5 (Mechanism of Crystallisation) has been completely revised, and a companion Chapter 6 (Kinetics of Crystallisation) has been written. The new Chapters 5 and 6 provide an introduction to the science of crystallisation, and a detailed survey has been made of the latest work in these rapidly expanding research areas. Chapters 7, 8 and 9 (Crystallisation Techniques, Industrial Crystallisation Processes and Crystallisation Equipment) are devoted to the unit operation of crystallisation. Here again, substantial changes have been made to include the latest developments in crystallisation technology. Chapter 10 (Crystalliser Operation and Design) is entirely new and deals with the recently developed techniques of crystal size distribution analysis, based on the particle population balance concept, and describes progress towards the fundamental design of crystallisers. Chapter 11 (Size Classification of Crystals) is a major extension of a former chapter, which dealt almost exclusively with sieving and screening. Other classification techniques are now covered, and the subject of fluid-particle suspension is dealt with in some detail, because of its particular relevance to agitated-vessel and fluidised-bed crystallisers. A number of new Appendix tables, dealing with crystal and solution properties, have been added. The aims and objectives of the book remain more or less the same — it is written primarily for chemical engineers and process chemists. The changes and additions reflect current developments in the chemical industry, and take into account the more recent research trends.

The response to the first edition was very gratifying, and I should like to thank all those who took the trouble to write to me. Many of their useful suggestions have been incorporated in this edition. I should also like to record my indebtedness to the research students and other members of my research team, past and present, at University College London, who have contributed so much, in so many ways, to the contents of this second edition. Especial thanks are due to my colleagues Dr. Alvin W. Nienow and Dr. John Garside, for their advice and helpful comments on the new sections, and to Miss Valerie H. Potter, Mrs. Jean S. Whitehouse and Miss Rae Schupack, who undertook the arduous typing task.

Once again I am greatly indebted to my wife for her constant encouragement and invaluable help in checking and assembling the final manuscript.

London,   J. W. MULLIN
1971

Preface to the First Edition

CRYSTALLISATION must surely rank as the oldest unit operation, in the chemical engineering sense. Sodium chloride, for example, has been manufactured by this process since the dawn of civilisation. Today there are few sections of the chemical industry that do not, at some stage, utilise crystallisation as a method of production, purification or recovery of solid material. Apart from being one of the best and cheapest methods available for the production of pure solids from impure solutions, crystallisation has the additional advantage of giving an end product that has many desirable properties. Uniform crystals have good flow, handling and packaging characteristics; they also have an attractive appearance, and this latter property alone can be a very important sales factor.

The industrial applications of crystallisation are not necessarily confined to the production of pure solid substances. In recent years large-scale purification techniques have been developed for substances that are normally liquid at room temperature. The petroleum industry, for example, in which distillation has long held pride of place as the major processing operation, is turning its attention most keenly to low-temperature crystallisation as a method for the separation of 'difficult' liquid hydrocarbon mixtures.

It is rather surprising that few books, indeed none in the English language, have been devoted to a general treatment of crystallisation practice, in view of its importance and extensive industrial application. One reason for this lack of attention could easily be that crystallisation is still referred to as more of an art than a science. There is undoubtedly some truth in this old adage, as anyone who has designed and subsequently operated a crystalliser will know, but it cannot be denied that nowadays there is a considerable amount of science associated with the art.

Despite the large number of advances that have been made in recent years in crystallisation technology, there is still plenty of evidence of the reluctance to talk about crystallisation as a process divorced from considerations of the actual substance being crystallised. To some extent this state of affairs is similar to that which existed in the field of distillation some decades ago when little attempt had been made to correlate the highly specialised techniques developed, more or less independently, for the processing of such commodities as coal tar, alcohol and petroleum products. The transformation from an 'art' to a 'science' was eventually made when it came to be recognised that the key factor which unified distillation design methods lay in the equilibrium physical properties of the working systems.

There is a growing trend today towards a unified approach to crystallisation problems, but there is still some way to go before crystallisation ceases to be the Cinderella of the unit operations. More data, particularly of the applied kind, should be published. In this age of prolific outputs of technical literature such a recommendation is not made lightly, but there is a real deficiency of this type of published information. There is, at the same time, a wealth of knowledge and experience retained in the process industries, much of it empirical but none the less valuable when collected and correlated.

The object of this book is to outline the more important aspects of crystallisation theory and practice, together with some closely allied topics. The book is intended to serve process chemists and engineers, and it should prove of interest to students of chemical engineering and chemical technology. While many of the techniques and operations have been described with reference to specific processes or industries, an attempt has been made to treat the subject matter in as general a manner as possible in order to emphasise the unit operational nature of crystallisation. Particular attention has been paid to the newer and more recently developed processing methods, even where these have not as yet proved adaptable to the large-scale manufacture of crystals.

My thanks are due to the Editors of *Chemical Engineering Practice* for permission to include some of the material and many of the diagrams previously published by me in Volume 6 of their 12-volume series. I am indebted to Professor M. B. Donald, who first suggested that I should write on this subject, and to many of my colleagues, past and present, for helpful discussions in connection with this work. I would also like to take this opportunity of acknowledging my indebtedness to my wife for the valuable assistance and encouragement she gave me during the preparation of the manuscript.

London,   J. W. MULLIN
1960

Contents

APPENDIX

Nomenclature and Units

At the present time there is a world-wide movement towards the use of the metric International System of Units (SI) in science and technology. This fact has been taken into account in the present edition, and many of the data in the original text have been recalculated. Metric units are not used exclusively in all these original passages, for the simple reason that a number of the graphs, charts and tables of data, reproduced from literature sources, were expressed in Imperial or other units, and it would have been impracticable to change them at this stage. In any case, it must be acknowledged that non-metric units are still very widely used in the chemical and allied industries, and will no doubt continue in use for many years to come.

In the new and rewritten sections of this edition, metric units are used exclusively, and most of these conform strictly to the SI. In other sections where non-metric units have been retained, the SI equivalent is usually quoted alongside for easy reference. The advantages in the use of SI units are readily seen in calculations, which are generally simplified, and in the ease with which data from various sources can be compared. It is to be hoped, in scientific reporting at least, that the change to the use of SI units is rapid.

For a general account of the use of SI units reference can be made to a recent B.S.I. publication[1]. Thus recent papers[2-4] have dealt with SI units in chemical engineering. The following brief notes are given for general information and quick reference.

BASIC AND DERIVED SI UNITS

The basic SI units of mass, length and time are the kilogram (kg), metre (m) and second (s). The basic unit of thermodynamic temperature is the kelvin (K). It should be noted that the term 'degree kelvin' ($^{\circ}$K) is no longer used, but temperatures and temperature differences may also be expressed in degrees Celsius ($^{\circ}$C). The unit for the amount of substance is the mole (mol), defined as the amount of substance which contains as many elementary units as there are atoms in $0 \cdot 012$ kg of carbon-12. Chemical engineers, however, are tending to use the kilogram-mole (kmol $= 10^3$ mol) as the preferred unit.

Several of the derived SI units have special names. The unit for energy, work and quantity of heat is the joule (J). 1 J = 1 N m. The unit of power is the watt (W). $1 \text{ W} = 1 \text{ J s}^{-1}$. The unit of pressure is the pascal (Pa).

$1 \, \text{Pa} = 1 \, \text{N m}^{-2}$. Up to the present moment, however, there is no general acceptance of the pascal for expressing pressures; many workers prefer to use multiples and submultiples of the bar ($1 \, \text{bar} = 10^5 \, \text{Pa} = 10^5 \, \text{N m}^{-2} \approx 1$ atmosphere).

The standard atmosphere (760 mm Hg) is defined as $1 \cdot 0133 \times 10^5$ Pa.

The prefixes for unit multiples and submultiples are:

10^{-18}	atto	a	10^1	deca	da
10^{-15}	femto	f	10^2	hecto	h
10^{-12}	pico	p	10^3	kilo	k
10^{-9}	nano	n	10^6	mega	M
10^{-6}	micro	μ	10^9	giga	G
10^{-3}	milli	m	10^{12}	tera	T
10^{-2}	centi	c			
10^{-1}	deci	d			

and rules for using these prefixes are described elsewhere[1, 2].

Conversion factors for some common units used in chemical engineering are listed below. An asterisk (*) denotes an exact relationship.

Length	*1 in	:	25·4 mm
	*1 ft	:	0·304 8 m
	*1 yd	:	0·914 4 m
	1 mile	:	1·609 3 km
	*1 Å (ångstrom)	:	10^{-10} m
Time	*1 min	:	60 s
	*1 h	:	3·6 ks
	*1 day	:	86·4 ks
	1 year	:	31·5 Ms
Area	*1 in^2	:	645·16 mm^2
	1 ft^2	:	0·092 903 m^2
	1 yd^2	:	0·836 13 m^2
	1 acre	:	4046·9 m^2
	1 hectare	:	10 000 m^2
	1 mile^2	:	2·590 km^2
Volume	1 in^3	:	16·387 cm^3
	1 ft^3	:	0·028 32 m^3
	1 yd^3	:	0·764 53 m^3
	1 UK gal	:	4546·1 cm^3
	1 US gal	:	3785·4 cm^3
Mass	1 oz	:	28·352 g
	1 grain	:	0·064 80 g
	*1 lb	:	0·453 592 37 kg
	1 cwt	:	50·802 3 kg
	1 ton	:	1016·06 kg
Force	1 pdl	:	0·138 26 N
	1 lbf	:	4·448 2 N
	1 kgf	:	9·806 7 N

Force *continued*	1 tonf	:	9·964 0 kN
	*1 dyn	:	10^{-5} N
Temperature difference	*1 degF (degR)	:	$\frac{5}{9}$ degC (K)
Energy (work, heat)	1 ft lbf	:	1·355 8 J
	1 ft pdl	:	0·042 14 J
	*1 cal (internat. table)	:	4·186 8 J
	1 erg	:	10^{-7} J
	1 Btu	:	1·055 06 kJ
	1 chu	:	1·899 1 kJ
	1 hp h	:	2·684 5 MJ
	*1 kW h	:	3·6 MJ
	1 therm	:	105·51 MJ
	1 thermie	:	4·185 5 MJ
Calorific value (volumetric)	1 Btu/ft^3	:	37·259 kJ/m^3
	1 chu/ft^3	:	67·067 kJ/m^3
	1 kcal/ft^3	:	147·86 kJ/m^3
	1 kcal/m^3	:	4·186 8 kJ/m^3
	1 therm/ft^3	:	3·726 0 GJ/m^3
Velocity	1 ft/s	:	0·304 8 m/s
	1 ft/min	:	5·080 0 mm/s
	1 ft/h	:	84·667 µm/s
	1 mile/h	:	0·447 04 m/s
Volumetric flow	1 ft^3/s	:	0·028 316 m^3/s
	1 ft^3/h	:	7·865 8 cm^3/s
	1 UK gal/h	:	1·262 8 cm^3/s
	1 US gal/h	:	1·051 5 cm^3/s
Mass flow	1 lb/h	:	0·126 00 g/s
	1 ton/h	:	0·282 24 kg/s
Mass per unit area	1 lb/in^2	:	703·07 kg/m^2
	1 lb/ft^2	:	4·882 4 kg/m^2
	1 ton/sq mile	:	392·30 kg/km^2
Density	1 lb/in^3	:	27·680 g/cm^3
	1 lb/ft^3	:	16·019 kg/m^3
	1 lb/UK gal	:	99·776 kg/m^3
	1 lb/US gal	:	119·83 kg/m^3
Pressure	1 lbf/in^2	:	6·894 8 kN/m^2
	1 tonf/in^2	:	15·444 MN/m^2
	1 lbf/ft^2	:	47·880 N/m^2
	1 kgf/m^2	:	9·806 7 N/m^2
	*1 standard atm	:	101·325 kN/m^2
	*1 at (1 kgf/cm^2)	:	98·066 5 kN/m^2
	*1 bar	:	10^5 N/m^2

Pressure *continued*	1 ft water	:	2·989 1 kN/m^2
	1 in water	:	249·09 N/m^2
	1 inHg	:	3·386 4 kN/m^2
	1 mmHg (1 torr)	:	133·32 N/m^2
Power (heat flow)	1 hp (British)	:	745·70 W
	1 hp (metric)	:	735·50 W
	1 erg/s	:	10^{-7} W
	1 ft lbf/s	:	1·355 8 W
	1 Btu/h	:	0·293 08 W
	1 Btu/s	:	1·055 1 kW
	1 chu/h	:	0·527 54 W
	1 chu/s	:	1·899 1 kW
	1 kcal/h	:	1·163 0 kW
	1 ton of refrigeration	:	3516·9 W
Moment of inertia	1 lb ft^2	:	0·042 140 kg m^2
Momentum	1 lb ft/s	:	0·138 26 kg m/s
Angular momentum	1 lb ft^2/s	:	0·042 140 kg m^2/s
Viscosity, dynamic	*1 poise (1g/cm s)	:	0·1 N s/m^2 (0·1 kg/m s)
	1 lb/ft h	:	0·413 38 mN s/m^2
	1 lb/ft s	:	1·488 2 N s/m^2
Viscosity, kinematic	*1 stokes (1 cm^2/s)	:	10^{-4} m^2/s
	1 ft^2/h	:	0·258 06 cm^2/s
Surface energy (surface tension)	1 erg/cm^2 (1 dyn/cm)	: :	10^{-3} J/m^2 (10^{-3} N/m)
Surface per unit volume	1 ft^2/ft^3	:	3·280 8 m^2/m^3
Surface per unit mass	1 ft^2/lb	:	0·204 82 m^2/kg
Mass flux density	1 lb/h ft^2	:	1·356 2 g/s m^2
Heat flux density	1 Btu/h ft^2	:	3·154 6 W/m^2
	*1 kcal/h m^2	:	1·163 W/m^2
Heat transfer coefficient	1 Btu/h ft^2 °F	:	5·678 4 W/m^2 K
Specific enthalpy (latent heat, etc.)	*1 Btu/lb	:	2·326 kJ/kg
Heat capacity (specific heat)	*1 Btu/lb °F	:	4·186 8 kJ/kg K
Thermal conductivity	1 Btu/h ft °F	:	1·730 7 W/m K
	1 kcal/h m °C	:	1·163 W/m K

The values of some common physical constants in SI units include:

Avagadro number, N_A	$6 \cdot 023 \times 10^{-23}$ mol^{-1}
Boltzmann constant, k	$1 \cdot 3805 \times 10^{-23}$ J/K
Planck constant, h	$6 \cdot 626 \times 10^{-34}$ J s
Stefan–Boltzmann constant, σ	$5 \cdot 6697 \times 10^{-8}$ W/m^2 K^4
Standard temperature and pressure (s.t.p.)	273·15 K and $1 \cdot 013 \times 10^5$ N/m^2
Volume of 1 kmol of ideal gas at s.t.p.	22·41 m^3
Gravitational acceleration	9·807 m/s
Universal gas constant, R	8·314 J/mol K

REFERENCES

1. *The Use of SI Units*, P.D. 5686: 1969. London; British Standards Institution
2. MULLIN, J. W., 'SI units in chemical engineering', *Chem. Engr, Lond.,* **1967,** 176
3. LEES, F. P., 'SI unit conversion table for chemical engineering', *Chem. Engr, Lond.,* **1968,** 341
4. MULLIN, J. W., 'Recent developments in the change-over to the International System of Units (S.I.)', *Chem. Engr, Lond.,* **1971,** 352

1

The Crystalline State

THE three general states of matter—gaseous, liquid and solid—represent very different degrees of atomic or molecular mobility. In the gaseous state, the molecules are in constant, vigorous and random motion; a mass of gas takes the shape of its container, is readily compressed and exhibits a low viscosity. In the liquid state, random molecular motion is much more restricted. The volume occupied by a liquid is limited; a liquid only takes the shape of the occupied part of its container, and its free surface is flat, except in those regions where it comes into contact with the container walls. A liquid exhibits a much higher viscosity than a gas and is less easily compressed. In the solid state, molecular motion is confined to an oscillation about a fixed position, and the rigid structure generally resists compression very strongly; in fact it will often fracture when subjected to a deforming force.

Some substances, such as wax, pitch and glass, which possess the outward appearance of being in the solid state, yield and flow under pressure, and they are sometimes regarded as highly viscous liquids. Solids may be crystalline or amorphous, and the crystalline state differs from the amorphous state in the regular arrangement of the constituent molecules, atoms or ions into some fixed and rigid pattern known as a lattice. Actually, many of the substances that were once considered to be amorphous have now been shown, by X-ray analysis, to exhibit some degree of regular molecular arrangement, but the term 'crystalline' is most frequently used to indicate a high degree of internal regularity, resulting in the development of definite external crystal faces.

As molecular motion in a gas or liquid is free and random, the physical properties of these fluids are the same no matter in what direction they are measured. In other words, they are *isotropic*. True amorphous solids, because of the random arrangement of their constituent molecules, are also isotropic. Most crystals, however, are *anisotropic*; their mechanical, electrical, magnetic and optical properties can vary according to the direction in which they are measured. Crystals belonging to the cubic system are the exception to this rule; their highly symmetrical internal arrangement renders them isotropic. Anisotropy is most readily detected by refractive index measurements, and the striking phenomenon of double refraction exhibited by a clear crystal of Iceland spar (calcite) is probably the best-known example.

1

LIQUID CRYSTALS

Before considering the type of crystal with which everyone is familiar, namely the solid crystalline body, it is worth while mentioning a state of matter which possesses the flow properties of a liquid yet exhibits some of the properties of the crystalline state.

Although liquids are usually isotropic, some 200 cases are known of substances that exhibit anisotropy in the liquid state at temperatures just above their melting point. These liquids bear the unfortunate name 'liquid crystals'; the term is inapt because the word 'crystal' implies the existence of a rigid space lattice. Lattice formation is not possible in the liquid state, but some form of molecular orientation can occur with certain types of molecules under certain conditions. Accordingly, the name 'anisotropic liquid' is preferred to 'liquid crystal'. The name 'mesomorphic state' was proposed by Friedel (1922) to indicate that anisotropic liquids are intermediate between the true liquid and crystalline solid states.

Among the better-known examples of anisotropic liquids are *p*-azoxyphenetole, *p*-azoxyanisole, cholesteryl benzoate, ammonium oleate and sodium stearate. These substances exhibit a sharp melting point, but they melt to form a turbid liquid. On further heating, the liquid suddenly becomes clear at some fixed temperature. On cooling, the reverse processes occur at the same temperatures as before. It is in the turbid liquid stage that

(a)   (b)

Figure 1.1. Isotropic and anisotropic liquids. (a) Isotropic: molecules in random arrangement; (b) anisotropic: molecules aligned into swarms

anisotropy is exhibited. The changes in physical state occurring with change in temperature for the case of *p*-azoxyphenetole are:

$$\text{solid} \underset{}{\overset{137°C}{\rightleftharpoons}} \text{turbid liquid} \underset{}{\overset{167°C}{\rightleftharpoons}} \text{clear liquid}$$

solid	turbid liquid	clear liquid
(anisotropic)	(anisotropic, mesomorphic)	(isotropic)

The simplest representation of the phenomenon is given by Bose's swarm theory, according to which molecules orientate into a number of groups in parallel formation (*Figure 1.1*). In many respects this is rather similar to the behaviour of a large number of logs floating down a river. Substances that can exist in the mesomorphic state are usually organic compounds, often aromatic, with elongated molecules. Properties such as double refraction and the production of interference colours in polarised light are attributed to the scattering of light at the boundaries of these swarms.

The mesomorphic state is conveniently divided into two main classes. The *smectic* (soap-like) state is characterised by an oily nature, and the flow of such liquids occurs by a gliding movement of thin layers over one another. Liquids in the *nematic* (thread-like) state flow like normal viscous liquids, but mobile threads can often be observed within the liquid layer. A third class, in which strong optical activity is exhibited, is known as the *cholesteric* state; some workers regard this state as a special case of the nematic. The name arises from the fact that cholesteryl compounds form the majority of known examples.

For further information on this subject, reference should be made to the original literature (see Bibliography).

CRYSTALLINE SOLIDS

The true solid crystal comprises a rigid lattice of molecules, atoms or ions, the locations of which are characteristic of the substance. The regularity of the internal structure of this solid body results in the crystal having a characteristic shape; smooth surfaces or faces develop as a crystal grows, and the planes of these faces are parallel to atomic planes in the lattice. Very rarely, however, do any two crystals of a given substance look identical; in fact, any two given crystals often look completely different in both size and external shape. In a way this is not very surprising, as many crystals, especially the natural minerals, have grown under different conditions. Few natural crystals have grown 'free'; most have grown under some restraint resulting in stunted growth in one direction and exaggerated growth in another.

This state of affairs prevented the general classification of crystals for centuries. The first advance in the science of crystallography came when Steno (1669) observed a unique property of all quartz crystals. He found that the angle between any two given faces on a quartz crystal was constant, irrespective of the relative sizes of these faces. This fact was confirmed later by other workers, and the Law of Constant Interfacial Angles was proposed by Haüy (1784): the angles between corresponding faces of all crystals of a given substance are constant. They may vary in size, and the development of the various faces (the crystal habit) may differ considerably, but the interfacial angles do not vary; they are characteristic of the substance.

Crystal angles are measured with an instrument known as a goniometer. The simple contact goniometer (*Figure 1.2*), consisting of an arm pivoted on a protractor, can only be used on moderately large crystals, and precision greater than $\pm 0.5°$ is rarely possible. This type of instrument is seldom used nowadays. The reflecting goniometer (*Figure 1.3*), which was developed by Wollaston (1809), is a more versatile and accurate apparatus. A crystal is mounted at the centre of a graduated turntable, a beam of light from an illuminated slit being reflected from one face of the crystal. The reflection is observed in a telescope and read on the graduated scale. The turntable is then rotated until the reflection from the next face of the crystal is observed in the telescope, and a second reading is taken from the scale. The difference α between the two readings is the angle between the normals to the two faces, and the interfacial angle is therefore $(180-\alpha)°$.

CRYSTAL SYMMETRY

Many of the geometric shapes that appear in the crystalline state are readily recognised as being to some degree symmetrical, and this fact can be used

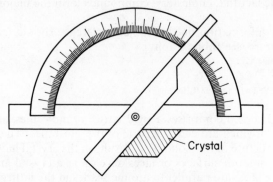

Figure 1.2. Simple contact goniometer

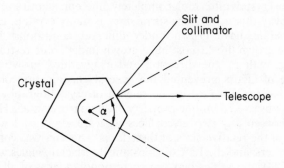

Figure 1.3. Reflecting goniometer

as a means of crystal classification. The three simple elements of symmetry which can be considered are:

1. Symmetry about a point (a *centre* of symmetry)
2. Symmetry about a line (an *axis* of symmetry)
3. Symmetry about a plane (a *plane* of symmetry)

It must be remembered, however, that while some crystals may possess a centre and several different axes and planes of symmetry, others may have no element of symmetry at all.

A crystal possesses a centre of symmetry when every point on the surface of the crystal has an identical point on the opposite side of the centre, equidistant from it. A perfect cube is a good example of a body having a centre of symmetry (at its mass centre).

If a crystal is rotated through 360° about any given axis, it obviously returns to its original position. If, however, the crystal *appears* to have reached its original position more than once during its complete rotation, the chosen axis is an axis of symmetry. If the crystal has to be rotated through 180° (360/2) before coming into coincidence with its original position, the axis is one of twofold symmetry (called a diad axis). If it has to be rotated through 120° (360/3), 90° (360/4) or 60° (360/6) the axes are of threefold symmetry (triad axis), fourfold symmetry (tetrad axis) and sixfold symmetry (hexad axis), respectively. These are the only axes of symmetry possible in the crystalline state.

A cube, for instance, has 13 axes of symmetry: 6 diad axes through opposite edges, 4 triad axes through opposite corners and 3 tetrad axes

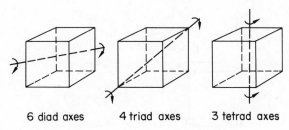

6 diad axes   4 triad axes   3 tetrad axes

Figure 1.4. The 13 axes of symmetry in a cube

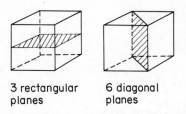

3 rectangular   6 diagonal
planes   planes

Figure 1.5. The 9 planes of symmetry in a cube

through opposite faces. One each of these axes of symmetry is shown in *Figure 1.4*.

The third simple type is symmetry about a plane. A plane of symmetry bisects a solid object in such a manner that one half becomes the mirror image of the other half in the given plane. This type of symmetry is quite common and is often the only type exhibited by a crystal. A cube has 9 planes of symmetry: 3 rectangular planes each parallel to two faces, and 6 diagonal planes passing through opposite edges, as shown in *Figure 1.5*.

It can be seen, therefore, that the cube is a highly symmetrical body, as it possesses 23 elements of symmetry (a centre, 9 planes and 13 axes). An octahedron also has the same 23 elements of symmetry; so, despite the difference in outward appearance, there is a definite crystallographic relationship between these two forms. *Figure 1.6* indicates the passage from the cubic (hexahedral) to the octahedral form, and vice versa, by a progressive and symmetrical removal of the corners. The intermediate solid forms shown (truncated cube, truncated octahedron and cubo-octahedron) are three of the 13 Archimedean semi-regular solids which are called *combination forms*, i.e. combinations of a cube and an octahedron. Crystals exhibiting combination forms are commonly encountered. The tetrahedron is also related to the cube and octahedron; in fact these three forms belong to the five regular solids of geometry. The other two (the regular dodecahedron and icosahedron) do not occur in the crystalline state. The rhombic dodecahedron, however, is frequently found, particularly in crystals of garnet. *Table 1.1*

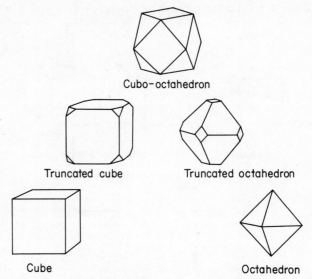

Cubo–octahedron

Truncated cube   Truncated octahedron

Cube   Octahedron

Figure 1.6. Combination forms of cube and octahedron

Table 1.1. PROPERTIES OF SOME REGULAR AND SEMI-REGULAR FORMS
FOUND IN THE CRYSTALLINE STATE

Form	Faces	Edges	Corners	Edges at a corner	Elements of symmetry		
					Centre	Planes	Axes
Regular solids							
Tetrahedron	4	6	4	3	*No*	6	7
Hexahedron (cube)	6	12	8	3	*Yes*	9	13
Octahedron	8	12	6	4	*Yes*	9	13
Semi-regular solids							
Truncated cube	14	36	24	3	*Yes*	9	13
Truncated octahedron	14	36	24	3	*Yes*	9	13
Cubo-octahedron	14	24	12	4	*Yes*	9	13

lists the properties of the six regular and semi-regular forms most often encountered in crystals. The Euler relationship is useful for calculating the number of faces, edges and corners of any polyhedron:

$$E = F + C - 2$$

This relationship states that the number of edges is two less than the sum of the number of faces and corners.

A fourth element of symmetry which is exhibited by some crystals is known by the names 'compound, or alternating, symmetry', or symmetry about a 'rotation–reflection axis' or 'axis of rotatory inversion'. This type of symmetry obtains when one crystal face can be related to another by performing two operations: (a) rotation about an axis, and (b) reflection in a plane at right angles to the axis, or inversion about the centre. *Figure 1.7*

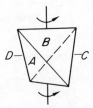

Figure 1.7. An axis of compound symmetry

illustrates the case of a tetrahedron, where the four faces are marked *A, B, C* and *D*. Face *A* can be transformed into face *B* after rotation through 90°, followed by an inversion. This procedure can be repeated four times, so the chosen axis is a compound axis of fourfold symmetry.

CRYSTAL SYSTEMS

There are only 32 possible combinations of the above-mentioned elements of symmetry, including the asymmetric state (no elements of symmetry), and these are called the 32 *point groups* or *classes*. All but one or two of these classes have been observed in crystalline bodies. For convenience these 32 classes are grouped into seven *systems*, which are known by the following names: regular (5 possible classes), tetragonal (7), orthorhombic (3), monoclinic (3), triclinic (2), trigonal (5) and hexagonal (7).

The first six of these systems can be described with reference to three axes, *x, y* and *z*. The *z* axis is vertical, and the *x* axis is directed from front to back and the *y* axis from right to left, as shown in *Figure 1.8a*. The angle between the axes *y* and *z* is denoted by α, that between *x* and *z* by β, and that between *x* and *y* by γ. Four axes are required to describe the hexagonal system; the *z* axis is vertical and perpendicular to the other three axes (*x, y* and *u*), which are coplanar and inclined at 60° (or 120°) to one another, as shown in *Figure 1.8b*. Some workers prefer to describe the trigonal system with reference to four axes. Descriptions of the seven crystal systems, together with some of the other names occasionally employed, are given in *Table 1.2*.

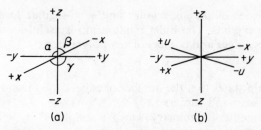

Figure 1.8. Crystallographic axes for describing the seven crystal systems: (a) *three axes* $\hat{yz} = \alpha$; $\hat{xz} = \beta$; $\hat{xy} = \gamma$; (b) *four axes (hexagonal system)* $xy = \hat{yu} = \hat{ux} = 60°$ $(120°)$

Table 1.2. THE SEVEN CRYSTAL SYSTEMS

System	Other names	Angles between axes	Length of axes	Examples
Regular	Cubic Octahedral Isometric Tesseral	$\alpha = \beta = \gamma = 90°$	$x = y = z$	Sodium chloride Potassium chloride Alums Diamond
Tetragonal	Pyramidal Quadratic	$\alpha = \beta = \gamma = 90°$	$x = y \neq z$	Rutile Zircon Nickel sulphate
Orthorhombic	Rhombic Prismatic Isoclinic Trimetric	$\alpha = \beta = \gamma = 90°$	$x \neq y \neq z$	Potassium permanganate Silver nitrate Iodine α-Sulphur
Monoclinic	Monosymmetric Clinorhombic Oblique	$\alpha = \beta = 90° \neq \gamma$	$x \neq y \neq z$	Potassium chlorate Sucrose Oxalic acid β-Sulphur
Triclinic	Anorthic Asymmetric	$\alpha \neq \beta \neq \gamma \neq 90°$	$x \neq y \neq z$	Potassium dichromate Copper sulphate
Trigonal	Rhombo- hedral	$\alpha = \beta = \gamma \neq 90°$	$x = y = z$	Sodium nitrate Ruby Sapphire
Hexagonal	None	z axis is perpendicular to the x, y and u axes, which are inclined at $60°$	$x = y = u \neq z$	Silver iodide Graphite Water (ice) Potassium nitrate

For the regular, tetragonal and orthorhombic systems, the three axes x, y and z are mutually perpendicular. The systems differ in the relative lengths of these axes: in the regular system they are all equal; in the orthorhombic system they are all unequal; and in the tetragonal system two are equal and the third is different. The three axes are all unequal in the monoclinic and triclinic systems; in the former, two of the angles are 90° and one angle is different, and in the latter all three angles are unequal and none is equal to 90°. Sometimes the limitation 'not equal to 30°, 60° or 90°' is also applied to the triclinic system. In the trigonal system three equal axes intersect at equal angles, but the angles are not 90°. The hexagonal system is described with reference to four axes. The axis of sixfold symmetry (hexad axis) is usually chosen as the z axis, and the other three equal-length axes, located in a plane at 90° to the z axis, intersect one another at 60° (or 120°).

Each crystal system contains several classes that exhibit only a partial symmetry; for instance, only one-half or one-quarter of the maximum number of faces permitted by the symmetry may have been developed. The

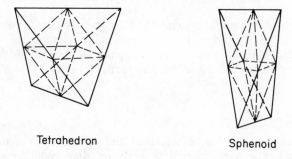

Tetrahedron   Sphenoid

Figure 1.9. Hemihedral forms of the octahedron and tetragonal bipyramid

holohedral class is that which has the maximum number of similar faces, i.e. possesses the highest degree of symmetry. In the *hemihedral* class only half this number of faces have been developed, and in the *tetartohedral* class only one-quarter have been developed. For example, the regular tetrahedron (4 faces) is the hemihedral form of the holohedral octahedron (8 faces) and the wedge-shaped sphenoid is the hemihedral form of the tetragonal bipyramid (*Figure 1.9*).

It has been mentioned above that crystals exhibiting combination forms are often encountered. The simplest forms of any crystal system are the prism and the pyramid. The cube, for instance, is the prism form of the regular system and the octahedron is the pyramidal form, and some combinations of these two forms have been indicated in *Figure 1.6*. Two simple combination forms in the tetragonal system are shown in *Figure 1.10*. *Figures 1.10a* and *b* are the tetragonal prism and bipyramid, respectively. *Figure 1.10c* shows a tetragonal prism that is terminated by two tetragonal pyramids, and *Figure 1.10d* the combination of two different tetragonal bipyramids. It frequently happens that a crystal develops a group of faces which intersect to form a series of parallel edges; such a set of faces is said to constitute a *zone*. In *Figure 1.10b*, for instance, the four prism faces make a zone.

The crystal system favoured by a substance is to some extent dependent on the atomic or molecular complexity of the substance. More than 80 per cent of the crystalline elements and very simple inorganic compounds belong

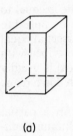  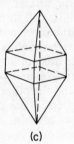 

(a)   (b)   (c)   (d)

Figure 1.10. Simple combination forms in the tetragonal system: (a) tetragonal prism; (b) tetragonal bipyramid; (c) combination of prism and bipyramid; (d) combination of two bipyramids

to the regular and hexagonal systems. As the constituent molecules become more complex, the orthorhombic and monoclinic systems are favoured; about 80 per cent of the known crystalline organic substances and 60 per cent of the natural minerals belong to these systems.

MILLER INDICES

All the faces of a crystal can be described and numbered in terms of their axial intercepts. The axes referred to here are the crystallographic axes (usually three), which are chosen arbitrarily; one or more of these axes may be axes of symmetry or parallel to them, but three convenient crystal edges can be used if desired. It is best if the three axes are mutually perpendicular, but this cannot always be arranged. On the other hand, some crystals require four axes for indexing purposes.

If, for example, three crystallographic axes have been decided upon, a plane that is inclined to all three axes is chosen as the standard or *parametral plane*. It is sometimes possible to choose one of the crystal faces to act as the parametral plane. The intercepts X, Y and Z of this plane on the axes x, y and z are called parameters a, b and c. The ratios of the parameters $a:b$ and $b:c$ are called the axial ratios, and by convention the values of the parameters are reduced so that the value of b is unity.

W. H. Miller (1839) suggested that each face of a crystal could be represented by the indices h, k and l, defined by

$$h = \frac{a}{X}, \quad k = \frac{b}{Y} \quad \text{and} \quad l = \frac{c}{Z}$$

For the parametral plane, the axial intercepts X, Y and Z are the parameters a, b and c, so the indices h, k and l are a/a, b/b and c/c, i.e. 1, 1 and 1. This is usually written (111). The indices for the other faces of the crystal are calculated from the values of their respective intercepts X, Y and Z, and these intercepts can always be represented by ma, nb and pc, where m, n and p are small whole numbers or infinity (Haüy's Law of Rational Intercepts).

The procedure for allotting face indices is indicated in *Figure 1.11*, where equal divisions are made on the x, y and z axes. The parametral plane ABC, with axial intercepts of $OA = a$, $OB = b$ and $OC = c$, respectively, is indexed (111) as described above. Plane DEF has axial intercepts $X = OD = 2a$, $Y = OE = 3b$ and $Z = OF = 3c$; so the indices for this face can be calculated as

$$h = a/X = a/2a = \tfrac{1}{2}$$
$$k = b/Y = b/3b = \tfrac{1}{3}$$
$$l = c/Z = c/3c = \tfrac{1}{3}$$

Hence $h:k:l = \tfrac{1}{2}:\tfrac{1}{3}:\tfrac{1}{3}$, and multiplying through by six, $h:k:l = 3:2:2$. Face DEF, therefore, is indexed (322). Similarly, face DFG, which has axial intercepts of $X = 2a$, $Y = -2b$ and $Z = 3c$, gives $h:k:l = \tfrac{1}{2}:-\tfrac{1}{2}:\tfrac{1}{3} = 3:-3:2$ or ($3\bar{3}2$). Thus the Miller indices of a face are inversely proportional to its axial intercepts.

The generally accepted notation for Miller indices is that the (hkl) represents a crystal face or lattice plane, while $\{hkl\}$ represents a crystallographic form comprising all faces that can be derived from hkl by symmetry operations of the crystal.

Figure 1.12 shows two simple crystals belonging to the regular system. As there is no inclined face in the cube, no face can be chosen as the parametral plane (111). The intercepts Y and Z of face A on the axes y and z are at infinity, so the indices h, k and l for this face will be a/a, b/∞ and c/∞, or (100). Similarly, faces B and C are designated (010) and (001), respectively.

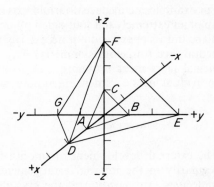

Figure 1.11. Intercepts of planes on the crystallographic axes

For the octahedron, face A is chosen arbitrarily as the parametral plane, so it is designated (111). As the crystal belongs to the regular system, the axial intercepts made by the other faces are all equal in magnitude, but not in sign, to the parametral intercepts a, b and c. For instance, the intercept of face B on the z axis is negative, so this face is designated ($11\bar{1}$). Similarly, face C is designated ($1\bar{1}1$), and the unmarked D face is ($1\bar{1}\bar{1}$).

Figure 1.13 shows some geometrical figures representing the seven crystal systems, and *Figure 1.14* indicates a few characteristic forms exhibited by crystals of some common substances.

Occasionally, after careful goniometric measurement, crystals may be found to exhibit plane surfaces which appear to be crystallographic planes, being symmetrical in accordance with the symmetry of the crystal, but which cannot be described by simple indices. These are called *vicinal faces*. A simple method for determining the existence of these faces is to observe the reflection of a spot of light on the face: four spot reflections, for example, would indicate four vicinal faces.

The number of vicinal faces corresponds to the symmetry of the face, and this property may often be used as an aid to the classification of the crystal.

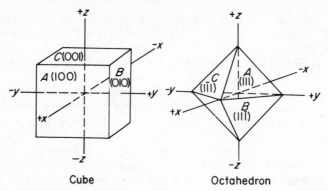

Cube   Octahedron

Figure 1.12. Two simple crystals belonging to the regular system, showing the use of Miller indices

For example, a cube face (fourfold axis of symmetry) may appear to be made up of an extremely flat four-sided pyramid with its base being the true (100) plane but its apex need not necessarily be at the centre of the face. An octahedral face (threefold symmetry) may show a three-sided pyramid. These vicinal faces most probably arise from the mode of layer growth on the individual faces commencing at point sources (see Chapters 5 and 6).

SPACE LATTICES

The external development of smooth faces on a crystal arises from some regularity in the internal arrangement of the constituent ions, atoms or molecules. Any account of the crystalline state, therefore, should include some reference to the internal structure of crystals. It is beyond the scope of this book to deal in any detail with this large topic, but a brief description will be given of the concept of the space lattice. For further information reference should be made to the specialised works listed in the Bibliography.

It is well known that some crystals can be split by cleavage into smaller crystals which bear a distinct resemblance in shape to the parent body. While there is clearly a mechanical limit to the number of times that this

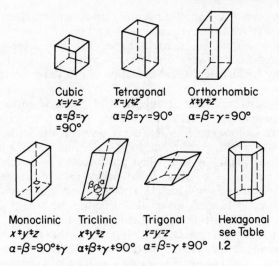

Cubic
$x=y=z$
$\alpha=\beta=\gamma$
$=90°$

Tetragonal
$x=y\ne z$
$\alpha=\beta=\gamma=90°$

Orthorhombic
$x\ne y\ne z$
$\alpha=\beta=\gamma=90°$

Monoclinic
$x\ne y\ne z$
$\alpha=\beta=90°\ne\gamma$

Triclinic
$x\ne y\ne z$
$\alpha\ne\beta\ne\gamma\ne90°$

Trigonal
$x=y=z$
$\alpha=\beta=\gamma\ne90°$

Hexagonal
see Table
1.2

Figure 1.13. *The seven crystal systems*

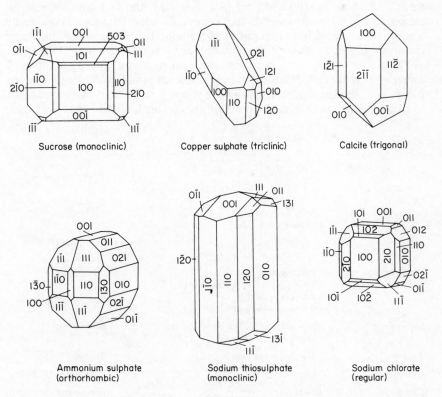

Sucrose (monoclinic)

Copper sulphate (triclinic)

Calcite (trigonal)

Ammonium sulphate
(orthorhombic)

Sodium thiosulphate
(monoclinic)

Sodium chlorate
(regular)

Figure 1.14. *Some characteristic crystal forms*

process can be repeated, eighteenth century investigators, Hooke and Haüy in particular, were led to the conclusion that all crystals are built up from a large number of minute units, each shaped like the larger crystal. This hypothesis constituted a very important step forward in the science of crystallography because its logical extension led to the modern concept of the space lattice.

A space lattice is a regular arrangement of points in three dimensions, each point representing a structural unit, e.g. an atom or a molecule. The whole structure is homogeneous, i.e. every point in the lattice has an environment identical with every other point's. For instance, if a line is drawn between any two points, it will, when produced in both directions, pass through other points in the lattice whose spacing is identical with that of the chosen pair. Another way in which this homogeneity can be visualised is to imagine an observer located within the structure; he would get the same view of his surroundings from any of the points in the lattice.

By geometrical reasoning, Bravais (1848) came to the conclusion that there were only 14 possible basic types of lattice that could give the above environmental identity. These 14 unit cells can be classified into seven groups based on their symmetry, and these seven groups correspond to the seven crystal systems listed in *Table 1.2*. The 14 Bravais lattices are given in *Table 1.3*. The three cubic lattices are illustrated in *Figure 1.15*; the first comprises eight elementary particles arranged at the corners of a cube, the second consists of a cubic structure with a ninth particle located at the centre of the cube, and the third of a cube with six extra particles, each located on a face of the cube.

The points in any lattice can be arranged to lie on a large number of different planes, called lattice planes, some of which will contain more points per unit area than others. The external faces of a crystal are parallel to lattice planes, and the most commonly occurring faces will be those which

Table 1.3. THE FOURTEEN BRAVAIS LATTICES

Type of symmetry	Lattice	Corresponding crystal system
Cubic	Cube Body-centred cube Face-centred cube	Regular
Tetragonal	Square prism Body-centred square prism	Tetragonal
Orthorhombic	Rectangular prism Body-centred rectangular prism Rhombic prism Body-centred rhombic prism	Orthorhombic
Monoclinic	Monoclinic parallelepiped Clinorhombic prism	Monoclinic
Triclinic	Triclinic parallelepiped	Triclinic
Rhomboidal	Rhombohedron	Trigonal
Hexagonal	Hexagonal prism	Hexagonal

correspond to planes containing a high density of points, usually referred to as a high reticular density (Law of Bravais). Cleavage also occurs along lattice planes. Bravais suggested that the surface energies, and hence the rates of growth, should be inversely proportional to the reticular densities, so that the planes of highest density will grow at the slowest rate and the low-density planes, by their high growth rate, will remove themselves.

Although there are only 14 basic lattices, interpenetration of lattices can occur in actual crystals, and it has been deduced that 230 combinations are possible which still result in the identity of environment of any given point. These combinations are the 230 space groups, which are further divided into the 32 point groups, or classes, mentioned above in connection

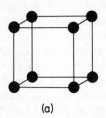
(a)

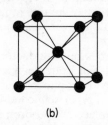
(b)

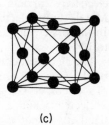
(c)

Figure 1.15. The three cubic lattices: (a) cube; (b) body-centred cube; (c) face-centred cube

with the seven crystal systems. The law of Bravais has been extended by Donnay and Harker into a more generalised form (the Bravais–Donnay–Harker Principle) by consideration of the space groups rather than the lattice types.

SOLID STATE BONDING

Four main types of crystalline solid may be specified according to the method of bonding in the solid state, viz. ionic, covalent, molecular and metallic. There are materials intermediate between these classes, but most crystalline solids can be classified as predominantly one of the basic types.

The *ionic crystals* (e.g. sodium chloride) are composed of charged ions held in place in the lattice by electrostatic forces, and separated from the oppositely charged ions by regions of negligible electron density. In *covalent crystals* (e.g. diamond) the constituent atoms do not carry effective charges; they are connected by a framework of covalent bonds, the atoms sharing their outer electrons. *Molecular crystals* (e.g. organic compounds) are composed of discrete molecules held together by weak attractive forces (e.g. π-bonds or hydrogen bonds). They are not linked by true chemical bonds.

Metallic crystals (e.g. copper) comprise ordered arrays of identical cations. The constituent atoms share their outer electrons, but these are so loosely held that they are free to move through the crystal lattice and confer 'metallic' properties on the solid. For example, ionic, covalent and molecular crystals are essentially non-conductors of electricity, because the electrons are all locked into fixed quantum states. Metals are good conductors because of the presence of mobile electrons.

Semiconducting crystals (e.g. germanium) are usually covalent solids with some ionic characteristics, although a few molecular solids (e.g. some polycyclic aromatic hydrocarbons such as anthracene) are known in which under certain conditions a small fraction of the valency electrons are free to move in the crystal. The electrical conductivity of semiconductors is electronic in nature, but it differs from that in metals. Metallic conductivity decreases when the temperature is raised, because thermal agitation exerts an impeding effect. On the other hand, the conductivity of a semiconductor increases with heating, because the number of electron–'hole' pairs, the electricity carriers in semiconductors, increases greatly with temperature. Metals have electrical resistivities in the ranges 10^{-6} to 10^{-4} ohm cm. Insulators cover the range 10^{10} to 10^{22} (diamond) and semiconductors 10^2 to 10^9.

The electrical conductivity of a semiconductor can be profoundly affected by the presence of impurities. For example, if x silicon atoms in the lattice of a silicon crystal are replaced by x phosphorus atoms, the lattice will gain x electrons and a negative (n-type) semiconductor results. On the other hand, if x silicon atoms are replaced by x boron atoms, the lattice will lose x electrons and a positive (p-type) semiconductor is formed. The impurity atoms are called 'donors' or 'acceptors' according to whether they give or take electrons to or from the lattice.

ISOMORPHISM AND POLYMORPHISM

Two or more substances that crystallise in almost identical forms are said to be *isomorphous* (Greek: 'of equal form'). This is not a contradiction of Haüy's law, because these crystals do show small, but quite definite, differences in their respective interfacial angles. Isomorphs are often chemically similar and can then be represented by similar chemical formulae; this statement is one form of Mitscherlich's Law of Isomorphism, which is now recognised only as a broad generalisation. One group of compounds which obey and illustrate Mitscherlich's law is represented by the formula $M_2'SO_4 \cdot M_2''' (SO_4)_3 \cdot 24H_2O$ (the alums), where M' represents a univalent radical (e.g. K or NH_4) and M''' represents a tervalent radical (e.g. Al, Cr or Fe). Many phosphates and arsenates, sulphates and selenates are also isomorphous.

Sometimes isomorphous substances can crystallise together out of a solution to form 'mixed crystals' or, as they are better termed, solid solutions. In such cases the composition of the homogeneous solid phase that is deposited follows no fixed pattern; it depends largely on the relative concentrations and solubilities of the substances in the original solvent. For instance, chrome alum, $K_2SO_4 \cdot Cr_2(SO_4)_3 \cdot 24H_2O$ (purple), and potash alum, $K_2SO_4 \cdot Al_2(SO_4)_3 \cdot 24H_2O$ (colourless), crystallise from their respective aqueous solutions as regular octahedra. When an aqueous solution containing both salts is crystallised, regular octahedra are again formed, but the colour of the crystals (which are now homogeneous solid solutions) can vary from almost colourless to deep purple, depending on the proportions of the two alums in the crystallising solution.

Another phenomenon often shown by isomorphs is the formation of over-growth crystals. For example, if a crystal of chrome alum (octahedral) is placed in a saturated solution of potash alum, it will grow in a regular manner such that the purple core is covered with a continuous colourless overgrowth. In a similar manner an overgrowth crystal of nickel sulphate, $NiSO_4 \cdot 7H_2O$ (green), and zinc sulphate, $ZnSO_4 \cdot 7H_2O$ (colourless), can be prepared.

There have been many 'rules' and 'tests' proposed for the phenomenon of isomorphism, but in view of the large number of known exceptions to these it is now recognised that the only general property of isomorphism is that crystals of the different substances shall show very close similarity. All the other properties, including those mentioned above, are merely confirmatory and not necessarily shown by all isomorphs.

A substance capable of crystallising into different, but chemically identical, crystalline forms is said to exhibit *polymorphism*. Dimorphous and trimor-phous substances are commonly known, e.g.

Carbon: graphite (hexagonal)
diamond (regular)

Silicon dioxide: cristobalite (regular)
tridymite (hexagonal)
quartz (trigonal)

The term *allotropy* instead of polymorphism is often used when the substance is an element.

The different crystalline forms exhibited by one substance may result from a variation in the crystallisation temperature or a change of solvent. Sulphur, for instance, crystallises in the form of orthorhombic crystals (α-S) from a carbon disulphide solution, and of monoclinic crystals (β-S) from the melt. In this particular case the two crystalline forms are interconvertible: β-sulphur cooled below $95\cdot5$ °C changes to the α form. This interconversion between two crystal forms at a definite transition temperature is called *enantiotropy* (Greek: 'change into opposite') and is accompanied by a change in volume. Ammonium nitrate (melting point $169\cdot2$ °C) exhibits four enantiotropic changes between -18 and 125 °C, as shown below:

	(I)	(II)	(III)	(IV)	(V)
liquid	$\rightleftharpoons$ cubic	$\rightleftharpoons$ trigonal	$\rightleftharpoons$ orthorhombic	$\rightleftharpoons$ orthorhombic	$\rightleftharpoons$ tetragonal
	$169\cdot6$°C	$125\cdot2$°C	$84\cdot2$°C	$32\cdot3$°C	-18°C

The transitions from forms II to III and IV to V result in volume increases; the changes from I to II and III to IV are accompanied by a decrease in volume. These volume changes frequently cause difficulty in the processing and storage of ammonium nitrate. The salt can readily burst a metal con-tainer into which it has been cast when change II to III occurs. The drying of ammonium nitrate crystals must be carried out within fixed temperature limits, e.g. 40–80 °C, otherwise the crystals can disintegrate when a transition temperature is reached.

When polymorphs are not interconvertible, the crystal forms are said to be *monotropic*; graphite and diamond are monotropic forms of carbon. The term *isopolymorphism* is used when each of the polymorphous forms of one substance is isomorphous with the respective polymorphous form of another substance. For instance, the regular and orthorhombic polymorphs of arsenious oxide, As_2O_3, are respectively isomorphous with the regular and orthorhombic polymorphs of antimony trioxide, Sb_2O_3. These two oxides are thus said to be isodimorphous.

ENANTIOMORPHISM

Two crystals of the same substance which are the mirror images of each other are said to be enantiomorphous (Greek: 'of opposite form'). These crystals have no planes of symmetry at all. Most enantiomorphous substances exhibit the property of optical activity, i.e. they are capable of rotating the plane of polarised light; one form will rotate it to the left (laevo-rotatory or L form) and the other to the right (dextro-rotatory or D form). Tartaric acid (*Figure 1.16*) and certain sugars are well-known examples of optically active substances.

Enantiomorphous substances are not necessarily optically active, but all known optically active substances are capable of being crystallised into

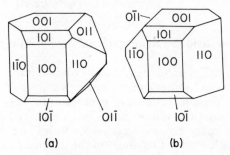

Figure 1.16. (a) Dextro- and (b) laevo-*tartaric acid crystals (monoclinic system)*

enantiomorphous forms. In some cases the solution of an optically active crystal is also optically active, which indicates that the actual molecules of the substance are enantiomorphous. In other cases solution or fusion destroys the optical activity, which indicates that enantiomorphism was confined to the crystal structure only.

Optical activity has been associated with compounds that possess one or more atoms around which different elements or groups are arranged asymmetrically, so that the molecule can exist in mirror image forms. In organic compounds the presence of an asymmetric carbon atom often favours optical activity. Tartaric acid offers a good example of this, and three possible arrangements of the tartaric acid molecule are shown in *Figure 1.17*. The (a) and (b) forms are mirror images of each other; both contain two asymmetric carbon atoms and both are optically active; one will be the D form and the other the L form. There are two asymmetric carbon atoms in formula (c) but this form (*meso*-tartaric acid) is optically inactive; the potential optical activity of one-half of the molecule is compensated by the opposite potential optical activity of the other.

The case of tartaric acid serves to illustrate another property, known as *racemism*. A mixture of crystalline D and L tartaric acids dissolved in water can, if mixed in the right proportions, produce an optically inactive solution. Crystallisation of this solution will yield crystals of optically inactive racemic

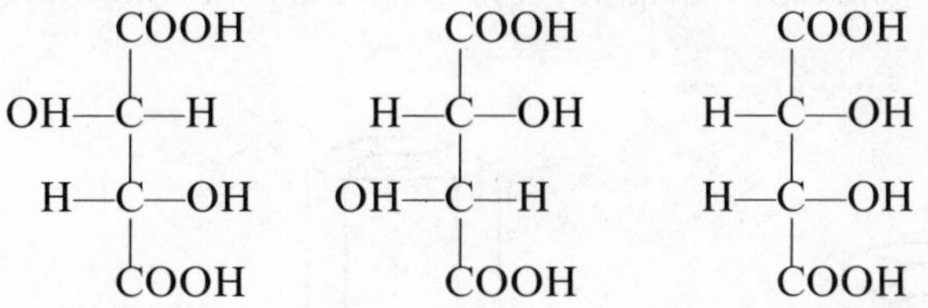

```
      COOH              COOH              COOH
       |                 |                 |
  OH—C—H            H—C—OH            H—C—OH
       |                 |                 |
  H—C—OH            OH—C—H            H—C—OH
       |                 |                 |
      COOH              COOH              COOH

       (a)               (b)               (c)
```

Figure 1.17. The tartaric acid molecule: (a) and (b) optically active forms; (c) meso-tartaric acid, optically inactive

acid which are different in form from the D and L crystals. There is, however, a difference between a racemic and a *meso-* form of a substance; the former can be resolved into D and L forms but the latter cannot.

A racemate can be resolved in a number of ways. Pasteur (1848) found that crystals of the sodium ammonium salt of racemic acid,

$$Na \cdot NH_4 \cdot C_4H_4O_6 \cdot H_2O$$

deposited from aqueous solution, consisted of two clearly different types, one being the mirror image of the other. The D and L forms were easily separated by hand picking. Bacterial attack was also shown by Pasteur to be effective in the resolution of racemic acid. *Penicillium glaucum* allowed to grow in a dilute solution of sodium ammonium racemate destroys the D form but, apart from being a rather wasteful process, the attack is not always completely selective. A racemate is best resolved by forming a salt or ester with an optically active base (usually an alkaloid) or alcohol. For example, a racemate of an acidic substance A with, say, the dextro form of an optically active base B will give

$$DLA + DB \rightarrow DA \cdot DB + LA \cdot DB$$

and the two salts $DA \cdot DB$ and $LA \cdot DB$ can be separated by fractional crystallisation.

CRYSTAL HABIT

Although crystals can be classified according to the seven general systems (*Table 1.1*), the relative sizes of the faces of a particular crystal can vary considerably. This variation is called a modification of habit. The crystals may grow more rapidly, or be stunted, in one direction; thus an elongated growth of the prismatic habit gives a needle-shaped crystal (acicular habit) and a stunted growth gives a flat plate-like crystal (tabular, platy or flaky habit). Nearly all manufactured and natural crystals are distorted to some degree, and this fact frequently leads to a misunderstanding of the term 'symmetry'. Perfect geometric symmetry is rarely observed in crystals, but crystallographic symmetry is readily detected by means of a goniometer.

Figure 1.18 shows three different habits of a crystal belonging to the hexagonal system. The centre diagram (b) shows a crystal with a predominant

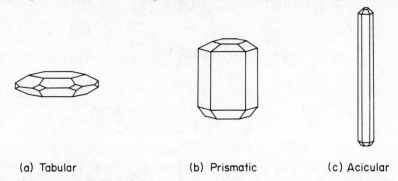

(a) Tabular   (b) Prismatic   (c) Acicular

Figure 1.18. Crystal habit illustrated on a hexagonal crystal

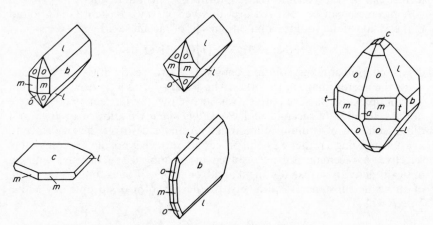

Figure 1.19. Some common habits of potassium sulphate crystals: a = {100}, b = {010}, c = {011}, l = {021}, m = {110}, o = {111}, t = {130}

prismatic habit. This combination-form crystal is terminated by hexagonal pyramids and two flat faces perpendicular to the vertical axis; these flat parallel faces cutting one axis are called *pinacoids*. A stunted growth in the vertical direction (or elongated growth in the directions of the other axes) results in a tabular crystal (a); excessively flattened crystals are usually called plates or flakes. An elongated growth in the vertical direction yields a needle or acicular crystal (c); flattened needle crystals are usually called blades. *Figure 1.19* shows some of the habits exhibited by potassium sulphate crystals grown from aqueous solution.

The relative growths of the faces of a crystal can be altered, and often controlled, by a number of factors. Rapid crystallisation, such as that produced by the sudden cooling or seeding of a supersaturated solution, often results in the formation of needle crystals; impurities in the crystallising solution may stunt the growth of a crystal in certain directions; and crystallisation from solutions of the given substance in difference solvents may result in a change of habit. The degree of supersaturation or supercooling of a solution or melt often exerts a considerable influence on the crystal habit, and so can the state of agitation of the system. These and other factors affecting the control of habit are discussed in Chapters 5 and 6.

DENDRITES

Rapid crystallisation from supercooled melts, supersaturated solutions and vapours frequently produces tree-like formations called dendrites, the growth of which is indicated in *Figure 1.20*. The main crystal stem grows quite

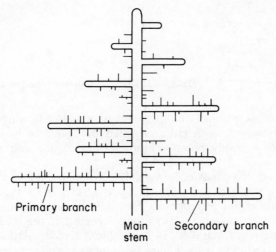

Primary branch

Main   Secondary branch
stem

Figure 1.20. Dendritic growth

rapidly in a supercooled system that has been seeded, and at a later stage primary branches grow at a slower rate out of the stem, often at right angles to it. In certain cases, small secondary branches may grow slowly out of the primaries. Eventually branching ceases and the pattern becomes filled in with crystalline material.

Most metals crystallise from the molten state in this manner, but because of the filling-in process the final crystalline mass may show little outward appearance of dendrite formation. The fascinating patterns of snow crystals are good examples of dendritic growth, and the frosting of windows often affords a visual observation of this phenomenon occurring in two dimensions.

The growth of a dendrite can be observed quite easily under a microscope by seeding a drop of a saturated solution on the slide.

Dendrites form most commonly during the early stages of crystallisation; at later stages a more normal uniform growth takes place and the pattern may be obliterated. Dendritic growth occurs quite readily in thin liquid layers, probably because of the high rate of evaporative cooling, whereas agitation tends to suppress this type of growth. Dendrite formation is favoured by substances that have high latent heats of crystallisation and low heat conductivities.

COMPOSITE CRYSTALS

Most crystalline natural minerals, and many crystals produced industrially, exhibit some form of aggregation or intergrowth, and prevention of the formation of these composite crystals is one of the problems of large-scale crystallisation. The presence of aggregates in a crystalline mass spoils the

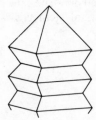

Figure 1.21. Parallel growth on a crystal of potash alum

appearance of the product and interferes with its free-flowing nature. More important, however, aggregation is often indicative of impurity because crystal clusters readily retain impure mother liquor and resist efficient washing.

Composite crystals may occur in simple symmetrical forms or in random clusters. The simplest form of aggregate results from the phenomenon known as *parallel growth*; individual forms of the same substance grow on the top of one another in such a manner that all corresponding faces and edges of the individuals are parallel. Potash alum, $K_2SO_4 \cdot Al_2(SO_4)_3 \cdot 24H_2O$, exhibits this type of growth; *Figure 1.21* shows a typical structure in which regular octahedra are piled on top of one another in a column symmetrical about the vertical axis. Parallel growth is often associated with isomorphs; for instance, parallel growths of one alum can be formed on the crystals of another, but this property is no longer regarded as an infallible test for isomorphism.

Another composite crystal frequently encountered is known as a *twin* or a *macle*; it appears to be composed of two intergrown individuals, similar in form, joined symmetrically about an axis (a twin axis) or a plane (a twin plane). A twin axis is a possible crystal edge and a twin plane is a possible crystal face. Many types of twins may be formed in simple shapes such as a V, +, L and so forth, or they may show an interpenetration giving the appearance of one individual having passed completely through the other (*Figure 1.22*). Partial interpenetration (*Figure 1.23*) can also occur. In some cases, a twin crystal may present the outward appearance of a form that possesses a higher degree of symmetry than that of the individuals, and this is known as *mimetic twinning*. A typical example of this behaviour is ortho-rhombic potassium sulphate, which can form a twin looking almost identical with a hexagonal bipyramid.

Parallel growth and twinning (or even triplet formation) are usually encountered when crystallisation has been allowed to take place in an undisturbed medium. Although twins of individuals belonging to most of the seven crystal systems are known, twinning occurs most frequently when the crystals belong to the orthorhombic or monoclinic systems. Certain impurities in the crystallising medium can cause twin formation even though an appreciable agitation is effected; this is one of the problems encountered in the commercial crystallisation of sugar.

The formation of crystal clusters, aggregates or conglomerates which

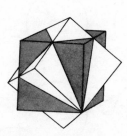

Figure 1.22. Interpenetrant twin of two cubes (e.g. fluorspar)

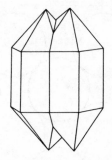

Figure 1.23. Partial interpenetrant twin (e.g. quartz)

possess no symmetrical properties is probably more frequently encountered in large-scale crystallisation than the formation of twins. Relatively little is still known about the growth of these irregular crystal masses, but among the factors that often favour their formation are poor agitation, the presence of certain impurities in the crystallising solution, seeding at high degrees of supersaturation and the presence of too many seed crystals, leading to conditions of overcrowding in the crystalliser.

IMPERFECTIONS IN CRYSTALS

Very few crystals are perfect. Indeed, in many cases they are not required to be, since lattice imperfections and other defects can confer some important chemical and mechanical properties on crystalline materials. Surface defects can also greatly influence the process of crystal growth. There are three main types of lattice imperfection: point (zero-dimensional), line (one-dimensional) and surface (two-dimensional).

Point Defects

The common point defects are indicated in *Figure 1.24*. *Vacancies* are lattice sites from which units are missing, leaving 'holes' in the structure. These units may be atoms, e.g. in metallic crystals, molecules (molecular crystals) or ions (ionic crystals). The *interstitials* are foreign atoms that occupy positions in the interstices between the matrix atoms of the crystal. In most cases the occurrence of interstitials leads to a distortion of the lattice.

More complex point defects can occur in ionic crystals. For example, a cation can leave its site and become relocated interstitially near a neighbouring cation. This combination of defects (a cation vacancy and an interstitial

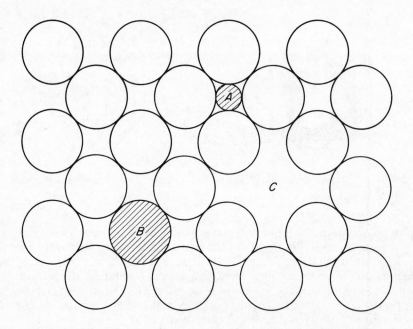

Figure 1.24. Representation of some common point defects: A, interstitial impurity; B, substitutional impurity; C, vacancy

cation) is called a *Frenkel imperfection*. A cation vacancy combined with an anion vacancy is called a *Schottky imperfection*.

A foreign atom that occupies the site of a matrix atom is called a *substitutional impurity*. Many types of semiconductor crystals contain controlled quantities of substitutional impurities. Germanium crystals, for example, can be grown containing minute quantities of aluminium (p-type semi-conductors) or phosphorus (n-type).

Line Defects

The two main types of line defect which can play an important role in the mode of crystal growth are the *edge* and *screw dislocations*. Both of these are responsible for slip or shearing in crystals. Large numbers of dislocations

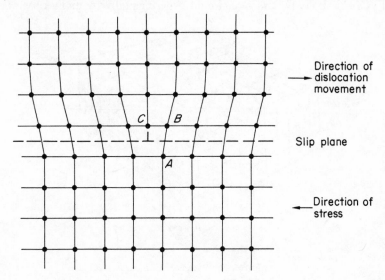

Figure 1.25. Movement of an edge dislocation through a crystal

occur in most crystals; they form readily during the growth process under the influence of surface and internal stresses.

Figure 1.25 shows in diagrammatic form the cross-sectional view of a crystal lattice in which the lower part of a vertical row of atoms is missing. The position of the dislocation is marked by the symbol ⊥; the vertical stroke of this symbol indicatès the extra plane of atoms and the horizontal stroke indicates the slip plane. The line passing through all the points ⊥, i.e. drawn vertical to the plane of the diagram, is called the edge dislocation line. In an edge dislocation, therefore, the atoms are displaced at right angles to the dislocation line.

The process of slip under the action of a shearing force may be explained as follows (see *Figure 1.25*). The application of a shear stress to a crystal causes atom *A* to move further away from atom *B* and closer to atom *C*. The bond between *A* and *B*, which is already strained, breaks and a new bond is formed between *A* and *C*. The dislocation thus moves one atomic distance to the right, and if this process is continued the dislocation will eventually reach the edge of the crystal. The direction and magnitude of slip are indicated by the *Burgers vector*, which may be one or more atomic spacings. In the above example, where the displacement is one lattice spacing, the Burgers vector is equal to 1.

A screw dislocation forms when the atoms are displaced along the dislocation line, rather than at right angles to it as in the case of the edge dislocation. *Figure 1.26* indicates this type of lattice distortion. In this example the Burgers vector is 1 (unit step height), but its magnitude may be any integral number.

Screw dislocations give rise to a particular mode of growth, first postulated by F. C. Frank (1949) and later found experimentally by many observers,

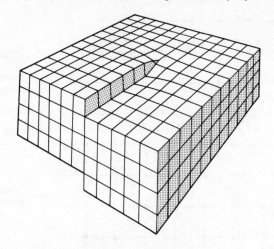

Figure 1.26. A screw dislocation

in which the attachment of growth units to the face of the dislocation results in the development of a spiral growth pattern over the crystal face (see Chapter 5).

Surface Defects

A variety of surface imperfections, or mismatch boundaries, can be produced in crystalline materials as a result of mechanical or thermal stresses or irregular growth. Grain boundaries, for example, can be created between individual crystals of different orientation in a polycrystalline aggregate.

When the degree of mismatching is small, the boundary can be considered to be composed of a line of dislocations. A low-angle *tilt boundary* is equivalent to a line of edge dislocations, and the angle of tilt is given by $\theta = b/h$ where b is the Burgers vector and h the average vertical distance between the dislocations (*Figure 1.27*). A *twist boundary* can be considered, when the degree of twist is small, as a succession of parallel screw dislocations. For a full account of this subject reference should be made to the specialised works (see Bibliography).

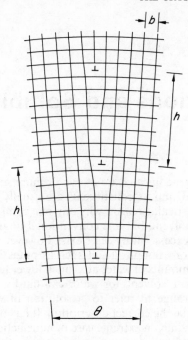

Figure 1.27. A simple tilt boundary

BIBLIOGRAPHY

BOWEN, H. J. M., *Properties of Solids and their Atomic Structures,* 1967. London; McGraw-Hill
BRAGG, W. L., *The Crystalline State,* 1933. London; Bell
BUCKLEY, H. E., *Crystal Growth,* 1952. London; Chapman and Hall
BUNN, C. W., *Chemical Crystallography,* 2nd Ed., 1961. Oxford; Clarendon Press
DONNAY, J. D. H. (Ed.), *Crystal Data: determination tables,* 2nd Ed., 1963. Washington; American
    Crystallographic Association
FLINT, Y., *Essentials of Crystallography,* 1966. Moscow; Peace Publishers
GRAY, G. W., *Molecular Structure and the Properties of Liquid Crystals,* 1962. London; Academic
    Press
HARTSHORNE, N. H. and STUART, A., *Practical Optical Crystallography,* 2nd Ed., 1969. London;
    Arnold
HIRTH, J. P. and LOETHE, J., *Theory of Dislocations,* 1968. New York; McGraw-Hill
KITAIGORODSKII, A. I., *Organic Crystal Chemistry,* 1961. New York; Consultants Bureau
PARTINGTON, J. R., *An Advanced Treatise on Physical Chemistry,* Vol. III, 'The Properties of
    Solids', 1952. London; Longmans Green
PHILLIPS, F. C., *Introduction to Crystallography,* 2nd Ed., 1956. London; Longmans Green
PORTER, M. W. and SPILLER, R. C. (Eds.), *The Barker Index of Crystals* (3 Vols.), 1951. Cam-
    bridge; Heffer
READ, W. T., *Dislocations in Crystals,* 1953. New York; McGraw-Hill
TUTTON, A. E. H., *Crystallography and Practical Crystal Measurement* (2 Vols.), 2nd Ed., 1922.
    London; Macmillan
WELLS, A. F., *Structural Inorganic Chemistry,* 3rd Ed., 1962. Oxford; Clarendon Press
WYCKOFF, R. W. G., *Crystal Structures* (5 Vols.), 2nd Ed., 1963–8. New York; Interscience
'Dislocations in Solids', *Discuss. Faraday Soc.,* No. 38 (1964)
VAN BUEREN, H. G., *Imperfections in Crystals,* 1960. New York; Interscience

2

Solutions and Solubility

A SOLUTION is a homogeneous mixture of two or more substances; it may be gaseous, liquid or solid, and its constituents are usually called solvents and solutes. There is no particular reason why any one component of a solution should be termed the solvent, but it is conventional to give this name to the component present in excess. Many cases exist, however, where considerable confusion can arise. For example, a salt such as potassium nitrate fuses in the presence of small amounts of water at a much lower temperature than the pure salt does. The term 'solvent' for water can hardly be justified in such cases. It may seem strange to refer to 'a solution of water in potassium nitrate', yet this would be the correct description. It has been suggested[1] that fusion is nothing more than an extreme case of liquefaction by solution, so it may be said that when a salt dissolves in water, the salt does, in fact, melt.

Owing to the widespread and often indiscriminate use of the word 'melt', it is difficult to give a precise definition of the term. Strictly speaking, a melt is the liquid phase of a pure substance that is solid at normal temperatures. In its general application the term also includes homogeneous liquid mixtures of two or more substances that solidify on cooling. Thus α-naphthol (m.p. 96 °C) in the liquid state is a melt. Homogeneous liquid mixtures consisting of, say, α-naphthol and β-naphthol (m.p. 122 °C), or α-naphthol, β-naphthol and naphthalene (m.p. 80 °C), would also be considered to be melts, whereas liquid mixtures containing, say, α-naphthol and benzene, or α-naphthol, β-naphthol and ethyl alcohol could be classified as solutions. It must be pointed out, however, that no rigid definition is possible; the KNO_3–H_2O system quoted above, and the many well-known cases of hydrated salts dissolving in their own water of crystallisation at elevated temperatures, would in all probability be considered to be melts.

SOLUBILITY DIAGRAMS

The composition of a solution can be expressed in a number of ways. One is to state the weight of solute present in a given volume of solution; for example, the concentration of a certain salt solution may be quoted as 20 g/l. Although this sort of expression may be convenient for the analytical laboratory, it can be most inconvenient in industrial practice, since it is necessary to know the density of the solution before the relative weights of solute and solvent can be determined. Volume measures can be rather misleading, as they are dependent upon temperature. Another method is to state the weight of solute present in a given weight of solution, but here again the relative weights of solute and solvent are not expressed precisely.

The solubility of a solute is most conveniently stated as the parts by weight per part (or 100 parts) by weight of solvent. To avoid confusion in the case of hydrated salts dissolved in water, the solute concentration should always refer to the *anhydrous* salt. No difficulty will then arise in cases where several hydrated forms can exist over the temperature range considered.

All the above methods of solubility expression can lead to the use of the term 'percentage concentration', and unless precisely defined this term can be very misleading. For instance, a 10 per cent aqueous solution of sodium sulphate could, without further definition, be taken to mean any one of the following:

10 g of Na_2SO_4 (anhyd.) in 100 g of water
10 g of Na_2SO_4 (anhyd.) in 100 g of solution
10 g of $Na_2SO_4 \cdot 10H_2O$ in 100 g of water
10 g of $Na_2SO_4 \cdot 10H_2O$ in 100 g of solution

To show how misleading this loose form of expression can be, let 10 g of Na_2SO_4 (anhyd.) in 100 g of water be the correct description of the solution concentration. This would then be equivalent to

9·1 g of Na_2SO_4 (anhyd.) in 100 g of solution
20·6 g of $Na_2SO_4 \cdot 10H_2O$ in 100 g of solution
26·0 g of $Na_2SO_4 \cdot 10H_2O$ in 100 g of water

Solubility data may also be recorded in terms of equivalent, molar or molal quantities; a normal solution (N) contains one g-equivalent of the solute per litre of solution, and a molar solution (M) contains one mole per litre. These expressions are particularly useful in laboratory practice, but both are temperature-dependent; the normality and molarity of a given solution decreases with an increase in temperature. A molal solution (m) contains one mole of solute per kg of solvent, and concentrations are often expressed in terms of molality when phase changes occur in the solute–solvent system over a given temperature range.

Concentrations expressed as moles of solute per mole of mixture are frequently used in industrial practice, especially for multicomponent liquid mixtures. The mole fraction x of a particular component in a mixture of several substances is given by

$$x_1 = \frac{m_1/M_1}{m_1/M_1 + m_2/M_2 + m_3/M_3 + \ldots} \tag{2.1}$$

where m is the mass of a particular component, and M its molecular weight. For any mixture the sum of all the mole fractions is unity. The term 'mole percentage' ($100\,x$) is also used.

The mean molecular weight, $\bar{M}$, of a mixture is given by

$$\bar{M} = \frac{m_1 + m_2 + m_3 + \ldots}{m_1/M_1 + m_2/M_2 + m_3/M_3 \ldots} \tag{2.2}$$

As a simple illustration of the use of these mass and mole compositions, take the case of an aqueous solution of ammonium chloride (mol. wt. = 53·5) containing 31 g of NH_4Cl per 100 g of solution; this can be expressed in the following ways:

(a) *Mass composition*

$$= 31 \times 100/(100-31) = 45 \text{ g } NH_4Cl/100 \text{ g water}$$

(b) *Molal composition*

$$NH_4Cl = 31/53·5 \qquad = 0·58 \text{ mol}$$
$$H_2O \quad = (100-31)/18 = \underline{3·83 \text{ mol}}$$
$$4·41 \text{ mol}$$

molality $= 0·58 \times 1000/(100-31) = 8·4$

(c) *Mole fractions*

$$NH_4Cl = 0·58/4·41 \qquad = 0·13$$
$$H_2O \quad = 3·83/4·41 \qquad = 0·87$$

In the majority of cases the solubility of a solute in a solvent increases with an increase in temperature, but there are a few well-known exceptions to this

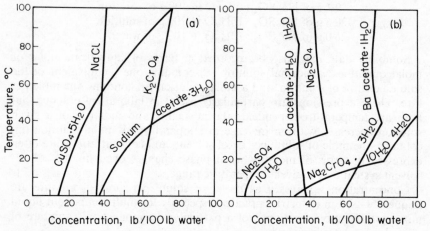

Figure 2.1. Solubility curves for some salts in water: (a) *smooth curves;* (b) *indicating occurrence of phase changes*

rule. Some typical solubility curves for various salts in water are shown in *Figure 2.1*, where all concentrations are expressed as pounds of anhydrous substance per 100 lb of water. In *Figure 2.1a* sodium chloride is a good example of a salt whose solubility increases only slightly with an increase in temperature, whereas sodium acetate shows a fairly rapid increase.

The solubility characteristics of a solute–solvent system have a considerable influence on the choice of a method of crystallisation. It would be useless, for instance, to cool a hot saturated solution of sodium chloride in the hope of depositing crystals in any quantity; cooling from, say, 90 to 20 °C would only produce about 7 lb of NaCl for every 100 lb of water present. The yield could be increased, however, by removing some of the water by evaporation, and this is what is done in practice. On the other hand, a direct cooling–crystallisation operation would be adequate for a salt such as copper sulphate; cooling from 90 to 20 °C would produce about 44 lb of $CuSO_4$ for every 100 lb of water present in the original solution. As the stable phase of copper sulphate at 20 °C is the pentahydrate, the actual crystal yield would be about 69 lb of $CuSO_4 \cdot 5H_2O$ for every 100 lb of water present initially.

Not all solubility curves are smooth, as can be seen in *Figure 2.1b*. A discontinuity in the solubility curve denotes a phase change. For example, the solid phase deposited from an aqueous solution of sodium sulphate below 32·4 °C will consist of the decahydrate, whereas the solid deposited above this temperature will consist of the anhydrous salt. The solubility curves for two different phases meet at the transition point, and a system may show a number of these points. For instance, three forms of ferrous sulphate may be deposited from aqueous solution, depending upon the temperature: $FeSO_4 \cdot 7H_2O$ up to 56 °C, $FeSO_4 \cdot 4H_2O$ from 56 to 64 °C and $FeSO_4 \cdot H_2O$ above 64 °C. In the case of sodium hydroxide no less than six hydrates and the anhydrous substance can be deposited from aqueous solution between -24 and 62 °C.

Referring back to the case of sodium sulphate (*Figure 2.1b*), it can be seen that above 32·4 °C, when the anhydrous salt is the stable form, the solubility decreases with an increase in temperature. This negative solubility effect, or inverted solubility as it is sometimes called, is also exhibited by substances such as anhydrous sodium sulphite, calcium sulphate (gypsum), calcium, barium and strontium acetates, calcium hydroxide, etc. These substances can cause trouble in certain types of crystalliser by causing a deposition of scale on the heat transfer surfaces.

The general trend of a solubility curve can be predicted from Le Chatelier's Principle, which, for the present purpose, can be stated as: when a system in equilibrium is subjected to a change in temperature or pressure, the system will adjust itself to a new equilibrium state in order to relieve the effect of the change. Most solutes dissolve in their near-saturated solutions with an absorption of heat (endothermic heat of solution); thus an increase in the temperature of these solutions results in an increase in the solubility. An inverted solubility effect occurs when the solute dissolves in its near-saturated solution with an evolution of heat (exothermic heat of solution). Strictly speaking, solubility is also a function of pressure, but the effect is quite negligible in the systems encountered in crystallisation from the liquid phase.

Many equations have been proposed for the correlation and prediction of solubility data, but none has been found to be of general applicability. In any case, an experimentally determined solubility is undoubtedly preferred to an estimated value. Nevertheless, there are two equations which are commonly used to express the influence of temperature on solubility, viz.

$$c = A + Bt + Ct^2 + \ldots \tag{2.3}$$

and

$$\log x = \frac{a}{T} + b \tag{2.4}$$

In equation (2.3), c = mass of solute per given mass of solvent, and t = temperature in degC or degF. Values of the constants A, B and C, etc., for a number of common solute–solvent systems may be found in the literature. The solubility in equation (2.4) is expressed as the mole fraction x of non-solvated solute in the solution, T is absolute temperature (K), and a and b

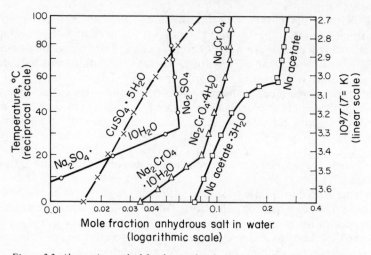

Figure 2.2. Alternative method for the graphical representation of solubility data

are constants. Equation (2.4) can be extremely useful. The graphical interpolation and extrapolation of solubility values or the estimation of transition points from the conventional solubility plots, such as those shown in Figure 2.1, can prove difficult and unreliable when only a few experimental values are available, especially when the points lie on one or more different curved lines. However, if equation (2.4) can be applied – and it does apply reasonably well to a large number of systems – these difficulties can be minimised.

Solubility data can be plotted, in accordance with equation (2.4), in the manner shown in Figure 2.2. Mole fractions x are recorded on the logarithmic abscissa and the values of $10^3/T$ (because $1/T$ in the range 273–373 K is a rather small quantity) are recorded on the right-hand linear ordinate scale. Alternatively, special log-reciprocal graph paper can be used and the temperature in degC can be plotted direct. The left-hand ordinate in Figure 2.2 is marked off on the reciprocal scale.

In Figure 2.1 the solubility of $CuSO_4$ over the temperature range 0–100 °C is represented by a smooth curve, and the solubilities of Na_2SO_4 and Na_2CrO_4 are represented by smooth curves that intersect at transition points. Several advantages of the log x versus $1/T$ plot shown in Figure 2.2

immediately become apparent. First, the data for the above three salts lie on a series of straight lines. Second, and probably not so important, the data for the highly soluble salts sodium chromate and sodium acetate can be recorded for the complete range 0–100 °C on the same graph as that used for the data of the less soluble salts. Third, the existence of transition points can be more clearly detected.

It is easier, for example, to produce the two straight lines for sodium sulphate in *Figure 2.2* to meet at 32·4 °C than it is to extend the two corresponding curves in *Figure 2.1*. The two straight lines for $CuSO_4$ intersect at about 67 °C, which indicates a phase transition at this temperature; this transition between two different crystalline forms of the pentahydrate was not detected in *Figure 2.1*. Incidentally, the transition $CuSO_4 \cdot 5H_2O \rightleftharpoons CuSO_4 \cdot 3H_2O$ occurs at 95·9 °C. Only two of the transitions for the sodium chromate system are indicated in *Figure 2.2*. There are actually three transition points in this system: $10H_2O \rightleftharpoons 6H_2O$ (19·6 °C), $6H_2O \rightleftharpoons 4H_2O$ (26·0 °C) and $4H_2O \rightleftharpoons$ anhydrous (64·8 °C).

The solubility data for sodium acetate are included in *Figure 2.2* to illustrate the fact that straight lines do not always result from this method of plotting. Curved lines are often obtained for highly soluble substances, or in regions where the temperature coefficient of solubility is high, or in cases where several hydrates can exist over a narrow range of temperature. It is possible, of course, that the curved portion of the sodium acetate line in the region of about 40–58 °C is really a series of straight lines representing hydrates other than the trihydrate, but no evidence to support this view is available.

THEORETICAL CRYSTAL YIELD

If the solubility data for a substance in a particular solvent are known, it is a simple matter to calculate the maximum yield of pure crystals that could be obtained by cooling or evaporating a given solution. The calculated yield will be a maximum, because the assumption has to be made that the final mother liquor in contact with the deposited crystals will be just saturated. Generally, some degree of supersaturation may be expected, but this cannot be estimated. The yield will refer only to the quantity of pure crystals deposited from the solution, but the actual yield of solid material may be slightly higher than that calculated, because crystal masses invariably retain some mother liquor even after filtration. When the crystals are dried they become coated with a layer of material that is frequently of a lower grade than that in the bulk of the crystals. Impure dry crystal masses produced commercially are very often the result of inadequate mother liquor removal.

Washing on a filter helps to reduce the amount of mother liquor retained by a mass of crystals, but there is always the danger of reducing the final yield by dissolution during the washing operation. If the crystals are readily soluble in the working solvent, another liquid in which the substance is relatively insoluble may be used. Alternatively, a wash consisting of a cold, near-saturated solution of the pure substance in the working solvent may be employed. The efficiency of washing depends largely on the shape and size of the crystals (see pp. 290–293).

The calculation of the yield for the case of crystallisation by cooling is quite straightforward if the initial concentration and the solubility of the substance at the lower temperature are known. The calculation can be complicated slightly if some of the solvent is lost, deliberately or accidentally, during the cooling process, or if the substance itself removes some of the solvent, e.g. by taking up water of crystallisation. All these possibilities are taken into account in the following equations, which may be used to calculate the maximum yields of pure crystals under a variety of conditions.

Let C_1 = initial solution concentration (kg anhydrous salt/kg solvent)
C_2 = final solution concentration (kg anhydrous salt/kg solvent)
W = initial weight of solvent (kg)
V = solvent lost by evaporation (kg per kg of original solvent)
R = ratio of molecular weights of hydrate and anhydrous salt
Y = crystal yield (kg)

Substance crystallises unchanged (e.g. anhydrous salt)

Total loss of solvent: $Y = WC_1$ (2.5)

No loss of solvent: $Y = W(C_1 - C_2)$ (2.6)

Partial loss of solvent: $Y = W[C_1 - C_2(1 - V)]$ (2.7)

Substance crystallises as a solvate

Total loss of free solvent: $Y = WRC_1$ (2.8)

No loss of solvent: $Y = \dfrac{WR(C_1 - C_2)}{1 - C_2(R - 1)}$ (2.9)

Partial loss of solvent: $Y = \dfrac{WR[C_1 - C_2(1 - V)]}{1 - C_2(R - 1)}$ (2.10)

Equation (2.10) can, of course, be used as the general equation for all cases.

Example

Calculate the theoretical yield of pure crystals that could be obtained from a solution containing 1000 kg of sodium sulphate (mol. wt. = 142) in 5000 kg of water by cooling to 10 °C. The solubility of sodium sulphate at 10 °C is 9·0 parts of anhydrous salt per 100 parts of water, and the deposited crystals will consist of the decahydrate (mol. wt. = 322). Assume that 2 per cent of the water will be lost by evaporation during the cooling process.

Solution

$$R = 322/142 = 2·27$$
$$C_1 = 0·2 \text{ kg Na}_2\text{SO}_4 \text{ per kg of water}$$
$$C_2 = 0·089 \text{ kg Na}_2\text{SO}_4 \text{ per kg of water}$$
$$W = 5000 \text{ kg of water}$$
$$V = 0·02 \text{ kg per kg of water present initially}$$

Substituting these values in equation (2.10):

$$Y = \frac{5000 \times 2·27[0·2 - 0·09(1 - 0·02)]}{1 - 0·09(2·27 - 1)}$$
$$\text{Yield} = 1432 \text{ kg Na}_2\text{SO}_4 \cdot 10\text{H}_2\text{O}$$

SATURATION AND SUPERSATURATION

A solution that is in equilibrium with the solid phase is said to be saturated with respect to that solid. However, it is relatively easy to prepare a solution containing more dissolved solid than that represented by saturation condition, and such a solution is said to be supersaturated. Uncontaminated

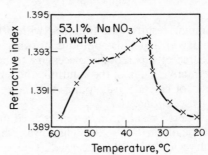

Figure 2.3. Showing the change in refractive index of an aqueous salt solution on cooling. (After H. A. MIERS[3]*)*

solutions in clean containers, cooled slowly without disturbance in a dust-free atmosphere, can readily be made to show appreciable degrees of super-saturation. The state of supersaturation is an essential feature of all crystal-lisation operations. Wilhelm Ostwald (1897) first introduced the terms 'labile' (unstable) and 'metastable' supersaturation; they refer to super-saturated solutions in which spontaneous deposition of the solid phase, in the absence of solid nuclei, will and will not occur, respectively.

MIERS[2-5] carried out extensive researches into the relationship between supersaturation and spontaneous crystallisation. He measured the refractive indices of concentrated aqueous salt solutions during a cooling process[3], and *Figure 2.3* shows one typical result: the change in refractive index of a 53·1 per cent solution of sodium nitrate in water with decrease in temperature.

A glass prism of known refractive index was immersed in the solution and the refractive index was measured by the method of total reflection within the prism with a goniometer. As the initially unsaturated solution was cooled, with stirring, from about 60 °C, the refractive index increased considerably until a few small crystals appeared in the solution. The refractive index continued to increase at a slightly slower rate and reached a maximum value at about 36 °C. At this point a copious separation of fine crystals occurred, accompanied by a sudden drop in the refractive index with no change in temperature. Further cooling reduced the refractive index to an approximately constant value at about 20 °C. A 53 per cent solution of sodium nitrate is saturated at about 50 °C, but the spontaneous deposition of crystals did not occur until the temperature of the solution had fallen to 36 °C.

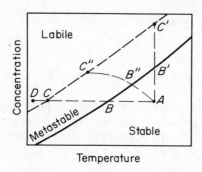

Figure 2.4. The solubility–supersolubility diagram

Similar results were obtained with other concentrations of sodium nitrate and with aqueous solutions of other salts, with and without stirring. The degree of supercooling necessary to bring about a sudden reduction of the refractive index of the solution, and simultaneously cause a copious separation of crystals, was found to be lower for an agitated solution than for a quiescent one.

Miers's results can be represented in another manner, as shown diagrammatically in *Figure 2.4*. The lower continuous line is the normal solubility curve for the salt concerned. Temperatures and concentrations at which spontaneous crystallisation occurs are represented by the upper broken curve, generally referred to as the supersolubility curve. This curve is not so well defined as the solubility curve and its position in the diagram depends on, among other things, the degree of agitation of the solution. Some workers prefer to picture the supersolubility 'curve' as a region or narrow band located in the supersaturated zone.

However, in spite of the fact that the supersolubility curve is ill-defined, there is no doubt that a region of metastability exists in the supersaturated region above the solubility curve. The diagram is therefore divided into three zones, one well-defined and the other two variable to some degree:

1. The stable (unsaturated) zone, where crystallisation is impossible.
2. The metastable (supersaturated) zone, between the solubility and supersolubility curves, where spontaneous crystallisation is improb-

able. However, if a crystal seed were placed in such a metastable solution, growth would occur on it.
3. The unstable or labile (supersaturated) zone, where spontaneous crystallisation is probable, but not inevitable.

If a solution represented by point A in *Figure 2.4* is cooled without loss of solvent (line ABC), spontaneous crystallisation cannot occur until conditions represented by point C are reached. At this point, crystallisation may be spontaneous or it may be induced by seeding, agitation or mechanical shock. Further cooling to some point D may be necessary before crystallisation can be induced, especially with very soluble substances such as sodium thiosulphate. Although the tendency to crystallise increases once the labile zone is penetrated, the solution may have become so highly viscous as to prevent crystallisation and would set to a glass.

Supersaturation can also be achieved by removing some of the solvent from the solution by evaporation. Line $AB'C'$ represents such an operation carried out at constant temperature. Penetration beyond the supersolubility curve into the labile zone rarely happens, as the surface from which evaporation takes place is usually supersaturated to a greater degree than the bulk of the solution. Crystals which appear on this surface eventually fall into the solution and seed it, often before conditions represented by point C' are reached in the bulk of the solution. In practice, a combination of cooling and evaporation is employed, and such an operation is represented by the line $AB''C''$ in *Figure 2.4*.

EXPRESSION OF SUPERSATURATION

The supersaturation, or supercooling, of a system may be expressed in a number of different ways, and considerable confusion can be caused if the basic units of concentration are not clearly defined. The temperature must also be specified.

Among the most common expressions of supersaturation are the concentration driving force, Δc, the supersaturation ratio, S, and a quantity sometimes referred to as the absolute or relative supersaturation, σ, or percentage supersaturation, 100σ. These quantities are defined by

$$\Delta c = c - c^* \tag{2.11}$$

$$S = \frac{c}{c^*} \tag{2.12}$$

$$\sigma = \frac{\Delta c}{c^*} = S - 1 \tag{2.13}$$

where c is the solution concentration, and c^* is the equilibrium saturation at the given temperature.

For one-component systems the term supercooling is used, and the corresponding terms are

$$\Delta \theta = \theta^* - \theta \quad \text{or} \quad \Delta T = T^* - T \tag{2.14}$$

$$\sigma' = \frac{\Delta\theta}{\theta^*} \quad \text{or} \quad \sigma'' = \frac{\Delta T}{T^*} \qquad (2.15)$$

where $\theta = {}^\circ C$ and $T = K$. It will be noted that $\Delta\theta = \Delta T$, but $\sigma' \neq \sigma''$.

Of the three expressions for supersaturation only Δc is dimensional, unless the composition is given in mole fractions. The magnitudes of the expressions depend on the units used to express concentration, as the following examples show.

Example 1

Potassium sulphate (mol. wt. = 174) at 20 °C. The equilibrium saturation $c^* = 109$ g of K_2SO_4/kg of water, which gives a solution density of 1·08 g/cm^3. Let the concentration of a supersaturated solution $c = 116$ g/kg, giving a solution density of 1·09 g/cm^3 at 20 °C. Then the following quantities may be calculated:

Solution composition	c	c^*	Δc	S	σ
g/kg water	116	109	7·0	1·06	0·06
g/kg solution	104	98·3	5·7	1·06	0·06
g/l solution ($= kg/m^3$)	113·3	106·1	7·2	1·07	0·07
mol/l solution ($= kmol/m^3$)	0·650	0·608	0·042	1·07	0·07
mol fraction of K_2SO_4	0·0119	0·0112	0·0007	1·06	0·06

The quantity that changes most in this example is Δc; neither S nor σ is very greatly affected. However, with very soluble substances considerable changes can occur in all expressions of supersaturation, as seen in the next example.

It is essential to quote the temperature when we express the supersaturation of a system, since the equilibrium saturation concentration is temperature dependent. In the case of potassium sulphate, for example, $S = 1·06$ means a concentration driving force $\Delta c = 7$ g/kg of water at 20 °C and 13 g/kg at 80 °C.

Example 2

Sucrose (mol. wt. = 342) at 20 °C, $c^* = 2040$ g/kg of water (density 1·33 g/cm^3). Let $c = 2450$ g/kg of water (density 1·36 g/cm^3). Then

Solution composition	c	c^*	Δc	S	σ
g/kg water	2450	2040	410	1·20	0·20
g/kg solution	710	671	39	1·06	0·06
g/l solution ($= kg/m^3$)	966	893	73	1·08	0·08
mol/l solution ($= kmol/m^3$)	2·82	2·61	0·21	1·08	0·08
mol fraction of sucrose	0·114	0·097	0·017	1·18	0·18

The situation becomes even more confused if the salt crystallises in the form of a hydrate, since solution compositions can be expressed in terms of either the anhydrous or hydrated salt.

It is not possible to say, with any firm conviction, which type of expression of solution composition or supersaturation is best; different expressions suit different circumstances. For mass balance calculations, e.g. estimation of crystal yield, concentration units such as kg of anhydrous salt/kg of solvent or kg of hydrate/kg of free solvent are generally most convenient. The advantage of the former is that it is not affected by any phase changes that might occur over the range of temperature used in the crystallisation process. The advantage of the latter is that mass balance calculations are simplified. Volumetric expressions such as kg/m^3 are useful for calculating flow rates and crystalliser volumes, but mass balances become very complicated because these units are temperature dependent and are affected by volume changes on crystallisation. For kinetic expressions there would appear to be a case for supersaturation expressed on a molar basis, but despite this the mass concentration basis is most generally used.

Conversion from one set of solubility units to another may be made through the equations

$$c_1 = \frac{c_2}{1-c_2} = \frac{c_3}{R-c_3} = \frac{c_4}{R+c_4(R-1)} \qquad (2.16)$$

where c_1 = kg anhydrous salt/kg water; c_2 = kg anhydrous salt/kg solution; c_3 = kg hydrate/kg solution; c_4 = kg hydrate/kg water; and R = ratio of molecular weights of hydrate and anhydrous salt.

MEASUREMENT OF SUPERSATURATION

If the concentration of a solution can be measured at a given temperature, and the corresponding equilibrium saturation concentration is known, then it is a simple matter to calculate the supersaturation (equations 2.11–2.13). Just as there are many methods of measuring concentration, so there are also many ways of measuring supersaturation, but not all of these are readily applicable to industrial crystallisation practice.

Solution concentration may be determined directly by analysis, or indirectly by measuring some property of the system that is a sensitive function of concentration. Properties frequently chosen for this purpose include density, viscosity, refractive index and electrical conductivity, and these can often be measured with high precision, especially if the actual measurement is made under carefully controlled conditions in the laboratory. However, for the operation of a crystalliser under laboratory or pilot plant conditions the demand is usually for an *in situ* method, preferably one capable of continuous operation. In these circumstances problems may arise from the temperature dependence of the property being measured. Nevertheless several of the above properties can be measured, more or less continuously, with sufficient accuracy for supersaturation determination. These techniques are described in Chapter 3. In general, density and refractive index are the least temperature-sensitive properties.

For industrial crystallisation, where temperature and feedstock conditions cannot be controlled with precision, very crude methods of supersaturation measurement may have to be employed. The most common method consists of a mass balance coupled with feedstock, exit liquor and crystal production rates taken over a suitable period, e.g. several hours, to smooth out fluctuations.

The supersaturation of a solution may be determined from a knowledge of its boiling point elevation by applying the principles of Dühring's rule (the boiling point of a solution is a linear function of the boiling point of the pure solvent at the same pressure. HOLVEN[6] reported that, over the range

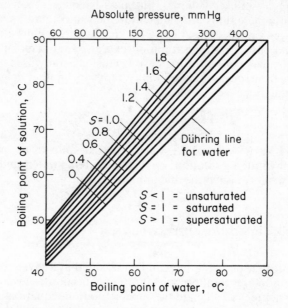

Figure 2.5. Duhring-type plot showing constant supersaturation lines (range $S = 0$ to $1·8$) for aqueous solutions of sucrose. (After A. L. HOLVEN[6])

of pressures normally encountered in sugar boiling practice, a plot of the solution boiling point for a given degree of supersaturation against the boiling point of water at the same absolute pressure yielded a straight line (*Figure 2.5*). He used these findings to develop an automatic method for recording and controlling the degree of supersaturation in sugar crystallisers.

DETERMINATION OF SOLUBILITY

The technique of solubility determination is basically simple, but it demands a high degree of practical skill and accuracy. A solution saturated with a given solid solute, as defined above, is one that is in equilibrium with the solid phase, and it is the achievement of true equilibrium which presents one of the biggest experimental difficulties. Prolonged and intimate contact is required between excess solid and solution at a constant temperature, usually for several hours. In certain cases contact for many days or even weeks may be necessary; viscous solutions and systems at relatively low temperatures require long contact times. Both the solute and solvent should be of the highest purity possible, unless for some reason solubility data for commercial substances are required, and the solute particles should be small enough to facilitate fairly rapid dissolution. Very small particles, however, cannot settle readily in viscous solutions, and it is generally found that a close-sieved sample around 30-mesh size (500 μm) is suitable for most purposes.

Once equilibrium has been attained, the mixture is allowed to stand for an hour or more, still at constant temperature, to enable any finely dispersed solids to settle. A sample of the clear supernatant liquid is carefully withdrawn by means of a warmed pipette, and a weighed quantity (*not* a measured volume) of the sample is analysed. One point on the solubility curve can then be plotted. The analysis of solutions can be effected by conventional volumetric or gravimetric techniques, but in certain cases colorimetric methods or measurements of refractive index, density, conductivity, etc., may yield a more rapid, and sometimes a more reliable, result.

There is, however, one further step to be taken in order to complete the information, and that is to determine the composition of the solid phase that was in equilibrium with the˙solution at the given temperature. The stable phase can change appreciably over quite short ranges of temperature, especially in hydrated systems. For example, in the determination of the solubility of sodium carbonate in water over the temperature range 0 to 100 °C it would be found that the stable solid phase is $Na_2CO_3 \cdot 10H_2O$ up to 32·0 °C, $Na_2CO_3 \cdot 7H_2O$ between 32·0 and 35·4 °C, and $Na_2CO_3 \cdot H_2O$ above 35·4 °C. In order to determine the composition of the solid phase, a sample is taken from the container in which the equilibrium state was achieved and dried carefully at the temperature of the experiment, before the analysis is commenced. The 'wet residue' and 'synthetic complex' methods described in Chapter 4 provide alternative procedures for the determination of solid phase composition.

A check can be made on a solubility determination at a given temperature by approaching equilibrium in two different ways: (a) from the unsaturated state and (b) from the supersaturated state. In the first method a quantity of solid, in excess of the amount required to saturate the solvent at the given temperature, is added to the solvent and the two are agitated until equilibrium is reached. In the second method the same quantities of solute and solvent are mixed, but the system is then heated above the required temperature, if solubility increases with temperature, so that most of the solid is dissolved. The solution is agitated for a long period at the required temperature and the excess solid is deposited. If the two solubility determinations agree, it can be reasonably assumed that the result represents the true equilibrium saturation concentration at the given temperature.

Constant temperature control during the experimental procedure is essential, although its limits vary according to the system under investigation and the required precision of the solubility. Far greater care has to be taken when the solubility changes appreciably with a change in temperature. In the determination of the solubility of, say, sodium chloride in water at 30 °C (*Figure 2.1a*), a variation of $\pm 0.5°$ in the experimental temperature would allow for a potential precision of about 0·1 per cent in the solubility, but the same temperature variation would only allow for a potential precision of about 5 per cent in the case of sodium sulphate at 30 °C (*Figure 2.1b*). It should be quite obvious that the thermometer used in the thermostat must be accurately calibrated with reference to a standard thermometer.

A simple apparatus for the determination of solubility consists of a thermostatically controlled water-bath (or oil-bath for temperatures above about 80 °C), a stoppered flask or test-tube of about 50 cm^3 capacity and a device for shaking the flask while it is immersed in the bath. Alternatively, the flask can be fitted with a stirrer; a very compact apparatus operated on this principle has been described in detail by PURDON and SLATER[7].

The sampling tube, or pipette, can be left standing in a stoppered tube immersed in the bath so that it attains the same temperature as the solution

(a)   (b)   (c)

Figure 2.6. Appearance of the illuminated slit during saturation temperature measurements: (a) unsaturated; (b) saturated; (c) supersaturated solutions in contact with the flat crystal face. (After L. A. DAUNCEY *and* J. E. STILL[9])

under investigation. A filter-tip consisting of a sintered glass end-piece, or more simply a piece of glass wool, can be provided on the sampling tube to prevent finely dispersed solids being drawn up with the sample of solution. Some sampling pipettes, e.g. the Landolt type, can be used as the weighing vessel. ZIMMERMAN[8] has made an extensive review of the literature up to 1950 on the subject of experimental solubility determination.

An apparatus for the direct and rapid measurement of the saturation temperature of solutions has been devised by DAUNCEY and STILL[9]. This new technique is based on an optical effect caused by the slight change in concentration, and therefore in refractive index, occurring in a layer of solution immediately in contact with a crystal that is either growing or dissolving. The saturation cell is a small Perspex container fitted with a stirrer (or the solution may be passed continuously through it), a calibrated thermometer and a holder for a medium-sized crystal. The cell is placed in a thermostatically controlled water-bath also made of Perspex. A beam of light from an optical slit is directed on to an edge of the crystal, and the appearance of the slit when viewed from behind the crystal will take the form of one of the three sketches shown in *Figure 2.6*. The light is bent into an obtuse angle when the solution is unsaturated (a), and into an acute angle when it is supersaturated (c). As soon as it is determined that the solution near the crystal face is unsaturated or supersaturated, the temperature controller is adjusted until view (b) is obtained

Dauncy and Still worked with ethylenediamine hydrogen tartrate and ammonium dihydrogen phosphate, and reported that points on the solubility curves could be plotted at the rate of 8–10 per hour. WISE and NICHOLSON[10] have adapted this method and applied it successfully in the determination of sugar solubilities. KELLY[11] has reviewed these and other techniques in some detail.

Another simple technique for determining saturation temperatures is as follows. About 40 cm^3 of the solution to be tested is cooled, e.g. under the tap, in a 50 cm^3 conical flask until the solute begins to crystallise out. The flask is then fitted with a magnetic stirrer and a thermometer graduated in 0·1 °C, and the temperature is steadily increased at a slow rate (~ 2 °C/h). The temperature at which the last crystal dissolves is recorded as the saturation temperature. This point is readily detected by observing the scintillating crystals with the aid of a lamp. The precision of the method is about ± 0.05 °C. An apparatus for making these measurements is described in Chapter 6 (see *Figure 6.4*).

If the temperature–solubility characteristics of the system are known, the saturation temperature measurement readily gives the solution concentration. Alternatively, the technique may be used for solubility determinations using solutions of known composition. This is particularly useful for checking the solubility of a solute in an impure process liquor.

EFFECT OF PARTICLE SIZE ON SOLUBILITY

If the particles of a solute suspended in a solvent are small enough, concentrations greater than the concentration represented by the normal solubility of the substance can temporarily be obtained. Numerous attempts have been made to correlate the relationship between particle size and solubility, the first being due to Ostwald (1900), who derived thermodynamically an equation that was later corrected by Freundlich (1909) in the form

$$\ln \frac{c_1}{c_2} = \frac{2M\sigma}{RT\rho}\left(\frac{1}{r_1} - \frac{1}{r_2}\right) \tag{2.17}$$

where c_1 and c_2 = solubilities of spherical particles of radius r_1 and r_2, respectively; R = gas constant; T = absolute temperature; ρ = density of the solid; M = molecular weight of the solid in solution; and σ = surface energy of the solid particle in contact with the solution.

If the normal equilibrium solubility of a substance, i.e. the solubility of large particles with flat surfaces ($r \to \infty$), is denoted by c^*, the solubility c_r of a particle of radius r can be expressed as

$$\ln \frac{c_r}{c^*} = \frac{2M\sigma}{RT\rho r} \tag{2.18}$$

The Ostwald–Freundlich relationship, which is also known as the Gibbs–Thomson or Gibbs–Kelvin equation, has been found[12] to hold fairly well for dilute aqueous solutions of gypsum ($CaSO_4 \cdot 2H_2O$) for particle sizes from 0·5 to 50 μm.

For a salt such as sodium chloride in water the solubility increase only becomes significant for particle sizes smaller than about 1 μm. At 25 °C, for example, $T = 298$ K, $M = 59$, $\rho = 2\cdot17 \times 10^3$ kg/m³, $\sigma = 0\cdot1$ J/m², $R = 8\cdot3 \times 10^3$ J/kmol K. Thus for a 1 μm crystal ($L = 0\cdot5 \times 10^{-6}$ m), $c/c^* = 1\cdot004$ (i.e. 0·4 per cent increase); for 0·1 μm, $c/c^* = 1\cdot05$ (5 per cent increase) and for 0·01 μm, $c/c^* = 1\cdot55$ (55 per cent increase). For a substance such as sucrose, with a much higher molecular weight, the critical size may be nearer 10 μm and the solubility increase for particles $\sim 0\cdot01$ μm may be tenfold.

The Ostwald–Freundlich equation involves a number of assumptions that are not strictly valid. For instance, both the density of the solid and the solid–liquid surface energy are assumed to be independent of particle size. In addition, the particles are considered to be spherical, and no account is taken of any dissociation of the solid in solution. JONES[13, 14] modified the equation to include cases of particles with various geometric shapes and to allow for degrees of ionisation or dissociation of the dissolved solid. Later[15] this modified equation was used to provide a general theory of supersaturation in which a critical size of the crystal nucleus necessary for growth to commence was specified. The Jones equation, which allowed for ionisation effects, was rather complex and solubilities could not be calculated directly from it. DUNDON and MACK[16] presented a simpler relationship, which can be written in a form similar to that of equation (2.17):

$$\ln \frac{c_1}{c_2} = \frac{2M\sigma}{RT\rho} \left(\frac{1}{r_1} - \frac{1}{r_2} \right) (1 - \alpha + n\alpha) \tag{2.19}$$

where α is the degree of dissocation, and n the number of ions formed from the dissociation of one molecule. If large particles are considered ($r_1 \rightarrow \infty$), $c_1 \rightarrow c^*$ and equation (2.19) becomes

$$\ln \frac{c_r}{c^*} = \frac{2M\sigma}{RT\rho r} (1 - \alpha + n\alpha) \tag{2.20}$$

HULETT[17, 18] investigated the effect of fine grinding on the solubilities in water of various substances and reported a 19 per cent increase in the solubility of gypsum for 0·4 μm particles and an 81 per cent increase for 0·1 μm particles of $BaSO_4$. DUNDON[19] measured the increases in the solubilities of many other salts and found an apparent relationship between the surface energy σ, molecular volume (mol. wt./density) and hardness of the crystalline substances (see *Table 2.1*).

Equations (2.17)–(2.20) suffer from the serious defect that they postulate a continual increase of solubility with reduction in particle size. To overcome this anomaly, KNAPP[20] considered that the particles carried a small surface electric charge so that the total surface energy would be the sum of the normal surface energy and the electric charge. From these considerations he derived, for the case of isolated charged spheres, an equation written in the form

$$c_r = c^* \exp\left(\frac{a}{r} - \frac{b}{r^4}\right) \tag{2.21}$$

where $a = 2\sigma M/RT\rho$ and $b = Q^2M/8\pi KRT\rho$, Q being the electric charge on the particle and K the dielectric constant of the solid. Equation (2.19) gives a curve of the type shown in *Figure 2.7*. As the particle size is reduced, the solubility increases above the normal solubility until some maximum value is reached. On account of the repulsion between the charged particles, any further reduction of particle size would result in a reduction of solubility. The work of DUNDON and MACK[16, 19] appears to afford some confirmation of Knapp's postulation. Solubility determination with $CaSO_4 \cdot 2H_2O$ in the size range 0·2–0·5 μm, for instance, showed a maximum at about 0·3 μm[19].

HARBURY[21] proposed a modified form of equation (2.20) to allow for the fact

Table 2.1. PARTICLE SIZE AND SOLUBILITY INCREASE (AFTER DUNDON[19])

Substance	Particle size μm	Solubility increase, %	Molecular volume	Surface energy, erg/cm²	Hardness (Mohs)
PbI_2	0·4	2	74·8	130	v. soft
$CaSO_4 \cdot 2H_2O$	0·2–0·5	4–12·5	74·2	370	1·6–2
Hg_2CrO_4	0·3	10	60·1	575	1·6–2
PbF_2	0·3	9	29·7	900	1·6–2
$SrSO_4$	0·25	26	46·4	1400	3–3·5
$BaSO_4$	0·1	80	52·0	1250	2·5–3·5
CaF_2	0·3	18	24·6	2500	2·5–4

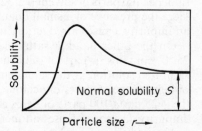

Figure 2.7. The effect of particle size on solubility.
(*After* L. F. KNAPP[20])

that the surface energy σ is not independent of particle radius at the very low values of r encountered in crystal nuclei (0·01–0·001 μm). He also suggested that σ should be replaced by a quantity σ', which, though of the same dimensions as σ, no longer represents the specific free surface energy of a macro crystal lattice. In point of fact, σ' was really a 'catch-all' for several correction factors; and at extremely low values of r, σ' is only a small fraction of σ. Harbury measured the maximum increases in the solubilities of a number of fairly soluble inorganic salts, such as KNO_3, $KClO_3$, $K_2Cr_2O_7$ and $Na_2SO_4 \cdot 10H_2O$ in water. Values of the ratio $c_{max}:c^*$ were found to be 2·75, 3·0, 6·0 and 13·0, respectively. He suggested that this ratio was a function of a quantity which he called the 'molar surface' of the substance, defined as $(M/\rho)^{\frac{1}{3}}$. For the above-mentioned inorganic salts it was found that

$$\ln \frac{c_{max}}{c^*} \simeq 3.3(M/\rho)^{\frac{1}{3}} \tag{2.22}$$

EFFECT OF IMPURITIES ON SOLUBILITY

Pure solutions are rarely encountered outside the analytical laboratory. Industrial solutions are almost invariably impure, and the presence of an impurity can have a considerable effect on the solubility characteristics of a system. If to a saturated binary solution of A (a solid solute) and B (a liquid solvent) a third component C (also soluble in B) is added, one of four conditions can result. First – this is comparatively rare – nothing may happen: the system remains in its original saturated state. Second, component C may combine or react chemically with A by forming a complex or compound, thus altering the whole nature of the system. These two cases will not be considered here.

The third and fourth possibilities, however, are extremely important. The presence of component C may make the solution undersaturated or supersaturated with respect to solute A. In the former case, A would be precipitated or 'salted-out'. In the latter, the solution would be capable of dissolving more A or, in other words, A would be 'salted-in'.

Take, for example, the case of a solution of salol (phenyl salicylate) in ethyl alcohol at 25 °C. At this temperature a saturated solution contains 54 parts of salol per 100 parts of ethyl alcohol. If, however, the aqueous ethanol azeotrope had been used as the solvent (95·5 per cent ethyl alcohol by weight), a saturated solution at 25 °C would contain only 31 parts of salol per 100 parts of solvent (i.e. 32·5 parts/100 parts of ethyl alcohol present). Here the presence of a small quantity of water, which may be considered as an impurity, causes a reduction in the solubility of salol.

On the other hand, a saturated solution of salicylic acid in benzene at 25 °C contains 1 part of salicylic acid per 100 parts of benzene, and a saturated solution at the same temperature in a solvent containing 95 parts of benzene per 5 parts of acetone contains 4 parts per 100 parts of solvent (i.e. 4·25 parts/100 parts of benzene present). In this case the presence of the 'impurity' (acetone) causes an increase in the solubility of the salicylic acid.

The addition of a soluble salt to a saturated aqueous solution of another salt will usually result in the precipitation of some of the latter if the two salts have an ion in common. Salting-out due to the common ion effect can be explained by the *solubility product* principle first proposed by Nernst (1889). In a saturated solution of a sparingly soluble salt, MR, dissociation may be assumed to be complete because of the very low electrolyte concentration. If one molecule of MR dissociates into n positive ions and n' negative ions according to the equation

$$M_n R_{n'} \rightleftharpoons nM^+ + n'R^-$$

then

$$(c_{M^+})^n (c_{R^-})^{n'} = k_s \tag{2.23}$$

where c_{M^+} and c_{R^-} are the ionic concentrations (e.g. g-ions/l) and k_s is the solubility product. If the solubility of a uni-univalent salt $(n = n' = 1)$ at a given temperature is denoted by s^* (e.g. mol/l) then

$$c_{M^+} = c_{R^-} = s^*$$

and

$$(c_{M^+})(c_{R^-}) = k_s \tag{2.24}$$
$$= s^{*2} \tag{2.25}$$

The addition of a soluble salt N^+R^- to a saturated solution of M^+R^- would result in a reduction in the solubility of salt M^+R^-. The lower solubility, s, produced by the addition of e g-ions/l of R^- ions is given by

$$s = \tfrac{1}{2}(e^2 - 4s^{*2})^{\frac{1}{2}} - \tfrac{1}{2}e$$
$$= \tfrac{1}{2}(e^2 - 4k_s)^{\frac{1}{2}} - \tfrac{1}{2}e \tag{2.26}$$

The common ion effect is of great importance in gravimetric analysis when an almost complete precipitation of one substance is essential.

Strictly speaking, the simple solubility product principle can only be applied to solutions of sparingly soluble salts (about 0·01 mol/l or less), and a more fundamental approach involves the use of the chemical potential and activity concepts.

If one molecule of a solute dissociates in solution according to the equation

$$M_nR_{n'} \rightleftharpoons nM^+ + n'R^-$$

then the chemical potential μ of the undissociated solid in equilibrium with the saturated solution must be equal to the sum of the chemical potentials of the ions in solution, thus

$$\mu_{M_nR_{n'}} = n\mu_{M^+} + n'\mu_{R^-}$$
$$= nRT \ln a_{M^+} + nRT \ln a_{R^-} \tag{2.27}$$

where a_{M^+} and a_{R^-} are the activities of the M^+ and R^- ions and R is the gas constant. So long as the solution is saturated with respect to MR, the chemical potential of the solid, $\mu_{M_nR_{n'}}$, will remain constant; therefore

$$nRT \ln a_{M^+} + nRT \ln a_{R^-} = \text{constant} \tag{2.28}$$

and

$$(a_{M^+})^n(a_{R^-})^{n'} = \text{constant} = K_s \tag{2.29}$$

where K_s is the activity product or the activity solubility product. For a uni-univalent salt

$$(a_{M^+})(a_{R^-}) = K_s \tag{2.30}$$

The activity, a, of an ion may be expressed in terms of ionic concentration, c, and the activity coefficient, f, by the relationship

$$a = cf \tag{2.31}$$

Therefore

$$(c_{M^+} f_{M^+})(c_{R^-} f_{R^-}) = K_s \tag{2.32}$$

For a sparingly soluble salt the activity coefficients f_{M^+} and f_{R^-} are almost equal to unity; thus equation (2.32) reduces to equation (2.24).

A number of cases that appear anomalous when the simple solubility product is used can be explained when activity coefficients are taken into account. For instance, the addition of a common ion usually decreases the solubility of a salt, but cases are known where large additions of a salt with a common ion result in increased solubility. The reason for this is that a large increase in ionic concentration can bring about a reduction in the activity coefficients. Thus from equation (2.32) an increase in c_{R^-} will result in a decrease in c_{M^+}, i.e. precipitation of MR, if f_{M^+} and f_{R^-} remain fairly constant, but an increase in c_{R^-} to a value which reduces both f_{M^+} and f_{R^-} must result in an increase in c_{M^+} if K_s is to remain constant. The addition of a salt without a common ion usually increases the solubility; this again is the result of the increased ionic concentration reducing the activity coefficients, but the salting-in effect is much less pronounced than the salting-out effect considered above.

For a complete account of these concepts and their application to aqueous solutions, reference should be made to specialised works [22-24].

The salting-out effect of an electrolyte produced when it is added to an aqueous solution of a non-electrolyte, at one given temperature, can often be represented by the equation

$$\log \frac{s}{s^*} = kC \tag{2.33}$$

where s^* and s are the equilibrium saturation concentrations (mol/l) of the non-electrolyte in pure water and a salt solution of concentration C (mol/l), respectively. The constant k is called the salting parameter, and it refers to one particular electrolyte and its effect on one particular non-electrolyte. This type of relationship will often apply with a reasonable accuracy for low non-electrolyte concentrations and electrolyte concentrations up to 4 or 5 mol/l.

Occasionally the presence of an electrolyte increases the solubility of a non-electrolyte in water (negative value of k), and this salting-in effect is exhibited by several salts with large anions or cations, which themselves are very soluble in water. Sodium benzoate and sodium p-toluenesulphonate are good examples of these 'hydrotropic' salts, and the phenomenon of salting-in is sometimes referred to as 'hydrotropism'. Values of the salting parameter k for three salts applied to benzoic acid in aqueous solution are: NaCl(0·17), KCl(0·14) and sodium benzoate ($-0·22$). LONG and MCDEVIT[25] have made a comprehensive review of salting-in and salting-out phenomena.

The salting effect of an impurity can also be considered in terms of a simple solubility ratio c/c^*, where c and c^* are the equilibrium saturation concentrations (e.g. parts/100 parts of solvent) of the given solute in the impure solution and pure solvent, respectively, at one given temperature. KELLY[26] used this approach and called the solubility ratio the 'solubility coefficient'. *Figure 2.8a* shows the influence of certain substances on the solubility of sucrose in water; the solubility coefficient of sucrose is plotted against the 'purity' of the sucrose solution. *Figure 2.8b* shows the effect of sucrose on the solubilities of several other substances. Kelly concluded that solutes that form a hydrate have a salting-out effect on sucrose within the temperature ranges over which the hydrates are stable, whereas solutes which

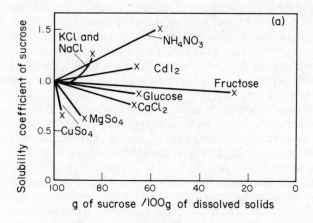

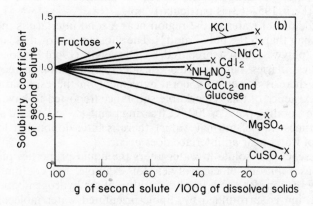

Figure 2.8. Influence of (a) second solute on the solubility coefficient of sucrose, (b) sucrose on the solubility coefficient of the second solute. (After F. H. C. KELLY[26])

do not form a hydrate have salting-in effect. Solutes that increased the solubility of sucrose also had their own solubilities increased and vice versa, although fructose and ammonium nitrate proved exceptions to this generalisation.

When considerable quantities of a soluble impurity are present in, or deliberately added to, a binary solution, the system is best analysed with reference to a ternary equilibrium diagram (see Chapter 4).

STRUCTURE OF SOLUTIONS

Water is a unique liquid. It is also the most abundant compound on earth ($\sim 10^{21}$ kg in the oceans with perhaps a similar quantity bound up as water of crystallisation in rocks and minerals) and it is an essential constituent of all living organisms. Its unusual properties, such as a high boiling point compared with its related hydrides, a high thermal conductivity, dielectric constant and surface tension, a low enthalpy of fusion, the phenomenon of maximum density (at 4 °C), etc., are usually explained by assuming that liquid water has a structure.

The two most acceptable models of the molecular arrangement are a quartz-like structure based on tetrahedra more closely packed than in ice, proposed by Bernal and Fowler (1933), and a clathrate-like structure with pentagonal dodecahedra arranged randomly, proposed by Pauling (1960). It is not possible at the present time to decide conclusively between these two models, but there is no doubt that liquid water does retain a loose local tetrahedral structure for short periods, and this structure is maintained by hydrogen bonds disposed tetrahedrally around each oxygen atom. Hydrogen bonded clusters readily form, but their lifetime is short (probably $\sim 10^{-11}$ s); and the name 'flickering clusters' based on the model of H. S. Frank and his co-workers (1957–63) is particularly apt.

The structure of water in the region near a solid interface is much more complex than that in the bulk liquid[33]. The substance known as 'anomalous water' or 'polywater', first prepared by Derjaguin (1966) by condensation in quartz capillaries, has attracted considerable attention in recent years. Although there is as yet no general agreement on the explanation for or even the existence of this remarkable substance (reported to have a density ~ 1.4 g/cm^3, boiling point > 200 °C, freezing point < -50 °C and a viscosity some 10 times that of normal water), there is little doubt that long-range ordering of water near an interface does occur.

The presence of a solute in water alters the liquid properties profoundly. In aqueous solutions of electrolytes, for example, the coulombic forces exerted by the ions lead to a local disruption of the hydrogen bonded structure. Each ion is surrounded by dipole orientated water molecules firmly bonded in what is known as the 'primary hydration sphere'. For monatomic and monovalent ions, four molecules of water most probably exist in this firmly fixed layer. For polyvalent ions such as Cr^{3+}, Fe^{3+} and Al^{3+}, six is a common number. The hydrated proton H_3O^+ most probably exists as $H_3O(H_2O)_3^+$.

The electrostatic effects of an ion, however, can extend far beyond the primary hydration sphere. This accounts for the very large so-called hydration numbers that have been reported for some ions (up to 700 for Na^+, for example). There is clearly a much larger region around the ion which contains loosely bound, but probably non-orientated, water. This assembly constitutes the 'secondary hydration sphere'. Other names for the two regions are 'inner and outer co-ordination spheres' and 'close and distant hydration'.

The complexities of ionic solutions arising from ion–solvent interactions, involving solvation and complex formation, have yet to be unravelled. This subject is a very active area of research at the present moment and the findings are of considerable importance in attempts to elucidate mechanisms of crystallisation (see Chapter 6). Detailed accounts of current theories of liquid structure are given in references 27–38a.

Supersaturated Solutions

The structure of a supersaturated solution is probably more complex than that of an unsaturated or saturated solution. As reported by KHAMSKII[39], a number of attempts have been made to find the distinguishing features of supersaturated solutions by investigating the dependences of various physical properties on concentration. In most cases, however, no evidence of discontinuity of the property–concentration curves at the equilibrium saturation point has been found. This has also been the author's experience with such physical properties as density, viscosity, refractive index and diffusivity in aqueous solutions of potassium sulphate, ammonium and potassium dihydrogen phosphate, and ammonium and potassium alum and potassium sulphate.

However, evidence has recently been reported of significant structural

Table 2.2. CONCENTRATION GRADIENTS DEVELOPED IN QUIESCENT AQUEOUS CITRIC ACID SOLUTIONS KEPT UNDER ISOTHERMAL CONDITIONS[40]

| Initial concentration at all three positions | | Concentration after time t | | | | | | Solution temp., | Time t, |
| | | Top† | | Middle† | | Bottom† | | | |
c	S	c	S	c	S	c	S	°C	h
2·247	1·055	2·244	1·053	2·253	1·057	2·263	1·061	28·2	70
2·568	1·393	2·531	1·372	2·546	1·381	2·604	1·412	22·6	71
2·624	1·173	2·500	1·131	2·616	1·169	2·652	1·182	30·0	92
2·336	1·185	2·303	1·168	2·333	1·183	2·336	1·185	25·2	336
‡1·553	0·714	1·553	0·714	1·553	0·714	1·553	0·714	28·5	158

c = solution concentration (g of citric acid monohydrate/g of 'free' water).
c^* = equilibrium saturation concentration (g of citric acid monohydrate/g of 'free' water).
S = supersaturation ratio = c/c^*.
† = vertical distance between the sample points = 20 cm.
‡ = unsaturated solution (one of many similar runs).

changes in highly concentrated solutions[40]. It has been shown that supersaturated aqueous solutions of citric acid kept quiescent at constant temperature develop concentration gradients with the highest concentrations in the lower regions (*Table 2.2*). This unusual behaviour is taken as a strong indication of the existence of molecular clusters in such solutions. This is perhaps not unexpected, because citric molecules by virtue of their —OH and —COOH groups:

$$
\begin{array}{ccc}
 & \text{OH} & \\
 & | & \\
\text{CH}_2\!\!-\!\!\!\!&\text{C}&\!\!\!\!-\!\!\text{CH}_2 \\
| & | & | \\
\text{COOH} & \text{COOH} & \text{COOH}
\end{array}
$$

are capable of extensive hydrogen bonding between themselves and with the solvent water molecules. Clusters of closely packed citric acid molecules (mol. wt. = 192) will possess a much higher density than closely packed water molecules, so some degree of 'sedimentation' might be expected. However, since such a system is not in thermodynamic equilibrium, after a finite time a degree of supersaturation higher than some critical value should be achieved at a low position in the system and eventually a phase transition should result (see Chapter 5).

CHOICE OF A SOLVENT

For the industrial crystallisation of inorganic substances from solution, the use of water as the solvent is almost exclusive. This fact is quite understandable, because, apart from the relative ease with which a very large number of chemical compounds dissolve in it, water is readily available, cheap and quite innocuous. For these reasons water is used whenever possible in the industrial crystallisation of organic compounds. Yet there are many cases in this particular field where for a variety of reasons some other solvent must be employed.

The selection of the best solvent for a given crystallisation operation is not always an easy matter. Many factors must be considered and some compromise must inevitably be made; several undesirable characteristics may have to be accepted to secure the aid of one important solvent property. There are several hundred organic liquids that are potentially capable of acting as crystallisation solvents, but outside the laboratory the list can be shortened to a few dozen selected from the following groups: acetic acid and its esters; lower alcohols and ketones; ethers; chlorinated hydrocarbons, e.g. chloroform and carbon tetrachloride; benzene, toluene, xylene and light petroleum fractions.

Occasionally a mixture of two or more solvents will be found to possess the best properties for crystallisation purposes. Common binary solvent mixtures that have proved useful include alcohol–water, alcohol–ketone, alcohol–ether, alcohol–carbon tetrachloride, alcohol–chloroform, alcohol–benzene, alcohol–toluene, alcohol–xylene, benzene–hexane and hexane–trichlorethylene. Sulphuric acid, of various strengths, can often be used to advantage as a solvent for certain organic compounds.

The standard reference works for solubility data include those of SEIDELL[41], STEPHEN and STEPHEN[42] and International Critical Tables[43]. Comprehensive tables of solubility data are given in Appendixes A3 and A4.

Broadly speaking, the factors that have to be considered in choosing a solvent can be grouped under the headings: solvent properties, economic aspects and industrial hazards. Several well-known publications[44-48] deal comprehensively with these aspects. The following brief notes indicate the main points in the selection process.

Solvent Power

The solute to be crystallised should be capable of being dissolved fairly easily in the solvent; it should also be capable of being deposited easily from the solution in the desired crystalline form after cooling, evaporation or dilution. It is important to remember that the crystal habit can often be changed by changing the solvent; for example, one solvent may favour the precipitation of needle crystals, whereas another may induce prisms.

There are many exceptions to the frequently quoted rule that 'like dissolves like', but this rough empiricism can serve as a useful guide. Solvents may be classified as being polar or non-polar; the former description is given to liquids which have high dielectric constants, e.g. water, acids, alcohols, and the latter refers to liquids of low dielectric constant, e.g. aromatic hydrocarbons. A non-polar solute (e.g. anthracene) is usually more soluble in a non-polar solvent (e.g. benzene) than in a polar solvent (e.g. water). However, close chemical similarity between solute and solvent should be avoided, because their mutual solubility will in all probability be high, and crystallisation may prove difficult or uneconomical.

The 'power' of a solvent is usually expressed as the mass of solute that can be dissolved in a given mass of pure solvent at one specified temperature. Water, for example, is a more powerful solvent at 20 °C for calcium chloride than, say, n-propanol (75 and 16 g/100 g of solvent, respectively). At the same temperature n-propanol is a more powerful solvent than water for, say, benzoic acid (42·5 and 0·29 g/100 g of solvent, respectively).

In cooling–crystallisation the temperature coefficient of solubility is another important factor to be considered. For example, at 20 °C water is a more powerful solvent for potassium sulphate (11 g/100 g of water) than for potassium chlorate (7 g/100 g), but the converse is true at 80 °C (K_2SO_4 = 21 g, $KClO_3$ = 39 g/100 g). Thus, on cooling the saturated solutions from 80 to 20 °C, about 82 per cent of the dissolved $KClO_3$ is deposited compared with about 47 per cent of the K_2SO_4.

Both the solvent power and the temperature coefficient of solubility must be considered in choosing a solvent for a cooling–crystallisation process; the former quantity decides the volume of the crystalliser, and the latter determines the solute yield. It frequently happens, especially in aqueous organic systems, that a low solubility is combined with a high temperature coefficient of solubility. For example, the solubilities of salicylic acid in water at 20 and 80 °C are 0·20 and 2·26 g/100 g, respectively. Therefore, on cooling from 80 to 20 °C, most of the dissolved solute (91 per cent) is deposited, and consequently the solute yield is high. However, on account of the low solubility even at the higher temperature, such a large crystalliser would be required to give a reasonable through-put that water could not be considered as a solvent for salicylic acid.

Potassium chromate is an example of a solute with a reasonably high solubility in water but a low temperature coefficient of solubility (61·7 and 72·1 g/100 g at 20 and 80 °C, respectively). The low yield on cooling (about 14 per cent) makes it necessary to effect crystallisation in some other manner, such as a combination of cooling and evaporation, thus increasing the cost of the operation.

Purity

No deleterious impurity, dissolved or suspended, should be introduced into the crystallising system. The solvent, therefore, should be as clean and as pure as possible. No colouring matter should be permitted to affect the appearance of the final crystals, and no residual odours should remain in the product after drying. This latter problem is often encountered after crystallisation from petroleum and coal tar solvents. If no previous experience has been obtained with a solvent, simple laboratory trials should be made; but due caution should be exercised, as laboratory filtration, washing and drying techniques generally prove to be far more efficient than the corresponding large-scale operations.

Chemical Reactivity

The solvent should be stable under all foreseeable operating conditions; it should neither decompose nor oxidise, and it must not attack any of the materials of construction of the plant. When organic solvents are being used, care must be taken in choosing the correct gasket materials; most common types of rubber, for example, swell and disintegrate after prolonged contact with hydrocarbons such as benzene and toluene.

The solute and solvent should not be capable of reacting together chemically; glacial acetic acid, for instance, would hardly be chosen if the solute were liable to be acetylated. Solvate formation, however, may be permitted under certain circumstances. Hydrated crystals are frequently desired as end-products; but should the anhydrous substance be required, the drying process may prove difficult and expensive. Methanol, ethanol, benzene and acetic acid are also known to form solvates with certain substances, and the loss of solvent on drying imposes an additional cost on the process.

The removal of a solvent from a solvate by drying at elevated temperatures generally results in the destruction of the crystalline nature of the product. Freeze drying, i.e. sublimation of the solvent, may prevent this undesirable feature, but the costs of the operation are invariably prohibitive. Occasionally the difficulties may be overcome by washing the crystalline mass with a liquid that acts as a solvent for the mother solvent and a non-solvent for the solute.

Handling and Processing

Highly viscous solvents are not usually conducive to efficient crystallisation, filtration and washing operations. In general, therefore, solvents of low viscosity are preferred. If the solvent recovery process involves distillation, a reasonably volatile solvent is desirable, the latent heat of vaporisation should be low and possible azeotrope formation should be avoided. On the other hand, the loss of a solvent with a high vapour pressure from filters and other processing equipment can be considerable, and may prove both costly and hazardous. Solvents with freezing points above about $-5\,°C$ present wintertime storage and transportation difficulties. Benzene (f.p. $5\,°C$) and glacial acetic acid (f.p. $16\,°C$) are good examples of commonly used solvents that suffer from this defect.

Inflammability and Toxicity

A large proportion of the organic solvents employed in crystallisation processes are inflammable, and their use necessitates stringent operating conditions. Two of the most important properties of an inflammable solvent are the flash-point and explosive limits; the former is defined as the temperature at which the mixture of air and vapour above the liquid can be ignited by means of a spark, and the latter usually refers to the percentages by volume in air between which the air–vapour mixture will explode in a confined space. Diethyl ether is an example of a solvent with a very low flash-point ($-30\,°C$) and wide range of explosive limits (1–50 per cent).

Rigidly enforced safety precautions should be taken on the plant, and operating personnel must be made fully aware of the potential dangers. Costly flame-proof lights and electrical equipment have to be installed, and all vessels and pipelines should be earthed as a precaution against the build-up of static electricity. Maintenance tools should be of the non-ferrous type, and operators should wear rubber-soled footwear to avoid the accidental production of a spark. Adequate fire-fighting equipment is necessary and safe solvent storage areas must be provided.

All organic solvents are toxic to a greater or lesser degree; the prolonged inhalation of almost any vapour will produce some harmful effect on a human being. Some solvents are acute poisons; some have a cumulative poisoning effect; and others produce narcosis or intoxication on inhalation, or dermatitis when contacted with the skin. Details on these aspects and on the maximum vapour concentrations permitted in working areas can be obtained from the specialised reference books.

Adequate ventilation should be provided in the plant, and all vessels and containers should be leak-free. Many of the precautions discussed above with regard to inflammable solvents should be enforced, particularly those concerning the operating personnel.

REFERENCES

1. GUTHRIE, F., 'Salt solutions and attached water', *Phil. Mag.*, **18** (1884), 22, 105
2. MIERS, H. A., 'Variation in angles of alum crystals', *Phil. Trans. R. Soc.*, **A202** (1904), 459
3. MIERS, H. A. and ISAAC, F., 'Refractive indices of crystallising solutions', *J. chem. Soc.*, **89** (1906), 413
4. MIERS, H. A. and ISAAC, F., 'The spontaneous crystallisation of binary mixtures', *Proc. R. Soc.*, **A79** (1907), 322
5. MIERS, H. A., 'The growth of crystals in supersaturated liquids', *J. Inst. Metals*, **37** (1927), 331
6. HOLVEN, A. L., 'Supersaturation in sugar boiling operations', *Ind. Engng Chem.*, **34** (1942), 1234
7. PURDON, F. F. and SLATER, V. W., *Aqueous Solution and the Phase Diagram*, 1946. London; Arnold
8. ZIMMERMAN, H. K., 'The experimental determination of solubilities', *Chem. Rev.*, **51** (1952), 25
9. DAUNCEY, L. A. and STILL, J. E., 'Apparatus for the direct measurement of the saturation temperatures of solutions', *J. appl. Chem., Lond.*, **2** (1952), 399
10. WISE, W. S. and NICHOLSON, E. B., 'Solubilities and heats of crystallisation of sucrose and methyl α-d-glucoside in aqueous solution', *J. chem. Soc.* **(1955)**, 2714
11. KELLY, F. H. C., 'The solubility of sucrose in impure solutions', in *Principles of Sugar Technology*, Vol. 2, ed. P. Honig, 1959. Amsterdam; Elsevier
12. JONES, W. J. and PARTINGTON, J. R., 'Experiments on supersaturated solutions', *J. chem. Soc.*, **107** (1915), 1019
13. JONES, W. J., 'Über die Grösse der Oberflächenenergie fester Stoffe', *Z. phys. Chem.*, **82** (1913), 448
14. JONES, W. J., 'Über die Beziehung zwischen geometrischer Form und Dampfdruck, Löslichkeit und Formenstabilität', *Annln Phys.*, **41** (1913), 441
15. JONES, W. J. and PARTINGTON, J. R., 'A theory of supersaturation', *Phil. Mag.*, **29** (1915), 35
16. DUNDON, M. L. and MACK, E., 'The solubility and surface energy of calcium sulphate', *J. Am. chem. Soc.*, **45** (1923), 2479
17. HULETT, G. A., 'Beziehungen zwischen Oberflächenspannung und Löslichkeit, *Z. phys. Chem.*, **37** (1901), 385
18. HULETT, G. A., 'The solubility of gypsum as affected by particle size and crystallographic surface', *J. Am. chem. Soc.*, **27** (1905), 49
19. DUNDON, M. L., 'The surface energy of several salts', *J. Am. chem. Soc.*, **45** (1923), 2658
20. KNAPP, L. F., 'The solubility of small particles and the stability of colloids', *Trans. Faraday Soc.*, **17** (1922), 457
21. HARBURY, L., 'Solubility and melting point as functions of particle size', *J. phys. Chem.*, **50** (1946), 190
22. GLASSTONE, S., *Textbook of Physical Chemistry*, pp. 954–974, 2nd Ed., 1953. London; Macmillan
23. HILDEBRAND, J. H. and SCOTT, R. L., *Solubility of Non-electrolytes*, 3rd Ed., 1950. New York; Reinhold
24. LEWIN, S., *The Solubility Product Principle*, 1960. London; Pitman
25. LONG, F. A. and MCDEVIT, W. F., 'Activity coefficients of non-electrolyte solutes in aqueous salt solutions', *Chem. Rev.*, **51** (1952), 119
26. KELLY, F. H. C., 'Phase equilibria in sugar solutions (Parts I–V, Ternary Systems, Parts VI–X, Quaternary and Quinary Systems)', *J. appl. Chem., Lond.*, **4** (1954), 401; **5** (1955), 66, 120, 170
27. KAVANAU, J. L., *Water and Water-Solute Interactions*, 1964. San Francisco; Holden Day
28. SAMOILOV, O. YA., *Structure of Aqueous Electrolyte Solutions and the Hydration of Ions* (trans. D. J. G. IVES), 1965. New York; Consultants Bureau
29. EISENBERG, D. and KAUZMANN, W., *The Structure and Properties of Water*, 1969. Oxford; Clarendon Press
30. DROST-HANSEN, W., 'Structure of water near solid interfaces', *Ind. Engng. Chem.*, **61** (11) (1969), 10
31. HAMER, W. J. (Ed.), *The Structure of Electrolyte Solutions*, 1957. New York; Wiley

32. HARNED, H. S. and OWEN, B. B., *The Physical Chemistry of Electrolyte Solutions*, 3rd Ed., 1958. New York; Reinhold
33. ROBINSON, R. A. and STOKES, R. H., *Electrolyte Solutions*, 2nd Ed., 1970. London; Butterworths
34. HUNT, J. P., *Metal Ions in Aqueous Solution*, 1963. New York; Benjamin
35. FRANKS, F. and IVES, D. J. G., 'Structural properties of alcohol-water mixtures', *Q. Rev. chem. Soc.*, **20** (1966), 1
36. 'Interactions in Ionic Solutions', *Discuss. Faraday Soc.*, No. 24 (1957)
37. NANCOLLAS, G. H., *Interactions in Electrolyte Solutions*, 1966. Amsterdam; Elsevier
38. 'The Structure and Properties of Liquids', *Discuss. Faraday Soc.*, No. 43 (1967)
38a. *Trace Inorganics in Water* (Symposium Proceedings), *Adv. Chem. Ser.*, No. 73 (1968)
39. KHAMSKII, E. V., *Crystallisation from Solutions* (transl. from Russian), p. 14, 1969. New York; Consultants Bureau
40. MULLIN, J. W. and LECI, C. L., 'Evidence of molecular cluster formation in supersaturated solutions of citric acid', *Phil. Mag.*, **19** (1969), 1075
41. SEIDELL, A., *Solubilities of Inorganic and Metal Organic Compounds*, 4th Ed. (ed. W. F. LINKE), Vol. I, 1958. New York; Van Nostrand. Vol. II, 1965, Washington; American Chemical Society
42. STEPHEN, H. and STEPHEN, T., *Solubilities of Inorganic and Organic Compounds* (in 5 parts) 1963. London; Pergamon
43. *International Critical Tables*, Vol. IV, 1933. New York; McGraw-Hill
44. DURRANS, T. H., *Solvents*, 7th Ed., 1957. London; Chapman and Hall
45. MELLAN, I., *Source Book of Industrial Solvents* (in 3 volumes), 1959. New York; Reinhold
46. SAX, N. I., *Dangerous Properties of Industrial Materials*, 1968. New York; Reinhold
47. WEISSBERGER, A. *et al.*, *Organic Solvents*, 2nd Ed., 1955. New York; Wiley
48. WADDINGTON, T. C. (Ed.), *Non-aqueous Solvent Systems*, 1965. New York; Academic Press

3

Physical and Thermal Properties

PHYSICAL PROPERTIES

DENSITY

THE concentration of commercial solutions is often expressed in terms of a density measurement. Density is defined as mass per unit volume, and the most common expression of this quantity encountered in the scientific literature has the unit g/cm^3. In industrial and engineering practice the units lb/ft^3, kg/m^3 and lb/gal (U.S. and Imperial) are frequently employed. It must be remembered, however, that the density of a solution is affected by temperature as well as by concentration.

The ratio of the density of a liquid at one temperature to the density of water at another is known as the *specific gravity* of the liquid; thus for complete definition all expressions of specific gravity must be accompanied by a temperature designation. For instance, the specific gravity of a solution quoted as 1·23 (20/20 °C) refers to the ratio of the densities of the solution and water both at 20 °C, while 1·23 (20/4 °C) refers to the solution at 20 °C and water at 4 °C. At 4 °C (39·2 °F) water exhibits its maximum density of 1·000 g/cm^3; thus a specific gravity expressed with reference to water at 4 °C is numerically equal to the density of the solution in g/cm^3 at the given temperature.

Density measurements are frequently used to determine solution concentration or supersaturation in crystallisers, by withdrawing small samples and determining the density by a standard technique. However, this method is rather tedious and some considerable time may elapse before a result can be obtained. A density meter that allows a continuous *in situ* method of measurement has been described by GARSIDE and MULLIN[1]. The basic principle of the instrument is that a plummet is suspended electromagnetically in the solution under test. The position of the plummet is sensed by search coils and it is prevented from sinking by an electromagnetic force from a solenoid located directly above. The search coils allow just sufficient current to flow in the solenoid to maintain the plummet in a central position in the test chamber, thereby affording a sensitive means for recording density changes ($\pm 0\cdot0001$ g/cm^3).

Liquid densities are most conveniently measured by the use of a hydrometer, which consists of a hollow cylinder or float, loaded with lead shot or mercury, on which is mounted a graduated stem. The depth to which the stem sinks in the liquid gives a measure of the density. As density varies with temperature, hydrometers are marked with the temperature at which they are calibrated. Liquid density can also be determined by the specific gravity bottle, pyknometer or Westphal balance methods, details of which are given in most textbooks of practical physics.

For reasons of convenience, several density scales have been developed for industrial use, and hydrometers graduated in these scales are readily available. In Great Britain, for instance, the *Twaddell* scale is widely employed, especially in the acid and alkali industries. It is used for liquids heavier than water (pure water at 4 °C = 0 °Tw), and the relationship between specific gravity and degrees Twaddell is given by

$$s.g. = 1 + (°Tw/200) \tag{3.1}$$

or

$$°Tw = 200 \, (s.g. - 1) \tag{3.2}$$

On account of its simplicity and easy convertibility into and from values of specific gravity, the Twaddell scale has also gained popularity in many parts of the world.

The *Beaumé* (or Baumé) scale was developed in the salt industry; pure water was taken as 0 °Bé, each degree representing 1 per cent of dissolved salt. Unfortunately, no reference temperature was originally specified, and as percentage composition can be interpreted in many ways, as many as 20 different Beaumé scales have been in use at one time or another. Several of these are still in use, but the two most widely used are the *Rational* and the *U.S.* scales. The relationship between these and specific gravity are:

(a) *liquids lighter than water*

Rational °Bé = (144·3/s.g.) − 144·3   [water = 0° at 15 °C]
U.S.   °Bé = (140/s.g.) − 130   [water = 10° at 60 °F]

(b) *liquids heavier than water*

Rational °Bé = 144·3 − (144·3/s.g.)   [water = 0° at 15 °C]
U.S.   °Bé = 145 − (145/s.g.)   [water = 0° at 60 °F]

Thus there is little difference between the two 'heavy' scales, but an appreciable difference between the two 'light' scales.

A Beaumé-type density scale was proposed by the American Petroleum Institute for liquids lighter than water. This is now known as the *A.P.I.* scale and is used in the petroleum industries of most countries. Water at 60 °F is 10 °A.P.I. The conversion between degrees A.P.I. and specific gravity is given by

$$A.P.I. = (141·5/s.g.) - 131·5 \tag{3.3}$$

Several other density scales are still in fairly common use in certain industries. These are best described in tabular form, as in *Table 3.1*; the terms 'light' and 'heavy' refer to the scale conversions for liquids lighter or heavier than water, respectively. The number of degrees on a particular scale is denoted by N.

The heavy Brix or Fisher scale quoted above is not widely used, but another Brix scale is used throughout the sugar industry. On this scale a

Table 3.1. INDUSTRIAL DENSITY SCALES

Scale	Reference temperature	Specific gravity	
		Light	Heavy
Balling	17·5 °C	200/(200 + N)	200/(200 − N)
Beck	12·5 °C	170/(170 + N)	170/(170 − N)
Brix	60 °F	400/(400 + N)	400/(400 − N)
Cartier	12·5 °C	136·8/(126·1 + N)	136·8/(126·1 − N)
Fisher	60 °F	as Brix scale	

Table 3.2. COMPARISON BETWEEN VARIOUS DENSITY SCALES USED IN INDUSTRIAL PRACTICE (REFERENCE TEMPERATURE = 60 °F)

Specific gravity	°Tw	°Bé		°A.P.I.	lb/gal (Imp.)	lb/gal (U.S.)	lb/ft³	kg/m³
		Rational	U.S.					
0·50	—	144·3	150·0	151·5	5·0	4·17	31·2	500
0·55	—	118·0	124·5	125·8	5·5	4·58	34·3	550
0·60	—	96·2	103·3	104·3	6·0	4·99	37·4	600
0·65	—	77·7	85·4	86·2	6·5	5·41	40·5	650
0·70	—	61·9	70·0	70·6	7·0	5·83	43·6	700
0·75	—	48·1	56·7	57·2	7·5	6·24	46·7	750
0·80	—	36·1	45·0	45·4	8·0	6·66	49·8	800
0·85	—	25·4	34·7	35·0	8·5	7·08	52·9	850
0·90	—	16·0	25·5	25·7	9·0	7·49	56·1	900
0·95	—	7·6	17·4	17·5	9·5	7·91	59·2	950
1·00	—	0·0	10·0	10·0	10·0	8·33	62·3	1000
1·00	0	0·0	0·0	10·0	10·0	8·33	62·3	1000
1·05	10	6·7	6·9	3·3	10·5	8·75	65·4	1050
1·10	20	13·1	13·2	—	11·0	9·16	68·5	1100
1·15	30	18·8	18·9	—	11·5	9·58	71·7	1150
1·20	40	24·1	24·2	—	12·0	10·00	74·8	1200
1·25	50	28·8	29·0	—	12·5	10·41	77·9	1250
1·30	60	33·2	33·5	—	13·0	10·83	81·0	1300
1·35	70	37·4	37·6	—	13·5	11·25	84·1	1350
1·40	80	41·2	41·4	—	14·0	11·66	87·3	1400
1·45	90	44·7	45·0	—	14·5	12·08	90·4	1450
1·50	100	48·1	48·3	—	15·0	12·50	93·5	1500

solution containing N per cent sugar by weight is said to have a density of N °Brix. A density scale known by the names Salinometer, Salometer or Salimeter is used in the salt industry. Water is 0 °Sal and a saturated aqueous solution of sodium chloride at 20 °C (26·5 parts of salt/100 parts of solution by weight) is 100 °Sal.

Table 3.2 compares some of the more common density scales used in industrial practice. The densities of a number of aqueous solutions are recorded in the Appendix (*Tables A.8 and A.9*).

VISCOSITY

The performance of many items of equipment encountered in crystal-lisation practice is often profoundly affected by the flow properties of the liquid media. Heat transfer, for example, may be severely impeded in 'thick' sluggish liquors or magmas; crystallisation may occur only with difficulty, and filtration and washing of the crystalline product may be impaired.

A measure of the resistance to flow exhibited by a fluid moving over itself is given by the property known as viscosity, which is defined as a shearing stress per unit area per unit velocity gradient within the fluid. The dimensions of viscosity, therefore, are: mass/length . time, and the common c.g.s. unit is the poise (1 g/cm s). As this unit is rather large, the centipoise (cP) is more frequently used. The SI unit for viscosity is kg/m s (or N s/m²):

$$1 \text{ cP} = 0·01 \text{ P} = 10^{-3} \text{ N s/m}^2 = 2·42 \text{ lb/ft . h}$$

The above viscosity, η, is often referred to as the 'dynamic' or 'absolute' or 'coefficient of' viscosity. The kinematic viscosity, v, of a fluid is equal to η/ρ, where ρ is its density. The dimensions of v are length²/time and the common c.g.s. unit is the stokes (1 cm²/s) or centistokes (cSt):

$$1 \text{ cSt} = 0·01 \text{ St} = 10^{-6} \text{ m}^2/\text{s} = 0·0388 \text{ ft}^2/\text{h}$$

The viscosity of a liquid decreases with increasing temperature, and for many liquids the relationship

$$\eta = A . \exp(-B/T) \tag{3.4}$$

holds reasonably well. A and B are constants, and the temperature T is in degrees absolute. Plots of log η against $1/T$, or log η against log T, usually yield fairly straight lines, and this fact may be used in estimating the viscosity of a liquid at some temperature if values of η are known at two other temperatures. *Table 3.3* gives some values of the viscosity of several common solvents over a range of temperature.

In general, solid electrolytes and non-electrolytes increase the viscosity of water, although a few exceptions to this rule are known. Occasionally the increase in viscosity is considerable, as shown by the system sucrose–water (*Table 3.4*). At low temperatures the viscosity of sucrose solutions increases greatly with an increase in concentration, while at high concentration the decrease in viscosity with increasing temperature is extremely rapid. *Figure*

3.1a shows an example of a solute that decreases the viscosity of the solvent; in this system (KI–water) a minimum viscosity is obtained. Several other potassium and ammonium salts also exhibit a similar behaviour. *Figure 3.1b* shows the effect of concentration and temperature on the system ethanol–water; this system exhibits a maximum viscosity. A comprehensive survey of the viscosity characteristics of aqueous solutions of electrolytes is made by STOKES and MILLS[3], who also give experimental data on a considerable

Table 3.3. ABSOLUTE VISCOSITIES (CENTIPOISE) OF SOME COMMON SOLVENTS

Solvent	Temperature, °C					
	0	20	40	60	80	100
Water	1·79	1·00	0·655	0·468	0·355	0·281
Acetone	0·389	0·322	0·261	—	—	—
Benzene	—	0·654	0·492	0·396	0·318	—
Toluene	0·76	0·587	0·471	0·380	0·310	0·250
o-Xylene	1·11	0·825	0·625	0·502	0·405	0·345
Carbon tetrachloride	1·37	0·975	0·746	0·595	—	—
Methanol	0·810	0·592	0·456	0·350	—	—
Ethanol	1·77	1·19	0·826	0·605	—	—
n-Propanol	4·20	2·56	1·40	0·925	0·645	0·443
n-Butanol	5·14	2·95	1·77	1·15	0·762	0·540

Table 3.4. ABSOLUTE VISCOSITIES (CENTIPOISE) OF AQUEOUS SOLUTIONS OF SUCROSE (AFTER E. HATSCHEK[2], FROM DATA OF BINGHAM AND JACKSON)

Temperature, °C	g sucrose per 100 g solution		
	20	40	60
0	3·804	14·77	238
10	2·652	9·794	109·8
20	1·960	6·200	56·5
30	1·504	4·382	33·78
40	1·193	3·249	21·28
50	0·970	2·497	14·01
60	0·808	1·982	9·83
70	0·685	1·608	7·15
80	0·590	1·334	5·40
90	—	1·123	4·15
100	—	0·960	3·34

number of systems. Viscosities of some aqueous solutions are recorded in the Appendix (*Table A.10*). Unfortunately, no completely reliable method is available for the prediction of the viscosities of solutions or liquid mixtures.

The viscosity characteristics of liquids can be altered considerably by the presence of finely dispersed solid particles, especially of colloidal size. Einstein (1911) suggested that the viscosity of a suspension of rigid spherical particles in a liquid, when the distance between the spheres is much greater than their diameter, would be governed by the equation

$$\eta_s = \eta_0(1 + 2{\cdot}5\phi) \tag{3.5}$$

where η_s = effective viscosity of the disperse system, η_0 = viscosity of the pure dispersion medium, and ϕ = ratio of the volume of the dispersed particles to the total volume of the disperse system. In other words, $\phi = 1 - \varepsilon$, where ε is the voidage of the system.

The Einstein equation applies reasonably well to lyophobic sols and very dilute suspensions. For moderately concentrated lyophilic sols modifications

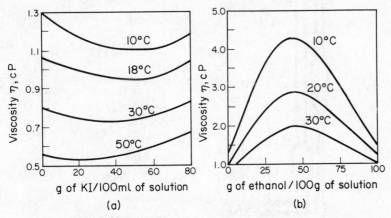

Figure 3.1. Aqueous solutions exhibiting (a) minimum, (b) maximum viscosities. (After E. HATSCHEK[2])

of the Einstein equation have been proposed, the simplest of which is due to Guth and Simha (1936):

$$\eta_s = \eta_0(1 + 2{\cdot}5\phi + 14{\cdot}1\phi^2) \tag{3.6}$$

Very high apparent viscosities can be recorded for concentrated suspensions of solid particles. A recent attempt[4] to derive a functional dependence of the effective viscosity on the concentration of uniform solid spheres in liquid suspension yielded the relationship

$$\eta_s = \frac{9}{8}\eta_0 \left\{ \frac{(\phi/\phi_m)^{\frac{1}{3}}}{(1 - \phi/\phi_m)^{\frac{1}{3}}} \right\} \tag{3.7}$$

for values of $\phi/\phi_m \to 1$, where ϕ_m is the maximum attainable volumetric concentration of solids in the system (usually about 0·6 for spherical particles). Equation (3.7) has met with experimental support in the region $(\phi/\phi_m) > 0.7$.

However, for solids concentrations such as those normally encountered in industrial crystallisers (say $\varepsilon \sim 0{\cdot}8$, $\phi \sim 0{\cdot}2$ and $\phi/\phi_m \sim 0{\cdot}3$ for granular crystals) the much simpler equation (3.6) predicts the order of magnitude of the apparent viscosity reasonably well.

Numerous instruments have been devised for the measurement of viscosity.

Many of these are based on the flow of fluid through a capillary tube, and Poiseuille's law can be applied in the form

$$\eta = \frac{\pi r^4 \Delta P}{8lV}$$ (3.8)

where ΔP = the pressure drop across the capillary of length l and radius r, and V = volume of fluid flowing in unit time. One simple type of U-*tube viscometer* is shown in *Figure 3.2*. The liquid under test is sucked into leg B until the level in this leg reaches mark z. The tube is arranged truly vertical, and the temperature of the liquid is measured and kept constant. The liquid

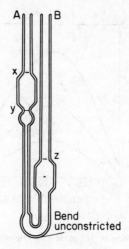

Figure 3.2 Simple U-tube viscometer

is then sucked up into leg A to a point 1 cm above mark x, and the time t (sec) for the meniscus to fall from x to y recorded. The mean of several measurements is taken. The kinematic viscosity of the liquid v (centistokes) can be calculated from the equation

$$v = kt$$ (3.9)

where k is a constant for the apparatus, determined by measurements on a liquid of known viscosity, e.g. water.

The falling-sphere method of viscosity determination also has many applications, and Stokes' law may be applied in the form

$$\eta = \frac{(\rho_s - \rho_l) d^2 g}{18u}$$ (3.10)

where d, ρ_s and u are the diameter, density and terminal velocity, respectively, of a solid sphere falling in the liquid of density ρ_l. A simple *falling-sphere viscometer* is shown in *Figure 3.3*. The liquid under test is contained in the inner tube of 3·2 mm diameter and about 450 mm length. The central portion of the tube contains two reference marks, 150 mm apart. The tube is held truly vertical, and the temperature of the liquid is measured and kept constant. A 1·5 mm diameter steel ball, free from rust and previously warmed to the test temperature, is allowed to fall down the tube, the time of its passage between the two reference marks being measured. A mean of several measurements should be taken. The viscosity of the liquid can then be calculated from equation (3.10).

A simple method devised by CLARK[5] for making a continuous measurement of viscosity has interesting possibilities for laboratory and pilot plant crystallisers. Basically, the liquor under test is forced continuously through a small-bore glass tube by a constant-speed gear pump and the pressure differential is recorded on a mercury manometer. Precisions of $\pm 1\%$ appear to be possible over the range 0 to 50 cP, provided that the necessary degree of temperature control can be exercised.

Several high-precision viscometers are based on the concentric-cylinder method. The liquid under test is contained in the annulus between two vertical coaxial cylinders; one cylinder can be made to rotate at a constant speed, and the couple required to prevent the other cylinder rotating can be

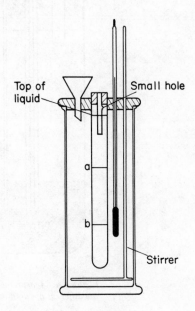

Top of liquid

Small hole

a

b

Stirrer

Figure 3.3. Falling-sphere viscometer

measured. For detailed information on practical viscometry reference should be made to specialised publications[2, 6, 7].

SURFACE TENSION

Of the many methods available for measuring the surface or interfacial tension of liquids[8, 9], the capillary rise and ring techniques are probably the most useful for general application.

Capillary Rise

The surface tension, σ, of a liquid can be determined from the height, h, of the liquid column in a capillary tube of radius r. If the liquid completely wets the tube (zero contact angle),

$$\sigma = \tfrac{1}{2}rh\,\Delta\rho g \tag{3.11}$$

where $\Delta\rho$ is the difference in density between the liquid and the gaseous atmosphere above it. The height, h, can be accurately measured with a cathetometer from the base of the liquid meniscus to the flat surface of the

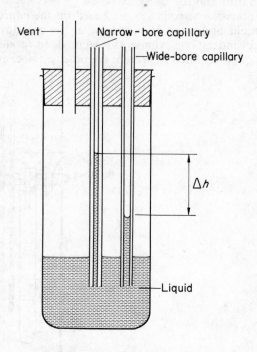

Figure 3.4. *Measurement of surface tension by the differential capillary rise method*

free liquid surface in a containing vessel. However, to minimise errors, this reference to a flat surface can be eliminated by measuring the difference in capillary rise in two tubes of different bore (*Figure 3.4*). Then

$$\sigma = \tfrac{1}{2}r_1 h_1 \Delta\rho g = \tfrac{1}{2}r_2 h_2 \Delta\rho g$$

i.e.

$$\sigma = \frac{\Delta h \Delta\rho\, r_1 r_2 g}{2(r_1 - r_2)} \tag{3.12}$$

The differential height Δh can be measured with a high precision.

Ring Method

The ring technique, and its many variations, are widely used in industrial laboratories. Several kinds of commercial apparatus incorporating a torsion balance are available under the name du Nouy tensometer. The method is simple and rapid, and is capable of measuring the surface tension of a pure liquid to a precision of 0·3% or better.

The force necessary to pull a ring (usually of platinum or platinum–iridium wire) from the surface of the liquid is measured. The surface tension is calculated from the pull and the dimensions of the ring after the appropriate correction factors have been applied[9].

The surface tensions of some common solvents at different temperatures are given in *Table 3.5*.

It is often possible to predict the surface tension of non-aqueous mixtures of solvents by assuming a linear dependence with mole fraction. Aqueous

Table 3.5. SURFACE TENSIONS OF SOME COMMON SOLVENTS AT DIFFERENT TEMPERATURES

Solvent	Surface tension, dyn/cm					
	0°	10°	20°	30°	40°	50 °C
Water	76·0	73·5	72·8	71·2	69·6	67·9
Benzene	31·6	30·2	28·9	27·2	26·3	25·0
Toluene	30·8	29·7	28·5	27·4	26·2	25·1
CCl$_4$	29·5	28·0	26·8	25·5	24·4	23·1
Acetone	25·5	24·4	23·3	22·3	21·2	—
Methanol	24·3	23·4	22·6	21·7	20·8	—
Ethanol	24·1	23·1	22·3	21·4	20·6	19·8

1dyn/cm = 10^{-3} N/m = 10^{-3} J/m^2.

solutions, however, generally show a pronounced non-linear behaviour and prediction is not recommended.

The surface tension of a liquid decreases with an increase in temperature, but the decrease is not always linear for aqueous solutions. Water is often taken as the standard reference liquid for surface tension measurements with a value of about 72 dyn/cm (0·072 N/m) at 25 °C and the non-linear temperature dependence is shown by the data in *Table 3.5*.

The addition of an electrolyte to water generally increases the surface tension very slightly, although an initial decrease is usually observed at low concentrations (<0·002 mol/l)[10]. Non-electrolytes generally decrease the surface tension of water. For example, saturated aqueous solutions of α-naphthol, adipic acid and benzoic acid at 22 °C are 48, 55 and 60 dyn/cm, respectively, whereas a saturated solution of potassium sulphate at the same temperature has a surface tension of 73 dyn/cm.

DIFFUSIVITY

Diffusion is of fundamental importance to the processes of crystallisation and dissolution. The solute, or crystalline component, moves from the region of high concentration (activity) to one of low concentration (activity), and this migration is accompanied by a counter flux of solvent molecules. The paucity of experimental data for diffusion in solid–liquid systems greatly hampers the prediction of mass transfer rates in crystallisation and dissolution, and recourse is frequently made to empirical methods. This practice, however, can lead to gross errors[11-13].

Two examples of a theoretical approach to the problem of the prediction of diffusion coefficients in fluid media are the equations postulated by Einstein (1905) and Eyring (1936). The former is based on kinetic theory and a modification of Stokes' law for the movement of a particle in a fluid, and is most conveniently expressed in the form

$$D = \frac{kT}{\phi r \eta} \tag{3.13}$$

where D = diffusivity (cm^2/s), T = absolute temperature (K), η = viscosity (g/s cm), r = molecular radius (cm), k = Boltzmann's constant and the factor ϕ has a numerical value between 4π and 6π depending on the solute/solvent molecular size ratio. Eyring's approach, based on reaction rate theory, treats a liquid as a disordered lattice structure with vacant sites into which molecules move, i.e. diffuse. For low solute concentration Eyring's equation may be expressed in a form identical with that of equation (3.13), but a different value of ϕ applies.

The usefulness of these equations, however, is strictly limited, because they both contain a term, r, which denotes the radius of the solute molecule. Values of this quantity are difficult to obtain. Consequently the most directly useful relationship that emerges is

$$D\eta/T = \text{constant} \tag{3.14}$$

which is of considerable value in predicting, for a given system, the effect of temperature and viscosity on the diffusion coefficient. This simple relation is often referred to as the Stokes–Einstein equation.

The very limited success of the theoretical equations has led to the development of many empirical and semi-empirical relationships for the prediction of diffusion coefficients. Amongst those devised for diffusion in liquids the following may be mentioned. For diffusion in aqueous solutions OTHMER and THAKAR[14] proposed a correlation which may be written in the form

$$D = \frac{14 \times 10^{-5}}{\eta^{1.1} V_1^{0.6}} \tag{3.15}$$

where V_1 is the molecular volume of the solute.

By correlating a large number of published experimental diffusivities WILKE and CHANG[15] arrived at the relationship

$$D = 7.4 \times 10^{-8} \frac{(\gamma M)^{\frac{1}{2}} T}{V_1^{0.6}} \tag{3.16}$$

where M and γ are the molecular weight and 'association parameter', respectively, for the solvent. For unassociated solvents, e.g. benzene, ether and heptane, $\gamma = 1$. For water, methanol and ethanol, $\gamma = 2.6$, 1.9 and 1.5, respectively.

Despite the widespread use of these and many other similar correlations, they are notoriously unreliable; deviations from experimental values as high as 30% are not unusual[11, 13]. Furthermore, these empirical relationships were devised from diffusion data on liquid/liquid systems, and there is little evidence to suggest that they are reliable for the prediction of the diffusion of solid solutes in liquid solutions, although an 'order of magnitude' estimation is generally possible. For example, the diffusivity of sodium chloride in water at 25 °C is 1.3×10^{-5} cm^2/s, while values calculated from equations (3.15) and (3.16) range from 1.7 to 2.6×10^{-5} cm^2/s. Similarly, for sucrose in

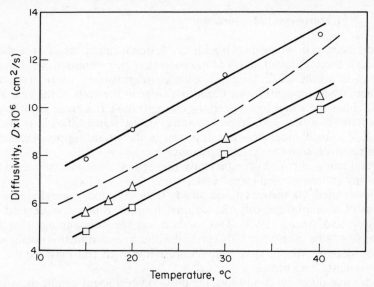

Figure 3.5. Diffusivities of saturated aqueous solutions of the hydroxybenzoic acids into water: o = ortho-, □ = meta-, △ = para-, – – – – – = *predicted from equation (3.16)*

water at 25 °C, $D = 5.2 \times 10^{-6}$ cm^2/s, while the predicted values range from 3.6 to 4.2×10^{-6} cm^2/s.

It is also important to note that these empirical correlations are meant to apply only to dilute solutions. Despite the fact that they all contain a term relating to viscosity which is a function of concentration, they usually fail to predict the rate of diffusion from a concentrated solution to a less concentrated one. For example, the diffusion coefficient for sucrose diffusing from a 1% aqueous solution into water at 25 °C is approximately five times the value for the diffusion between 61.5 and 60.5% solutions, whereas over the concentration range 1 to 60% the viscosity exhibits a fortyfold increase.

One deficiency of these empirical equations, viz. their inability to discriminate between isomers, was pointed out by MULLIN and COOK[16], who measured the diffusivities of *o-*, *m-* and *p-*hydroxybenzoic acid in water. The data measured at 20 °C are compared with values predicted by equation (3.16) in *Figure 3.5*, where the deviation is about ±30% between the estimated values and those measured for the m- and p- isomers, with a difference of about 60% between the o- and m-.

Clearly equation (3.16), and the other empirical relationships, fail to take some property of the system into account, and it is suggested[16] that this quantity is the 'size' of the diffusing component. For the case of the hydroxybenzoic acids the differences in diffusivity can be accounted for by considering the different hydrogen bonding tendencies of the three isomers.

Experimental Measurement

In a diffusion cell, where two liquids are brought into contact at a sharp boundary, three different states of diffusion may be recognised. In the case of 'free' diffusion, concentrations change progressively away from the interface; when concentrations begin to change at the ends of the cell, 'restricted' diffusion is said to occur. If the concentration at a given point in the cell remains constant with respect to time, 'steady-state' diffusion is taking place, and, as in all other steady-state processes, a constant supply of material to and removal from the system is required.

Several comprehensive accounts have been given of the methods used for measuring diffusion coefficients under these three conditions[3, 17-21]. The techniques used for restricted and steady-state diffusion generally involve the use of a diaphragm cell, the original form of which was devised by Northrop and Anson (1929). This method has the disadvantage that the cell has first to be calibrated with a system with a known diffusion coefficient, and for systems with relatively slow diffusivities each run may require an inconveniently long time.

Under free-diffusion conditions, the diffusion coefficient can be measured by analysis at the termination of the experiment or by continuous or intermittent analysis while diffusion continues. Care has to be taken not to disturb the system, and a variety of techniques using radioactivity, electrical conductance, light absorption and schlieren measurements have been devised. Probably the most widely employed methods for measuring diffusion coefficients in liquids are interferometric, resulting from the original work of Gouy (1880) and the later theoretical analysis of Kegeles and Gosting (1947). This technique has been used in several recently reported investigations.

Two kinds of diffusivity can be recorded, viz. the *differential*, D, and the *integral*, D. The differential diffusivity is a value for one particular concentration, c, and driving force, $c_1 - c_2$, where $c = (c_1 + c_2)/2$ and $c_1 - c_2$ is sufficiently small for D to remain unchanged over the concentration range. However, the diffusion coefficient is usually concentration dependent, and in most cases it is the integral diffusivity that is normally measured. This is an average value over the concentration range c_1 to c_2.

The differential diffusivity is of considerable theoretical importance, and it is only through this quantity that experimental measurements by different techniques can be compared. On the other hand, it is the integral coefficient that is generally required for mass transfer assessment, since this coefficient represents the true 'average' diffusivity over the concentration range involved in the mass transfer process.

For example, in the dissolution of a solid into a liquid, the solute diffuses from the saturated solution at the interface to the bulk solution. In crystallisation the solute diffuses from the supersaturated bulk solution to the saturated solution at the interface. The relevant diffusivity that should be used in an analysis of these two processes, therefore, is the integral diffusivity, which covers the range of concentration from equilibrium saturation to that in the bulk solution.

Integral diffusivities may be measured directly by the diaphragm cell technique[21]; but unless the concentration on both sides of the diaphragm (see *Figure 3.6*) is maintained constant throughout the experiment, the diffusivity measured is the rather complex double-average known as the 'diaphragm cell integral diffusivity', $\bar{D}_d$, defined by an integrated form of Fick's law of diffusion:

$$D_d = \frac{1}{\beta t} \ln \left(\frac{c_{1_0} - c_{2_0}}{c_{1_t} - c_{2_t}} \right) \qquad (3.17)$$

where c_0 and c_t are the initial and final (at time t) concentrations, 1 and 2 refer to the lower and upper cells, and β is the cell constant (cm^{-2}), which can be calculated from the cell dimensions:

$$\beta = \frac{A}{L} \left(\frac{1}{V_1} + \frac{1}{V_2} \right) \qquad (3.18)$$

where A and L = area and thickness of diaphragm, and V_1 and V_2 = volume of the cell compartments.

The method is extremely time-consuming and the process of converting integral to differential values is both tedious and inaccurate. It is more practicable, therefore, to measure the differential diffusivity, D, at intervals over the whole concentration range and to calculate the required integral diffusivity, $\bar{D}$, by means of the relationship

$$\bar{D} = \frac{1}{c_1 - c_2} \int_{c_2}^{c_1} D \, dc \qquad (3.19)$$

This procedure has been described in detail elsewhere[22], and the author and his co-workers have reported diffusivities measured by an interferometric technique for a number of aqueous salt solutions including ammonium dihydrogen phosphate (ADP)[11], potassium sulphate[23], ammonium and potassium chloride[22], ammonium and potassium alums[24] and nickel ammonium sulphate[25]. *Figure 3.7* shows the measured differential and calculated integral diffusivities for the systems KCl–H$_2$O and NH$_4$Cl–H$_2$O. Diffusivities for a number of common salt systems are given in the Appendix (*Table A.11*).

The diffusivity of a strong electrolyte at infinite dilution is called the *Nernst limiting value* of the diffusion coefficient, D^0, which can be calculated[19] from

$$D^0 = \frac{RT(v_1 - v_2)}{F^2 v_1 |z_1|} \cdot \frac{\lambda_1^0 \lambda_2^0}{\lambda_1^0 + \lambda_2^0} \qquad (3.20)$$

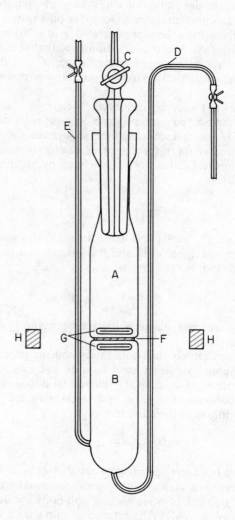

Figure 3.6. Diaphragm diffusion cell (after
F. A. L. DULLIEN *and* L. W. SHEMILT[21]*).*
A, light liquid compartment; B, heavy liquid
compartment; C, stop-cock; D, E, capillaries;
F, sintered glass diaphragm; G, polythene-
coated iron stirrers; H, rotating magnets

where λ_1^0 and λ_2^0 are the limiting conductivities, and v_1 and v_2 are the number of cations and anions of valency z_1 and z_2, respectively. Using the conditions of neutrality:

$$v_1 z_1 + v_2 z_2 = 0$$

so equation (3.20) may also be written

$$D^0 = \frac{RT}{F^2} \cdot \frac{|z_1| + |z_2|}{|z_1 z_2|} \cdot \frac{\lambda_1^0 \lambda_2^0}{\lambda_1^0 + \lambda_2^0} \tag{3.21}$$

F is the Faraday constant ($9 \cdot 6487 \times 10^4$ C mol^{-1}), R is the gas constant ($8 \cdot 3143$ J K^{-1} mol^{-1}), giving values of RT/F^2 of $2 \cdot 4381$, $2 \cdot 6166$ and $2 \cdot 7951 \times$

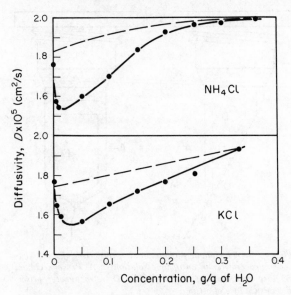

Figure 3.7. Measured differential ($\bullet$) and calculated integral (broken line) diffusivities for the systems NH$_4$Cl–H$_2$O and KCl–H$_2$O at 20 °C

10^{-7} Ω mol^{-1} s^{-1} at 0, 20 and 40 °C, respectively. These may be used with the values of λ^0 (cm^2 Ω^{-1} mol^{-1}) in the Appendix (Table A.13) to give values of D^0 in cm^2/s.

By the application of reaction rate theory[26] to both viscosity and diffusion it can be shown that

$$\eta = A \exp(-E_V/RT) \tag{3.22}$$

and

$$D = B \exp(-E_D/RT) \tag{3.23}$$

Equation (3.22) is the same as equation (3.4) and, as previously described, a plot of $\log \eta$ against $1/T$ should yield a straight line. In a similar manner a linear relationship should exist between $\log D$ and $1/T$; and the slopes of these plots give the respective energies of activation E_V (viscosity) and E_D (diffusion).

REFRACTIVE INDEX

Refractometric measurements can often be used for the rapid measurement of solution concentration. Several standard instruments (Abbé, Pulfrich, etc.) are available commercially and the techniques for their use are well documented[8]. A sodium lamp source is most usually used for illumination, and an instrument reading to the fourth decimal place is normally adequate for

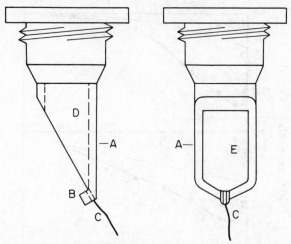

Figure 3.8. A technique for illuminating the prism of a dipping-type refractometer in an opaque solution[27]: A, Perspex collar; B, fibre optic holder; C, fibre optic; D, refractometer prism; E, polished face of prism

crystallisation work. It is advisable that calibration curves be measured, in terms of temperature and concentration, prior to the study with the actual system.

If a dipping-type refractometer[8] is used, a semi-continuous measurement may be made of the change in concentration as the system crystallises. However, if nucleation is heavy or if large numbers of crystals are present, it may be difficult to provide sufficient illumination for the prism because of the light scattering. One solution to this problem[27] is to use a fibre optic (a light wire) fitted into a collar around the prism illuminated from an external source (*Figure 3.8*). In this way undue heating of the solution is also avoided.

ELECTROLYTIC CONDUCTIVITY

The electrolytic conductance of an aqueous solution can often be measured with high precision and thus afford a useful means of determining concentration. A detailed account of the methods used in this area is given by ROBINSON and STOKES[19], but most published work is concerned with dilute systems.

In the author's experience, conductivity measurements are of limited use in crystallisation work because of the unreliability of measurement in near-saturated or supersaturated solutions. The temperature dependence of electrical conductivity usually demands a very high precision of temperature control. However, TORGESEN and HORTON[28] have successfully operated conductance cells for the control of ADP crystallisation and give a full description of their methods. They controlled the temperature to $\pm 0.002\,°C$.

CRYSTAL HARDNESS

Crystals vary in hardness not only from substance to substance but also from face to face on a given substance. One of the standard tests for hardness in non-metallic compounds and minerals is the scratch test, which gave rise to the Mohs scale. Ten 'degrees' of hardness are designated by common

Table 3.6. MOHS SCALE OF HARDNESS

Mohs hardness number	Reference substance	Formula	Approximate Vickers hardness, kgf/mm^2
1	talc	$3MgO \cdot 4SiO_2 \cdot H_2O$	50
2	gypsum	$CaSO_4 \cdot 2H_2O$	80
3	calcite	$CaCO_3$	130
4	fluorite	CaF_2	200
5	apatite	$CaF_2 \cdot 3Ca_3(PO_4)_2$	320
6	orthoclase	$K_2O \cdot Al_2O_3 \cdot 6SiO_2$	500
7	quartz	SiO_2	800
8	topaz	$(AlF)_2 \cdot SiO_4$	1 300
9	corundum	Al_2O_3	2 000
10	diamond	C	10 000

minerals in such an order that a given mineral will scratch the surface of any of the preceding members of the scale (see Table 3.6).

The hardness of metals is generally expressed in terms of their resistance to indentation. A hard indenter is pressed into the surface under the influence of a known load and the size of the resulting indentation is measured. A widely used instrument is the Vickers indenter, which gives a pyramidal indentation, and the results are expressed as a Vickers hardness number, $V(kgf/mm^2)$. Other tests include the Rockwell, which uses a conical indenter, and the Brinell, which uses a hard steel ball[29].

The relation between Mohs hardness, M, and the indentation hardness of solids is not a clear one. However, if diamond ($M = 10$) is omitted from the Mohs scale, the relationship

$$\log V = 0.2\,M + 1.5 \tag{3.24}$$

may be used for rough approximation purposes only. This equation does not apply for values of M between 9 (~ 2000 kgf/mm^2) and 10 ($\sim 10\,000$ kgf/mm^2). A few typical surface hardnesses (Mohs) of some common substances are:

sodium	0.5	aluminium	2–3
potassium	0.5	gold	2.5–3
lead	1.5	brass	3–4
magnesium	2	glass	3–4

The scratch test is not really suitable for specifying the hardness of substances commonly crystallised from aqueous solutions, because their Mohs values lie in a very short range, frequently between 1 and 3 for inorganic salts and below 1 for organic substances. For a reliable measurement of hardness of these soft crystals the indentation test is preferred. RIDGWAY[30] has measured mean values of the Vickers hardness (kgf/mm^2) for several crystalline substances:

sodium thiosulphate	($Na_2S_2O_3 \cdot 5H_2O$)	: 18
potassium alum	($KAl(SO_4)_2 \cdot 12H_2O$)	: 56
ammonium alum	($NH_4Al(SO_4)_2 \cdot 12H_2O$):	58
potassium dihydrogen phosphate	(KH_2PO_4)	:150
sucrose	($C_{12}H_{22}O_{11}$)	: 64

He has also determined the hardnesses of different faces of the same crystal:

ammonium dihydrogen phosphate	($NH_4H_2PO_4$)	(100)	: 69
		(110)	: 73
potassium sulphate	(K_2SO_4)	(100)	: 95
		(110)	:100
		(120)	:130

Hardness appears to be closely related to density (proportional to) and to atomic or molecular volume (inversely proportional), but few reliable data are available to develop reliable prediction methods. Much more work is clearly necessary in this area.

THERMAL PROPERTIES

Before the heating or cooling requirements for a crystallisation operation can be determined, a knowledge of many thermal properties of solid, liquid and gaseous systems may be required. A solution can be crystallised by cooling or evaporation, or by a combination of both processes. In the case of cooling, the heat that has to be removed from the system is the sum of the sensible heat and the heat of crystallisation (exothermic in most cases). Before the heating requirements for concentration processes can be calculated, the latent heat of vaporisation of the solvent and the heat of crystallisation of the solute must be known. In the case of crystallisation from the melt, a value of the latent heat of fusion, among other heat quantities, is needed.

The supersaturation of a solution, and thus its crystallisation, can also be achieved by the addition of a 'diluent' in the form of a liquid that is miscible in all proportions with the original solvent. The solute should be sparingly soluble in the diluent. This method of crystallisation is referred to as 'salting-out' or 'dilution crystallisation'. Water is the most widely used diluent in the crystallisation of organic compounds from alcoholic solution, but alcohols can be used as diluents in the crystallisation of inorganic salts from aqueous solution. When two solvents are mixed, heat is usually liberated; hence, a knowledge of the heat of mixing is required for the heat balance calculation of a salting-out process.

UNITS OF HEAT

The five heat energy units in common use are the gram-calorie (cal), the kilogram-calorie (kcal), the British thermal unit (Btu), the centigrade heat

Table 3.7. EQUIVALENT VALUES OF THE COMMON HEAT ENERGY UNITS

cal	kcal	Btu	chu	J
1	0·001	0·003 97	0·002 21	4·187
1 000	1	3·97	2·21	4 187
252	0·252	1	0·556	1 055
453·6	0·453 6	1·8	1	1 898
0·238 8	$2·388 \times 10^{-4}$	$9·478 \times 10^{-4}$	$5·275 \times 10^{-4}$	1

unit (chu) and finally the SI unit, the joule (J). This last unit is gradually replacing the others. The old definitions of the first four units are based on the heat energy required to raise the temperature of a unit mass of water by one degree:

1 cal = 1 g of water raised through 1 °C (or K)
1 kcal = 1 kg of water raised through 1 °C (or K)
1 chu = 1 lb of water raised through 1 °C (or K)
1 Btu = 1 lb of water raised through 1 °F (or R)

The new definitions of these units are linked to the basic SI unit, the joule. (1 J = 1 W s = 1 N m.) *Table 3.7* indicates the equivalent values of these heat units.

HEAT CAPACITY

The amount of heat energy associated with a given temperature change in a given system is a function of the chemical and physical states of the system. A measure of this heat energy can be expressed in terms of the quantity known as the heat capacity. The term 'specific heat' is often used synonymously with heat capacity, but strictly speaking the specific heat of a substance is the ratio of its heat capacity to the heat capacity of an equal mass of water, usually at the reference temperature of 15 °C. Heat capacities may be expressed on a mass or molal basis. As a one-degree temperature range is considered, the Celsius (centigrade), Kelvin, Fahrenheit or Rankine scales can be employed. The equivalent heat capacity units are:

Mass heat capacity, c

$$
\begin{aligned}
1 \text{ cal/g } °C \text{ (or K)} \quad &= 1 \text{ Btu/lb } °F \text{ (or } °R) \\
&= 1 \text{ chu/lb } °C \text{ (or K)} \\
&= 4 \cdot 187 \text{ kJ/kg K}
\end{aligned}
$$

Molal heat capacity, C

$$
\begin{aligned}
1 \text{ cal/mol } °C \text{ (or K)} &= 1 \text{ Btu/lb-mol. } °F \text{ (or } °R) \\
&= 1 \text{ chu/lb-mol. } °C \text{ (or K)} \\
&= 4 \cdot 187 \text{ kJ/kmol K}
\end{aligned}
$$

For gases two heat capacities have to be considered, at constant pressure, C_p, and at constant volume, C_v. The value of the ratio of these two quantities, $C_p/C_v = \gamma$, varies from about 1·67 for monatomic gases (e.g. He) to about 1·3 for triatomic gases (e.g. CO_2). For liquids and solids there is little difference between C_p and C_v, i.e. $\gamma \sim 1$, and it is usual to find C_p values only quoted in the literature.

Solids

Values of the heat capacities of solid substances near normal atmospheric temperature can be estimated with a reasonable degree of accuracy by combining two empirical rules. The first of these, due to Dulong and Petit, applies to solid elemental substances and may be written

mass heat capacity × atomic weight = atomic heat ≃ 6·2

The second rule, due to Kopp, applies to solid compounds and may be expressed by

molal heat capacity = sum of the atomic heats of the constituent atoms

In applying these rules, the following exceptions to the approximation 'atomic heat ≃ 6·2' must be noted:

$$
\begin{array}{llll}
C = 1 \cdot 8 & H = 2 \cdot 3 & B = 2 \cdot 7 & Si = 3 \cdot 8 \\
O = 4 \cdot 0 & F = 5 \cdot 0 & S = 5 \cdot 4 & [H_2O] = 9 \cdot 8
\end{array}
$$

The last substance, $[H_2O]$, refers to water as ice or as water of crystallisation in solid substances. Obviously a reliable measured value of a heat capacity is preferable to an estimated value, but in the absence of the former Kopp's rule can prove extremely useful. A few calculated and observed values of C_p are compared in *Table 3.8*.

Many inorganic solids have values of mass heat capacity, c_p, in the range $0.1-0.3$ cal/g °C and many organic solids have values in the range $0.2-0.5$.

Table 3.8. ESTIMATED (KOPP'S RULE) AND OBSERVED VALUES OF C_p FOR SEVERAL SOLID SUBSTANCES AT ROOM TEMPERATURE

Solid	Formula	Calculation	C_p, cal/mol °C Calc.	Obs.
Sodium chloride	NaCl	$6.2+6.2$	12.4	12.4
Magnesium sulphate	$MgSO_4 \cdot 7H_2O$	$6.2+5.4+4(4.0)+7(9.8)$	96.2	89.5
Iodobenzene	C_6H_5I	$6(1.8)+5(2.3)+6.2$	28.5	24.6
Naphthalene	$C_{10}H_8$	$10(1.8)+8(2.3)$	36.4	37.6
Potassium sulphate	K_2SO_4	$2(6.2)+5.4+4(4.0)$	33.8	30.6
Oxalic acid	$C_2H_2O_4 \cdot 2H_2O$	$2(1.8)+6(2.3)+4(4.0)$ $+2(9.8)$	43.8	43.5

In general, the heat capacity increases slightly with an increase in temperature. For example, the values of c_p at 0 and 100 °C for sodium chloride are 0.21 and 0.22 cal/g °C, respectively, and the corresponding values for anthracene are 0.30 and 0.35.

Liquids

Although several methods have been proposed[13] for the estimation of the heat capacity of a liquid, none is completely reliable. However, one recent method[31] worthy of mention is based on the additivity of the heat capacity contributions $[C_p]$ of the various atomic groupings in the molecules of organic liquids. *Table 3.9* gives some values of $[C_p]$, and the following

Table 3.9. CONTRIBUTIONS OF VARIOUS ATOMIC GROUPS TO THE HEAT CAPACITY OF ORGANIC LIQUIDS AT 20 °C (AFTER A. I. JOHNSON AND C. J. HUANG[31])

Group	$[C_p]$	Group	$[C_p]$
C_6H_5-	30.5	—OH	11.0
CH_3-	9.9	$-NO_2$	15.3
$-CH_2-$	6.3	$-NH_2$	15.2
—CH	5.4	—CN	13.9
—COOH	19.1	—Cl	8.6
—COO— (esters)	14.5	—Br	3.7
C=O (ketones)	14.7	—S—	10.6
—H (formates)	3.6	—O— (ethers)	8.4

examples illustrate the use of the method—the heat capacity values (cal/mol °C) in parentheses denote values obtained experimentally at 20 °C:

methyl alcohol $(CH_3 \cdot OH)$ $9.9 + 11.0 = 20.9$ (19.5)

toluene $(C_6H_5 \cdot CH_3)$ $30.5 + 9.9 = 40.4$ (36.8)

isobutyl acetate $CH_3 \cdot COO \cdot CH \Big\langle \begin{matrix} CH_3 \\ CH_2 \cdot CH_3 \end{matrix}$

$$= 3(9.9) + 14.5 + 5.4 + 6.3 = 55.9 \ (53.3)$$

The heat capacity of a substance in the liquid state is generally higher than that of a substance in the solid state. A large number of organic liquids have mass heat capacity values in the range 0·4–0·6 cal/g °C at about room temperature. The heat capacity of a liquid usually increases with increasing temperature; for example, the values of c_p for ethyl alcohol at 0, 20 and 60 °C are 0·54, 0·56 and 0·68 cal/g °C, those for benzene at 20, 40 and 60 °C are 0·40, 0·42 and 0·45. Water is an exceptional case; it has a very high heat capacity and exhibits a minimum value at 30 °C. The values of C_p for water at 0, 15, 30 and 100 °C are 1·008, 1·000, 0·9987 and 1·007 cal/g °C, respectively.

Liquid Mixtures

There is no reliable method for predicting the heat capacity of liquid mixtures, but in the absence of experimental data the following equation may be used for the molal heat capacity of a mixture of two or more liquids:

$$C_{p_{\text{mixt}}} = x_A C_{pA} + x_B C_{pB} + \ldots \tag{3.25}$$

where x denotes the mole fraction of the given component of the mixture. For example, the value of c_p for methanol (mol. wt. = 32·0) at 20 °C is 0·58 cal/g °C. From equation (3.25) it can be calculated that an aqueous solution containing 75 mol per cent of methanol (mean mol. wt. of mixture = 28·5) has a mass heat capacity of 0·64 cal/g °C. The experimental value at 20 °C is 0·69.

For dilute aqueous solutions of inorganic salts a rough estimate of the mass heat capacity can be made by ignoring the heat capacity contribution of the dissolved substance, i.e.

$$c_p = 1 - X \tag{3.26}$$

where X = g solute/g water, and c_p = cal/g °C. Thus solutions containing 5 g NaCl, 10 g KCl and 15 g CuSO$_4$ per 100 g of solution would, by this method, be estimated to have mass heat capacities of 0·95, 0·90 and 0·85 cal/g °C, respectively. Experimental values (25 °C) for these solutions are 0·94, 0·91 and 0·83. This estimation method cannot be applied to aqueous solutions of non-electrolytes or acids.

Another rough estimation method for the heat capacity of aqueous solutions is based on the empirical relationship

$$|c_p| = \left|\frac{1}{\rho}\right| \tag{3.27}$$

where ρ = density of the solution in g/cm^3. For example, at 30 °C a 2 per cent aqueous solution of sodium carbonate by weight has a density of $1\cdot016\ g/cm^3$ and a heat capacity of $0\cdot98$ cal/g °C $(1/\rho = 0\cdot985)$, while a 20 per cent solution has a density of $1\cdot21$ and a heat capacity of $0\cdot86$ $(1/\rho = 0\cdot83)$.

THERMAL CONDUCTIVITY

The thermal conductivity, κ, of a substance is defined as the rate of heat transfer by conduction across a unit area, through a layer of unit thickness, under the influence of a unit temperature difference, the direction of heat

Table 3.10. THERMAL CONDUCTIVITIES OF SOME PURE LIQUIDS

Temperature, °C	Thermal conductivity, κ (W/m K)					
	Water	Acetone	Benzene	Methanol	Ethanol	CCl$_4$
10	0·59	0·16	0·14	0·21	0·18	0·11
40	0·62	0·15	0·15	0·19	0·17	0·099
95	0·67	—	—	—	—	—

transmission being normal to the reference area. Fourier's equation for steady conduction may be written as

$$\frac{dq}{dt} = -\kappa A \frac{d\theta}{dx} \tag{3.28}$$

where q, t, A, θ and x are units of heat, time, area, temperature and length (thickness), respectively.

The units of κ, therefore, may be expressed as cal/s cm^2 (°C/cm), which may be contracted to cal/s cm °C, or Btu/h ft^2 (°F/ft), which may be contracted to Btu/h ft °F. The SI unit for thermal conductivity is W/m K, and the conversion factors are

$$1\ cal/s\ cm\ °C = 418\cdot7\ W/m\ K = 241\cdot9\ Btu/h\ ft\ °F$$

An increase in the temperature of a liquid usually results in a slight decrease in thermal conductivity, but water is a notable exception to this generalisation. *Table 3.10* gives values of κ for a few pure liquids. In the reference literature thermal conductivity data, particularly on solutions, are sparse and often conflicting, and no reliable method of estimation has yet been devised. For aqueous methanol and ethanol solutions, however, the following relationship due to BATES, HAZZARD and PALMER[32] may be used with a reasonable degree of accuracy up to about 80 °C:

$$\kappa_{mixt} = \frac{\kappa_1 \sinh m_1\delta + \kappa_2 \sinh m_2\delta}{\sinh (100\delta)} \qquad (3.29)$$

where κ_1, κ_2 = thermal conductivities of water and alcohol, m_1, m_2 = mass fractions of water and alcohol, and δ = a constant (0·90 for methanol, 0·94 for ethanol).

FREEZING AND BOILING POINTS

When a non-volatile solute is dissolved in a solvent, the vapour pressure of the solvent is lowered. Consequently, at any given pressure, the boiling point of a solution is higher and the freezing point lower than those of the pure solvent. For dilute ideal solutions, i.e. such as obey Raoult's law, the boiling point elevation and freezing point depression can be calculated by an equation of the form

$$\Delta T = \frac{mK}{M} \qquad (3.30)$$

where m = mass of solute dissolved in a given mass of pure solvent (usually 1000 g) and M = molecular weight of the solute. When ΔT refers to the freezing point depression, $K = K_f$, the cryoscopic constant; when ΔT refers to the boiling point elevation, $K = K_b$, the ebullioscopic constant. Values of K_f and K_b for several common solvents are given in *Table 3.11*; these, in effect, give the depression in freezing point, or elevation in boiling point, in °C when 1 mol of solute is dissolved, without dissociation or association, in 1000 g of solvent.

The cryoscopic and ebullioscopic constants can be calculated from values

Table 3.11. CRYOSCOPIC AND EBULLIOSCOPIC CONSTANTS FOR SOME COMMON SOLVENTS

Solvent	Freezing point, °C	Boiling point, °C	K_f	K_b
Acetic acid	16·7	118·1	3·9	3·1
Acetone	−95·5	56·3	2·7	1·7
Aniline	− 6·2	184·5	5·9	3·2
Benzene	5·5	80·1	5·1	2·7
Carbon disulphide	−108·5	46·3	3·8	2·4
Carbon tetrachloride	− 22·6	76·8	32·0	4·9
Chloroform	− 63·5	61·2	4·8	3·8
Cyclohexane	6·2	80	20·0	2·8
Nitrobenzene	5·7	211	7·0	5·3
Methyl alcohol	−97·8	64·7	2·6	0·8
Phenol	42·0	181	7·3	3·0
Water	0·0	100·0	1·86	0·52

of the latent heats of fusion and vaporisation, respectively, by the equation

$$K = \frac{RT^2}{1000l} \tag{3.31}$$

When $K = K_f$, T refers to the freezing point, T_f (K), and l to the latent heat of fusion, l_f (cal/g). When $K = K_b$, T refers to the boiling point, T_b, and $l = l_{vb}$, the latent heat of vaporisation at the boiling point. The gas constant $R = 1\cdot987$ cal/mol K.

Equation (3.31) cannot be applied to concentrated solutions or to aqueous solutions of electrolytes. In these cases the freezing point depression cannot readily be estimated. The boiling point elevation, however, can be predicted with a reasonable degree of accuracy by means of an empirical rule. Dühring (1878) observed that the boiling point of a solution is a linear

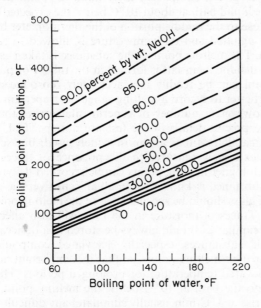

Figure 3.9. Dühring plot for aqueous solutions of sodium hydroxide

function of the boiling point of the pure solvent. Therefore, if the boiling points of solutions of different concentrations are plotted against those of the solvent at different pressures, a family of straight lines (not necessarily parallel) will be obtained. A typical Dühring plot for aqueous solutions of sodium hydroxide is given in Figure 3.9, from which it can be seen, for example, that at a pressure at which water boils at 180 °F (83 °C), a solution containing 50 per cent by weight of NaOH would boil at about 260 °F (127 °C), i.e. a boiling point elevation of about 80 °F (44 °C).

MELTING POINT

The melting point of a solid organic substance is frequently adopted as a criterion of purity; but before any reliance can be placed on the test, it is absolutely necessary for the experimental procedure to be standardised. Several types of melting point apparatus are available commercially, such as heating blocks on which samples of the material are placed and observed while the temperature of the surface of the block is recorded, but the most widely used method, and the one most frequently recommended, consists of heating a powdered sample of the material in a glass capillary tube located close to the bulb of a thermometer in an agitated bath of liquid.

The best type of glass tube is 0·9–1·2 mm internal diameter, about 100 mm long, with walls about 0·1 mm thick. The tube is sealed at one end, and the powdered sample is scraped into the tube and knocked or vibrated down to the closed end to give a compacted layer about 2–4 mm deep. It is then inserted into the liquid bath at about 10 °C below the expected melting point and attached close to the middle portion of the thermometer bulb. The bath is continuously agitated and the temperature is allowed to rise steadily at about 3 °C/min. The melting point of the substance is taken as the temperature at which a definite meniscus is formed in the tube. For pure substances the melting point can be readily and accurately reproduced; for impure substances it is better to record a melting range of temperature.

There are, however, a number of experimental errors that can be encountered if the precise details of the test are not specified. For instance, soft soda glass capillary tubes are most widely used, but several organic substances are extremely sensitive to the presence of traces of alkali. In these cases much higher melting points can be recorded if Pyrex glass tubes are used. If a substance is known to be alkali-sensitive, e.g. acetylsalicylic acid, the type of glass should be specified before the melting point is used as a test for purity. Traces of moisture in the tube will also affect the melting point; thus the capillaries should always be stored in a desiccator.

Some organic substances, especially acetylated compounds, begin to decompose near their melting points, so that it is important not to keep the sample at an elevated temperature for prolonged periods. The insertion of the capillary into the bath at 10 °C below the melting point, allowing the temperature to rise at 3 °C/min, usually eliminates any difficulties; with some unstable compounds an insertion at 5 °C below the melting point may, however, be necessary. For reproducible results to be obtained, the sample should be in a finely divided state, at least finer than 150-mesh ($\sim$ 100 μm). The thermometer, preferably divided in $\frac{1}{10}$ °C, should be accurately calibrated over its whole range and, if in constant use, should be checked regularly. The well-known standardisation temperatures are the freezing and boiling points of water, but other standards that can be used are the melting points of pure organic substances, such as those indicated in *Table 3.12*.

The liquid used in the heating bath depends on the working temperature; water is quite suitable for melting points from about room temperature to about 70 °C; liquid paraffin is widely used but readily darkens at temperatures in excess of about 200 °C. Phosphoric acid and glycerol, also used,

Table 3.12. MELTING POINTS OF PURE ORGANIC COMPOUNDS USED FOR
CALIBRATING THERMOMETERS

Substance	Melting point, °C
Phenyl salicylate (salol)	42
p-Dichlorbenzene	53
Naphthalene	80
m-Dinitrobenzene	90
Acetamide	114
Benzoic acid	122
Urea	133
Salicylic acid	159
Succinic acid	183
Anthracene	217
p-Nitrobenzoic acid	242
Anthraquinone	285

likewise tend to darken. Sulphuric acid or solutions of potassium sulphate in H_2SO_4 are frequently recommended for heating-bath media, but the potential hazards hardly need any emphasis. In recent years silicone fluids have become available which, though expensive, do not darken and have a prolonged working life.

LATENT HEAT

When a substance undergoes a phase change, a quantity of heat is transferred between the substance and its surrounding medium. This enthalpy change is called the latent heat, and the following types may be distinguished:

solid I $\rightleftharpoons$ solid II (latent heat of transition)
solid $\rightleftharpoons$ liquid (latent heat of fusion)
solid $\rightleftharpoons$ gas (latent heat of sublimation)
liquid $\rightleftharpoons$ gas (latent heat of vaporisation)

Only the last three of these represent significant quantities of heat energy; heats of transition can for most industrial purposes be ignored. For example, the transformation of monoclinic to orthorhombic sulphur is accompanied by an enthalpy change of about 0·4 cal/g, whereas the fusion of orthorhombic sulphur is accompanied by an enthalpy change of about 17 cal/g.

Latent heats, like the other thermal properties, can be expressed on a mass or molar basis, e.g. cal/g or cal/mol; but to avoid confusion, all latent heats in this section will be expressed on a molar basis. The relationship between the old-established and the newer SI units for latent heat are:

1 cal/g $= 1$ chu/lb $= 1·8$ Btu/lb $= 4187$ J/kg
1 cal/mol $= 1$ chu/lb-mol $= 1·8$ Btu/lb-mol $= 4187$ J/kmol $= 4·187$ J/mol

The relationship between any latent heat, l, and the pressure–volume–temperature conditions of a system is given by the Clapeyron equation

$$\frac{\mathrm{d}p}{\mathrm{d}T} = \frac{l}{T\Delta v} \tag{3.32}$$

where $\mathrm{d}p/\mathrm{d}T$ = rate of change of vapour pressure with absolute temperature and Δv = volume change accompanying the phase change.

A typical temperature–pressure phase diagram for a one-component system is shown in *Figure 3.10*. The sublimation curve AX indicates the increase of the vapour pressure of the solid with an increase in temperature. This is expressed quantitatively by the Clapeyron equation written as

$$\frac{\mathrm{d}p}{\mathrm{d}T} = \frac{l_s}{T(v_g - v_s)} \tag{3.33}$$

where v_g and v_s = the molar specific volumes of the vapour and solid, respectively, and l_s = latent heat of sublimation. The vaporisation curve XB

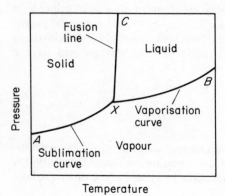

Figure 3.10. Temperature–pressure diagram for a single-component system

indicates the increase of the vapour pressure of the liquid with an increase in temperature:

$$\frac{\mathrm{d}p}{\mathrm{d}T} = \frac{l_v}{T(v_g - v_l)} \tag{3.34}$$

where v_l = the specific volume of the liquid, and l_v = latent heat of vaporisation. Curve XB is not a continuation of curve AX; this fact can be confirmed by calculating their slopes $\mathrm{d}p/\mathrm{d}T$ from equations (3.33) and (3.34) at point X.

The fusion line XC indicates the effect of pressure on the melting point of the solid; it can either increase or decrease with an increase in pressure, but the effect is so small that line XC deviates only slightly from the vertical. When the fusion line deviates to the right, as in *Figure 3.10*, the melting point increases with an increase in pressure, and the substance contracts on freezing. Most substances behave in this manner. When the line deviates to the left, the melting point decreases with increasing pressure, and the substance expands on freezing. Water and type metals are among the few examples of this behaviour that can be quoted. The equation for the fusion line is

$$\frac{\mathrm{d}p}{\mathrm{d}T} = \frac{l_f}{T(v_1 - v_s)} \tag{3.35}$$

where l_f is the latent heat of fusion.

The latent heats of sublimation, vaporisation and fusion are related by

$$l_s = l_f + l_v \tag{3.36}$$

but this additivity equation is applicable only at one specific temperature. The variation of a latent heat with temperature can be calculated from the Clausius equation

$$\frac{\mathrm{d}l}{\mathrm{d}T} - \frac{l}{T} = c_2 - c_1 \tag{3.37}$$

When $l = l_v$, c_1 and c_2 are the heat capacities of the liquid, just on the point of vaporisation, and of the saturated vapour, respectively. Equation (3.37) can also be used for calculating l_f and l_s, the appropriate values of c being inserted.

The specific volumes v_1 and v_s are much smaller than v_g; equations (3.33) and (3.34) can therefore be simplified to

$$\frac{\mathrm{d}p}{\mathrm{d}T} = \frac{l}{Tv_g} \tag{3.38}$$

where $l = l_v$ or l_s. From the ideal gas laws $v_g = RT/p$, so that equation (3.38) may be written

$$\frac{\mathrm{d}p}{p} = \frac{l}{RT^2} \mathrm{d}T \tag{3.39}$$

This is the Clausius–Clapeyron equation. If the latent heat is considered to be constant over a small temperature range, T_1 to T_2, equation (3.39) may be integrated to give

$$\ln \frac{p_2}{p_1} = l \frac{(T_2 - T_1)}{RT_1 T_2} \tag{3.40}$$

Equation (3.40) can be used to estimate latent heats of vaporisation and sublimation if vapour pressure data are available, or to estimate vapour pressures from a value of the latent heat. Analysis of sublimation problems, for instance, is frequently difficult owing to the scarcity of published vapour pressure and latent heat data. If two values of vapour pressure are available, however, a considerable amount of information can be derived from equation (3.40), as illustrated by the following example[33].

Example

Suppose that the only data available on solid anthracene are that its vapour pressures at 210 and 145 °C are 30 and 1 mmHg, respectively; the vapour pressure at 100 °C is required.

Solution

Equation (3.40) can be used twice — first to calculate a value of l_s, then that of the required pressure:

$$\ln\left(\frac{30}{1}\right) = \frac{l_s(483-418)}{1\cdot99 \times 483 \times 418}$$

$$l_s = 21\,000 \text{ cal/mol}$$

Substituting this value of l_s in equation (3.40) again,

$$\ln\left(\frac{30}{p_{100^\circ C}}\right) = \frac{21\,000(483-373)}{1\cdot99 \times 483 \times 373}$$

therefore,

$$p_{100^\circ C} = 0\cdot048 \text{ mmHg}$$

Heat of Vaporisation

There are several methods available for the estimation of latent heats of vaporisation at the atmospheric boiling point of the liquid, l_{vb}. The well-known rules due to Trouton (1884) and Kistiakowsky (1923) are only suitable for nonpolar liquids, but a more recent method due to GIACALONE[34] is fairly reliable for both polar and non-polar liquids. These three empirical rules may be summarised

$$l_{vb} = 21\,T_b \qquad \text{(Trouton)} \qquad (3.41)$$

$$= (8\cdot75 + 4\cdot571\log_{10} T_b)T_b \quad \text{(Kistiakowsky)} \qquad (3.42)$$

$$= \left(\frac{RT_cT_b}{T_c - T_b}\right)\ln P_c \qquad \text{(Giacalone)} \qquad (3.43)$$

where l_{vb} = latent heat of vaporisation (cal/mol) at boiling point, T_b = boiling point (K) of the liquid at 760 mmHg, T_c = critical temperature (K) of the liquid, and P_c = critical pressure (atm) of the liquid. (1 mmHg = 133·3 N/m²; 1 atm = 760 mmHg = $1\cdot013 \times 10^5$ N/m².)

Table 3.13 illustrates the use of these three equations and compares calculated

Table 3.13. COMPARISON BETWEEN ESTIMATED AND EXPERIMENTAL VALUES OF THE LATENT HEAT OF VAPORISATION OF A LIQUID AT ITS BOILING POINT

Liquid	Atm. boiling point, K	Critical temperature, K	Critical pressure, atm	Latent heat of vaporisation at boiling point, cal/mol			
				Eq. (3.41)	Eq. (3.42)	Eq. (3.43)	Experimental
Benzene	353	563	48·6	7 413	7 200	7 300	7 350
Naphthalene	491	750	39·2	10 310	10 340	10 320	10 250
Water	273	647	218·0	5 733	5 460	9 440	9 708
Ethanol	351	516	62·7	7 371	7 130	9 050	9 380

and experimental values of l_{vb}. It can be seen that equations (3.41) and (3.42) are completely unreliable for polar liquids. Despite its limitations, Trouton's rule has the merit of simplicity and when coupled with equation (3.40) provides a rapid estimation method for the boiling point of a non-polar liquid at pressures both above and below atmospheric; the latent heat of vaporisation is assumed to be a constant. A combination of equations (3.40) and (3.41) gives

$$\ln \frac{p_2}{p_1} = \frac{21 T_2(T_2 - T_1)}{R T_2 T_1} \tag{3.44}$$

Example

Benzene boils at 80 °C at 760 mmHg. Estimate its boiling point at a pressure of 200 mmHg.

Solution

From equation (3.44)

$$\ln \left(\frac{760}{200} \right) = \frac{21 \times 353 \times (353 - T_1)}{1 \cdot 99 \times 353 \times T_1}$$

$$T_1 = 314 \text{ K } (41 \text{ °C})$$

The observed boiling point of benzene at 200 mmHg is 42·0 °C.

The latent heat of vaporisation of a liquid at some temperature T_1 can be calculated from the latent heat at another temperature T_2 by means of the equation, due to WATSON[35],

$$l_{v_1} = l_{v_2} \left(\frac{T_c - T_1}{T_c - T_2} \right)^{0 \cdot 38} \tag{3.45}$$

For example, the latent heat of vaporisation of benzene at its boiling point (353 K) is 7350 cal/mol, its critical temperature 563 K. From equation (3.45) the corresponding value at 25 °C (298 K) can be calculated as 8030 cal/mol, which compares with an experimental result of 8060 cal/mol.

HEATS OF SOLUTION AND CRYSTALLISATION

Heat is usually absorbed from the surrounding medium when a solute dissolves in a solvent without reaction (heat of solution), or the solution temperature falls if the dissolution occurs adiabatically. When a solute crystallises out of its solution, heat is usually liberated (heat of crystallisation) and the solution temperature is increased. The reverse cases, viz. heat evolution on dissolution and heat absorption on crystallisation, are also encountered, especially with solutes that exhibit an inverted solubility characteristic. The dissolution of an anhydrous salt in water at a temperature at which the hydrated salt is the stable crystalline form frequently leads to the formation of heat energy, owing to the exothermic nature of the hydration process:

$$AB + xH_2O \rightarrow AB \cdot xH_2O$$
$$\text{(anhydrous)} \qquad\qquad \text{(hydrate)}$$

Table 3.14 lists the heats of solution of anhydrous and hydrated magnesium sulphate and sodium carbonate in water to illustrate the effect of water of crystallisation.

The enthalpy changes associated with dissolution (ΔH_{soln}) and crystallisation (ΔH_{cryst}) are generally recorded as the number of heat units liberated by the system when the process takes place isothermally. According to this system of nomenclature, if an adiabatic operation is considered, the expression $\Delta H_{soln} = +q$ (heat units per unit mass of solute) means that the solution temperature will increase; $\Delta H_{soln} = -q$ means that it will fall.

The magnitude of the heat effect accompanying the dissolution of a solute in a given solute in a solvent or undersaturated solution depends on the quantities of solute and solvent involved, the initial and final concentrations

Table 3.14. HEATS OF SOLUTION OF ANHYDROUS AND HYDRATED SALTS IN WATER AT 18 °C AND INFINITE DILUTION

Salt	Formula	Heat of solution, kcal/mol
Magnesium sulphate	$MgSO_4$	+21·1
	$MgSO_4 \cdot H_2O$	+14·0
	$MgSO_4 \cdot 2H_2O$	+11·7
	$MgSO_4 \cdot 4H_2O$	+ 4·9
	$MgSO_4 \cdot 6H_2O$	+ 0·55
	$MgSO_4 \cdot 7H_2O$	− 3·18
Sodium carbonate	Na_2CO_3	+ 5·57
	$Na_2CO_3 \cdot H_2O$	+ 2·19
	$Na_2CO_3 \cdot 7H_2O$	−10·18
	$Na_2CO_3 \cdot 10H_2O$	−16·22

1 kcal/mol = 4·187 kJ/mol.

and the temperature at which the dissolution occurs. The standard reference temperature is now generally taken as 25 °C, but the older reference temperature of 18 °C is still encountered.

In crystallisation practice it is usual to take the heat of crystallisation as being equal in magnitude, but opposite in sign, to the heat of solution, i.e.

$$\Delta H_{cryst} = -\Delta H_{soln} \qquad (3.46)$$

This assumption, of course, is not strictly correct, but the error involved is small. Heats of solution are generally recorded as the enthalpy change associated with the dissolution of a unit quantity of solute in a large excess of pure solvent, i.e. the heat of solution at infinite dilution, ΔH_{soln}^{∞}. For most practical purposes the term 'infinite dilution' is taken to mean <0·01 mole fraction of solute in the solution.

The heat of crystallisation is numerically equal to the heat of solution only when the latter refers to the dissolution of the solute in an almost saturated solution at the specific temperature. The temperature correction may be neglected, but the heat of dilution, ΔH_{dil}, should be taken into account if an accurate value of the heat of crystallisation is required, i.e.

$$\Delta H_{cryst} = -\Delta H_{soln}^{\infty} + \Delta H_{dil} \qquad (3.47)$$

Few values of heats of dilution are available in the literature, especially for the concentration ranges usually required, but this quantity is usually only a small fraction of the heat of solution. Furthermore, as the dilution of most aqueous salt solutions is exothermic, i.e. the concentration is endothermic, the true value of the heat of crystallisation will be slightly less than that obtained by taking the negative value of the heat of solution alone. Therefore the calculated quantity of heat to be removed from a crystallising solution will be slightly greater than the true value; this small error acts as a factor of safety in the design of the heat transfer equipment.

ENTHALPY–CONCENTRATION DIAGRAMS

The heat effects accompanying a crystallisation operation are most frequently determined by making a heat balance over the system, and for a reasonable degree of accuracy many calculations may be necessary, involving a knowledge of heat capacities and heats of crystallisation, dilution, vaporisation and so on. Much of the burden of calculation, however, can be eased by the use of a graphical technique. Merkel (1929) and Bošnjaković (1932) demonstrated a convenient method for representing enthalpy data for solutions on an enthalpy–concentration (H–x) diagram. McCABE[36] drew attention to the use of the H–x chart for the analysis of several chemical engineering operations, and this approach is now widely used for distillation, evaporation and refrigeration processes, to name but a few.

With regard to applying it to crystallisation there are two difficulties. First, enthalpy–concentration diagrams are available, in the literature at least, for only a very few aqueous–inorganic systems. Second, the construction of an H–x chart is laborious and would normally be undertaken only if many calculations were to be performed, e.g. on a system of commercial importance. Nevertheless, once an H–x chart is available, its use is simple, and a great deal of information can be obtained rapidly. If the concentration x of one component of a binary mixture is expressed as a mass fraction, the enthalpy is expressed as a number of heat units per unit mass of mixture, e.g. Btu/lb, cal/g or J/kg. Molar units are less frequently used.

The basic rule governing the use of an H–x chart is that an adiabatic mixing, or separation, process is represented by a straight line. In *Figure 3.11*, points A and B represent the concentrations and enthalpies x_A, H_A and x_B, H_B of two mixtures of the same system. If A is mixed adiabatically with B, the enthalpy and concentration of the resulting mixture is given by point C on the straight line AB. The exact location of point C, which depends on the masses m_A and m_B of the two initial mixtures, can be determined by the mixture rule or lever-arm principle:

$$m_A(x_C - x_A) = m_B(x_B - x_C) \tag{3.48}$$

or

$$x_C = \frac{m_B x_B + m_A x_A}{m_A + m_B} \tag{3.49}$$

Similarly, if mixture A were to be removed adiabatically from mixture C,

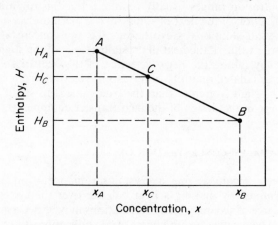

Figure 3.11. An adiabatic mixing process represented on an H–x diagram

the enthalpy and concentration of residue B can be located on the straight line through points A and C by means of the equation

$$x_B = \frac{m_C x_C - m_A x_A}{m_C - m_A} \tag{3.50}$$

An H–x diagram for the system NaOH–water at atmospheric pressure is shown in *Figure 3.12*; this chart[36, 37] is constructed on the basis that the enthalpy of pure water at 32 °F is zero. The curved isotherms refer to homogeneous solutions only. The lower right-hand region of the diagram below the saturation curve represents saturated solutions in equilibrium with the various hydrates of NaOH. A simple example will demonstrate the use of this diagram.

Example

150 lb of water at 50 °F is added to 100 lb of a 45 per cent solution of NaOH. If the mixing is carried out adiabatically, estimate the temperature of the resulting mixture.

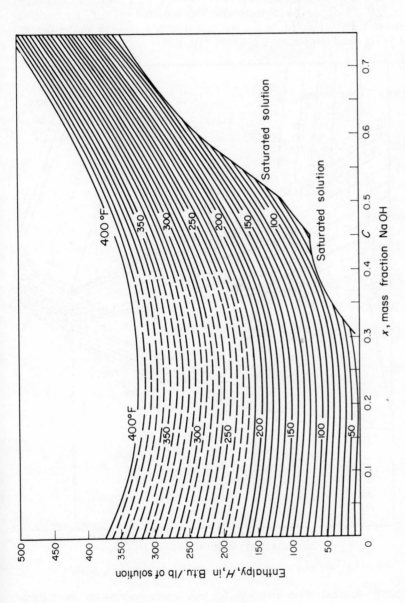

Figure 3.12. Enthalpy–concentration diagram for the system sodium hydroxide–water. (From W. L. McCABE and J. C. SMITH[37], by courtesy of McGraw-Hill)

Solution

Locate on *Figure 3.12* point *A* at $x_A = 0$ on the 50 °F isotherm and point *B* at $x_B = 0.45$ on the 60 °F isotherm. The required point *C* representing the final mixture is located, by the mixture rule, at $x_C = 0.18$ on the straight line *AB*. The final temperature is about 110 °F.

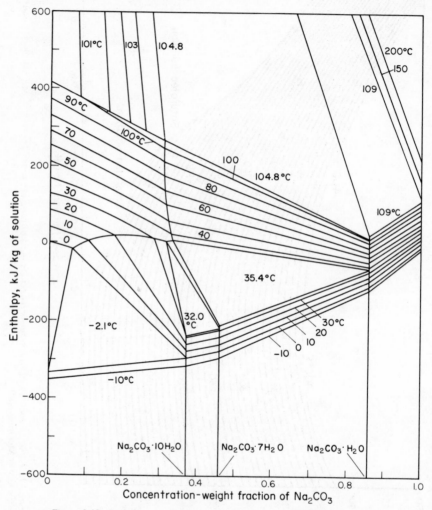

Figure 3.13. *Enthalpy–concentration diagram for the system* Na_2CO_3–H_2O

Enthalpy–concentration diagrams for salt–water systems are more complex than the one described above (*Figure 3.12*), because enthalpies of liquid and solid phases have to be recorded. Charts for aqueous solutions of sodium carbonate, sodium sulphate (both in SI units) and magnesium sulphate (in Imperial units) are given in *Figures 3.13–3.15*. In *Figure 3.15*, for example, the isotherms in the region above curve *pabcdq* represent enthalpies and concentrations of unsaturated aqueous solutions of $MgSO_4$, and the very slight curvature of these isotherms, compared with those in *Figure 3.12*, indicates that the heat of dilution of $MgSO_4$ solutions is very small. Point *p* (zero enthalpy) represents pure water at 32 °F, point *n* the enthalpy of pure

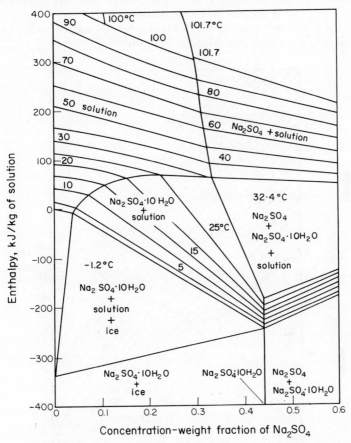

Figure 3.14. Enthalpy–concentration diagram for the system Na_2SO_4–H_2O

ice at the same temperature. The portion of the diagram below curve *pabcdq*, which represents liquid–solid systems, can be divided into five polythermal regions:

pae   solutions of $MgSO_4$ in equilibrium with pure ice
abfg   equilibrium mixtures of $MgSO_4 \cdot 12H_2O$ and saturated solution
bcih   equilibrium mixtures of $MgSO_4 \cdot 7H_2O$ and saturated solution
cdlj   equilibrium mixtures of $MgSO_4 \cdot 6H_2O$ and saturated solution
dqrk   equilibrium mixtures of $MgSO_4 \cdot H_2O$ and saturated solution

In between these five regions lie four isothermal triangular areas, which represent the following conditions:

aef (25 °F)   mixtures of ice, cryohydrate *a* and $MgSO_4 \cdot 12H_2O$

bfh (37·5 °F) mixtures of solid $MgSO_4 \cdot 12H_2O$ and $MgSO_4 \cdot 7H_2O$ in a 21 per cent $MgSO_4$ solution

cji (118·8 °F) mixtures of solid $MgSO_4 \cdot 7H_2O$ and $MgSO_4 \cdot 6H_2O$ in a 33 per cent $MgSO_4$ solution

dkl (154·4 °F) mixtures of solid $MgSO_4 \cdot 6H_2O$ and $MgSO_4 \cdot H_2O$ in a 37 per cent $MgSO_4$ solution

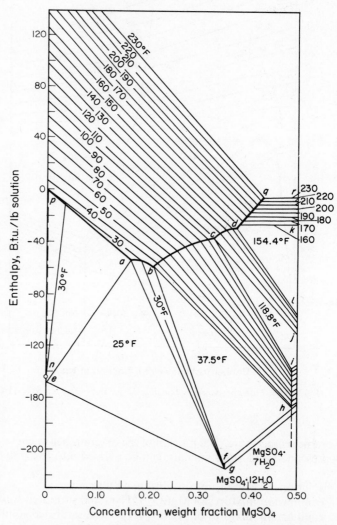

Figure 3.15. Enthalpy–concentration diagram for the system $MgSO_4$–H_2O. (From W. L. MCCABE[38], *by courtesy of* McGraw-Hill)

The short vertical lines *fg* and *ih* represent the compositions of solid $MgSO_4\cdot$ $12H_2O$ (0·359 mass fraction $MgSO_4$) and $MgSO_4\cdot7H_2O$ (0·49 mass fraction). The following example demonstrates the use of *Figure 3.15*.

Example

Calculate (a) the quantity of heat to be removed and (b) the theoretical crystal yield when 5000 lb of a 30 per cent solution of $MgSO_4$ by weight at

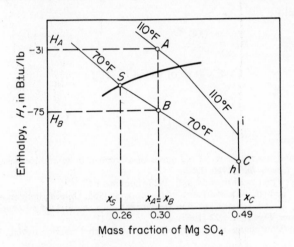

Figure 3.16. Graphical solution of Example (*enlarged portion of Figure 3.15 — not to scale*)

110 °F is cooled to 70 °F. Evaporation and radiation losses may be neglected.

Solution

Figure 3.16 indicates the relevant section – not to scale – of the *H–x* diagram in *Figure 3.15*.

(a) Initial solution, *A* $\quad\quad x_A = 0\cdot30,\ H_A = -31$ Btu/lb
Cooled system, *B* $\quad\quad\quad x_B = 0\cdot30,\ H_B = -75$ Btu/lb
Enthalpy change $\quad\quad\quad\quad\quad\quad \Delta H = -44$ Btu/lb
Heat to be removed $\quad\quad 44 \times 5000 = 220\,000$ Btu

(b) The cooled system *B*, located in the region *bcih* in *Figure 3.15*, comprises $MgSO_4\cdot7H_2O$ crystals in equilibrium with solution *S* on curve *bc*. The actual proportions of solid and solution can be calculated by the mixture rule.

Solution composition $\quad\quad\quad\quad x_S = 0\cdot26$
Crystalline phase composition $\quad\quad x_C = 0\cdot49$

Thus

$$\frac{m_S}{m_C} = \frac{x_C - x_B}{x_B - x_S}$$

$$= \frac{0\cdot49 - 0\cdot30}{0\cdot30 - 0\cdot26} = 4\cdot75$$

and

$$m_B = m_S + m_C = 5000 \text{ lb}$$

Therefore, yield

$$m_C = \frac{5000}{5\cdot75}$$

$$= 870 \text{ lb of } MgSO_4 \cdot 7H_2O$$

REFERENCES

1. GARSIDE, J. and MULLIN, J. W., 'Continuous measurement of solution concentration in a crystalliser', *Chemy Ind.*, **1966**, 2007
2. HATSCHEK, E., *The Viscosity of Liquids*, 1928. London; Bell
3. STOKES, R. H. and MILLS, R., *Viscosity of Electrolytes*, 1965, Oxford; Pergamon
4. FRANKEL, N. A. and ACRIVOS, A., 'On the viscosity of concentrated suspension of solid spheres', *Chem. Engng Sci.*, **22** (1967), 847
5. CLARK, R. C., 'Continuous monitoring of viscosity of stirred reaction media', *Chemy Ind.*, **1966**, 489
6. *Determination of the Viscosity of Liquids*, B.S. 188, 1957. London; British Standards Institution
7. DINSDALE, A. and MOORE, F., *Viscosity and its Measurement*, 1962. London; Institute of Physics
8. FINDLAY, A., *Practical Physical Chemistry*, 8th Ed. (ed. J. A. KITCHENER), 1954. London; Longmans Green
9. HARKINS, W. D. and ALEXANDER, A. E., in *Physical Methods of Organic Chemistry*, 3rd Ed. (ed. A. WEISSBERGER), p. 757, 1959. New York; Interscience
10. HARNED, H. S. and OWEN, B. O., *The Physical Chemistry of Electrolytic Solutions*, 3rd Ed., 1958. New York; Reinhold
11. MULLIN, J. W. and COOK, T. P., 'Diffusivity of ammonium dihydrogen phosphate in aqueous solutions', *J. appl. Chem., Lond.*, **13** (1963), 423
12. NIENOW, A. W., 'Diffusivity in the liquid phase', *Brit. Chem. Engng*, **10** (1965), 827
13. REID, R. C. and SHERWOOD, T. K., *The Properties of Gases and Liquids*, 2nd Ed., 1966. New York; McGraw-Hill
14. OTHMER, D. F. and THAKAR, M. S., 'Correlating diffusion coefficients in liquids', *Ind. Engng Chem.*, **45** (1953), 589
15. WILKE, C. R. and CHANG, P., 'Correlation of diffusion coefficients in dilute solutions', *A. I. Ch. E. Jl*, **1** (1955), 264
16. MULLIN, J. W. and COOK, T. P., 'Diffusion and dissolution of the hydroxybenzoic acids in water', *J. appl. Chem., Lond.*, **15** (1965), 145
17. GEDDES, A. L. in *Techniques of Organic Chemistry* (ed. A. WEISSBERGER), Vol. I, Pt. I, 1949. New York; Interscience
18. JOST, W., *Diffusion in Solids, Liquids and Gases*, 1952. New York; Academic Press
19. ROBINSON, R. A. and STOKES, R. H., *Electrolyte Solutions*, 2nd Ed., 1970. London; Butterworths

20. TYRRELL, H. J. V., *Diffusion and Heat Flow in Liquids*, 1961. London; Butterworths
21. DULLIEN, F. A. L. and SHEMILT, L. W., 'Diffusion coefficients for the liquid system ethanol-water', *Can. J. chem. Engng*, **39** (1961), 242
22. MULLIN, J. W., GARSIDE, J. and UNAHABHOKHA, R., 'Diffusivities of ammonium and potassium alums in aqueous solutions', *J. appl. Chem., Lond.*, **15** (1965), 502
23. NIENOW, A. W., UNAHABHOKHA, R. and MULLIN, J. W., 'Diffusion and mass transfer of ammonium and potassium chloride in aqueous solution', *J. appl. Chem., Lond.*, **18** (1968), 154
24. MULLIN, J. W. and NIENOW, A. W., 'Diffusion coefficients of potassium sulphate in water', *J. chem. Engng Data*, **9** (1964), 526
25. MULLIN, J. W. and OSMAN, M. M., 'Diffusivity, density, viscosity and refractive index of nickel ammonium sulphate aqueous solutions', *J. chem. Engng Data*, **12** (1967), 516
26. GLASSTONE, S., LAIDLER, K. J. and EYRING, H., *The Theory of Rate Processes*, 1941. New York; McGraw-Hill
27. LECI, C. L. and MULLIN, J. W., 'Refractive index measurements in liquids rendered opaque by the presence of suspended solids', *Chemy Ind.*, **1968**, 1517
28. TORGESEN, J. L. and HORTON, A. T., 'Electrolytic conductance of ammonium dihydrogen phosphate solutions in the saturated region', *J. phys. Chem., Ithaca*, **67** (1963), 376
29. BOWDEN, F. P. and TABOR, D., *The Friction and Lubrication of Solids*, Part II, pp. 320–349, 1964. Oxford; Clarendon
30. RIDGWAY, K., School of Pharmacy, University of London (private communication)
31. JOHNSON, A. I. and HUANG, C. J., 'Estimation of heat capacities of organic liquids', *Can. J. Technol.*, **33** (1955), 421
32. BATES, O. K., HAZZARD, G. and PALMER, G., 'Thermal conductivity of liquids', *Ind. Engng Chem. analyt. Edn*, **10** (1938), 314
33. MULLIN, J. W., 'Sublimation in theory and practice', *Ind. Chemist*, **31** (1955), 540
34. GIACALONE, A., *Gazz. chim. ital.*, **81** (1951), 180 (reported in reference 13)
35. WATSON, K. M., 'Thermodynamics of the liquid state', *Ind. Engng Chem.*, **35** (1943), 398
36. MCCABE, W. L., 'The enthalpy-concentration chart – a useful device for chemical engineering calculations', *Trans. Am. Inst. chem. Engrs*, **31** (1935), 129
37. MCCABE, W. L. and SMITH, J. C., *Unit Operations of Chemical Engineering*, 2nd Ed., 1968. New York; McGraw-Hill
38. MCCABE, W. L., 'Crystallisation', in *Chemical Engineers' Handbook*, 4th Ed. (ed. J. H. PERRY), 1963. New York; McGraw-Hill

4

Phase Equilibria

THE amount of information which the simple solubility diagram can yield is strictly limited. For a more complete picture of the behaviour of a given system over a wide range of temperature, pressure and concentration, a phase diagram must be employed. This type of diagram represents graphically, in two or three dimensions, the equilibria between the various phases of a system. The Phase Rule developed by J. Willard Gibbs (1876) relates the number of components, C, phases, P, and degrees of freedom, F, of a system by means of the equation

$$P+F = C+2$$

These three terms are defined as follows.

The number of *components* of a system is the minimum number of chemical compounds required to express the composition of any phase. In the system water–copper sulphate, for instance, five different chemical compounds can exist, viz. $CuSO_4 \cdot 5H_2O$, $CuSO_4 \cdot 3H_2O$, $CuSO_4 \cdot H_2O$, $CuSO_4$ and H_2O; but for the purpose of applying the Phase Rule there are considered to be only two components, $CuSO_4$ and H_2O, because the composition of each phase can be expressed by the equation

$$CuSO_4 + x\, H_2O \rightleftharpoons CuSO_4 \cdot x\, H_2O$$

Again, in the system represented by the equation

$$CaCO_3 \rightleftharpoons CaO + CO_2$$

three different chemical compounds can exist, but there are only two components because the composition of any phase can be expressed in terms of the compounds CaO and CO_2.

A *phase* is a homogeneous part of a system. Thus any heterogeneous system comprises two or more phases. Any mixture of gases or vapours is a one-phase system. Mixtures of two or more completely miscible liquids or solids are also one-phase systems, but mixtures of two partially miscible liquids or a heterogeneous mixture of two solids are two-phase systems, and so on.

The three variables that can be considered in a system are temperature, pressure and concentration. The number of these variables that may be changed in magnitude without changing the number of phases present is called the number of *degrees of freedom*. In the equilibrium system water–ice–water vapour $C = 1$, $P = 3$, and from the Phase Rule, $F = 0$. Therefore in this system there are no degrees of freedom: no alteration may be made in either temperature or pressure (concentration is obviously not a variable in a one-component system) without a change in the number of phases. Such a system is called 'invariant'.

For the system water–water vapour $C = 1$, $P = 2$ and $F = 1$; thus only one variable, pressure or temperature, may be altered independently without changing the number of phases. Such a system is called 'univariant'. The one-phase water vapour system has two degrees of freedom; thus both temperature and pressure may be altered independently without changing the number of phases. Such a system is called 'bivariant'.

Summarising, it may be said that the physical nature of a system can be expressed in terms of phases, and that the number of phases can be changed by altering one or more of three variables: temperature, pressure or concentration. The chemical nature of a system can be expressed in terms of components, and the number of components is fixed for any given system.

ONE-COMPONENT SYSTEMS

The two variables that can affect the phase equilibria in a one-component, or unary, system are temperature and pressure. The phase diagram for such a system is therefore a temperature–pressure equilibrium diagram. *Figure 4.1* illustrates it for the case of sulphur. This system is chosen because

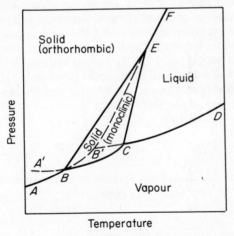

Figure 4.1. Phase diagram for sulphur (not to scale)

it brings out several important points. The diagram (not drawn to scale) indicates the equilibrium relationships between vapour, liquid and two solid forms of sulphur. The area enclosed by the curve *ABEF* is the region in which orthorhombic sulphur is the stable solid form. The areas enclosed by curves *ABCD* and *FECD* indicate the existence of vapour and liquid sulphur, respectively. The 'triangular' area *BEC* represents the region in which monoclinic sulphur is the stable solid form. Curves *AB* and *BC* are the vapour pressure curves for orthorhombic and monoclinic sulphur, respectively, and these curves intersect at the transition point *B*.

Curve *BE* indicates the effect of pressure on the transition temperature for the change orthorhombic S $\rightleftharpoons$ monoclinic S. Point *B*, therefore, is a triple point representing the temperature and pressure (95·5 °C and 0·0038 mm Hg, i.e. 0·51 N/m^2) at which orthorhombic sulphur and sulphur vapour can coexist in stable equilibrium. Curve *EF* indicates the effect of pressure on the melting point of orthorhombic sulphur; point *E* is a triple point representing the temperature and pressure (151 °C and 1290 atm, i.e. 1·31 × 10^8 N/m^2) at which orthorhombic and monoclinic sulphur and liquid sulphur are in stable equilibrium. Curve *CD* is the vapour pressure curve for liquid sulphur, and curve *CE* indicates the effect of pressure on the melting point of monoclinic sulphur. Point *C*, therefore, is another triple point (115 °C and 0·018 mmHg, i.e. 2·4 N/m^2) representing the equilibrium between monoclinic and liquid sulphur and sulphur vapour.

The broken lines in *Figure 4.1* represent metastable conditions. If orthorhombic sulphur is heated rapidly beyond 95·5 °C, the change to the monoclinic form does not occur until a certain time has elapsed; curve *BB'*, a continuation of curve *AB*, is the vapour pressure curve for metastable orthorhombic sulphur above the transition point. Similarly, if monoclinic sulphur is cooled rapidly below 95·5 °C, the change to the orthorhombic form does not take place immediately, and curve *BA'* is the vapour pressure curve for metastable monoclinic sulphur below the transition point. Likewise, curve *CB'* is the vapour pressure curve for metastable liquid sulphur below the 115 °C transition point, and curve *B'E* the melting point curve for metastable orthorhombic sulphur. Point *B'*, therefore, is a fourth triple point (110 °C and 0·013 mmHg, i.e. 1·7 N/m^2) of the system.

Only three of the four possible phases orthorhombic (solid), monoclinic (solid), liquid and vapour can coexist in stable equilibrium at any one time, and then only at one of the three 'stable' triple points. This in fact can be deduced from the Phase Rule:

$$3 + F = 1 + 2$$
$$F = 0$$

Another important one-phase system is water with a triple point of −0·0075 °C and 4·6 mmHg, i.e. 610 N/m^2.

ENANTIOTROPY AND MONOTROPY

A pure substance capable of existing in two different crystalline forms is called dimorphous. The transformation from one form to the other can be reversible or irreversible: in the former case the two crystalline forms are said to be enantiotropic; in the latter, monotropic. These phenomena, already described in Chapter 1, can be demonstrated with reference to the pressure–temperature phase diagram.

Figure 4.2a shows the phase reactions exhibited by two enantiotropic solids,

α and β. *AB* is the vapour pressure curve for the α form, *BC* that for the β form, and *CD* that for the liquid. Point *B*, where the vapour pressure curves of the two solids intersect, is the transition point; the two forms can coexist in equilibrium under these conditions of temperature and pressure. Point *C* is a triple point at which vapour, liquid and β solid can coexist. This point can be considered to be the melting point of the β form.

If the α solid is heated slowly, it changes into the β solid and finally melts. The vapour pressure curve *ABC* is followed. Conversely, if the liquid is cooled slowly, the β form crystallises out first and then changes into the α form. Rapid heating or cooling, however, can result in a different behaviour. The vapour pressure of the α form can increase along curve *BB'*, a continuation of *AB*, the α form now being metastable. Similarly, the liquid vapour pressure can fall along curve *CB'*, a continuation of *DC*, the liquid being

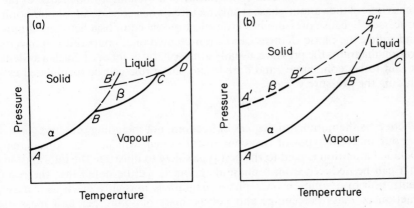

Figure 4.2. *Pressure–temperature diagrams for dimorphous substances:* (a) *enantiotropy;* (b) *monotropy*

metastable. Point *B'*, therefore, is a metastable triple point at which the liquid, vapour and α solid can coexist in metastable equilibrium.

The type of behaviour described above is well illustrated by the case of sulphur (*Figure 4.1*), where the orthorhombic and monoclinic forms are enantiotropic; the transition point occurs at a lower temperature than does the triple point.

Figure 4.2b shows the pressure–temperature curves for a monotropic substance. *AB* and *BC* are the vapour pressure curves for the α solid and liquid, respectively, and *A'B'* is that for the β solid. In this case the vapour pressure curves of the α and β forms do not intersect, so there is no transition point. The solid form with the higher vapour pressure at any given temperature (β in this case) is the metastable form. Curves *BB'* and *BB"* are the vapour pressure curves for the liquid and metastable α solid, so *B'* is a metastable triple point. If this system did exhibit a true transition point, it would lie at point *B"*; but as this represents a temperature higher than the melting point of the solid, it cannot exist.

A typical case of monotropy is the change from white to red phosphorus.

Benzophenone is another example of a monotropic substance: the stable melting point is 49 °C, whereas the metastable form melts at 29 °C.

TWO-COMPONENT SYSTEMS

The three variables that can affect the phase equilibria of a two-component, or binary, system are temperature, pressure and concentration. The behaviour of such a system should, therefore, be represented by a space model with three mutually perpendicular axes of pressure, temperature and concentration. Alternatively, three diagrams with pressure–temperature, pressure–concentration and temperature–concentration axes, respectively, can be employed. However, in most crystallisation processes the main interest lies in the liquid and solid phases of a system; a knowledge of the behaviour of the vapour phase is only required when considering sublimation processes. Because pressure has little effect on the equilibria between liquids and solids, the phase changes can be represented on a temperature–concentration diagram; the pressure, usually atmospheric, is ignored. Such a system is said to be 'condensed', and a 'reduced' phase rule can be formulated excluding the pressure variable:

$$P + F' = C + 1$$

where F' is the number of degrees of freedom, not including pressure.

Four different types of two-component system will now be considered. Detailed attention is paid to the first type solely to illustrate the information that can be deduced from a phase diagram. It will be noted that the concentration of a solution on a phase diagram is normally given as a mass fraction or mass percentage and not as 'mass of solute per unit mass of solvent', as recommended for the solubility diagram (Chapter 2). Mole fractions and mole percentages are also suitable concentration units for use in phase diagrams.

SIMPLE EUTECTIC

A typical example of a system in which the components do not combine to form a chemical compound is shown in *Figure 4.3*. Curves *AB* and *BC* represent the temperatures at which homogeneous liquid solutions of naphthalene in benzene begin to freeze or to crystallise. The curves also represent, therefore, the temperatures above which mixtures of these two components are completely liquid. The name 'liquidus' is generally given to this type of curve. In aqueous systems of this type one liquidus is the freezing point curve, the other the normal solubility curve. Line *DBE* represents the temperature at which solid mixtures of benzene and naphthalene begin to melt, or the temperature below which mixtures of these two components are completely solid. The name 'solidus' is generally given to this type of line. The melting or freezing points of pure benzene and naphthalene are given by points *A* (5·5 °C) and *C* (80·2 °C), respectively. The upper area enclosed by the liquidus, *ABC*, represents the homogeneous liquid phase, i.e. a solution of naphthalene in benzene; that enclosed by the solidus, *DBE*, indicates solid mixtures of benzene and naphthalene. The small and large 'triangular' areas *ABD* and *BCE* represent mixtures of solid benzene and solid naphthalene, respectively, and benzene–naphthalene solution.

If a solution represented by point *x* is cooled, pure solid benzene is deposited when the temperature of the solution reaches point *X* on curve *AB*. As solid benzene separates out, the solution becomes more concentrated in naphthalene and the equilibrium temperature of the system falls, following curve *AB*. If a solution represented by point *y* is cooled, pure solid naphthalene is deposited when the temperature reaches point *Y* on the solubility

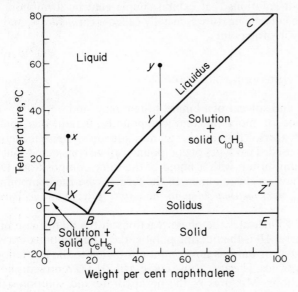

Figure 4.3. Phase diagram for the simple eutectic system naphthalene–benzene

curve; the solution becomes more concentrated in benzene and the equilibrium temperature follows curve *CB*. Point *B*, common to both curves, is the eutectic point ($-3.5\,^{\circ}$C and 0·189 mass fraction of naphthalene), and this is the lowest freezing point in the whole system. At this point a completely solidified mixture of benzene and naphthalene of fixed composition is formed; it is important to note that the eutectic is a physical mixture, not a chemical compound. Below the eutectic temperature all mixtures are solid.

If the solution *y* is cooled below the temperature represented by point *Y* on curve *BC* to some temperature represented by point *z*, the composition of the system as a whole remains unchanged. The physical state of the system has been altered, however; it now consists of a solution of benzene and naphthalene containing solid naphthalene. The composition of the solution, or mother liquor, is given by point *z* on the solubility curve, and the proportions of solid naphthalene and solution are given by the ratio of the lengths zZ and zZ', i.e.

$$\frac{\text{mass of solid C}_{10}\text{H}_8}{\text{mass of solution}} = \frac{zZ}{zZ'}$$

A process involving both cooling and evaporation can be analysed in two steps. The first is as described above, i.e. the location of points z, Z and Z'; this represents the cooling operation. If benzene is evaporated from the system, z no longer represents the composition; thus the new composition point z' (not shown in the diagram) is located along line ZZ' between points z and Z. Then the ratio $z'Z/z'Z'$ gives the proportions of solid and solution.

The systems $KCl-H_2O$ and $(NH_4)_2SO_4-H_2O$ are good examples of aqueous salt solutions that exhibit simple eutectic formation. In aqueous systems the eutectic mixture is usually called a *cryohydrate*, and the eutectic point a 'cryohydric point'.

COMPOUND FORMATION

The solute and solvent of a binary system may, and frequently do, combine to form one or more different compounds. In aqueous solutions these compounds are called 'hydrates'; for non-aqueous systems the term 'solvate' is sometimes used. Two types of compound can be considered: one can coexist in stable equilibrium with a liquid of the same composition, and the other cannot behave in this manner. In the former case the compound is said to have a *congruent melting point*; in the latter, to have an *incongruent melting point*.

Figure 4.4 illustrates the phase reactions in the manganese nitrate–water system. Curve AB is the freezing point curve. The solubility curve $BCDEFG$ for $Mn(NO_3)_2$ in water is not continuous owing to the formation of several different hydrates. The area above curve $ABCDEFG$ represents homogeneous liquid solutions. Mixtures of the hexahydrate and solution exist in areas BCH and ICD. The tetrahydrate is the stable phase in region DEJ and the dihydrate in EKF. The rectangular areas under FH, IJ and KL represent completely solidified systems (ice and hexahydrate, hexa- and tetrahydrates, and tetra- and dihydrates, respectively). Point B is a eutectic or cryohydric point with the co-ordinates $-36\,°C$ and 0.405 mass fraction of $Mn(NO_3)_2$.

Point C in *Figure 4.4* indicates the melting point ($25.8\,°C$) and composition (0.624 mass fraction) of the hexahydrate. Thus when a solution of this composition is cooled to $25.8\,°C$ it solidifies to the hexahydrate, i.e. no change in composition occurs. Point C, therefore, is a congruent point. Similarly, point E is the congruent point for the tetrahydrate (melting point $37.1\,°C$, composition 0.713). Points D and F are the other two eutectic points of the system. Point G is the transition point at which the dihydrates decomposes into the monohydrate and water, i.e. it is the incongruent melting point of the dihydrate. The vertical broken line at 0.834 mass fraction represents the composition of the dihydrate.

The behaviour of manganese nitrate solutions on cooling can be traced in the same manner as that described above for simple eutectic systems. The solution concentrations and the proportions of solid and solution can similarly be deduced graphically. The process of isothermal evaporation in congruent melting systems presents an interesting phenomenon. For example, the mixture represented by point X in *Figure 4.4* represents a slurry of ice and solution; but when sufficient water is removed to bring the system composition into the region to the right of curve AB, it becomes a homogeneous liquid solution. When more water is removed, so that region BCG is entered, the system partially solidifies again, depositing crystals of the

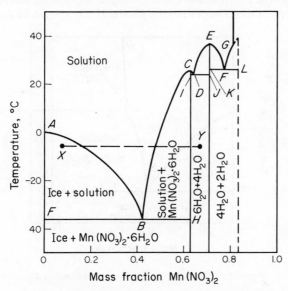

Figure 4.4. Phase diagram for the system $Mn(NO_3)_2$–H_2O

hexahydrate. On further evaporation, once the composition exceeds 62·4 per cent of $Mn(NO_3)_2$, e.g. at point Y, the system solidified completely to a mixture of the hexa- and tetrahydrates. The reverse order of behaviour occurs on isothermal hydration.

The formation of eutectics and solvates with congruent points is observed in many organic, aqueous inorganic and metallic systems. The case illustrated above is a rather simple example. Some systems form a large number of solvates and their phase diagrams can become rather complex. Ferric chloride, for example, forms four hydrates, and the $FeCl_3$–H_2O phase diagram exhibits five cryohydric points and four congruent points.

A solvate that is unstable in the presence of a liquid of the same composition is said to have an incongruent melting point. Such a solvate melts to form a solution and another compound, which may or may not be a solvate. For instance, the hydrate $Na_2SO_4 \cdot 10H_2O$ melts at 32·4 °C and immediately breaks down into the anhydrous salt and water; hence, this temperature is the incongruent melting point of the decahydrate. The terms 'meritectic point' and 'transition point' are also used instead of the expression 'incongruent melting point'.

Figure 4.5 illustrates the behaviour of the system sodium chloride–water. The various areas are marked on the diagram. *AB* is the freezing point curve and *BC* is the solubility curve for the dihydrate. Point *B* (-21 °C) is a eutectic or cryohydric point at which a solid mixture of ice and $NaCl \cdot 2H_2O$ of fixed composition (0·29 mass fraction of NaCl) is deposited. At point *C* (0·15 °C) the dihydrate decomposes into the anhydrous salt and water; this is, therefore, the incongruent melting point, or transition point, of $NaCl \cdot 2H_2O$. The vertical line commencing at 0·619 mass fraction of NaCl represents the composition of the dihydrate. If this system had a congruent

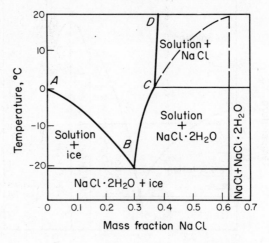

Figure 4.5. Phase diagram for the system $NaCl–H_2O$

melting point, which it does not have, this line would meet the peak of the extension of curve *BC* (e.g., see *Figure 4.4*).

Many aqueous and organic systems exhibit eutectic and incongruent points. Several cases are known of an inverted solubility effect after the transition point (see *Figure 2.1b*); the systems $Na_2SO_4–H_2O$ and $Na_2CO_3–H_2O$ are particularly well-known examples of this phenomenon.

SOLID SOLUTIONS

Many binary systems when submitted to a cooling operation do not at any stage deposit one of the components in the pure state: both components are deposited simultaneously. The deposited solid phase is, in fact, a solid solution. Only two phases can exist in such a system: a homogeneous liquid solution and a solid solution. Therefore, from the reduced phase rule, $F' = 1$, so an invariant system cannot result. One of three possible types of equilibrium diagram can be exhibited by systems of this kind. In the first type, illustrated in *Figure 4.6a*, all mixtures of the two components have freezing or melting points intermediate between the melting points of the pure components. In the second type shown in *Figure 4.6b*, a minimum is produced in the freezing and melting point curves. In the third, rare, type of diagram (not illustrated) a maximum is exhibited in the curves.

Figure 4.6a shows the temperature–concentration phase diagram for the system naphthalene–β-naphthol, which forms a continuous series of solid solutions. The melting points of pure naphthalene and β-naphthol are 80 and 120 °C, respectively. The upper curve is the liquidus or freezing point curve, the lower the solidus or melting point curve. Any system represented by a point above the liquidus is completely molten, and any point below the solidus represents a completely solidified mass. A point within the area enclosed by the liquidus and solidus curves indicates an equilibrium mixture of liquid and solid solution. Point X, for instance, denotes a liquid of composition L in equilibrium with a solid solution of composition S, and point Y a liquid L' in equilibrium with a solid S'.

The phase reactions occurring on the cooling of a given mixture can be traced as follows. If a homogeneous liquid represented by point A (60 per cent β-naphthol) is cooled slowly, it starts to crystallise when point L (105 °C) is reached. The composition of the first crystals is given by point S (82 per cent β-naphthol). As the temperature is lowered further, more crystals are deposited but their composition changes successively along curve SS', and the liquid composition changes along curve LL'. When the

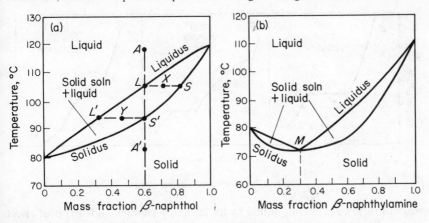

Figure 4.6. Solid solutions: (a) continuous series (naphthalene–β-naphthol; (b) minimum melting point (naphthalene–β-naphthylamine)

temperature is reduced to 94 °C (points L' and S'), the system solidifies completely. The over-all composition of the solid system at some temperature represented by, say, point A' is the same as that of the original homogeneous melt, assuming that no crystals have been removed during the cooling process, but the system is no longer homogeneous because of the successive depositions of crystals of varying composition. The changes occurring when a solid mixture A' is heated can be traced in a manner similar to the cooling operation.

Figure 4.6b shows the relatively uncommon, but not rare, type of binary

system in which a common minimum temperature is reached by both the
upper liquidus and lower solidus curves. These two curves approach and
touch at point M. The example shown in *Figure 4.6b* is the system naphthalene–β-naphthylamine. Freezing and melting points of mixtures of this
system do not necessarily lie between the melting points of the pure components. Three sharp melting points are observed: 80 °C (pure naphthalene),
110 °C (pure β-naphthylamine) and 72·5 °C (mixture M, 0·3 mass fraction β-
naphthylamine). Although the solid solution deposited at point M has a
definite composition, it is not a chemical compound. The components of
such a minimum melting point mixture are rarely, if ever, present in stoi-
chiometric proportions. Point M, therefore, is not a eutectic point: the
liquidus curve is completely continuous; it only approaches and touches the
solidus at M. The phase reactions occurring when mixtures of this system
are cooled can be traced in the same manner as that described for the continuous series solid solutions.

THERMAL ANALYSIS

Equilibrium in solid–liquid systems may be determined by the solubility
methods discussed in Chapter 2. If these are not convenient or applicable,
another technique, known as thermal analysis, may be employed. A phase

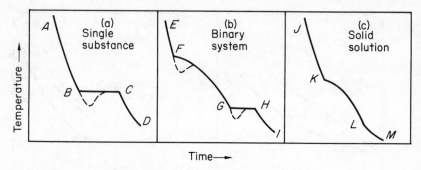

Figure 4.7. Some typical cooling curves

reaction is always accompanied by an enthalpy change, and this heat effect
can readily be observed if a cooling curve is plotted for the system. In many
cases a simple apparatus can be used; a 6 × 1 in glass boiling tube, fitted with
a stirrer and a thermometer graduated in $\frac{1}{10}$ °C, suspended in a shielding
vessel or refrigerant bath, will suffice. The temperature of the system is
recorded at regular intervals of, say, 1 min.

A smooth cooling curve is followed until a phase reaction takes place, and
then the accompanying heat effect causes an arrest or change in slope.
Figure 4.7a shows a typical example for a pure substance. AB is the cooling
curve for the homogeneous liquid phase. At point B the substance starts to
freeze and the system remains at constant temperature, the freezing point,
until solidification is complete at point C. The solid then cools at a rate

indicated by curve CD. It is possible, of course, for the liquid phase to cool below the freezing point, and some systems may exhibit appreciable degrees of supercooling. The dotted curve in *Figure 4.7a* denotes the sort of path followed if supercooling occurs. Seeding of the system will minimise these effects.

Figure 4.7b shows the type of cooling curve obtained for a binary system in which eutectic or compound formation occurs. The temperature of the

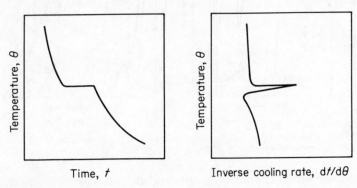

Figure 4.8. Detection of the arrest point for a single substance: (a) on a temperature–time curve, (b) on an inverse rate curve

homogeneous liquid phase falls steadily along curve EF until, at point F, deposition of the solid phase commences. The rate of cooling changes along curve FG as more and more solid is deposited. The composition of the remaining solution changes until the composition of the eutectic is reached, then crystallisation or freezing continues at constant temperature (line GH), i.e. the eutectic behaves as a single pure substance. The completely solidified system cools along curve HI. Supercooling, denoted by the dotted lines, may be encountered at both arrest points if the system is not seeded.

Figure 4.7c shows a typical cooling curve for a binary mixture that forms a series of solid solutions. The first arrest, K, in the curve corresponds to the onset of freezing, and this represents a point on the liquidus. The second arrest, L, occurs on the completion of freezing and represents a point on the solidus. It will be noted that no constant-temperature freezing point occurs in such a system.

The discontinuities may not always be clearly defined on a cooling curve (temperature, θ, versus time, t, plot). In such cases the arrest points can often be greatly exaggerated by plotting an inverse rate curve (θ versus $dt/d\theta$, i.e. the inverse of the cooling rate). A typical plot is shown in *Figure 4.8*.

Equilibria in solid solutions are better studied by a heating than by a cooling process. This is the basis of the thaw–melt method first proposed by Rheinboldt (1925). An intimate mixture, of known composition, of the two pure components is prepared by melting, solidifying and then crushing to a fine powder. A small sample of the powder is placed in a melting-point tube, attached close to the bulb of a thermometer graduated in $\frac{1}{10}$ °C, and immersed in a stirred bath. The temperature is raised slowly and regularly at

112   PHASE EQUILIBRIA

a rate of about 1 °C in 5 min. The 'thaw point' is the temperature at which liquid first appears in the tube; this is a point on the solidus. The 'melt point' is the temperature at which the last solid particle melts; this is a point on the liquidus. Only pure substances and eutectic mixtures have sharp melting points. The thaw–melt method is particularly useful if the system is

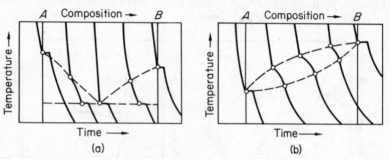

Figure 4.9. Construction of equilibrium diagrams from 'thaw–melt' data: (a) eutectic system; (b) solid solution

prone to supercooling, and it has the added advantage of requiring only small quantities of test material.

The construction of equilibrium diagrams from cooling or thaw–melt data is indicated in *Figure 4.9*. In practice, however, a large number of different mixtures of the two components *A* and *B*, covering the complete range from pure *A* to pure *B*, would be tested. The liquidus curves are drawn through the first-arrest points, the solidus curves through the second-arrest points. Only at 100 per cent *A*, 100 per cent *B* and the eutectic point do the liquidus and solidus meet.

DIFFERENTIAL THERMAL ANALYSIS

A method widely used for observing phase changes and measuring the associated changes in enthalpy is the technique known as differential thermal analysis (DTA). A small test sample, often only a few milligrams, is heated next to a sample of reference material in an identical container. The reference material, chosen for its similarity to the test sample, must not exhibit any phase change over the temperature range under consideration.

When the test sample undergoes a phase change, there will be a heat release or absorption. For example, if it melts it will absorb heat and its temperature will lag behind that of the reference material (*Figure 4.10a*). The difference in temperature between the two samples is detected by a pair of matched thermocouples and recorded as a function of time. The area between the differential curve (*Figure 4.10b*) and the base line is a function of the enthalpy associated with the phase change.

Instruments are now commercially available which record the energy required to keep the test sample at the same temperature as the reference sample when it undergoes a phase change, and in this way give a direct measure of the enthalpy change. Detailed accounts of the various techniques of thermal analysis are given in the relevant publications listed in the Bibliography.

DILATOMETRY

The dilatometric methods for detecting phase changes utilise volume changes in the same way as the calorimetric methods utilise thermal effects. Dilatometry is widely used in the analysis of melts, and particularly of fats and waxes. The techniques and equipment are usually quite simple.

Solids absorb heat on melting and usually evolve heat in changing from one polymorphic form to another. Most substances expand on melting and contract when they undergo polymorphic transformation. Consequently, dilatometric (specific volume–temperature) curves bear a close resemblance to calorimetric (enthalpy–temperature) curves. The melting dilation corresponds to the heat of fusion, and the coefficient of cubical expansion, a, corresponds to the specific heat capacity, c_p. The ratio c_p/a is virtually a constant independent of temperature.

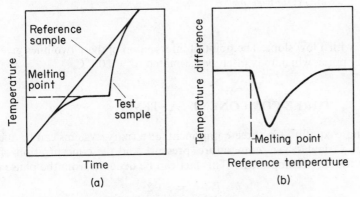

Figure 4.10. Differential thermal analysis: (a) comparative heating curves, (b) differential temperature curve for a single substance

A high-precision dilatometer used for fats and waxes is shown in Figure 4.11. Mercury, or some other suitable liquid, is used as the confining fluid and its thread in the capillary, C, communicates with the reservoir, R. Volume changes in the sample, S, are measured by weighing the mercury in the reservoir before and after. The small expansion bulb, B, is warmed to expel any air that enters the end of the capillary when the flask is detached. Owing to the high density of mercury, and the accuracy with which weighings can be made, volume changes as small as 10^{-5} cm^3/g have been detected.

Melting points can be determined with great precision by dilatometry. A plot of dilation versus temperature usually gives two straight lines – one for the solid dilation, which generally has a steep slope, and one for the

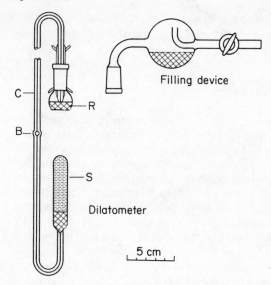

Filling device

Dilatometer

5 cm

Figure 4.11. Gravimetric dilatometer and filling device. (After A. E.
BAILEY; *see Bibliography)*

liquid, with a low slope. The point of intersection of these two lines gives the
melting point, which may often be estimated to ± 0.01 °C.

THREE-COMPONENT SYSTEMS

The phase equilibria in three-component, or ternary, systems can be affected
by four variables, viz. temperature, pressure and the concentration of any
two of the three components. This fact can be deduced from the phase rule:

$$P + F = 3 + 2$$

which indicates that a one-phase ternary system will have four degrees of
freedom. It is impossible to represent the effects of the four possible variables
in a ternary system on a two-dimensional graph. For solid–liquid systems,
however, the pressure variable may be neglected, and the effect of temperature will be considered later.

The composition of a ternary system can be represented graphically on a
triangular diagram. Two methods are in common use. The first utilises the
equilateral triangle, and the method of construction is shown in *Figure 4.12*.
The apexes of the triangle represent the pure components *A*, *B* and *C*. A
point on a side of the triangle stands for a binary system, *AB*, *BC* or *AC*; a
point within the triangle represents a ternary system *ABC*. The scales may
be constructed in any convenient units, e.g. weight or mole percent, weight
or mole fraction, etc., and any point on the diagram must satisfy the equation
$A + B + C = 1$ or 100. The quantities of the components *A*, *B* and *C* in a

given mixture *M* (*Figure 4.12d*) are represented by the perpendicular distance from the sides of the triangle.

Special triangular graph paper is required if the equilateral diagram is to be used, and for this reason many workers prefer to employ the right-angled triangular diagram which can be drawn on ordinary linear graph paper.

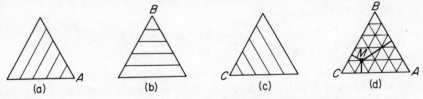

Figure 4.12. Construction of the equilateral triangular diagram

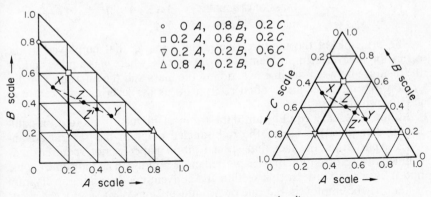

Figure 4.13. Construction of the right-angled triangular diagram

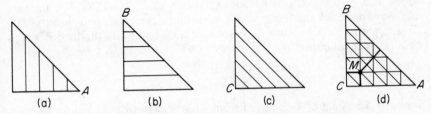

Figure 4.14. Composition plots on triangular diagrams

The construction of the right-angled isosceles triangle is shown in *Figure 4.13*. Again, as in the case of the equilateral triangle, each apex represents a pure component *A*, *B* or *C*, a point on a side a binary system, and a point within the triangle a ternary system; in all cases $A + B + C = 1$ or 100. The quantities of *A*, *B* and *C* in a given mixture *M* (*Figure 4.13d*) are represented by the perpendicular distances to the sides of the triangle. If two compositions, *A* and *B*, *B* and *C*, or *A* and *C* are known, the composition of the third component is fixed on both triangular diagrams.

Two actual plots are shown in *Figure 4.14* to illustrate the interpretation of these diagrams. For clarity the C scale has been omitted from the right-angled diagram; the C values can be obtained from the expression $C = 1 - (A + B)$. The 'mixture rule' is also illustrated in *Figure 4.14*. When any two mixtures X and Y are mixed together, the composition of the final mixture Z is represented by a point on the diagram located on a straight line drawn between the points representing the initial mixtures. The position of Z is located by the expression

$$\frac{\text{mass of mixture } X}{\text{mass of mixture } Y} = \frac{\text{distance } YZ}{\text{distance } XZ}$$

For example, if one part of a mixture X (0·1A, 0·5B, 0·4C) is mixed with one part of a mixture Y (0·5A, 0·3B, 0·2C), the composition of the final mixture Z (0·3 A, 0·4B, 0·3C) is found on the line XY where $XZ = YZ$. Again, if 3 parts of Y are mixed with 1 part of X, the mixture composition Z' (0·4A, 0·35B, 0·25C) is found on the line XY where $XZ' = 3(YZ')$. The mixture rule also applies to the removal of one or more constituents from a system. Thus, one part of a mixture X removed from 2 parts of a mixture Z would yield one part of a mixture Y given by:

$$\frac{\text{mass of original } Z}{\text{mass of } X \text{ removed}} = \frac{YX}{YZ} = \frac{2}{1}$$

Similarly, one part of X removed from 4 parts of Z' would yield 3 parts of a mixture Y given by

$$\frac{\text{mass of original } Z'}{\text{mass of } X \text{ removed}} = \frac{YX}{YZ'} = \frac{4}{1}$$

The principle of the mixture rule is the same as that employed in the operation of lever-arm problems, i.e. $m_1 l_1 = m_2 l_2$, where m is a mass and l is the distance between the line of action of the mass and the fulcrum. For this reason, the mixture rule is often referred to as the lever-arm or centre of gravity principle.

Although ternary equilibrium data are most frequently plotted on equilateral diagrams, the use of the right-angled diagram has several advantages. Apart from the fact that special graph paper is not required, it is claimed that information may be plotted more rapidly on it, and some people find it easier to read. In this section the conventional equilateral diagram will mostly be employed, but one or two illustrations of the use of the right-angled diagram will be given.

EUTECTIC FORMATION

Equilibrium relationships in three-component systems can be represented on a temperature–concentration space model as shown in *Figure 4.15a*. The ternary system *ortho*-, *meta*- and *para*-nitrophenol, in which no compound formation occurs, is chosen for illustration purposes. The three components will be referred to as O, M and P, respectively. Points O', M' and P' on the vertical edges of the model represent the melting points of the pure components *ortho-* (45 °C), *meta-* (97 °C) and *para-* (114 °C). The vertical faces of the prism represent the temperature–concentration diagrams for the three binary systems *O–M*, *O–P* and *M–P*. These diagrams are each similar to that shown in *Figure 4.3* described in the section on binary eutectic systems. In this case, however, the solidus lines have been omitted for clarity.

The binary eutectics are represented by points *A* (31·5 °C; 72·5 per cent *O*, 27·5 per cent *M*), *B* (33·5 °C; 65·5 per cent *O*, 24·5 per cent *M*) and *C* (61·5 °C; 54·8 per cent *M*, 45·2 per cent *P*). Curve *AD* within the prism represents the effect of the addition of the component *P* to the *O–M* binary eutectic *A*. Similarly, curves *BD* and *CD* denote the lowering of the freezing points of the binary eutectics *B* and *C*, respectively, on the addition of the third component. Point *D*, which indicates the lowest temperature at which

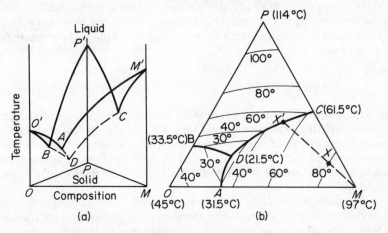

Figure 4.15. *Eutectic formation in the three-component system* o-, m- *and* p-*nitrophenol:* (a) *temperature–concentration space model;* (b) *projection on a triangular diagram*

solid and liquid phases can coexist in equilibrium in this system, is a ternary eutectic point (21·5 °C; 57·7 per cent *O*, 23·2 per cent *M*, 19·1 per cent *P*). At this temperature and concentration the liquid freezes invariantly to form a solid mixture of the three components. The section of the space model above the freezing point surfaces formed by the liquidus curves represents the homogeneous liquid phase. The section below these surfaces down to a temperature represented by point *D* denotes solid and liquid phases in equilibrium. Below this temperature the section of the model represents a completely solidified system.

Figure 4.15b is the projection of the curves *AD*, *BD* and *CD* in *Figure 4.15a* on to the triangular base. The apexes of the triangle represent the pure components *O*, *M* and *P* and their melting points. Points *A*, *B* and *C* on the sides of the triangle indicate the three binary eutectic points, point *D* the ternary eutectic point. The projection diagram is divided by curves *AD*, *BD* and *CD* into three regions which denote the three liquidus surfaces in the space model. The temperature falls from the apexes and sides of the triangle towards the eutectic point D, and several isotherms showing points on the liquidus surfaces are drawn on the diagrams. The phase reactions occurring when a given ternary mixture is cooled can now be traced.

A molten mixture with a composition as in point X starts to solidify when the temperature is reduced to 80 °C. Point X lies in the region $ADCM$, so pure *meta-* is deposited on decreasing temperature. The composition of the remaining melt changes along line MXX' in the direction away from point M representing the deposited solid phase (the mixture rule). At X', where line MXX' meets curve CD, the temperature is about 50 °C, and at this point a second component (*para-*) also starts to crystallise out. On further cooling, *meta-* and *para-* are deposited and the liquid phase composition changes in the direction $X'D$. When melt composition and temperature reach point D, the third component (*ortho-*) crystallises out, and the system solidifies without any further change in composition. A similar reasoning may be applied to the cooling, or melting, of systems represented by points in the other regions of the diagrams.

AQUEOUS SOLUTIONS

There are many different types of phase behaviour encountered in ternary systems consisting of water and two solid solutes. Only a few of the simpler cases will be considered here; attention will be devoted to a brief survey of systems in which there is (a) no chemical reaction, (b) formation of a solvate, e.g. a hydrate, (c) formation of a double salt, and (d) formation of a hydrated double salt.

At one given temperature the composition of, and phase equilibria in, a ternary aqueous solution can be represented on an isothermal triangular diagram. The construction of these diagrams has already been described. Polythermal diagrams can also be constructed, but in the case of complex systems the charts tend to become congested and rather difficult to interpret.

No Compound Formed

This simplest case is illustrated in *Figure 4.16* for the system KNO_3–$NaNO_3$–H_2O at 50 °C. Neither salt forms a hydrate, nor do they combine chemically. Point A represents the solubility of KNO_3 in water at the stated temperature (46·2 g/100 g of solution) and point C the solubility of $NaNO_3$ (53·2 g/100 g). Curve AB indicates the composition of saturated ternary solutions that are in equilibrium with solid KNO_3, curve BC those in equilibrium with solid $NaNO_3$. The upper area enclosed by ABC represents the region of unsaturated homogeneous solutions. The three 'triangular' areas are constructed by drawing straight lines from point B to the two apexes of the triangle; the compositions of the phases within these regions are marked on the diagram. At point B the solution is saturated with respect to both KNO_3 and $NaNO_3$, and from the reduced phase rule $F' = 1$. This means that point B is univariant, or invariant when the temperature is fixed.

The effect of isothermal evaporation on such a system can be shown as follows. If water is evaporated from an unsaturated solution represented by point X_1 in the diagram, the solution concentration will increase, following line X_1X_2. Pure KNO_3 will be deposited when the concentration reaches point X_2. If more water is evaporated to give a system of composition X_3, the composition of the solution will be represented by point X'_3 on the saturation curve AB; and when composition X_4 is reached, by point B: any further removal of water will cause the deposition of $NaNO_3$. All solutions in contact with solid will thereafter have a constant composition B, which is referred to as the *drying-up point* of the system. After the complete evaporation of water the composition of the solid residue is indicated by point X_5 on the base line.

Similarly, if an unsaturated solution, represented by a point located to the right of B in the diagram, were evaporated isothermally, only $NaNO_3$ would

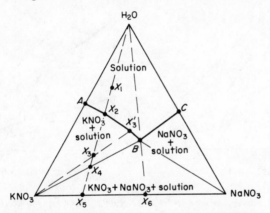

Figure 4.16. Phase diagram for the system KNO_3–$NaNO_3$–H_2O at 50 °C

be deposited until the solution composition reached the drying-up point B, when KNO_3 would also be deposited. The solution composition would thereafter remain constant until evaporation was completed. If water is removed isothermally from a solution of composition B, the composition of the deposited solid is given by point X_6 on the base line, and it remains unchanged throughout the remainder of the evaporation process.

The effect of the addition of one of the salts to the system KNO_3–$NaNO_3$–H_2O at 50 °C is shown in *Figure 4.17a*. This time the equilibria are plotted on a right-angled triangular diagram simply to demonstrate the use of this type of chart. Points A and C, as in *Figure 4.16*, refer to the solubilities at 50 °C of KNO_3 and $NaNO_3$, respectively. Curves AB and BC indicate the saturated ternary solutions in equilibrium with solid KNO_3 or $NaNO_3$, and show, for instance, that the solubility of KNO_3 in water is depressed when $NaNO_3$ is present in the system, and vice versa.

Take, for example, a binary system $NaNO_3$–H_2O represented by point Y_1 (0·7 mass fraction of $NaNO_3$ and 0·3 H_2O). As this point lies in the 'triangular' region to the right of curve BC, the system consists of a saturated solution of $NaNO_3$, with a composition given by point C, and excess solid $NaNO_3$. If a quantity of KNO_3 is added to this binary system, the temperature being kept constant at 50 °C so that the new composition is represented by point Y_2 (0·64 $NaNO_3$, 0·1 KNO_3, 0·26 H_2O), the composition of the ternary saturated solution in contact with the excess solid $NaNO_3$ present is given by Y_2' (0·46 $NaNO_3$, 0·15 KNO_3, 0·39 H_2O) on the line drawn from the apex N through Y_2 to meet curve BC. As more KNO_3 is added, the solution concentration alters, following curve CB. At point B the solution becomes saturated with respect to both $NaNO_3$ and KNO_3; its concentration is 0·4 $NaNO_3$, 0·29 KNO_3, 0·31 H_2O. If after this point further quantities of KNO_3 are

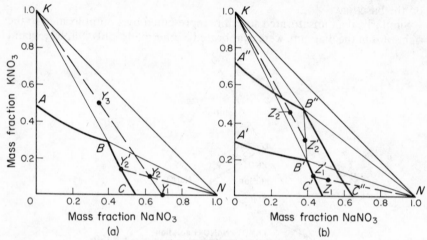

Figure 4.17. Phase diagrams for the system KNO_3–$NaNO_3$–H_2O : *(a) at 50 °C ; (b) at 25 and 100 °C*

added to bring the system concentration up to some point Y_3, no more KNO_3 dissolves; the solution composition remains at point B.

The interpretation of these phase diagrams is aided by remembering the rule of mixtures – i.e. on the removal or addition of any component from or to a system, the composition of the system changes along a straight line drawn from the original composition point to the apex representing the pure given component. In *Figure 4.17a* the right-angled apex represents pure water, the top apex K pure KNO_3 and the other acute apex N pure $NaNO_3$.

The effect of temperature on the system KNO_3–$NaNO_3$–H_2O is shown in *Figure 4.17b*. Two isotherms, $A'B'C'$ and $A''B''C''$, for 25 and 100 °C, respectively, are drawn on this diagram. The lower left-hand area enclosed by $A'B'C'$ represents homogeneous unsaturated solutions at 25 °C, the larger area enclosed by $A''B''C''$ unsaturated solutions at 100 °C. The line $B'B''$ shows the locus of the drying-up points between 25 and 100 °C. To illustrate the effect of temperature changes in the system, let point Z_1 refer to the composition (0·5 $NaNO_3$, 0·1 KNO_3, 0·4 H_2O) of a certain quantity of the ternary mixture. From the position of Z_1 in the diagram it can be seen that at 100 °C the system would be a homogeneous unsaturated solution, but at 25 °C it would consist of pure undissolved $NaNO_3$ in a saturated aqueous solution of $NaNO_3$ and KNO_3. Thus pure $NaNO_3$ would crystallise out of the solution Z_1 on cooling from, say, 100 to 25 °C, in fact at about 50 °C. Despite the phase changes, of course, the over-all system composition remains at Z_1 until one or more components are removed. At 25 °C the composition of the solution in contact with the crystals of $NaNO_3$ is given by the intersection of the line from N through Z_1 with curve $B'C'$, i.e. at point Z_1' (0·43 $NaNO_3$, 0·11 KNO_3, 0·46 H_2O). The quantity of $NaNO_3$ which would crystallise out at 25 °C is given by the mixture rule

$$\frac{\text{mass of crystals deposited}}{\text{mass of saturated solution}} = \frac{\text{length } Z_1 Z_1'}{\text{length } Z_1 N}$$

where N represents the $NaNO_3$ apex of the triangle.

When a pure solute is to be crystallised from a ternary two-solute system by cooling, there is usually a temperature limit below which the desired solute becomes 'contaminated' with the other solute. This can be demonstrated by considering a system represented by point Z_2 in *Figure 4.17b*. The composition at Z_2 is 0·3 $NaNO_3$, 0·45 KNO_3, 0·25 H_2O; at 100 °C the system is a homogeneous unsaturated solution. At 25 °C, however, this point lies in the region where both solid $NaNO_3$ and KNO_3 are in equilibrium with a saturated solution of both salts, its composition being given by point B'. If it is desired to cool solution Z_2 in order to yield only KNO_3 crystals, then the temperature limitation is found by drawing a straight line from the KNO_3 apex K through point Z_2 and producing it to meet the drying-up line $B'B''$ at Z_2'. Point Z_2' occupies the position of an invariant point on an isotherm; by referring to *Figure 4.17a* it can be seen that it corresponds approximately to point B on the 50 °C isotherm. Thus solution Z_2 must not be cooled below 50 °C if only KNO_3 crystals are to be deposited.

Solvate Formation

When one of the solutes in a ternary system is capable of forming a compound, with the solvent, the phase diagram will contain more regions to consider than in the simple case described above. A common example of solvate formation is the production of a hydrated salt in a ternary aqueous system. *Figure 4.18* shows the isothermal diagrams for the system $NaCl$–Na_2SO_4–H_2O at two temperatures, 17·5 and 25 °C, at which different phase equilibria are exhibited. Sodium sulphate combines with water, under certain conditions, to form $Na_2SO_4 \cdot 10H_2O$. Sodium chloride, however, does not form a hydrate at the temperatures being considered. *Figure 4.18a* shows the case where the decahydrate is stable in the presence of $NaCl$, and *Figure 4.18b* that of the decahydrate being dehydrated by the $NaCl$ under certain conditions.

Points A and C in *Figure 4.18a* represent the solubilities of $NaCl$ (26·5 per cent w/w) and Na_2SO_4 (13·8 per cent) in water at 17·5 °C, curves AB and BC the ternary solutions in equilibrium with solid $NaCl$ and $Na_2SO_4 \cdot 10H_2O$, respectively. Point D shows the composition of the hydrate $Na_2SO_4 \cdot 10H_2O$.

For convenience, the following symbols are used on the diagram to mark the phase regions: S = solution; H = hydrate $Na_2SO_4 \cdot 10H_2O$; SO_4 = Na_2SO_4 and Cl = NaCl. The solution above curve ABC is unsaturated. The lowest triangular region represents a solid mixture of Na_2SO_4, $Na_2SO_4 \cdot 10H_2O$ and NaCl. Point B is the drying-up point of the system.

In *Figure 4.18b*, points A and D denote the solubilities of NaCl (26·6 per cent w/w) and Na_2SO_4 (21·6 per cent) in water at 25 °C, point E the composition of $Na_2SO_4 \cdot 10H_2O$. In this diagram there are three curves, AB, BC and CD, which give the composition of the ternary solutions in equilibrium with NaCl, Na_2SO_4 and $Na_2SO_4 \cdot 10H_2O$. The various phase regions are indicated on the diagram. If NaCl is added to a system in the region CDE, i.e. to an equilibrium mixture of solid $Na_2SO_4 \cdot 10H_2O$ in a solution of

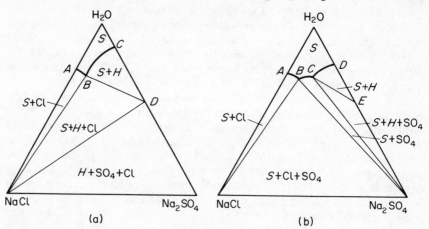

Figure 4.18. Phase diagrams for the system $NaCl-Na_2SO_4-H_2O$: (a) at 17·5 °C; (b) at 25 °C

NaCl and Na_2SO_4, the solution concentration will change along curve DC. When point C is reached, the NaCl can only dissolve by dehydrating the $Na_2SO_4 \cdot 10H_2O$, and anhydrous Na_2SO_4 is deposited. Further addition of NaCl will result in the complete removal of the decahydrate from the system, the solution concentration following curve CB; under these conditions the excess solid phase consists of anhydrous Na_2SO_4. At the drying-up point B the solution is saturated with respect to both NaCl and Na_2SO_4.

The effects of isothermal evaporation, salt additions and cooling can be traced from *Figure 4.18* in a manner similar to that outlined for *Figures 4.16* and *4.17*.

Double Compound Formation

Cases are encountered in ternary systems where the two dissolved solutes combine in fixed proportions to form a definite double compound. *Figure 4.19* shows two possible cases for a hypothetical aqueous solution of two salts A and B. Point C on the AB side of each triangle represents the composition of the double salt; points L and O show the solubilities of salts A and B in water at the given temperature. Curves LM and NO denote ternary solutions saturated with salts A and B, respectively, curve MN ternary solutions in equilibrium with the double salt C. The significance of the various areas is marked on the diagrams.

The isothermal dehydration of solutions in *Figure 4.19a* can be traced in the manner described for *Figures 4.16* and *4.17*. Point M is the drying-up point for solutions located to the left of broken line WR, point N that for solutions to the right of this line. A solution on line WM behaves as a solution of a single salt in water; when its composition reaches point M, a mixture of salt A and double salt C crystallises out in the fixed ratio of the lengths PC/AP. Similarly, a solution on line WN yields a mixture of B and C, in the ratio CQ/QB, when its composition reaches point N. A solution represented by a point on line WR also behaves as a solution of a single salt; when its composition reaches point R, the double compound C crystallises out and neither of salts A and B is deposited at any stage. Point R, therefore, is the third drying-up point of the system.

The phase diagram in *Figure 4.19b* shows a different case. There are only two drying-up points, M and N, in this system, the first for solutions located

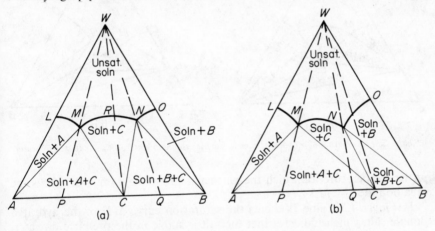

Figure 4.19. Formation of a double salt: (a) stable in water; (b) decomposed by water

to the left, the second for solutions to the right of line WN. Each solution on lines WM and WN behaves as a solution of single salt in water. The line WC does not cross the saturation curve MN of the double salt but cuts the saturation curve for salt B, indicating that the double salt is not stable in water; it is decomposed and salt B is deposited.

Hydrated Double Salt

Figure 4.20a shows the phase diagram for the case of a hydrated double salt that is stable in water. The best-known examples of this type of system are the alums ($M_2^I SO_4 \cdot M_2^{III}(SO_4)_3 \cdot 24H_2O$, where M^I and M^{III} represent mono- and tervalent cations, e.g. M^I = Na, K, NH$_4$, Cs, Rb, Tl or hydroxylamine; M^{III} = Al, Ti, V, Cr, Mn, Fe, Co or Ga; the sulphate radical may be replaced by selenate) and the Tutton salts ($M_2^I M^{II}(SO_4)_2 \cdot 6H_2O$, where M^I and M^{II} represent mono- and bivalent ions, respectively, e.g. M^I = NH$_4$, K, Rb, Cs or Te; M^{II} = Ni, Mn, Mg, Fe, Co, Zn or Cu).

In the case depicted salt A forms a hydrate of composition H. Its saturation curve is LM. Salt B is anhydrous and its saturation curve is ON. Point W represents water. Salts A and B combine together to form a hydrated double salt of composition denoted by point C within the triangular diagram. MN is the saturation curve for the hydrated double salt. The compositions of the phases in the eight separate regions are indicated in the diagram. The only region in which the pure hydrated double salt will crystallise out of solution,

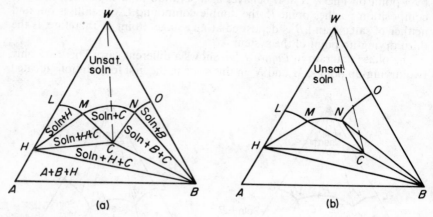

Figure 4.20. Formation of a hydrated double salt: (a) stable in water, (b) decomposed by water

at the temperature for which the particular phase diagram is drawn, is the area bounded by MNC.

In *Figure 4.20a* line WC cuts the saturation curve MN of the hydrated double salt, which indicates that the salt is stable in the presence of water. In *Figure 4.20b* line WC does not cross curve MN, which indicates that the hydrated double salt decomposes in the presence of water. This is a comparatively rare behaviour, but an example is the case of $MgSO_4 \cdot Na_2SO_4 \cdot 4H_2O$ (astrakanite) at 25 °C.

EQUILIBRIUM DETERMINATIONS

The solubility methods discussed in Chapter 2 are those most frequently used in the determination of equilibria in multicomponent systems, but for the complex cases the composition of the solid phase in equilibrium with the saturated solution is best analysed by the *wet-residue* method developed by

Schreinemakers (1893). This method is illustrated below with reference to a ternary system.

A quantity of solvent containing excess solute is kept at a constant temperature until equilibrium is achieved. The clear supernatant solution is then analysed. Most of the solution is decanted off the remaining solute, and the composition of a sample of the wet solid determined. Ternary systems can, as described above, be represented on a triangular diagram, so the solution composition gives a point on the solubility curve and the wet-residue composition one within the triangle. By virtue of the properties of a triangular diagram and the mixture rule, the solubility and wet-residue points and the point representing the solid phase must lie on a straight line. Therefore the point at which a line drawn through the solubility and wet-residue points meets the periphery of the triangle gives the composition of the solid phase.

An alternative procedure to the wet-residue method which is capable of yielding more rapid results is known as the *synthetic complex* method. Several mixtures of the solutes of known composition are prepared and a known quantity of solute is added to each sample. Thus a number of 'complex' points can be plotted within the triangular diagram. The samples are then allowed to achieve equilibrium at constant temperature by conventional methods, and the clear supernatant solution is analysed. Again, a line drawn through a solution point and its corresponding 'complex' point, extended to the periphery of the triangle, gives the composition of the solid phase.

FOUR-COMPONENT SYSTEMS

A one-phase, four-component or quaternary system has five degrees of freedom. Therefore the phase equilibria in these systems may be affected by the five variables: pressure, temperature and the concentrations of any three of the four components. To represent quaternary systems graphically, one or more of the above variables must be excluded. The effect of pressure on solid–liquid systems may be ignored, and if only one temperature is considered an isothermal space model can be constructed. If the concentration of one of the components is excluded, usually the liquid solvent, a two-dimensional graph can be drawn, but this simplification will be described later.

THREE SALTS AND WATER

The first, simple, type of quaternary system to be considered here consists of three solid solutes, A, B and C, and a liquid solvent, S. No chemical reaction takes place between any of the components, e.g. water and three salts with a common ion. The isothermal space model for this type of system can be constructed in the form of a tetrahedron (*Figure 4.21a*) with the solvent at the top apex and the three solid solutes on the base triangle. The four triangular faces of the tetrahedron represent the four ternary systems $A–B–C$, $A–B–S$, $A–C–S$ and $B–C–S$. The three faces, excluding the base, have the appearance of the 'two salts and water' diagram shown in *Figure 4.18a.*

A point on an edge of the tetrahedron represents a binary system, a point within it a quaternary. On the faces ABS, BCS and ACS the solubility curves meet at points L, M and N, respectively, which represent the solvent saturated with two solutes. They are the starting points for the three curves LO, MO

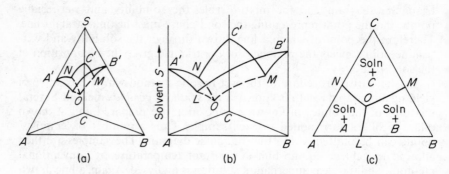

Figure 4.21. Isothermal representation of a quaternary system of the 'three salts in water' type: (a) tetrahedral space model; (b) triangular prism space model; (c) Jänecke projection

and NO, which denote solutions of three solutes in the solvent; point O represents the solution which, at the given temperature, is saturated with respect to all three solutes. All these curves form three curved surfaces within the space model. The section between these surfaces and the apex of the tetrahedron indicates unsaturated solution, that between the surfaces and the triangular base complex mixtures of liquid and solid.

Figure 4.21b shows another way in which systems of this type can be represented as a space model. Here it takes the form of a triangular prism where the apexes of the triangular base represent the three solid components and the vertical scale the liquid solvent. The interpretation of this model is similar to that just described for the tetrahedron; the same symbols have been used.

For a complete picture of the phase behaviour of quaternary systems a space model is essential; yet, because of its time-consuming construction, a two-dimensional 'projection' is frequently employed. Such a projection, named after E. Jänecke (1906), is shown in *Figure 4.21c.* In this type of isothermal diagram the solvent is excluded. The curved surfaces $A'LON$, $B'MOL$ and $C'NOM$ in *Figures 4.21a* and *4.21b*, which represent solutions in equilibrium with solutes A, B and C, respectively, are projected on to the triangular base and become areas $ALON$, $BMOL$ and $CNOM$ in *Figure 4.21c.* Curves LO, MO and NO denote solutions in equilibrium with two solutes, viz. A and B, B and C, A and C, respectively, while point O represents a solution in equilibrium with the three solutes. For this type of system the projection diagram can be plotted in terms of mass or mole fractions or percentages.

RECIPROCAL SALT PAIRS

The second, and more important, type of quaternary system that will be considered is one consisting of two solutes and a liquid solvent where the two solutes inter-react and undergo double decomposition. This behaviour is frequently encountered in aqueous solutions of two salts that do not have a common ion. Typical examples of double decomposition reactions of commercial importance are

$$KCl + NaNO_3 \rightleftharpoons NaCl + KNO_3$$
$$NaNO_3 + \tfrac{1}{2}(NH_4)_2SO_4 \rightleftharpoons NH_4NO_3 + \tfrac{1}{2}Na_2SO_4$$
$$KCl + \tfrac{1}{2}Na_2SO_4 \rightleftharpoons NaCl + \tfrac{1}{2}K_2SO_4$$
$$NaCl + \tfrac{1}{2}(NH_4)_2SO_4 \rightleftharpoons NH_4Cl + Na_2SO_4$$
$$NaNO_3 + \tfrac{1}{2}K_2SO_4 \rightleftharpoons KNO_3 + \tfrac{1}{2}Na_2SO_4$$

The four salts in each of the above systems form what is known as a 'reciprocal salt pair'. Although all four may be present in aqueous solution, the composition of any mixture can be expressed in terms of three salts and water. Thus, from the phase rule point of view, an aqueous reciprocal salt pair system is considered to be a four-component system.

Reciprocal salt pair solutions may be represented on an isothermal space model, in the form of either a square-based pyramid or a square prism.

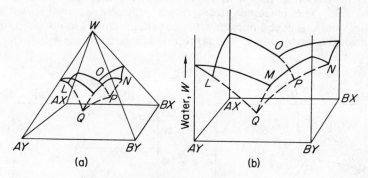

Figure 4.22. Isothermal representation of a quaternary system of the 'reciprocal salt pair' type: (a) square-based pyramid; (b) square prism

Figure 4.22a indicates the pyramidal model; the four equilateral triangular faces stand for the four ternary systems $AX–AY–W$, $AY–BY–W$, $BY–BX–W$ and $AX–BX–W$ (W = water) for the salt pair represented by the equation

$$AX + BY \rightleftharpoons AY + BX$$

The apex of the pyramid denotes pure water, its base the anhydrous quaternary system $AX–AY–BX–BY$. Points L, M, N and O on the four triangular faces of the pyramid indicate the equilibria between two salts and water. Point P, which represents a solution of three salts AX, BX and BY in water saturated with all three salts, is a quaternary invariant point. So is Q, which shows the equilibrium between salts AX, AY and BY and water. Curves OP, NP and LQ, MQ, which join these quaternary invariant points P and Q to the corresponding ternary invariant points on the triangular faces of the pyramid, represent solutions of three salts in water saturated with two salts, and so does curve PQ, joining the two quaternary invariant points.

The square-prism space model (*Figure 4.22b*) illustrates another way in which a quaternary system of the reciprocal salt pair type may be represented. The vertical axis stands for the water content, and the points on the diagram are the same as those marked on *Figure 4.22a*. In both diagrams all surfaces formed between the internal curves represent solutions of three salts in water saturated with one salt, all internal curves solutions of three salts in water saturated with two salts, and the two points P and Q solutions of three salts in water saturated with the three salts. The section above the internal curved surfaces denotes unsaturated solutions, the section below them mixtures of liquid and solid.

JÄNECKE'S PROJECTION

In order to simplify the interpretation of the phase equilibria in reciprocal salt pair systems, the water content may be excluded. The curves of the space model can then be projected on to the square base to give a two-dimensional graph, called a Jänecke projection, as described above. A typical projection is shown in *Figure 4.23a*; the lettering is that used in *Figure 4.22*. The enclosed areas, which represent saturation surfaces, indicate solutions in equilibrium with one salt, the curves solutions in equilibrium with two salts, points P and Q solutions in equilibrium with three salts.

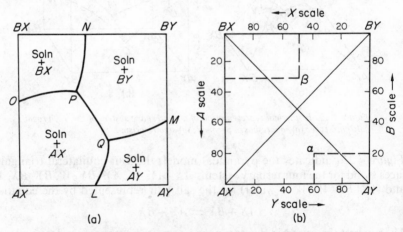

Figure 4.23. Interpretation of the Jänecke diagram for reciprocal salt pairs: (a) projection of the surfaces of saturation on to the base; (b) method of plotting

Molar or ionic bases must be used in this type of diagram for reciprocal salt pairs. The four corners of the square represent 100 mol of the pure salts AX, BX, BY and AY. Any point inside the square denotes 100 mol of a mixture of these salts; its composition can always be expressed in terms of three salts. The scales in *Figure 4.23b* are marked in ionic percentages of A, B, X and Y. Take, for example, 100 mol of a mixture expressed as

| Salt | moles of compounds | moles of ions | | | |
		A	B	X	Y
AX	20	20		20	
AY	60	60			60
BX	20		20	20	
	100	*80*	*20*	*40*	*60*

The totals of the $A+B$ ions (e.g. the basic radicals) and the $X+Y$ ions (e.g. the acidic radicals) must always equal 100. Thus point α, indicating this mixture, can be plotted: the square is divided by the two diagonals into four right-angled triangles, and point α lies in triangles $AX.AY.BX$ and $AX.AY.BY$. Therefore the composition of the above mixture could also have been expressed in terms of salts AX (40 mol), AY (40 mol) and BY (20 mol). In a similar manner, it can be shown that point β which lies within the two triangles $AX.BX.BY$ and $BX.BY.AY$ represents 100 mol of a mixture with a composition expressed either by 50 BY, 30 AX and 20 BX, or by 50 BX, 30 AY and 20 BY.

Although it is usually more convenient to plot ionic percentages on the square, it is quite in order to plot mole percentages of the salts direct. The numerical scales marked on *Figure 4.23b* must now be ignored. If point α is considered to lie in triangle $AX.AY.BX$, representing a mixture 20 AX, 60 AY and 20 BX, the compositions of the two salts at opposite ends of the diagonal AY and BX are used for plotting purposes. Thus point α is located by 60 along the horizontal AY scale and 20 up the vertical BX scale. If α is taken to lie in triangle $AX.AY.BY$, the composition is represented by 40 AX, 20 BY, 40 AY, and the AY and BX compositions are used for plotting. A similar reasoning may be applied to the plotting of point β in triangles $AX.BX.BY$ and $AY.BY.BX$.

Figure 4.24 shows Jänecke diagrams for solutions of a given reciprocal salt pair at different temperatures. These two simple cases will be used to demonstrate some of the phase reactions that can be encountered in such systems. Both diagrams are divided by the saturation curves into four areas which are actually the projections of the surfaces of saturation (e.g., see *Figure 4.22b*). Salts AX and BY can coexist in solution in stable equilibrium; the solutions are given by points along curve PQ. Salts BX and AY, however, cannot coexist in solution because their saturation surfaces are separated from each other by curve PQ. Thus AX and BY are called the *stable salt pair*, or the *compatible salts*, BX and AY the *unstable salt pair*, or the *incompatible salts*. In *Figure 4.24a* the AX–BY diagonal cuts curve PQ which joins the two quaternary invariant points, while in *Figure 4.24b* curve $P'Q'$ is not cut by either diagonal. These are two different cases to consider.

Point P represents a solution saturated with salts AX, BX and BY, Q one saturated with salts AX, BY and AY. In *Figure 4.24a* both P and Q lie in their 'correct' triangles, i.e. $AX.BX.BY$ and $AX.BY.AY$, respectively, and solutions represented by P and Q are said to be congruently saturated. In *Figure 4.24b* point Q' lies in its 'correct' triangle, $AX.BY.AY$, but P' lies in the

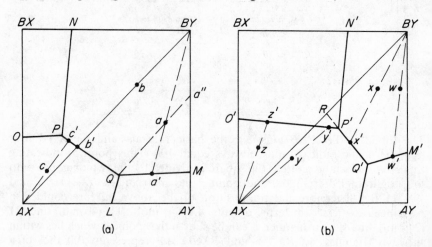

Figure 4.24. *Jänecke projections for aqueous solutions of a reciprocal salt pair, showing (a) two congruent points, (b) congruent and incongruent points*

'wrong' triangle, the same as Q'. Point Q', therefore, is *congruent* and point P' is *incongruent*.

Isothermal Evaporation

The phase reactions occurring on the removal of water from a reciprocal salt pair system will first be described with reference to *Figure 4.24a*. Point a which lies on the BY saturation surface represents a solution saturated with salt BY. When water is removed isothermally from this solution, the pure salt BY is deposited and the solution composition (i.e. the composition of the salts in solution, the water content being ignored) moves from a towards a' along the straight line drawn from BY through a to meet curve QM. When a sufficient quantity of water has been removed, the solution composition reaches point a' and here the solution is saturated with two salts, BY and AY.

Further evaporation results in the deposition of AY as well as BY; the composition of the solid phase being deposited is given by point M. The over-all composition of deposited solid therefore moves from BY towards a'' on the line $BY.AY$. The solution composition, being depleted in solid M, moves away from point M towards Q. On reaching point Q, three salts AX, AY and BY are deposited. The composition of the solid phase deposited is also given by point Q; the over-all composition of the solid phase, assuming that none has been removed from the system, by point a''. The solution composition, the water content being ignored, and the composition of the deposited solid phase remain constant at point Q for the rest of the evaporation process, and the over-all solids content changes along line $a''a$, composition a representing the completely dry complex. Point Q is a quaternary drying-up point for all solutions represented by points within triangle $AX.AY.BY$.

The isothermal evaporation of solution b on the diagonal can be traced as follows. If point b lies on the saturation surface, it represents a solution saturated with salt BY. While salt BY is being deposited, the solution composition changes along the diagonal from b towards b'. At b' the solution becomes saturated with salts AX and BY. This ternary system $(AX-BY-H_2O)$ thereafter dries up, without change in composition, at point b'. Point b', therefore, is a ternary drying-up point.

If point c lies on the saturation surface, it represents a solution saturated with salt AX. When this solution is evaporated isothermally, AX is deposited and the solution composition changes along line cc'. At c' salt BY also crystallises out and the composition of the solid phase deposited is given by b', the point at which the diagonal crosses line PQ. The solution composition, therefore, changes along line $c'P$, and at P the three salts AX, BY and BX are co-deposited: point P is the quaternary drying-up point for all solutions represented by points within triangle $AX.BX.BY$.

The isothermal evaporation of a solution denoted by point w in *Figure 4.24b* can be traced in the same manner as that described for point a in *Figure 4.24a*. Q' is the drying-up point. The evaporation of solution x can be traced as follows. At x the solution is saturated with salt BY, and this salt is deposited until the solution composition reaches x', where the solution is saturated with the two salts AX and BY. The composition of the solid phase being deposited at this stage is given by point R on the diagonal. As evaporation proceeds, the solution composition changes from point x' along line $x'Q'$, i.e. in a direction away from point R, and at Q' the solution is saturated with the three salts AX, AY and BY. Both solution and deposited solids thereafter have a constant composition until evaporation is complete: Q' is the quaternary drying-up point.

Point Q' is also the drying-up point for a solution represented by point y. The solution composition changes along line yy' while salt AX crystallises out, and then from y' towards P' while the two salts AX and BX of composition O' are deposited. At P' the solution is saturated with the three salts AX, BX and BY, the composition of the solid phase deposited at this point being given by R. On further evaporation, the solution composition remains constant at P' while salts AX and BY are deposited and salt BX is dissolved. When all BX has dissolved, the solution composition changes from P' towards Q', and the solution finally dries up at Q'.

Point P', therefore, is incongruent. It is not a true drying-up point except for the case where the original complex lies within the triangle representing the three salts of which it is the saturation point, i.e. AX, BY and BX. Point z may be taken as an example of this case. On evaporation, the solution composition changes from z to z' while salt AX is deposited, from z' towards P' while salts AX and BX are deposited. The composition of the solid phase at this latter stage is given by point O'. At P' this solution is saturated with salts AX, BX and BY. Further evaporation results in the deposition of AX and BY and the dissolution of BX. The solution dries up at point P'.

Representation of Water Content

So far in the discussion of Jänecke's projection for reciprocal salt pair systems the water content has been ignored. This is not too serious, because

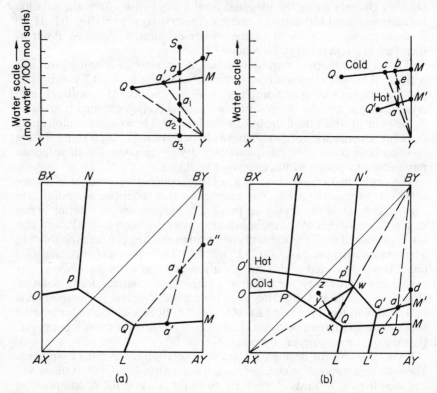

Figure 4.25. Representation of water content: (a) isothermal evaporation; (b) crystallisation by cooling

much information can be obtained from the projection before consideration of the quantity of water present. One way in which the water content can be represented is shown in *Figure 4.25a*; the plan shows the projection of the saturation surfaces, the elevation indicates the water contents. To avoid unnecessary complication, the elevation only shows the horizontal view of the particular saturation curve concerned in the problem.

The isothermal evaporation of water from a complex a was considered in

Figure 4.24a, where point *a*, representing the composition of the given complex, was taken to lie on the saturation surface. In *Figure 4.25a* the isothermal dehydration of an unsaturated solution *S* is considered, the dissolved salt having the same composition *a* as that in *Figure 4.24a*. Point *S*, therefore, is located on the elevation vertically above point *a* in the plan. The exact position of *S* is determined by the water content of the given solution, i.e. distance Sa_3 on the water scale denotes the moles of water per 100 mol of salt content. Line Sa_3, called the *water line*, represents the course of the isothermal dehydration. Points *Q* and *M* are similarly located on the elevation, according to their corresponding water contents, vertically above points *Q* and *M* on the plan. Point *a′* lies on curve *QM* vertically above *a′* in the plan. Point *T* on the elevation represents the water content of a saturated solution of pure salt *BY*, the salt to be deposited.

Three construction lines can now be drawn on the elevation. Line *Ta′* cuts the water line at point *a*. The *BY* saturation surface is assumed for simplicity to be plane, so *Taa′* is a line on this surface. The *Y* corner of the elevation represents pure salts *AY* and *BY* and all their mixtures. The line drawn from *a′* to *Y* (*BY* on plan) cuts the water line at a_1, that from *Q* to *Y* (*a″* on plan) at a_2.

When water is removed isothermally from the unsaturated solution *S*, the water content falls along the water line Sa_3. When point *a* is reached, the solution is saturated with salt *BY*, and pure *BY* starts to crystallise out. The quantity of water to be removed to achieve this condition is determined from the water scale readings on the elevation diagram, i.e. *Sa* mol of water has to be removed from a system containing 100 mol of salts dissolved in Sa_3 mol of water. Salt *BY* is deposited while the water content falls from *a* to a_1, and at point a_1 the solution (of composition *a′*) becomes saturated with salts *BY* and *AY*. Both salts are deposited while the water content falls from a_1 to a_2, and the over-all deposited solids content changes along line *BY/a″* on the plan. At point a_2 the solution (composition *Q*) is saturated with respect to the three salts *AX*, *AY* and *BY*, and further evaporation from a_2 to a_3 proceeds at constant solution composition *Q*. The solids composition changes along line *a″a* on the plan.

CRYSTALLISATION BY COOLING

The graphical procedure described above, viz. the drawing of a plan and elevation, provides a simple pictorial representation of the phase reactions occurring in a given system at two different temperatures. *Figure 4.25b* shows two isotherms labelled 'hot' and 'cold', respectively; they are in fact the curves from *Figure 4.24*, plotted on one diagram, and the same lettering is used. By way of example, two different cooling operations will be considered.

Point *a* on curve *Q′M′* represents a hot solution saturated with the two salts *AY* and *BY*. When it is cooled to the lower temperature, point *a* lies in the *BY* field of the projection. Line *BY/a* is drawn on the plan to meet curve *QM* at *b*, but point *b* represents the solution composition only if point *a* lies on the *BY* saturation surface in the 'cold' projection, i.e. if pure *BY* was crystallising out. To find the actual solution composition and the composition of the deposited solid phase, point b is projected from the plan onto curve QM in the elevation.

Point Y on the elevation diagram represents salts AY or BY or any mixture of them. Line Ya is drawn on the elevation and then produced to meet curve QM at c. It can be seen that in this case points b and c do not coincide. This means that the deposited solid phase is not pure salt BY but some mixture of BY and AY. Point c is projected from the elevation onto the plan, and line cad is drawn. Thus the final solution composition is given by point c, and the over-all solid phase composition by point d.

If pure salt BY was required to be produced during the cooling operation, the water content of the system would have to be adjusted accordingly. Solution point c has to move to become coincident with point d, and solid point d has to move to BY on the plan. In this case, therefore, water has to be added to the system, e.g. to the hot solution before cooling. The quantity of water required per 100 mol of salts is given by the vertical distance ae on the elevation.

A different sequence of operations is shown in another section of *Figure 4.25b*. Point w on curve $P'Q'$ represents a solution saturated with salts AX and BY at the higher temperature. At the lower temperature, however, point w lies in the BY field of the diagram. If the correct amount of water is present in the system, pure BY crystallises out on cooling, and the solution composition is given by point x located on line BY/w produced to meet curve PQ. A cyclic process can now be planned.

The pure salt BY is filtered off and a quantity of a solid mixture, e.g. of composition z, is added to solution x. The quantity of solid z to be added, calculated by the mixture rule, must be the amount necessary to give complex y, the composition of which is chosen so that, on being heated to the higher temperature, it lies in the AX field, yields the original solution w and deposits the pure salt AX. Thus the sequence of operations is

1. Cool solution w to the lower temperature
2. Filter off solid BY
3. Add solid mixture z to the mother liquor x to give complex y
4. Heat the complex to the higher temperature
5. Filter off solid AX
6. Cool mother liquor w, and so on

Of course, the water contents at each stage in the cycle must be adjusted so that the solutions deposit only one pure salt at a time. The quantities of water to be added or removed can be estimated graphically on the elevation diagram in the manner described above for solution a.

Only the simplest type of reciprocal salt pair diagram has been considered here. Many systems form hydrates or double salts; in others the stable salt pair at one temperature may become the unstable pair at another. For information on these more complicated systems reference should be made to specialised works on the Phase Rule. Purdon and Slater's publication (see Bibliography) is particularly noteworthy in this respect; many detailed graphical solutions of problems of commercial importance are given, and the analysis of five-component aqueous systems is discussed.

BIBLIOGRAPHY

BAILEY, A. E., *Melting and Solidification of Fats and Waxes*, 1950. New York; Interscience

BLASDALE, W. C., *Equilibria in Saturated Salt Solutions*, 1927. New York; Chemical Catalog Co.

BOWDEN, S. T., *The Phase Rule and Phase Reactions*, 1950. London; Macmillan

FINDLAY, A. and CAMPBELL, A. N., *The Phase Rule and its Applications*, 9th Ed., 1951. London; Longmans

HAASE, T. and SCHÖNERT, H., *Solid-Liquid Equilibrium*, 1969. Oxford; Pergamon

HILL, L. M., 'Phase Rule: Application to the Separation of Salt Solutions', in *Thorpe's Dictionary of Applied Chemistry*, 4th Ed., Vol. IX, 438, 1949. London; Longmans

MACKENZIE, R. C. (Ed.), *Differential Thermal Analysis*, 2 vols., 1969. New York; Academic Press

PURDON, F. F. and SLATER, V. W., *Aqueous Solution and the Phase Diagram*, 1946. London; Arnold

RICCI, J. E., *The Phase Rule and Heterogeneous Equilibrium*, 1951. New York; Van Nostrand

SCHWENKER, R. F. and GARN, P. D. (Eds.), *Thermal Analysis*, 2 vols., 1969. New York; Academic Press

TYRRELL, H. J. V. and BEEZER, A. E., *Thermometric Titrimetry*, 1968. London; Chapman and Hall

WETMORE, F. E. W. and LEROY, D. J., *Principles of Phase Equilibria*, 1951. New York; McGraw-Hill

5

Mechanism of Crystallisation

THE deposition of a solid crystalline phase from liquid and gaseous solutions, pure liquids and pure gases can only occur if some degree of supersaturation or supercooling has first been achieved in the system. The attainment of the supersaturated state is essential for any crystallisation operation, and the degree of supersaturation, or deviation from the equilibrium saturated condition, is the prime factor controlling the deposition process. Any crystallisation operation can be considered to comprise three basic steps: (1) achievement of supersaturation or supercooling, (2) formation of crystal nuclei, (3) growth of the crystals. All three processes may be occurring simultaneously in different regions of a crystallisation unit. The ideal crystallisation, of course, would consist of a strictly controlled step-wise procedure, but the complete cessation of nucleation cannot normally be guaranteed in a growing mass of suspended and circulating crystals.

The supersaturation of a system may be achieved by cooling, evaporation, the addition of a precipitant or as a result of the chemical reaction between two homogeneous phases. The state of supersaturation has already been discussed in Chapter 2. The other two factors, nucleation and growth, will be considered in this and the next chapter. Accounts of modern theories and research trends in these areas may be found in standard works, papers presented at recent symposia and annual reviews of the literature[1-38].

NUCLEATION

The condition of supersaturation alone is not sufficient cause for a system to begin to crystallise. Before crystals can grow there must exist in the solution a number of minute solid bodies known as centres of crystallisation, seeds embryos or nuclei. Nucleation may occur spontaneously or it may be induced artificially; these two cases are frequently referred to as homogeneous and heterogeneous nucleation, respectively. It is not always possible, however, to decide whether a system has nucleated of its own accord or whether it has done so under the influence of some external stimulus.

At the present time there is no general agreement on nucleation nomenclature, so to avoid confusion the terminology to be used in this and subsequent chapters will be defined here. The term 'primary' will be reserved for all cases of nucleation, homogeneous or heterogeneous, in systems that do not contain crystalline matter. On the other hand, nuclei are often generated in the vicinity of crystals present in a supersaturated system; this phenomenon will be referred to as 'secondary' nucleation. Thus we have:

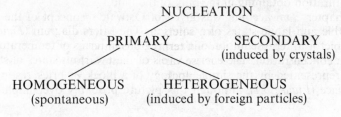

NUCLEATION

PRIMARY   SECONDARY
                           (induced by crystals)

HOMOGENEOUS   HETEROGENEOUS
(spontaneous)   (induced by foreign particles)

HOMOGENEOUS NUCLEATION

Exactly how a crystal nucleus is formed within a homogeneous fluid system is not known with any degree of certainty. To take a simple example, the condensation of a supersaturated vapour to the liquid phase is only possible after the appearance of microscopic droplets, called condensation nuclei, on the condensing surface. However, as the vapour pressure at the surface of these minute droplets is exceedingly high, they evaporate rapidly even though the surrounding vapour is supersaturated. New nuclei form while old ones evaporate, until eventually stable droplets are formed either by coagulation or under conditions of very high vapour supersaturation.

The formation of crystal nuclei is an even more difficult process. Not only have the constituent molecules to coagulate, resisting the tendency to redissolve, but they also have to become orientated into a fixed lattice. The number of molecules in a stable crystal nucleus can vary from about ten to several thousand; water (ice) nuclei, for instance, may contain about 100 molecules. The actual formation of such a nucleus can hardly result from the simultaneous collision of the required number of molecules; this would constitute an extremely unlikely event.

Most probably the mechanism of nucleation is as follows. The formation of liquid nuclei from a vapour, for example, where a stable nucleus contains n molecules, could occur by the bimolecular addition of molecules according to the scheme

$$A \quad + A \rightleftharpoons A_2$$
$$A_2 \quad + A \rightleftharpoons A_3$$
$$\vdots \qquad \vdots \qquad \vdots$$
$$A_{n-1} + A \rightleftharpoons A_n \text{ (critical cluster)}$$

Further molecular additions to the critical cluster would result in nucleation and subsequent growth of the droplets. Ions or molecules in a solution can also interact with one another to form short-lived clusters. Short chains may be formed initially, or flat monolayers, and eventually the lattice structure is built up. The construction process, which occurs very rapidly, can only continue in local regions of very high supersaturation, and many of these 'sub-nuclei' fail to achieve maturity; they simply redissolve because they are extremely unstable. If, however, the nucleus grows beyond a certain critical size, as explained below, it becomes stable under the average conditions of supersaturation obtaining in the bulk of the fluid.

In Chapter 2 an account was given of Ostwald's concept of the stable, metastable and labile states of a solution. The Miers diagram (*Figure 2.4*) gives a pictorial representation of these regions in terms of temperature and solution concentration. The energy levels of these various states of stability can be represented by the simple analogy of a block or brick resting on a flat surface (*Figure 5.1*), although this picture probably oversimplifies the

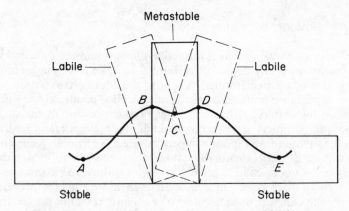

Figure 5.1. Demonstration of the stable, metastable and labile states

situation. The energy level (potential energy) in each of the various positions of the brick is denoted here by the height of the centre of gravity above the arbitrary datum of the flat surface. Cases *A* and *E* represent the lowest energy state, or state of maximum stability, which may be likened to the case of a saturated solution. Case *C* is also a stable state, but represents a higher energy level than *A* or *E*. A brick in this position cannot withstand any great displacement without reverting to the more stable positions *A* or *E*, so it may be considered to be metastable. This condition is similar, therefore, to the state of metastability in a supersaturated solution. Cases *B* and *D* represent unstable states, and also the highest energy level of the system. Any displacement would make the brick adjust itself into a more stable position. These cases may be taken to represent a labile supersaturated solution that would tend to nucleate spontaneously.

The structure of the assembly of molecules or ions which we call a critical nucleus is not known, and it is too small to observe directly. It could be a miniature crystal, nearly perfect in form. On the other hand, it could be a rather diffuse body with molecules or solvated ions in a state not too different from that in the bulk fluid, with no clearly defined surface. The classical theory of nucleation, stemming from the work of GIBBS[39], VOLMER[40], BECKER and DÖRING[41] and others, is based on the condensation of a vapour to a liquid, and this may readily be extended to crystallisation from the melt.

When a group of freely moving molecules becomes aggregated into a more condensed state, i.e. one in which the molecular movement is much more restricted, a quantity of energy is released. For example, when a vapour condenses to a liquid, the latent heat associated with the change of state is liberated. In a given system, therefore, the transitions from the gaseous to the liquid and then to the solid state represent a step-wise decrease in the degree of molecular mobility and likewise a decrease in the free energy of the system.

On the other hand, the formation of a liquid droplet or solid particle within a homogeneous fluid demands the expenditure of a certain quantity of energy in the creation of the liquid or solid surface. Therefore the total quantity of work, W, required to form a stable crystal nucleus is equal to the sum of the work required to form the surface, W_S (a positive quantity), and the work required to form the bulk of the particle, W_V (a negative quantity):

$$W = W_S + W_V \tag{5.1}$$

For the formation of a spherical liquid droplet in a supersaturated vapour, for instance, equation (5.1) can be written as

$$W = a\sigma - v\Delta p \tag{5.2}$$

where σ is the surface energy of the droplet per unit area, Δp is the difference in pressure between the vapour phase and the interior of the liquid droplet, and a and v are the surface area and volume, respectively, of the droplet. If the spherical droplet has radius r, then $a = 4\pi r^2$, $v = \frac{4}{3}\pi r^3$ and $\Delta p = 2\sigma/r$, so equation (5.2) can be written

$$W = 4\pi r^2\sigma - \tfrac{4}{3}\pi r^3 \cdot \frac{2\sigma}{r} \tag{5.3}$$

$$= \tfrac{4}{3}\pi r^2\sigma \tag{5.4}$$

From equations (5.3) and (5.4) it can be seen that the work of formation of a droplet equals one-third of that required to form the surface of the droplet. This fact was first deduced by GIBBS[39].

The increase in the vapour pressure of a liquid droplet as its size decreases can be estimated from the Gibbs–Thomson (Kelvin) formula, which may be written in the form

$$\ln\frac{p_r}{p^*} = \frac{2M\sigma}{RT\rho r} \tag{5.5}$$

where p_r and p^* are the vapour pressures over a liquid droplet of radius r and a flat liquid surface, respectively: in other words, p^* is the equilibrium saturation vapour pressure of the liquid. M is the molecular weight, ρ the density of the droplet, T the absolute temperature and R the gas constant.

A formula similar to equation (5.5) has already been discussed (p. 43) with respect to the increase in solubility of a particle as its size is reduced; in that particular case the concentration ratio c_r/c^* was used in place of the term p_r/p^* in equation (5.5). Both these terms give a measure of the supersaturation, S, of the system, so the Gibbs–Thomson equation may be rewritten in a more general form as

$$\ln S = \frac{2M\sigma}{RT\rho r} \tag{5.6}$$

or

$$r = \frac{2M\sigma}{RT\rho \ln S} \tag{5.7}$$

If this value of r is substituted in equation (5.4), we get

$$W = \frac{16\pi\sigma^3 M^2}{3(RT\rho \ln S)^2} \tag{5.8}$$

Equation (5.8) is an extremely important relationship: it gives a measure of the work of nucleation in terms of the degree of supersaturation of the system. It can be seen, for example, that when the system is only just saturated ($S = 1$, $\ln S = 0$), the amount of energy required for nucleation is infinite, so a saturated solution cannot nucleate spontaneously. However, equation (5.8) also suggests that any supersaturated solution can nucleate spontaneously because there is some finite work requirement associated with the process — it is merely a question of supplying the required amount of energy to the system. The necessary quantity of energy may be quite excessive, but it still remains a fact that spontaneous nucleation is theoretically possible at any degree of supersaturation.

The free energy changes associated with the process of homogeneous nucleation may be considered as follows. The over-all excess free energy, ΔG, between a small solid particle of solute and the solute in solution is equal to the sum of the surface excess free energy, ΔG_S, i.e. the excess free energy between the surface of the particle and the bulk of the particle, and the volume excess free energy, ΔG_V, i.e. the excess free energy between a very large particle ($r = \infty$) and the solute in solution. ΔG_S is a positive quantity, the magnitude of which is proportional to r^2. In a supersaturated solution G_V is a negative quantity proportional to r^3. Thus

$$\begin{aligned} \Delta G &= \Delta G_S + \Delta G_V \\ &= 4\pi r^2 \sigma + \tfrac{4}{3}\pi r^3 \Delta G_v \end{aligned} \tag{5.9}$$

where ΔG_v is the free energy change of the transformation per unit volume. The two terms on the right-hand side of equation (5.9) are of opposite sign and depend differently on r, so the free energy of formation, ΔG, passes through a maximum (see *Figure 5.2*). This maximum value, ΔG_{crit} corresponds to the critical nucleus, r_c, and for a spherical cluster is obtained by maximising equation (5.9) setting $d\Delta G/dr = 0$:

$$\frac{d\Delta G}{dr} = 8\pi r\sigma + 4\pi r^2 \Delta G_v = 0 \tag{5.10}$$

therefore

$$r_c = \frac{-2\sigma}{\Delta G_v} \tag{5.11}$$

where ΔG_v is a negative quantity. From equations (5.9) and (5.11) we get

$$\Delta G_{\text{crit}} = \frac{16\pi\sigma^3}{3(\Delta G_v)^2} = \frac{4\pi\sigma r_c^2}{3} \qquad (5.12)$$

The behaviour of a newly created crystalline lattice structure in a super-saturated solution depends on its size; it can either grow or redissolve, but the process which it undergoes should result in the decrease in the free energy of the particle. The critical size r_c, therefore, represents the minimum size of a stable nucleus. Particles smaller than r_c will dissolve, or evaporate if the

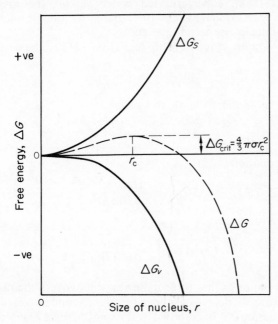

Figure 5.2. Free energy diagram for nucleation explaining the existence of a 'critical nucleus

particle is a liquid in a supersaturated vapour, because only in this way can the particle achieve a reduction in its free energy. Similarly, particles larger than r_c will continue to grow.

Although it can be seen from the free energy diagram why a particle of size greater than the critical size is stable, it does not explain how the amount of energy, ΔG_{crit}, necessary to form a stable nucleus is produced. This may be explained as follows. The energy of a fluid system at constant temperature and pressure is constant, but this does not mean that the energy level is the same in all parts of the fluid. There will be fluctuations in the energy about the constant mean value, i.e. there will be a statistical distribution of energy, or molecular velocity, in the molecules constituting the system, and in those supersaturated regions where the energy level rises temporarily to a high value nucleation will be favoured.

The rate of nucleation, J, e.g. the number of nuclei formed per unit time per unit volume, can be expressed in the form of the Arrhenius reaction velocity equation commonly used for the rate of a thermally activated process:

$$J = A \exp(-\Delta G / kT) \tag{5.13}$$

where k is the Boltzmann constant, the gas constant per molecule (1.3805×10^{-23} J K^{-1} = R/N, where R is the gas constant = 8.314 J K^{-1} mol^{-1} and N = the Avogadro number = 6.023×10^{23} mol^{-1}). For the nucleation of water droplets from supersaturated water vapour the pre-exponential factor, A, has been evaluated[42] as 10^{25}, giving the rate of nucleation, J, as the number of nuclei per second per cm^3.

The Gibbs–Thomson relationship (equation 5.6) may be written

$$\ln \frac{c}{c^*} = \ln S = \frac{2\sigma v}{kTr} \tag{5.14}$$

where v is the molecular volume; this gives

$$-\Delta G_v = \frac{2\sigma}{r} = \frac{kT \ln S}{v} \tag{5.15}$$

Hence, from equation (5.12)

$$\Delta G_{\text{crit}} = \frac{16\pi\sigma^3 v^2}{3(kT \ln S)^2} \tag{5.16}$$

and from equation (5.13)

$$J = A \exp\left[-\frac{16\pi\sigma^3 v^2}{3k^3 T^3 (\ln S)^2}\right] \tag{5.17}$$

This equation indicates that three main variables govern the rate of nucleation: temperature, T; degree of supersaturation, S; and interfacial tension, σ.

A plot of equation (5.17), as shown by the solid curve in *Figure 5.3*, indicates

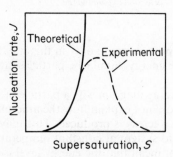

Figure 5.3. Effect of supersaturation on the nucleation rate

the extremely rapid increase in the rate of nucleation once some critical level of supersaturation is exceeded. Equation (5.17) may be rearranged to give

$$\ln S = \left[\frac{16\pi\sigma^3 v^2}{3k^3 T^3 \ln(A/J)}\right]^{\frac{1}{2}} \tag{5.18}$$

and if, arbitrarily, the critical supersaturation, S_{crit}, is chosen to correspond to a nucleation rate, J, of, say, one nucleus per second per unit volume, then equation (5.18) becomes

$$\ln S_{crit} = \left[\frac{16\pi\sigma^3 v^2}{3k^3\,T^3\,\ln A} \right]^{\frac{1}{2}} \tag{5.19}$$

Equation (5.19) may be used, e.g. by taking a value of the pre-exponential factor $A = 10^{25}\ cm^{-3}s^{-1}$, to estimate the solid–liquid surface energy term, σ, from nucleation data (see Chapter 6). On the other hand, the dominant effect of the degree of supersaturation can be shown by calculating the time required for the spontaneous appearance of nuclei in supercooled water vapour

Supersaturation, S	Time
1·0	∞
2·0	10^{62} years
3·0	10^3 years
4·0	0·1 s
5·0	10^{-13} s

Therefore, in this case, a 'critical' degree of supersaturation exists in the region of $S \sim 4\!\cdot\!0$, but it is also clear that nucleation would have occurred at any value of $S > 1\!\cdot\!0$ if sufficient time had elapsed.

For the case of a non-spherical nucleus the factor $16\pi/3$ in equations (5.12) and (5.16)–(5.19) must be replaced by the appropriate value (e.g. 32 for a cube).

Similar expressions to the above may be derived for homogeneous nucleation from the melt in terms of supercooling[43, 44]. The volume free energy ΔG_v is given by

$$\Delta G_v = \frac{L\Delta T}{T^*} \tag{5.20}$$

where T^* is the solid–liquid equilibrium temperature expressed in kelvins, $\Delta T = T^* - T$ is the supercooling and L is the latent heat of fusion. The radius of a critical nucleus is given by

$$r_c = \frac{2\sigma T^*}{L\Delta T} \tag{5.21}$$

and the rate of nucleation, from equation (5.13), may be expressed by

$$J = A \exp\left[-\frac{16\pi\sigma^3}{3kT^*L^2T_r(\Delta T_r)^2} \right] \tag{5.22}$$

where T_r is the reduced temperature defined by $T_r = T/T^*$ and $\Delta T_r = \Delta T/T^*$, i.e. $\Delta T_r = 1 - T_r$. Equation (5.22), like equation (5.17), indicates the dominant effect of supercooling on the nucleation rate.

For a wide range of substances, including metals, organic compounds and alkali halides[45], the critical homogeneous nucleation temperature expressed in kelvins is approximately 0·8–0·85 T^*, although recent work[46, 47] with hydrocarbons $> C_{15}$ has indicated values of $T_r \sim 0.95$.

The size of the critical nucleus is dependent on temperature, since the volume free energy, ΔG_v, is a function of the supercooling, ΔT, (equations 5.20 and 5.21) giving

$$r_c \propto (\Delta T)^{-1} \tag{5.23}$$

and from equation (5.12)

$$\Delta G_{crit} \propto (\Delta T)^{-2} \tag{5.24}$$

These relationships are shown in *Figure 5.4*, where it can be seen that the size of a critical nucleus increases with temperature. Similar relationships can

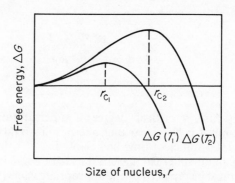

Figure 5.4. Effect of temperature on the size and free energy of formation of a critical nucleus ($T_1 < T_2$)

be derived for the effect of pressure in the case of condensation from a vapour.

Melts frequently demonstrate abnormal nucleation characteristics, as noted in the early work of TAMMAN[48]. The rate of nucleation usually follows an exponential curve (solid curve in *Figure 5.3*) as the supercooling is increased, but reaches a maximum and subsequently decreases (broken curve in *Figure 5.3*). Tamman suggested that this behaviour was caused by the sharp increase in viscosity with supercooling which restricted molecular movement and inhibited the formation of ordered crystal structures. It was left to the more recent studies of TURNBULL and FISHER[43] to quantify this behaviour with a modified form of equation (5.17):

$$J = A' \exp\left[-\frac{16\pi\sigma^3 v^2}{3k^3 T^3 (\ln S)^2} + \frac{\Delta G'}{kT} \right] \tag{5.25}$$

which includes a 'viscosity' term. When $\Delta G'$, the activation energy for molecular motion across the embryo–matrix interface, is exceptionally large (e.g. for highly viscous liquids and glasses) the other exponential term is small because under these circumstances S is generally very large. $\Delta G'$ then becomes the dominant factor in this rate equation and a decrease in nucleation rate is predicted.

Previously reported experimental observations of this reversal of the nucleation rate have been confined to melts, but it should also be expected in highly viscous solutions and such a behaviour has recently been observed in aqueous solutions of citric acid[49] (*Figure 5.5*).

The nucleation process has been discussed above in terms of the so-called classical theories stemming from the thermodynamic approach of Gibbs and

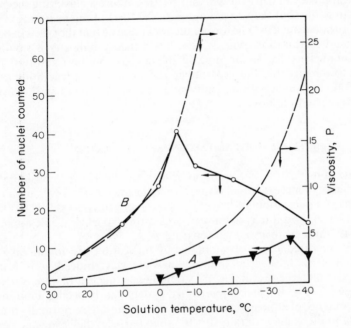

Figure 5.5. Spontaneous nucleation in supercooled citric acid solution:
A, 4·6 kg of citric acid monohydrate per kg of 'free' water ($T^* = 62\,°C$);
B, 7·0 kg/kg ($T^* = 85\,°C$)[49]

Volmer, with the modifications of Becker, Döring and later workers. The main criticism of these theories is their dependence on the quantity known as surface energy, σ, e.g. in the Gibbs–Thomson equation, and this term is probably meaningless when applied to small molecular aggregates of nucleic size. Other theories have been developed, such as that of STRANSKII and KAISHEV[50], which has been outlined in some detail by KHAMSKII[9], but these by and large lead to similar conclusions to the classical theories and will not be discussed here.

A more empirical approach to the nucleation process has been developed by CHRISTIANSEN and NIELSEN[14, 51], who proposed a relationship between the induction period, τ (the time interval between mixing two reacting solutions and the appearance of the precipitate) and the initial concentration, c, of the supersaturated solution:

$$\tau = k\,c^{1-p} \tag{5.26}$$

where k is a constant and p is the number of molecules in a critical nucleus. It is suggested that the induction period, which may range from microseconds to days depending on the supersaturation, represents the time needed for the assembly of a critical nucleus.

The so-called classical theories of homogeneous nucleation and the empirical theory of Christiansen and Nielsen all use a clustering mechanism of reacting ions, but they do not agree on the effect of supersaturation on the size of a critical nucleus. The former theories indicate that the size is dependent on the supersaturation, whereas the latter theory indicates a smaller but constant nucleus size. So far these differences have not been resolved, largely owing to the fact that the experimental investigation of true homogeneous nucleation is fraught with difficulty since the production of an impurity-free system is virtually impossible.

HETEROGENEOUS AND SECONDARY NUCLEATION

The rate of nucleation of a solution or melt can be affected considerably by the presence of mere traces of impurities in the system. However, an impurity that acts as a nucleation inhibitor in one case may not necessarily be effective in another; indeed it may act as an accelerator. No general rule applies and each case must be investigated separately.

Many reported cases of spontaneous nucleation are found on careful examination to have been induced in some way. Indeed, it is generally accepted that true examples of spontaneous nucleation are rarely encountered. For example, a supercooled system can be seeded unknowingly by the presence of atmospheric dust. Aqueous solutions as normally prepared in the laboratory may contain 10^6–10^8 solid particles (heteronuclei) per cm^3. It is virtually impossible to achieve a solution completely free of foreign bodies, although careful filtration may reduce the numbers to $< 10^3$ cm^{-3} and render the solution more or less immune to 'spontaneous' nucleation.

Cases are often reported of large volumes of a given system nucleating 'spontaneously' at smaller degrees of supercooling than small volumes. The most probable explanation is that the larger samples stand a greater chance of being contaminated with active heteronuclei. The size of the solid foreign bodies is important and there is evidence to suggest that the most active heteronuclei in liquid solutions lie in the range 0·1 to 1 μm.

Atmospheric dust frequently contains particles of the product itself, especially in industrial plants or in laboratories where samples of the crystalline material have been handled. Quite often some inert amorphous material in the dust acts as a seed; deliberate addition of small quantities of fine particles such as kieselguhr, silica and powdered glass has been found helpful for inducing crystallisation in specific cases.

Heteronuclei play an important role in atmospheric water condensation or ice formation[52]. Atmospheric nuclei have been classified as 'giant' (10 to 1 μm) which remain airborne for limited periods only, 'large' (1 to 0·2 μm) and 'Aitken' (0·2 to 0·005 μm). Particles smaller than about 10^{-3} μm are not normally found in air because they readily aggregate. Aitken nuclei are so called because they are active at the supersaturations produced in an Aitken counter, an apparatus in which a known volume of air is rapidly expanded; water droplets, formed on the particles, settle and are counted microscopically. Aitken nuclei, which occur to the extent of 10^4 to 10^5 cm^{-3} in the atmosphere, result from industrial smokes and vapours, ocean salts, land dusts, etc.

Probably the best method for inducing crystallisation is to inoculate or seed a supersaturated solution with small particles of the material to be crystallised. Deliberate seeding is frequently employed in industrial crystallisation to effect a control over the product size and size distribution. Nucleation in the presence of crystals is by definition (p. 137) 'secondary' and possible mechanisms of this type of nucleation are discussed on pp. 180 ff.

Seed crystals do not necessarily have to consist of the material being crystallised in order to be effective; isomorphous substances will frequently induce crystallisation. For example, phosphates will often nucleate solutions of arsenates; sodium tetraborate decahydrate can nucleate sodium sulphate decahydrate; phenol can nucleate m-cresol; and so on. The success of silver iodide, as an artificial rain-maker, is generally attributed to the striking similarity of the AgI and ice crystal lattices. However, there are many cases where lattice similarity does not exist and it is possible that other factors have to be considered; evidence has been produced, for example, to indicate that nucleation by seeding is dependent on the surface charge of the nucleating substrate[53].

In laboratory and large-scale crystallisations the first sign of nucleation often appears in one given region of the vessel, usually in regions where there is a local high degree of supersaturation, such as near a cooling surface or at the surface of the liquid. It is not uncommon to find some particular spot on the vessel wall or on the stirrer acting as a crystallisation centre. The most reasonable explanation of this phenomenon is that minute cracks and crevices in the surface retain tiny crystals from a previous batch which seed the system when it becomes supercooled. It is possible, of course, for a metal or glass surface to be in a condition in which it acts as a catalyst for nucleation.

As the presence of a suitable foreign body or 'sympathetic' surface can induce nucleation at degrees of supercooling lower than those required for spontaneous nucleation, the over-all free energy change associated with the formation of a critical nucleus under heterogeneous conditions, $\Delta G'_{crit}$, must be less than the corresponding free energy change, ΔG_{crit}, associated with homogeneous nucleation, i.e.

$$\Delta G'_{crit} = \phi \Delta G_{crit} \qquad (5.27)$$

where the factor ϕ is less than unity.

It has been indicated above that the interfacial energy, σ, is one of the important factors controlling the nucleation process. *Figure 5.6* shows a typical interfacial energy diagram for three phases in contact; in this case, however, the three phases are not the more familiar solid, liquid and gas, but two solids and a liquid. The three interfacial energies are denoted by σ_{cl} (between the solid crystalline phase, c, and the liquid, l), σ_{sl} (between another foreign solid surface, s, and the liquid) and σ_{cs} (between the solid crystalline phase and the foreign solid surface). Resolving these forces in a horizontal direction

$$\sigma_{sl} = \sigma_{cs} + \sigma_{cl} \cos \theta \qquad (5.28)$$

or

$$\cos \theta = \frac{\sigma_{sl} - \sigma_{cs}}{\sigma_{cl}} \qquad (5.29)$$

The angle θ, the angle of contact between the crystalline deposit and the foreign solid surface, corresponds to the angle of wetting in liquid–solid systems.

The factor ϕ in equation (5.27) can be expressed[40] as

$$\phi = \frac{(2 + \cos \theta)(1 - \cos \theta)^2}{4} \qquad (5.30)$$

Thus, when $\theta = 180°$, $\cos \theta = -1$ and $\phi = 1$, equation (5.27) becomes

$$\Delta G'_{crit} = \Delta G_{crit} \qquad (5.31)$$

When θ lies between 0 and 180°, $\phi < 1$; therefore

$$\Delta G'_{crit} < \Delta G_{crit} \qquad (5.32)$$

When $\theta = 0$, $\phi = 0$, and

$$\Delta G'_{crit} = 0 \qquad (5.33)$$

The three cases represented by equations (5.31)–(5.33) can be interpreted as follows. For the case of complete non-affinity between the crystalline solid and the foreign solid surface (corresponding to that of complete non-wetting in liquid–solid systems), $\theta = 180°$, and equation (5.31) applies, i.e.

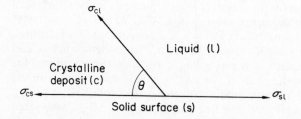

Figure 5.6. Surface energies at the boundaries between three phases (two solids, one liquid)

the over-all free energy of nucleation is the same as that required for homogeneous or spontaneous nucleation. For the case of partial affinity (cf. the partial wetting of a solid with a liquid), $0 < \theta < 180°$, and equation (5.32) applies, which indicates that nucleation is easier to achieve because the over-all excess free energy required is less than that for homogeneous nucleation. For the case of complete affinity (cf. complete wetting) $\theta = 0$, and the free energy of nucleation is zero. This case corresponds to the seeding of a super-saturated solution with tiny crystals of the required solute, i.e. no nuclei have to be formed in the solution. *Figure 5.7* indicates the relationship between ϕ and θ.

It has frequently been suggested that the heterogeneous nucleation of a solution may occur by seeding from embryos retained in cavities, e.g. in foreign bodies or the walls of the retaining vessel, under conditions in which the embryos would normally be unstable on a flat surface. This problem has been analysed by TURNBULL[54] for different types of cavity. The maximum diameter of a cylindrical cavity which will retain a stable embryo is given by

$$d_{max} = \frac{4\sigma_{cl}\cos\theta}{\Delta G_v} \qquad (5.34)$$

where ΔG_v is the volume free energy for the phase transformation. If the system is heated, this reducing the supersaturation or supercooling and eliminating all embryos in cavities larger than d_{max}, and subsequently cooled,

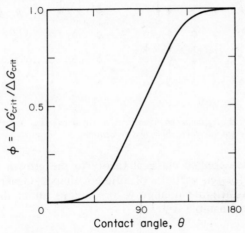

Figure 5.7. *Ratio of free energies of homogeneous and heterogeneous nucleation as a function of the contact angle*

the embryos retained in the cavities smaller than d_{max} will grow to the mouth of the cavity. They will then act as nuclei only if the cavity size $d_{max} \geqslant 2r_c$, where r_c is the size of a critical nucleus (equation 5.11 or 5.21).

TURNBULL and VONNEGUT[55] have developed a theory of nucleation catalysis in which the catalytic power of a solid surface is related to the lattice matching at the crystal/catalyst surface. The strain induced in the crystal nucleus depends on the lattice mismatch, δ, between the two solids, defined by $\delta = \Delta a/a$, where a is the lattice parameter of the unstrained crystal and Δa is the difference in lattice parameters between the crystal and the catalyst. If the nucleus can be strained by an amount ε in the interfacial plane and completely accommodate δ, the interface is said to be coherent (*Figure*

5.8a). The energy barrier to heterogeneous nucleation approaches a minimum under these conditions. When $\delta > \varepsilon \geqslant 0$, the interface is incoherent and is made up of regions of good fit interspersed with dislocations (see *Figure 5.8b*). The energy of a coherent interface is zero, but any coherency strain creates a strain energy in the crystal nucleus, proportional to δ^2, and this energy barrier reduces the volume free energy term, ΔG_v, in equation (5.12). Hence, the smaller the value of δ the smaller is the nucleation barrier, and therefore the more potent is the solid as a nucleation catalyst. Good agreement has been reported[55] between theory and practice in the case of the nucleation of ice from water in the presence of various solid catalysts. Broadly speaking, nuclei tend to form coherently with catalysts when the disregistry δ is less than about 0·02. On the other hand, oriented overgrowth (epitaxy) can occur with values of δ up to about 0·2.

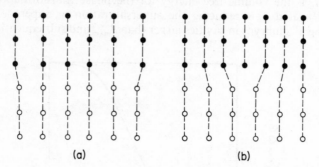

(a)   (b)

Figure 5.8. Atomic configurations at (a) a coherent and (b) an incoherent interface, showing a dislocation

FLETCHER[56] has applied classical theory to the growth of a crystalline phase on small foreign particles of different shapes. General relationships were devised to facilitate qualitative predictions about the behaviour of heteronuclei in metastable systems.

SPINODAL DECOMPOSITION

The existence of concentration fluctuations in a multicomponent fluid system is an implicit assumption in the GIBBS[39] theory of homogeneous nucleation. Two types of phase transition (nucleation) have been postulated, viz. composition fluctuations large in degree and infinitesimal in spatial extent (e.g. an infinitesimal droplet with properties approaching those of the bulk supercooled phase) or infinitesimal in degree and large in extent (e.g. continuous changes of phase). Classical nucleation theory, based on the former postulate, requires the further assumption that a sharp interface exists between the nucleating (stable) and supercooled (unstable) phases. The latter mode of transition, known as *spinodal decomposition,* does not require this assumption; a diffuse interface may be considered to exist between the phases.

The underlying theory for spinodal decomposition rests on Gibbs' derivation for the limit of stability of a fluid phase with respect to continuous changes of phase, represented by

$$\left.\frac{\partial^2 G}{\partial c^2}\right|_{T,P} = 0 \qquad (5.35)$$

where G is the Gibbs free energy per mole of solution and c is the solution concentration. On a phase diagram the locus of such points, representing the limit of stability, is referred to as the *spinodal* (see *Figure 5.9*). Thus, for spinodal

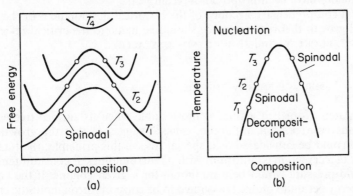

Figure 5.9. (a) Free energy–composition–temperature surface, showing the location of the spinodal; (b) temperature–composition graph of the spinodal

decomposition to occur, a spontaneous phase transition is necessary and the condition

$$(\partial^2 G/\partial c^2) \leqslant 0 \qquad (5.36)$$

should apply. Within the spinodal region any phase separation can lower the free energy of the system and no nucleation step is required. Outside this boundary, nucleation is essential to effect a phase change.

CAHN and HILLIARD[57] have applied this criterion to the work of formation of a critical nucleus in a two-component system, and were able to deduce the hitherto unknown properties of a critical nucleus for spinodal decomposition.

CRYSTAL GROWTH

As soon as stable nuclei, i.e. particles larger than the critical size, have been formed in a supersaturated or supercooled system, they begin to grow into crystals of visible size. Many attempts have been made to explain the mechanism and rate of crystal growth, and these may be broadly classified under the three general headings of 'surface energy', 'adsorption layer' and 'diffusion' theories.

The surface energy theories are based on the postulation of Gibbs (1878) and Curie (1885) that the shape a growing crystal assumes is that which has a minimum surface energy. This approach, although not completely abandoned, has largely fallen into disuse. The diffusion theories originated by Noyes and Whitney (1897) and Nernst (1904) presume that matter is deposited continuously on a crystal face at a rate proportional to the difference in concentration between the point of deposition and the bulk of the solution. The mathematical analysis of the operation is similar to that used for other diffusional and mass transfer processes. Volmer (1922) suggested that crystal growth was a discontinuous process, taking place by adsorption, layer by layer, on the crystal surfaces. Several notable modifications of this adsorption layer theory have been proposed in recent years.

For a comprehensive account of the historical development of the many crystal growth theories, reference should be made to the critical reviews by WELLS[58], BUCKLEY[1] and STRICKLAND-CONSTABLE[17].

SURFACE ENERGY THEORIES

An isolated droplet of a fluid is most stable when its surface free energy, and thus its area, is a minimum. GIBBS[39] suggested that the growth of a crystal could be considered as a special case of this principle: the total free energy of a crystal in equilibrium with its surroundings at constant temperature and pressure would be a minimum for a given volume. If the volume free energy per unit volume is assumed to be constant throughout the crystal, then

$$\sum_1^n a_i g_i = \text{minimum} \tag{5.37}$$

where a_i is the area of the ith face of a crystal bounded by n faces, and g_i the surface free energy per unit area of the ith face. Therefore, if a crystal is allowed to grow in a supersaturated medium, it should develop into an 'equilibrium' shape, i.e. the development of the various faces should be in such a manner as to ensure that the whole crystal has a minimum total surface free energy for a given volume.

Of course, a liquid droplet is very different from a crystalline particle; in the former the constituent atoms or molecules are randomly dispersed, whereas in the latter they are regularly located in a lattice structure. Gibbs was fully aware of the limitations of his simple analogy, but CURIE[59] found it a useful starting point for an attempt to evolve a general theory of crystal growth.

WULFF[60] showed that the equilibrium shape of a crystal is related to the free energies of the faces; he suggested that the crystal faces would grow at rates proportional to their respective surface energies. LAUE[61] has modified Wulff's theory, pointing out that all possible combinations of faces must be considered to determine which of the over-all surface free energies represents a minimum. However, the surface energy and the rate of growth of a face should be inversely proportional to the reticular or lattice density of the respective lattice plane, so that faces having low reticular densities would grow rapidly and eventually disappear. In other words, high index faces grow faster than low.

The velocity of growth of a crystal face is measured by the outward rate of movement in a direction perpendicular to that face. In order to maintain constant interfacial angles in the crystal (Haüy's law), the successive displacements of a face during growth or dissolution must be parallel to each other. Except for the special case of a geometrically regular crystal, the velocity of growth will vary from face to face. *Figure 5.10a* shows the ideal case of a crystal that maintains its geometric pattern as it grows; such a crystal is called 'invariant'. The three equal *A* faces grow at an equal rate; the smaller *B* faces grow faster; while the smallest face, *C*, grows fastest of all. A similar, but reverse, behaviour may be observed when a crystal of this type dissolves in a solvent; the *C* face dissolves at a faster rate than the other faces, but the sharp outlines of the crystal are soon lost once dissolution commences.

In practice, a crystal does not always maintain geometric similarity during growth; the smaller, faster-growing faces are often eliminated, and

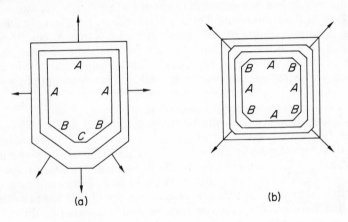

(a)   (b)

Figure 5.10. Velocities of crystal growth faces: (a) invariant crystal; (b) overlapping

this mode of crystal growth is known as 'overlapping'. *Figure 5.10b* shows the various stages of growth of such a crystal. The smaller *B* faces, which grow much faster than the *A* faces, gradually disappear from the pattern.

So far there is no general acceptance of the surface energy theories of crystal growth, since there is little quantitative evidence to support them. These theories, however, still continue to attract attention. For example, an interesting extension has recently been proposed by HERRING[62], who concluded that the thermodynamically stable form of a crystal may not always be a polyhedron but may have curved surfaces. But here again it is extremely difficult to visualise experimental techniques that would enable supporting evidence to be obtained. In any case, the main defect of the Gibbs–Curie–

Wulff theories is their failure to explain the well-known effects of super-saturation and solution movement on the crystal growth rate.

The concept of a crystal growth mechanism based on the existence of an adsorbed layer of solute atoms or molecules on a crystal face was first suggested by VOLMER[40]. Other workers who have contributed to, and modified, this theory include BRANDES[63], STRANSKII[64] and KOSSEL[65], although Kossel's theory, based on a sequence of repeatable steps, deviates to a large degree from Volmer's original postulation. Several comprehensive surveys of these and other modern theories of crystal growth are available[1, 17]. The brief account given here will serve merely to indicate the important features of layer growth and the role of crystal imperfections in the growth process.

Volmer's theory, or as some prefer to call it, the Gibbs–Volmer theory, is based on thermodynamic reasoning. When units of the crystallising substance arrive at the crystal face they are not immediately integrated into the lattice, but merely lose one degree of freedom and are free to migrate over the crystal face (surface diffusion). There will, therefore, be a loosely adsorbed layer of integrating units at the interface, and a dynamic equilibrium is established between this layer and the bulk solution. The adsorption layer, or 'third phase', as it is sometimes called, plays an important role in crystal growth (p. 161), secondary nucleation (p. 180) and precipitation phenomena. The thickness of the adsorption layer probably does not exceed 10 nm (100 Å) and may even be nearer 1 nm (10 Å).

Atoms, ions or molecules will link into the lattice in positions where the attractive forces are greatest, i.e. at the 'active centres', and under ideal conditions this step-wise build-up will continue until the whole plane face is completed (*Figure 5.11a* and *b*). Before the crystal face can continue to grow, i.e. before a further layer can commence, a 'centre of crystallisation' must come into existence on the plane surface, and in the Gibbs–Volmer theory it is suggested that a monolayer island nucleus, usually called a two-dimensional nucleus, is created (*Figure 5.11c*).

Expressions for the energy requirement of two-dimensional nucleation and the critical size of a two-dimensional nucleus may be derived in a similar manner to those for homogeneous (three-dimensional) nucleation discussed earlier. The over-all excess free energy of nucleation may be written

$$\Delta G = a\sigma + v\Delta G_v \qquad (5.38)$$

where a and v are the area and volume of the nucleus, and if this is a circular disc of radius r and height h, then

$$\Delta G = 2\pi rh\sigma + \pi r^2 h\Delta G_v \qquad (5.39)$$

and, maximising to find the critical size, r_c,

$$\frac{d\Delta G}{dr} = 2\pi h\sigma + 2\pi rh\Delta G_v = 0 \qquad (5.40)$$

whence

$$r_c = -\frac{\sigma}{\Delta G_v} \tag{5.41}$$

In other words, the critical radius of a two-dimensional nucleus is half that of a three-dimensional nucleus (equation 5.11) formed under similar environmental conditions.

Similarly,

$$\Delta G_{crit} = -\frac{\pi h \sigma^2}{\Delta G_v} \tag{5.42}$$

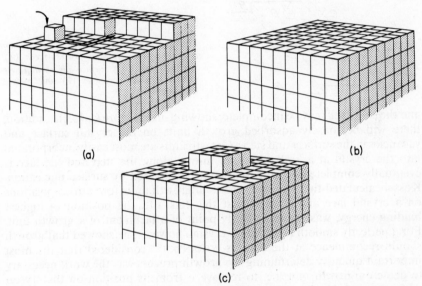

Figure 5.11. A mode of crystal growth without dislocations: (a) migration towards desired location; (b) completed layer; (c) surface nucleation

where ΔG_v is a negative quantity; so from equation (5.15)

$$\Delta G_{crit} = \frac{\pi h \sigma^2 v}{kT \ln S} \tag{5.43}$$

In a similar manner to that described earlier, the rate of two-dimensional nucleation, J', can be expressed in the form of the Arrhenius reaction velocity equation:

$$J' = B.\exp\left(-\Delta G_{crit}/kT\right) \tag{5.44}$$

or

$$J' = B.\exp\left[-\frac{\pi h \sigma^2 v}{k^2 T^2 \ln S}\right] \tag{5.45}$$

Comparing equations (5.16) and (5.43), it can be seen that the ratio of the energy requirements of three- to two-dimensional nucleation (sphere:

disc) is $16\sigma v/3hkT \ln S$. By inserting typical values for an inorganic salt, e.g. $\sigma = 10^{-1}$ J/m², $v = 2 \times 10^{-29}$ m³, $h = 5 \times 10^{-10}$ m, $kT = 4 \times 10^{-21}$ J, it can be calculated that the ratio is about 50:1 for a supersaturation of $S = 1 \cdot 1$ and about $1 \cdot 2 : 1$ for $S = 10$. In general, it may be said that a reasonably high degree of local supersaturation is necessary for two-dimensional nucleation to occur, but lower than that required for the formation of three-dimensional nuclei under equivalent conditions.

The Kossel model[65] of a growing crystal face is shown in *Figure 5.12*. The otherwise flat surface is divided into two regions by a monatomic *step*

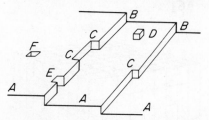

Figure 5.12. Kossel's model of a growing crystal surface showing flat surfaces (A), steps (B), kinks (C), surface-adsorbed growth units (D), edge vacancies (E) and surface vacancies (F)

and the step itself may be incomplete, showing one or more *kinks*. In addition, there will be loosely adsorbed growth units on the crystal surface and vacancies in the surfaces and steps. Growth units are most easily incorporated into the crystal at a kink; the kink moves along the step and the face is eventually completed. A fresh step could be created by surface nucleation. Kossel calculated the binding energies of growth units for various positions on a crystal face and showed that the kink site, the position of highest binding energy, was the most likely point of attachment of a growth unit. For a perfectly smooth heteropolar crystal, however, he showed that growth should recommence at the corners. STRANSKII[64] considered that the most important quantity determining the growth process was the work necessary to detach a growth unit, i.e. to remove it from its position on the crystal surface to infinity. In most cases the work of detachment of a molecule is lower than that of a single ion.

A crystal should grow fastest when its faces are entirely covered with kinks, and it is possible to estimate this theoretical maximum growth rate. It is unlikely, however, that the number of kinks would remain at this high value for any length of time; it is well known, for example, that broken crystal surfaces rapidly 'heal' and then proceed to grow at a much slower rate. However, many crystal faces readily grow at quite fast rates at relatively low supersaturations, far below those needed to induce surface nucleation. VOLMER and SCHULTZ[66], for example, showed that crystals of iodine could be grown from the vapour at 1 per cent supersaturation at rates some 10^{1000} times greater than those predicted by classical theory! It must be concluded, therefore, that the Kossel model, and its dependence on surface nucleation, is unreasonable for growth at moderate to low supersaturation.

A solution to the dilemma came in 1949 when FRANK[67] postulated that few crystals ever grow in the ideal layer-by-layer fashion without some imperfection occurring in the pattern. Most crystals contain dislocations, which cause steps to be formed on the faces (see Chapter 1). Of these the screw dislocation is considered to be important for crystal growth, since it obviates the necessity for surface nucleation. Once a screw dislocation has been formed, the crystal face can grow perpetually 'up a spiral staircase'. *Figure 5.13 a–c* indicates the successive stages in the development of a growth spiral starting from a screw dislocation. The curvature of the spiral cannot exceed a certain maximum value, determined by the critical radius for a two-dimensional nucleus under the conditions of supersaturation in the medium in which the crystal is growing. Quite often very complex spirals develop, especially when several screw dislocations grow together. Many examples of these are shown in the books by VERMA[68] and READ[69].

As a plane face never appears under conditions of spiral growth, surface nucleation is not necessary and the crystal grows as if the surface were covered with kinks. Growth continues uninterrupted at near the maximum theoretical rate for the given level of supersaturation. The behaviour of a crystal face with many dislocations is practically the same as that of a crystal face containing just one. BURTON, CABRERA and FRANK[70] developed a kinetic theory of growth in which the curvature of the spiral near its origin was related to the spacing of successive turns and the level of supersaturation. By the application of Boltzmann statistics they predicted kink populations, and by assuming that surface diffusion is an essential step in the process they were able to calculate the growth rate at any supersaturation. The apparently anomalous results of Volmer and Schultz could thus be explained. The Burton–Cabrera–Frank (BCF) relationship may be written.

$$R = A\sigma^2 \tanh(B/\sigma) \qquad (5.46)$$

where R = crystal growth rate, $\sigma = S-1$ and $S = c/c^*$ (see p. 37). A and B

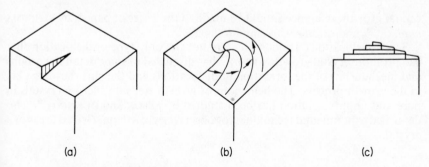

(a)   (b)   (c)

Figure 5.13. Development of a growth spiral starting from a screw dislocation

are complex temperature-dependent constants which include parameters depending on step spacings.

At low supersaturations the BCF equation approximates to $R \propto \sigma^2$, but at high supersaturations $R \propto \sigma$. In other words, it changes from a parabolic to a linear growth law as the supersaturation increases. The volume diffusion model proposed by CHERNOV[71] gives the same result. The general form of these expressions is shown in *Figure 5.14*. Actual experimental growth rate data are discussed in Chapter 6.

It should be pointed out that the BCF theory was derived for crystal growth from the vapour; and while it should also apply to growth from solutions (and melts), it is difficult to quantify the relationships because of the more complex nature of these systems. Viscosities, for example, are higher and diffusivities lower in solutions ($\sim 10^{-3}$ N s/m^2 (1 cP) and 10^{-9} m^2/s) than in vapours ($\sim 10^{-5}$ N s/m^2 and 10^{-4} m^2/s). In addition, the dependence of diffusivity on solute concentration can be complex (Chapter 3). Transport phenomena in ionic solutions can be complicated especially if the different ions exhibit complex hydration characteristics. Furthermore little is known about surface diffusion in adsorbed layers, and ion dehydration in or near these layers must present additional complicating factors.

For a comprehensive account of the relationships between the various surface and bulk diffusion models of crystal growth, and their relevance to

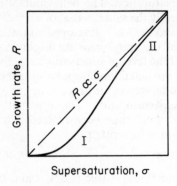

Figure 5.14. The Burton–Cabrera–Frank (BCF) supersaturation–growth relationship (I, $R \propto \sigma^2$; II, an approach to $R \propto \sigma$)

crystal growth, reference should be made to the series of papers by BENNEMA et al.[72, 73], and GILMER, GHEZ and CABRERA[74].

Screw dislocations, incidentally, are not the only type of dislocation that can promote crystal growth. ALBON[75] has discussed in some detail the origins and mechansims of the formation of dislocations and the part they can play in the growth process. The behaviour of growth steps on sucrose crystals in pure and impure solution has been studied by ALBON and DUNNING[76], who described experimental techniques for observing growth on selected areas of a crystal.

KINEMATIC THEORY

Two simultaneous processes are involved in the layer growth of crystals, viz. the generation of steps at some source on the crystal face followed by the movement of layers across the face. The characteristics of step motion and step bunching have been considered by several authors[71, 77-80] in an attempt to develop what has been called[77] a 'kinematic theory' of crystal growth.

Growth, and the reverse process of dissolution, may be considered in terms of the progression of steps across a crystal face (Figure 5.15). The step velocity, u, depends on the proximity of the other steps since all steps are competing for growth units. Thus

$$u = q/n \qquad (5.47)$$

where q is the step flux (the number of steps passing a given point per unit time) and n is the step density (the number of steps per unit length in a given region). The distance between steps, $\lambda = n^{-1}$. The slope of the surface, p, with reference to the close packed surfaces, i.e. the flat ledges, is given by

$$p = \tan \theta = hn \qquad (5.48)$$

and the face growth rate, v, normal to the reference surface by

$$v = hq = hnu \qquad (5.49)$$

where h is the step height.

If the steps are far apart ($\theta \to 0$), and the diffusion fields do not interfere with one another, the velocity of each step, u, will be a maximum. As the step spacings decrease and the slope increases, u decreases to a minimum at $hn = 1$ ($\theta = 45°$). Looking at it another way: as the slope θ increases, the

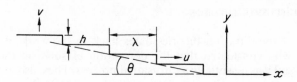

Figure 5.15. Two-dimensional diagrammatic representation of steps on a crystal face

face growth velocity v ($= u \tan \theta$) increases, approaches a flat maximum and then decreases to zero. The shape of this $v(p)$ curve, which is affected by the presence of impurities, is an important characteristic of the growth process[71].

For the two-dimensional case depicted in Figure 5.15 another velocity, $c = dx/dt$, may be defined which represents the motion of 'kinematic waves' (regions on the crystal surface with a constant slope p and velocity v). These waves do not contain the same monomolecular steps all the time, as the step velocity $u = v/p$ can be greater or less than c. When two kinematic waves of different slope meet, a discontinuity in slope occurs, giving rise to 'shock waves' across the surface.

Another problem that can be treated on the basis of the kinematic theory is that of step bunching. The steps that flow across a face are usually randomly spaced and of different height and velocity. Consequently they pile-up or bunch. Growth, and dissolution, can be characterised by the relationship between the step flux, q, and step density, n. Two general forms of this relationship can be considered depending upon whether $d^2q/dn^2 < 0$ (Type I) or $d^2q/dn^2 > 0$ (Type II). The former is analogous to the flow of traffic along a straight road and the latter to flood water on a river (see Figure 5.16).

FRANK[77], CABRERA and co-workers[78, 79] and CHERNOV[71] have analysed step bunching in terms of shock waves, using a continuum theory for step

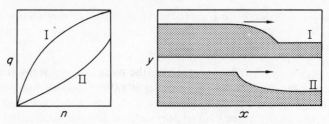

Figure 5.16. (a) Step flux density curves: type I, $d^2q/dn^2 < 0$; type II, $d^2q/dn^2 > 0$. (b) Surface profiles arising from bunches with type I and type II kinetics, respectively

motion. SCHWOEBEL and SHIPSEY[80] have adopted a probabilistic approach in which steps capture atoms, diffusing on the surface, with certain probabilities. The capture probability is assumed to depend on the directions from which the adsorbed atoms approach the step, and coalescence of steps or the stabilisation of step spacings can be explained.

DIFFUSION THEORIES

The origin of the diffusion theories dates back to the work of NOYES and WHITNEY[81] who considered that the deposition of solid on the face of a growing crystal was essentially a diffusional process. They also assumed that crystallisation was the reverse of dissolution, and that the rates of both processes were governed by the difference between concentration at the solid surface and in the bulk of the solution. An equation for crystallisation was proposed in the form

$$\frac{dm}{dt} = k_m A(c - c^*) \qquad (5.50)$$

where m = mass of solid deposited in time t; A = surface area of the crystal; c = solute concentration in the solution (supersaturated); c^* = equilibrium saturation concentration; and k_m = coefficient of mass transfer.

On the assumption that there would be a thin stagnant film of liquid adjacent to the growing crystal face, through which molecules of the solute would have to diffuse, NERNST[82] modified equation (5.50) to the form

$$\frac{dm}{dt} = \frac{D}{\delta} A(c - c^*) \qquad (5.51)$$

where D = coefficient of diffusion of the solute, and δ = length of the diffusion path.

The thickness δ of the stagnant film would obviously depend on the relative solid–liquid velocity, i.e. on the degree of agitation of the system. Film thicknesses varying from 20 to 150 μm have been measured on stationary crystals in stagnant aqueous solution, but MARC[83] found that the film thickness was virtually zero in vigorously stirred solutions. As this would imply an almost infinite rate of growth in agitated systems, it is obvious that the concept of film diffusion alone is not sufficient to explain the mechanism of crystal growth.

It was also shown by Marc that crystallisation is not necessarily the reverse of dissolution. A substance generally dissolves at a faster rate than it crystal-lises at, under the same conditions of temperature and concentration. Another important finding was made by MIERS[84], who determined, by refractive index measurements, the solution concentrations near the faces of crystals of sodium chlorate growing in aqueous solution; he showed that the solution in contact with a growing crystal face is not saturated but supersaturated.

In the light of these facts, a considerable modification was made to the diffusion theory of crystal growth by BERTHOUD[85] and VALETON[86], who suggested that there were two steps in the mass deposition, viz. a diffusion process, whereby solute molecules are transported from the bulk of the fluid phase to the solid surface, followed by a first-order 'reaction' when the solute molecules arrange themselves into the crystal lattice. These two stages, occurring under the influence of different concentration driving forces, can be represented by the equations

$$\frac{dm}{dt} = k_d A(c - c_i) \qquad \text{(diffusion)} \qquad (5.52)$$

and

$$\frac{dm}{dt} = k_r A(c_i - c^*) \qquad \text{(reaction)} \qquad (5.53)$$

where k_d = a coefficient of mass transfer by diffusion; k_r = a rate constant

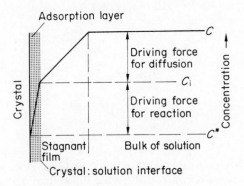

Figure 5.17. Concentration driving forces in crys-tallisation from solution

for the surface reaction; and c_i = solute concentration in the solution at the crystal–solution interface.

A pictorial representation of these two processes is shown in *Figure 5.17*, where the various concentration driving forces can be seen. It must be clearly understood, however, that this is only diagrammatic: the driving forces will rarely be of equal magnitude, and the concentration drop across the stagnant film is not necessarily linear. Furthermore, there appears to be some confusion in recent crystallisation literature between this hypothetical film and the more fundamental 'boundary layers' (see p. 202). Some of the characteristics of the adsorption layer are discussed on pp. 154 and 225.

Equations (5.52) and (5.53) are not easy to apply in practice because they involve interfacial concentrations that are difficult to measure. It is usually more convenient to eliminate the term c_i by considering an 'over-all' concentration driving force, $c - c^*$, which is quite easily measured. A general equation for crystallisation based on this over-all driving force can be written as

$$\frac{dm}{dt} = K_G A(c - c^*)^n \qquad (5.54)$$

where K_G is an over-all crystal growth coefficient. The exponent n is usually referred to as the 'order' of the over-all crystal growth process, but the use of this term should not be confused with its more conventional use in chemical kinetics, where the order always refers to the power to which a *concentration* should be raised to give a factor proportional to the rate of an elementary reaction. In crystallisation work the exponent, which is applied to a *concentration difference*, has no fundamental significance and cannot give any indication of the number of elementary species involved in the growth process.

If $n = 1$ and the surface reaction (equation 5.53) is also first-order, the interfacial concentration, c_i, may be eliminated from equations (5.52) and (5.53) to give

$$\frac{dm}{dt} = \frac{A(c - c^*)}{1/k_d + 1/k_r} \qquad (5.55)$$

i.e.

$$\frac{1}{K_G} = \frac{1}{k_d} + \frac{1}{k_r} \qquad (5.56)$$

or

$$K_G = \frac{k_d k_r}{k_d + k_r} \qquad (5.57)$$

For cases of extremely rapid surface reaction, i.e. large k_r, $K_G \approx k_d$ and the crystallisation process is controlled by the diffusional operation. Similarly, if the value of k_d is large, i.e. if the diffusional resistance is low, $K_G \approx k_r$, and the process is controlled by the surface reaction. It is worth pointing out that whatever the relative magnitude of k_d and k_r they will always contribute to K_G.

The diffusional step (equation 5.52) is generally considered to be linearly dependent on the concentration driving force, but the validity of the assumption of a first-order surface reaction (equation 5.53) is highly questionable. Many inorganic salts crystallising from aqueous solution give an over-all growth rate order, n, in the range 1·5 to 2[87, 88]. The rate equations, therefore, may be written

$$R_G = \frac{1}{A} \cdot \frac{dm}{dt} = k_d(c - c_i) \qquad \text{(diffusion)} \qquad (5.58)$$

$$= k_r(c_i - c^*)^z \qquad \text{(reaction)} \qquad (5.59)$$

$$= K_G(c - c^*)^n \qquad \text{(over-all)} \qquad (5.60)$$

The reverse process of dissolution may be represented by the over-all relationship

$$R_D = K_D(c^* - c)^{n'} \qquad (5.61)$$

where n' is generally, but not necessarily, unity. From equation (5.58)

$$c_i = c - R_G/k_d \qquad (5.62)$$

so equation (5.59), representing the surface integration step, may be written

$$R_G = k_r\left(\Delta c - \frac{R_G}{k_d}\right)^z \qquad (5.63)$$

where $\Delta c = c - c^*$ and $z \geqslant 1$. If $z = 1$,

$$R_G = \left[\frac{k_d k_r}{k_d + k_r}\right]\Delta c \qquad (5.64)$$

as in equation (5.57). However, if $z \neq 1$, the surface integration step is dependent on the concentration driving force in a non-linear manner. For example, if $z = 2$, equation (5.63) can be solved to give

$$R_G = k_d\left[\left(1 + \frac{k_d}{2k_r\Delta c}\right) - \sqrt{\left\{\left(1 + \frac{k_d}{2k_r\Delta c}\right)^2 - 1\right\}}\right]\Delta c \qquad (5.65)$$

However, apart from such simple cases, equation (5.63) cannot be solved explicitly for R_G and the relationship between the coefficients K_G, k_d and k_r remains obscure.

A quantitative measure of the degree of diffusion or surface integration control may be made through the concept of *effectiveness factors*. A surface integration effectiveness factor, η_r, may be defined[89] by

$$\eta_r = \frac{\text{measured over-all growth rate}}{\begin{array}{c}\text{growth rate expected when the crystal surface is exposed}\\\text{to conditions in the bulk solution}\end{array}}$$

$$= \frac{R_G}{k_r(c - c^*)^z} \qquad (5.66)$$

As the diffusion step becomes less important, $\eta_r \to 1$, and the surface integration step increasingly dominates the growth process. It can be shown that

$$\eta_r = (1 - h\eta_r)^z \qquad (5.67)$$

where

$$h = k_r(c - c^*)^{z-1}k_d^{-1}$$

Effectiveness factors may be used in conjunction with growth and dissolution rate data to calculate how closely measured growth rates approach the surface integration rate. Similarly, changes in relative crystal/solution velocity alter the relative magnitude of the diffusion and surface integration rates and the approach to true surface integration control at high velocities can be estimated[89].

It might be thought possible that the diffusional and surface reaction coefficients could be quantified by making certain assumptions. For example, if it is assumed that the diffusional mass transfer coefficient, k_d, in the crystallisation process is the same as that measured for crystal dissolution in near-saturated solutions under the same concentration driving force, temperature, etc., then values of k_r can be predicted.

Such calculations have been made[87, 88], but the validity of the assumption that the diffusion step in crystal growth can be related to the diffusion step in dissolution has not yet been proved. It is possible, for example, that dissolution is not a simple one-step process. Indeed some form of 'surface reaction' step has been measured for the dissolution of zinc in mercury[90] and lead sulphate in water[91]. Obviously more studies on reciprocal growth and dissolution need to be made before a coherent growth theory can be produced.

In any case, the problem is undoubtedly much more complex than the simple two-step process envisaged above. For an ionising solute crystallising from aqueous solution, for example, the following processes may all be taking place simultaneously:

1. Bulk diffusion of solvated ions through the diffusion boundary layer
2. Bulk diffusion of solvated ions through the adsorption layer
3. Surface diffusion of solvated or unsolvated ions
4. Partial or total desolvation of ions
5. Integration of ions into the lattice
6. Counter-diffusion of released water through the adsorption layer
7. Counter-diffusion of water through the boundary layer

In theory any one of these processes could become rate-controlling, and a rigorous solution of the problem is virtually unattainable. Furthermore the thicknesses of the different layers and films cannot be known with any certainty. Adsorbed molecular layers probably do not exceed 10^{-2} μm; partially disordered solution near the interface may account for another 10^{-1} μm; and the diffusion boundary layer is probably not much thicker than about 10 μm (see p. 202).

The individual face constants k_d and k_r are not only difficult if not impossible to determine, but can vary from face to face on the same crystal. It is even possible for k_d to vary over one given face: although it is true, as MIERS[84] found, that the solution in contact with growing crystal face is always supersaturated, the degree of supersaturation can vary at different points over the face[92]. From refractive index measurements BERG[93] showed, surprisingly, that the concentration was highest at the corners and lowest at the centre of the face, but whether this condition arose from the mode of crystal growth or whether it actually caused a particular mode of growth would be extremely difficult to decide.

The diffusion theories of crystal growth cannot yet be reconciled with the adsorption layer and dislocation theories. It is acknowledged that the diffusion theories have grave deficiencies (they cannot explain layer growth or the faceting of crystals, for example), yet crystal growth rates are conveniently measured and reported in diffusional terms. The utilisation of the mathematics of mass transfer processes makes this the preferred approach, from the chemical engineer's point of view at any rate, despite its many limitations.

Not only is it difficult to decide between basic theories for any given situation, it is also possible that the growth mechanism may change if the environment, e.g. temperature or concentration, changes. BRICE[94] suggests that no single growth mechanism can explain all data and that at least six different relations between growth rate and supersaturation are possible, depending upon the type of interface involved.

DIFFUSION AND REACTION STEPS

If a crystallisation process were entirely diffusion-controlled or surface reaction controlled, it should be possible to predict the growth rate by fundamental reasoning. In the case of diffusion-controlled growth, for example, the molecular flux, F (mol/s cm^2), is related to the concentration gradient, dc/dx, by

$$F = D(dc/dx) \tag{5.68}$$

where x is the length of the diffusion path and D is the diffusion coefficient. Therefore the rate of diffusion, dn/dt (mol s^{-1}), to a spherical surface, distance r from the centre, is given by

$$\frac{dn}{dt} = 4\pi r^2 D \frac{dc}{dr} \tag{5.69}$$

At any instant dn/dt is a constant, so equation (5.69) may be integrated to give

$$4\pi D \int_{c_1}^{c_2} dc = \frac{dn}{dt} \int_{r_1}^{r_2} \frac{dr}{r^2}$$

i.e.

$$\frac{dn}{dt} = \frac{4\pi D(c_2 - c_1)}{\dfrac{1}{r_1} - \dfrac{1}{r_2}} \tag{5.70}$$

If $c_1 = c^*$ (equilibrium saturation) at $r_1 = r$ (the surface of the sphere) and $c_2 = c$ (the bulk liquid concentration) at $r_2 = \infty$ (i.e. $r_2 \gg r_1$), then

$$\frac{dn}{dt} = 4\pi r D(c - c^*) \tag{5.71}$$

The linear particle growth rate, dr/dt, is obtained by multiplying both sides of this equation by the molecular volume, v, and dividing by the crystal surface area, $4\pi r^2$, to yield the general equation for the diffusion-controlled linear growth rate as

$$\frac{dr}{dt} = \frac{Dv(c-c^*)}{r} \tag{5.72}$$

The same relationship may be used for the reverse process of dissolution. For dissolution into a pure solvent ($c = 0$) integration of equation (5.72) gives

$$r^2 = r_0^2 - 2\,Dvc^*t \tag{5.73}$$

where r_0 is the initial size at time $t = 0$. The time for complete dissolution ($r = 0$) is thus given by

$$t = \frac{r_0^2}{2\,Dvc^*} \tag{5.74}$$

Substitution of typical values into equation (5.74) leads to some interesting observations. Small crystals of reasonably soluble salts may dissolve in fractions of a second, but those of sparingly soluble substances can take very long periods of time. For example, a 1 μm crystal ($r = 5 \times 10^{-7}$ m) of lead chromate ($D \approx 10^{-9}$ m^2/s, $v \approx 5 \times 10^{-5}$ m^3/mol, $c^* \approx 10^{-4}$ mol/m^3) would take about 7 h to dissolve in water at room temperature. A 10 μm crystal would take about 30 days. Tiny crystalline fragments of relatively insoluble substances may therefore remain undissolved in unsaturated solutions and act as nuclei in subsequent crystallisation operations. The properties of precipitates attributed to the past history of the system may well be associated with this behaviour.

For diffusion-controlled growth, with the growth rate speeded up by convection around falling particles, NIELSEN[95] recommends multiplication of equation (5.72) by a factor $F = (1+A)^{0.285}$, where $A = 2r^3 g\Delta\rho/9D\eta$ (g = acceleration of gravity, $\Delta\rho$ = relative density, η = viscosity). Nielsen also proposes growth relationships for cases of screw dislocation, mononuclear and polynuclear growth.

However, pure diffusion-controlled growth for all sizes of particles is most unlikely. From diffusional considerations TURNBULL[96] derived the mass flux, N, to a growing particle of radius r as

$$N = K\left(\frac{dr}{dt}\right) = \frac{D\kappa c}{D+\kappa r} \tag{5.75}$$

where D = diffusivity, k is a constant and κ is an interface transfer coefficient defined by $N = \kappa(c_r - c^*)$. Concentrations c, c_r and c^* refer to the bulk solution, particle surface and equilibrium saturation, respectively. Integration of equation (5.75) gives

$$\frac{r^2}{2D} + \frac{r}{\kappa} = Ktc \tag{5.76}$$

For $r \to 0$ this becomes

$$r \approx \kappa Ktc \tag{5.77}$$

indicating that the growth of very small nuclei should be interface-controlled.
    For large values of r

$$r \rightarrow \sqrt{(2DKtc)} \tag{5.78}$$

indicating diffusion control.

A further complicating factor in the use of growth rate expressions such as equation (5.72) is the fact that for very small particles the solubility is generally dependent on the particle size (Gibbs–Thomson effect, equation 5.14). In a recent analysis MATZ[97] has considered the combined effects of diffusion, integration and size-solubility. The over-all growth rate expression may be written as

$$R(x) = \tfrac{1}{2}R_\infty \left[\frac{1 - 3x^2 + 2x^3}{x^3 + Bx^4} \right] \tag{5.79}$$

$R(x)$ is the mean over-all linear growth rate (dr/dt) of a particle of dimensionless size $x = r/r_c$, where r is the particle radius and r_c is the size of a critical nucleus (*Figure 5.2*). The dimensionless factor $B = R_\infty/R^*$, where R_∞ is the growth rate of a large particle $(r \rightarrow \infty)$ determined from surface integration kinetics by

$$R_\infty = VNAFv \exp\left(-\frac{E}{RT}\right) \ln\left(\frac{c}{c^*}\right) \tag{5.80}$$

where V = molecular volume (cm^3/molecule), N = number of molecules per unit area at the crystal face (cm^{-2}), A = a dimensionless accommodation coefficient for the crystal, F = a dimensionless geometrical factor, v = frequency of oscillation of the lattice units (s^{-1}) and E = activation energy for crystal growth.

The growth rate for a diffusional process $R_D = R^*/x$ and R^* is expressed by a modified form of equation (5.72):

$$R^* = \frac{D(c - c^*)}{\rho \delta_0 r_c} \tag{5.81}$$

where ρ is the crystal density and $\delta_0 = \delta/r$, where δ is the thickness of the diffusion boundary layer.

For the growth of sodium chloride crystals from aqueous solution at 52 °C (325 K) and a supersaturation $S = 1.007$ ($c = 320.3$ and $c^* = 319.1$ g/l of solution) the following data may be used[97]. $N = 6.32 \times 10^{14}$ cm^{-2}, $r_c = 1.66 \times 10^{-6}$ cm, $\rho = 2.17$ g cm^{-3}, $\delta_0 = 0.04$, $A = 0.15$, $F = 0.167$, $V = 4.49 \times 10^{-23}$ cm^3 molecule^{-1}, $E = 5.4$ kcal mol^{-1}, $v = 10^{-13}$ s^{-1} and $R = 1.986$ cal mol^{-1} K^{-1}. Substituting these values in equations (5.80) and (5.81), we get $R_\infty = 6.04 \times 10^{-3}$ cm s^{-1}, $R^* = 2.26 \times 10^{-1}$ cm s^{-1}, and $B = R_\infty/R^* = 2.67 \times 10^{-2}$, and by use of equation (5.79) the variation of growth rate with size, $R(x)$, may be calculated (*Figure 5.18*).

The growth rate rapidly increases up to a maximum $(4.05 \times 10^{-5}$ m s$^{-1})$ at $x \approx 10$, i.e. at a particle diameter $\simeq 0.3$ μm. The growth rate than decreases until at $x \approx 10^4$ it is around the value normally expected for large crystals.

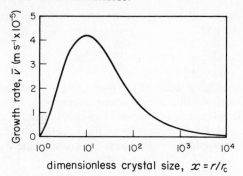

dimensionless crystal size, $x = r/r_c$

Figures 5.18. Predicted values of the over-all growth rates of NaCl crystals from aqueous solution[97]

For example, for a 1 mm crystal $x = r/r_c = 3.01 \times 10^4$ and $R(x) = 6.8 \times 10^{-6}$ m s^{-1}, which compares well with measured values.

CRYSTALLISATION FROM MELTS

The rate of crystallisation from a melt depends on the rate of heat transfer from the crystal face to the bulk of the liquid. As the process is generally accompanied by the liberation of heat of crystallisation, the surface of the crystal will have a slightly higher temperature than the supercooled melt.

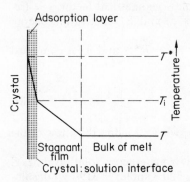

Figure 5.19. Temperature gradients near the face of a crystal growing in a melt

These conditions are shown in *Figure 5.19*, where the melting point of the substance is denoted by T^* and the temperature of the bulk of the super-cooled melt by T. The over-all degree of supercooling, therefore, is $T^* - T$. The temperature at the surface of the crystal, the solid–liquid interface, is denoted by T_i, so the driving force for heat transfer across the 'stagnant or effective' film of liquid close to the crystal face is $T_i - T$. The rate of heat transfer, dq/dt, can be expressed in the form of the equation

$$\frac{dq}{dt} = hA(T_i - T) \tag{5.82}$$

where A is the area of the growing solid surface and h is a film coefficient of heat transfer defined by

$$h = \frac{\kappa}{\delta'} \qquad (5.83)$$

where κ is the thermal conductivity and δ' is the effective film thickness for heat transfer. There is a distinct similarity between the form of equation (5.82) for heat transfer and equation (5.52) for mass transfer by diffusion. Agitation of the system will reduce the effective film thickness, increase the film coefficient of heat transfer and tend to increase the interfacial temperature, T_i, to a value near to that of the melting point, T^*.

TAMMAN[48] showed that the rate of crystallisation of a supercooled melt achieves a maximum value at a lower degree of supercooling, i.e. at a temperature higher than that required for maximum nucleation. The nucleation and crystallisation rate curves are dissimilar: the former has a relatively sharp peak (*Figure 5.3*), the latter usually a rather flat one. Tamman suggested that the maximum rate of crystallisation would occur at a melt temperature, T, given by

$$T = T^* - \left(\frac{\Delta H_{cryst}}{c_m} \right) \qquad (5.84)$$

where ΔH_{cryst} is the heat of crystallisation and c_m is the mean specific heat capacity of the melt.

The crystal growth rate (e.g. mass per unit time) may be expressed as a function of the over-all temperature driving force (cf. equation 5.54) by

$$\frac{dm}{dt} = K'_G A (T^* - T)^{n'} \qquad (5.85)$$

where A is the crystal surface area, K'_G is an over-all mass transfer coefficient for growth and exponent n' generally has a value in the range 1·5 to 2·5. The reverse process of melting, like that of dissolution, is often assumed to be first-order with respect to the temperature driving force, but this is not always the case[17], i.e.

$$-\frac{dm}{dt} = K_M A (T - T^*)^x \qquad (5.86)$$

where $x \geqslant 1$. K_M is an over-all mass transfer coefficient for melting.

Melting is a simultaneous heat and mass transfer process, i.e.

$$\frac{dq}{dt} = U_M A \Delta T = -\frac{dm}{dt} \Delta H_f \qquad (5.87)$$

therefore

$$-\frac{dm}{dt} = \frac{U_M \Delta T}{\Delta H_f} \qquad (5.88)$$

where $\Delta T = T - T^*$, q is a heat quantity, ΔH_f is the heat of fusion and U_M is an over-all heat transfer coefficient for melting.

The surface area of the melting solid $(A = \beta L^2)$ is related to the mass $(m = \alpha\rho L^3)$ by

$$A = \beta\left(\frac{m}{\alpha\rho}\right)^{\frac{2}{3}}$$ (5.89)

where L is a linear dimension, ρ = density, and α and β = volume and surface shape factors, respectively. Hence, equation (5.88) becomes

$$-\frac{dm}{dt} = \beta\left(\frac{m}{\alpha\rho}\right)^{\frac{2}{3}}\frac{U_M \Delta T}{\Delta H_f}$$ (5.90)

and, assuming the U_M, ΔT, α and β remain constant,

$$\int m^{-\frac{2}{3}}\,dm = -\frac{\beta U_M \Delta T}{(\alpha\rho)^{\frac{2}{3}}\Delta H_f}\int dt$$

or

$$\Delta(m^{\frac{1}{3}}) = -\gamma t$$ (5.91)

where $\gamma = \beta U_M \Delta T/3\Delta H_f\,(\alpha\rho)^{\frac{2}{3}}$, or in terms of the change in particle size, ΔL,

$$\Delta L = -\gamma'_t$$ (5.92)

where $\gamma' = \beta U_M \Delta T/3\Delta H_f\alpha\rho$.

Equations of the type of (5.91) and (5.92) find application in the assessment of melting processes and crystallisation from melts[98].

An absolute rate theory for the interfacial kinetics of crystal growth from the melt, based on a plausible model of the activated state, has been developed by KIRWAN and PIGFORD[99]. The theory predicts growth rates of pure materials, but not liquid metals, to within an order of magnitude, and correlates growth rates of crystals from their binary melts. In the experimental study[99] liquid compositions near the faces of crystals growing from binary melts were measured by a micro-interferometric technique.

For a full account of the basic theories of melting and crystal growth from the melt, reference should be made to the comprehensive monographs by BRICE[3] and UBBELOHDE[18].

CRYSTAL MORPHOLOGY AND STRUCTURE

The morphology of a crystal depends on the growth rates of the different crystallographic faces. Some faces grow very fast and have little or no effect on the growth form; the ones that have most influence are the slow-growing faces. The growth of a given face is governed by the crystal structure and defects on the one hand, and by the environmental conditions on the other.

A number of attempts have been made to predict the equilibrium growth form of a crystal. According to the Bravais rule (Chapter 1), the important faces governing the crystal morphology are those with the highest reticular densities and the greatest interplanar distances, d_{hkl}. Or, in simpler terms, the slowest-growing and most influential faces are the closest-packed and have the lowest Miller indices. The Gibbs–Curie–Wulff surface energy theory of crystal growth suggests that the growth form or 'equilibrium shape' should be such that the crystal has a minimum total surface free energy per unit volume.

A more recent morphological theory is that of HARTMAN and PERDOK[100, 101], who consider the bond energies involved in the integration of growth units into the lattice. In this theory crystal growth is considered as the formation of strong bonds between crystallising particles. A strong bond is defined as a bond in the first co-ordination sphere of a particle. The two-dimensional crystal shown in *Figure 5.20* is bounded by straight edges that are parallel to uninterrupted chains of strong bonds. Such a straight edge is

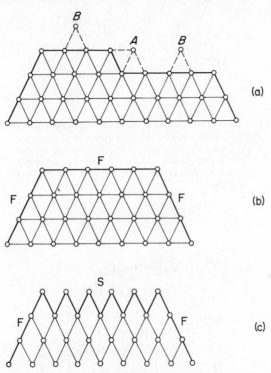

Figure 5.20. (a) Two-dimensional crystal. Each circle represents a growth unit of Kossel's repeatable step. (b) and (c) Projection of a three-dimensional crystal along a PBC. Each circle represents a PBC. An F-face results when neighbouring PBCs are linked together by strong bonds, otherwise an S-face results

formed when the probability of a particle being integrated is greater for site *A* than site *B*. In the case illustrated the particle at site *A* is bonded to the crystal with one strong bond more than at site *B*. The uninterrupted chains of strong bonds have been called *periodic bond chains* (PBC); and as the number of strong bonds per unit cell is limited, there exists a maximum length for the period of a PBC and, hence, a limited number of PBCs.

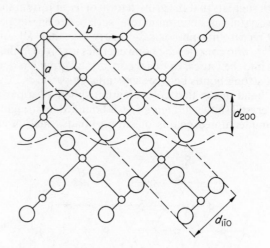

Figure 5.21. Projection of the rutile structure parallel to
[001]. Small circles = Ti, large circles = O. In a slice
$d_{1\bar{1}0}$ neighbouring PBCs [001] are bonded, so that
{110} is an F-form. In a layer d_{200} neighbouring chains
are not bonded, so that {100} is an S-form and less
important[101]

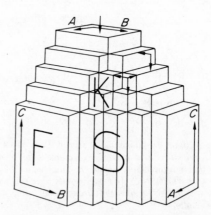

Figure 5.22. Crystal with three PBCs parallel
to [100] (A), [010] (B), and [001] (C). The
F-faces are (100), (010) and (001). The S-faces
are (110), (101) and (011). The K-face is
(111)[101]

The minimum thickness of a growth layer is the elementary 'slice', d_{hkl}, and faces that grow slice after slice are called flat or F-faces. The condition for a slice to exist is that two neighbouring parallel periodic bond chains be bonded together with strong bonds (*Figure 5.20b*). If this is not so, no slice exists, i.e. no layer growth can occur. Such faces are called stepped or S-faces (*Figure 5.20c*).

The structure of rutile, TiO_2 (*Figure 5.21*), shows PBCs of

in the direction [001]. These chains are bonded within d_{110} but not within d_{200}, so $\{110\}$ is an F-form, while $\{100\}$ is an S.

If no PBC exists within a layer, d_{hkl}, the face is called a kinked or K-face, which needs no nucleation for growth since it corresponds to a generalised type of Kossel's repeatable step (*Figure 5.22*). In terms of crystal structure dependent growth, therefore, the growth form should be bounded by F-faces only, although not all F-faces need be present.

The Hartman–Perdok approach is applied by making projections of the crystal structure parallel to a PBC and tabulating all the bonds. The packing of the chains determines the F-faces, provided that the chains are bonded by strong bonds. Sometimes it is easier to recognise the slices, and in that case the PBC may be found as the intersection of two slices. HARTMAN[101], BENNEMA[102], TROOST[103], and SIMON and BIENFAIT[104] have reported successful applications of this approach.

REFERENCES

The references to this Chapter will be found at the end of Chapter 6 (p. 225).

6

Crystallisation Kinetics

NUCLEATION

EXTERNAL INFLUENCES

NUCLEATION can often be induced by agitation, mechanical shock, friction and extreme pressures within solutions and melts, as shown by the early experiments of YOUNG[105] and BERKELEY[106]. The effects of external influences such as electric and magnetic fields, spark discharges, ultra-violet light, X-rays, γ-rays, sonic and ultrasonic irradiation have also been studied in recent years, but so far none of these methods has found any significant application in large-scale crystallisation practice. TIPSON[107], KAPUSTIN[108] and KHAMSKII[9] have summarised the findings of a large number of workers on this aspect of the subject.

Cavitation in an under-cooled liquid can cause nucleation, and this probably accounts for a number of the above reported effects. HUNT and JACKSON[109] have demonstrated, by a novel experimental technique, that nucleation occurs when a cavity collapses rather than when it expands. Extremely large pressures ($\sim 10^5$ bar, i.e. 10^{10} N/m^2) can be generated by the collapse of a cavity. The change in pressure lowers the crystallisation temperature of the liquid and nucleation results. It is suggested that nucleation caused by scratching the side of the containing vessel also takes place through the medium of cavitation.

Agitation is frequently used to induce crystallisation. Stirred water, for example, will allow only about $\frac{1}{2}$ °C of supercooling before spontaneous nucleation occurs, whereas undisturbed water will allow over 5 °C. Actually, very pure water, free from all extraneous matter, has been supercooled[110] some 39 °C. Most agitated solutions nucleate spontaneously at lower degrees of supercooling than quiescent ones. In other words, the supersolubility curve (*Figure 2.4*) tends to approach the solubility curve more closely in agitated solutions, and the width of the metastable zone is reduced, as is shown by the results in *Figure 6.1*.

However, the influence of agitation on the nucleation process is probably very complex. It is generally agreed that mechanical disturbances can enhance nucleation, but it has been shown by MULLIN and RAVEN[111] that an increase in the intensity of agitation does not always lead to an increase in nucleation. In other words, gentle agitation causes nucleation in solutions that are otherwise stable, and vigorous agitation considerably enhances nucleation, but the transition between the two conditions may not be continuous; a portion of the curve (see *Figure 6.2*) may have a reverse slope indicating a region where an increase in agitation actually *reduces* the tendency to nucleate. This phenomenon, observed with aqueous solutions of ammonium dihydrogen phosphate, magnesium sulphate and sodium nitrate,

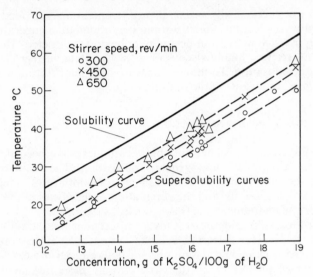

Figure 6.1. Effect of agitation on the metastable limits of K_2SO_4 solutions (cooling rate 0·6 °C/min)

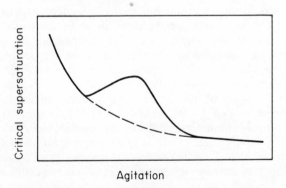

Figure 6.2. Influence of agitation on nucleation, showing a region where increased agitation can reduce the tendency to nucleate[111].

may be explained by assuming that agitation effects can lead to the disruption of sub-nuclei (molecular clusters) in the solution. Evidence for the existence of structured molecular clusters in supersaturated aqueous solutions has been provided by several workers[112–114].

Excessive cooling does not aid nucleation, as is shown by the work of

TAMMAN[48]. There is an optimum temperature for nucleation of a given system (see *Figures 5.3* and *5.5*) and any reduction below this value decreases the tendency to nucleate. As indicated by the classical nucleation equation (equation 5.17)

$$J = A \cdot \exp\left[-B(\ln S)^{-2}\right] \qquad (6.1)$$

nucleation can theoretically occur at any temperature, provided that the system is supercooled, but under normal conditions the temperature range over which massive nucleation occurs may be quite restricted. Therefore, if a system has set to a highly viscous or glass-like state, further cooling will not cause crystallisation. To induce nucleation the temperature would have to be increased to a value in the optimum region.

HOMOGENEOUS NUCLEATION

It is only in recent years that suitable techniques have been devised for studying the kinetics of homogeneous nucleation. The main difficulties have been the preparation of systems free from impurities, which might act as nucleation catalysts, and the elimination of the effects of the retaining vessel walls — glass frequently acts as a nucleation catalyst.

The first successful attempt to study homogeneous nucleation was made by VONNEGUT[115], who dispersed a liquid system into a large number of discrete droplets, exceeding the number of heteronuclei present. A significant number of droplets were therefore entirely mote-free and could be used for the study of true homogeneous nucleation. This technique has been applied to supercooled melts[116, 117], water[118] and aqueous salt solutions[119, 120].

The dispersed droplet method has many attendant experimental difficulties. Concentrations and temperatures must be measured with some precision for critical supersaturations to be determined; the tiny droplets (<1 mm) must be dispersed into an inert medium, e.g. an oil, which will not act as a nucleation catalyst; and any nuclei that form in the droplets have to be observed microscopically. WHITE and FROST[119] devised a droplet technique for the study of nucleation in ammonium nitrate solutions, and this was later developed by MELIA and MOFFIT[120] to study nucleation in aqueous solutions of ammonium chloride and bromide (see *Figure 6.3*). PRICE and GORNICK[121] have presented a theoretical and computational analysis for the suspension of melt droplets undergoing cooling at a constant rate.

The reliability of homogeneous nucleation studies is difficult to judge. Values of the kinetic constant, or collision factor as it is sometimes called, (the pre-exponential factor, A, in equation 6.1) reported by Melia and Moffitt lie in the range 10^3 to 10^5 cm^{-3} s^{-1}, which is well outside the range predicted from the Gibbs–Volmer theory ($\sim 10^{25}$). Doubts may be expressed, therefore, as to whether true homogeneous nucleation was observed or not. Of course, it is possible that the surface energy term, σ, which appears in constant B of equation (6.1) to the third power (see equation 5.17), cannot be assumed to be constant over a range of temperature. This point will be discussed in some detail later in the section on Precipitation.

An interesting technique, based on the observation of the phenomenon of crystalloluminescence, was reported by GARTEN and HEAD[112]. They showed that crystalloluminescence occurs only during the formation of a three-dimensional nucleus in solution, and that each pulse of light emitted lasting less than 10^{-7} s corresponds to a single nucleation event. Nucleation

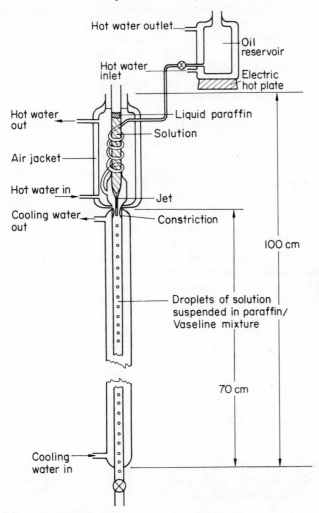

Figure 6.3. Apparatus for the study of nucleation in droplets of aqueous solution suspended in a heavy oil[120]

rates thus measured were close to those predicted from classical theory (equation 6.1) with collision factors in the range 10^{25} to 10^{30} cm^{-3} s^{-1}. It was suggested that true homogeneous nucleation occurs only at high supersaturations, and in their work on the precipitation of sodium chloride in the presence of lead impurities very high local supersaturations ($S > 14$)

were achieved. It would appear, therefore, that homogeneous nucleation (the formation of a true three-dimensional nucleus) can occur together with, or even instead of, heterogeneous nucleation if local supersaturation conditions are favourable.

The nucleation process is envisaged as the development by diffusion in the solution of a molecular cluster, as a disordered quasi-liquid, which after attaining critical size suddenly 'clicks' into crystalline form. As a result of this high-speed rearrangement, the surface of the newly formed crystalline particle may be expected to contain large numbers of imperfections that would encourage further rapid crystalline growth. As a nucleus appears to be generated in $< 10^{-7}$ s, its steady build-up as a crystalline body by diffusion is ruled out (a diffusion coefficient for NaCl of 10^{-5} cm^2/s gives a formation time more than ten times greater than the pulse period). These observations, therefore, may be taken as strong evidence for the existence and development of molecular clusters in supersaturated solutions.

From their work with sodium chloride Garten and Head suggested that a critical nucleus can be as small as about ten molecules. A different order of magnitude was proposed by OTPUSHCHENNIKOV[123], who measured the sizes of critical nuclei by observing the behaviour of ultrasonic waves in melts kept just above their freezing point. For phenol, naphthalene and azobenzene, for example, he suggested that about 1000 molecules constitute a stable nucleus. In contrast to this, the work of ADAMSKI[24] with relatively insoluble barium salts led to the conclusion that a critical nucleus was about 10^{-15} g, and as small as this mass may appear it still represents several million molecules. It is obvious, therefore, that there are still some widely diverging views on the question of the size of a crystal nucleus.

HETEROGENEOUS NUCLEATION

The lack of success of the classical nucleation theories in explaining the behaviour of real systems has led a number of authors to suggest that most primary nucleation in industrial crystallisers is heterogeneous rather than homogeneous and that empirical relationships such as

$$J = k_n \Delta c_{max}^m \qquad (6.2)$$

are the only ones that can be justified. J is the nucleation rate, k_n the nucleation rate constant, m the 'order' of the nucleation process and Δc_{max} the maximum allowable supersaturation. As explained earlier (p. 162) in connection with the empirical growth relationships, the 'order', m, has no fundamental significance. It does not give any indication of the number of elementary species involved in the nucleation process.

However, equation (6.2) is not entirely empirical, as demonstrated by NÝVLT[125] and NIELSEN[14], since it can be derived from the classical nucleation relationship (equation 6.1). The nucleation rate may be expressed in terms of the rate at which supersaturation is created by cooling, viz.

$$J = qb \qquad (6.3)$$

where $b = -\mathrm{d}\theta/\mathrm{d}t$ and q is the mass of solid deposited per unit mass of free solvent present when the solution is cooled by $1\,°C$. q is a function of the concentration change and of the crystallising species. In general,

$$q = \varepsilon \frac{\mathrm{d}c^*}{\mathrm{d}\theta} \tag{6.4}$$

where $\varepsilon = R/[1-c(R-1)]$. R is the ratio of the molecular weights of hydrate: anhydrous salt and c is the solution concentration expressed as mass of anhydrous solute per unit mass of solvent.

The maximum allowable supersaturation, Δc_{max}, may be expressed in terms of the maximum allowable undercooling, $\Delta\theta_{max}$:

$$\Delta c_{max} = \left(\frac{\mathrm{d}c^*}{\mathrm{d}\theta}\right) \Delta\theta_{max} \tag{6.5}$$

so equation (6.2) can be rewritten as

$$\varepsilon\left(\frac{\mathrm{d}c^*}{\mathrm{d}\theta}\right) b = k_n \left[\left(\frac{\mathrm{d}c^*}{\mathrm{d}\theta}\right) \Delta\theta_{max}\right]^m \tag{6.6}$$

or, taking logarithms,

$$\log b = (m-1) \log\left(\frac{\mathrm{d}c^*}{\mathrm{d}\theta}\right) - \log\varepsilon + \log k_n + m\log\Delta\theta_{max} \tag{6.7}$$

which indicates that the dependence of $\log b$ on $\log \Delta\theta_{max}$ is linear and the slope of the line is the 'order' of the nucleation process, m.

A simple apparatus[126], based on an earlier one devised by NÝVLT[15], used for determining solubility and nucleation characteristics, is shown in *Figure 6.4*. It consists essentially of a 50 ml Erlenmeyer flask fitted with a magnetic

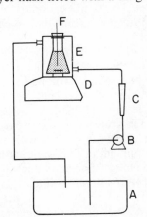

Figure 6.4. Apparatus for measuring metastable limits in agitated solutions: A, cooling water-bath; B, pump; C, flow meter; D, magnetic stirrer; E, Perspex water jacket; F, thermometer

stirrer and a thermometer graduated to $0\cdot1\,°C$, a cylindrical Perspex vessel, 90 mm diam. and 110 mm deep, to act as a water jacket for the flask, a 100 W lamp for illumination and heating, a pump, a Rotameter and a thermostatically controlled water-bath. About 40 ml of nearly saturated solution of known concentration is placed in the flask and sealed with the rubber stopper containing the thermometer. The flask is first cooled by shaking it under a cold-water tap until nuclei appear, and then slowly heated (at about 0·2 °C/h on approaching the saturation temperature) by the 100 W lamp. The rate of heating is easily controlled by varying the distance between the lamp and the flask. The stirring rate is maintained constant and the temperature at which the last crystalline particle disappears is taken as the saturation temperature.

The nucleation temperature is measured in a similar way. The flask containing the solution of known saturation temperature is placed in the jacket supplied with water at about 4 to 5 °C higher than the saturation temperature. A steady rate of cooling is maintained by controlling the water temperature and flow rate in the jacket. The temperature at which nuclei first appear and the time taken to cool from the saturation temperature are recorded. The difference between the saturation and nucleation temperatures is the maximum allowable undercooling, $\Delta\theta_{max}$, corresponding to a particular cooling rate.

For aqueous solutions the 'order' of the nucleation process, m, ranges from about 2 to 9, the lower values in the main being obtained with low molecular weight salts[125].

SECONDARY NUCLEATION

A supersaturated solution nucleates much more readily when crystals of the solute are present, as shown by the work of Miers in the early part of the century (see Chapter 2). The effects produced by the presence of a seed crystal have been called self-nucleation[127], avalanching[128], vegetative nucleation[129], ancillary nucleation[130], secondary nucleation[120, 131] and the breeding of nuclei. The term 'secondary' nucleation will be used here to distinguish the phenomenon from the so-called 'primary' nucleation discussed earlier (see p. 137).

Until recently, relatively few experimental studies had been reported on secondary nucleation, but in the few past years this has become a very active area of research. Among the early papers on this subject may be mentioned the work of TING and McCABE[127], who demonstrated that solutions of magnesium sulphate nucleated in a more reproducible manner at moderate supersaturations in the presence of seed crystals, and similar observations were made in studies with copper sulphate[132]. GYULAI[128] reported the development of fine wadding-like fibrous crystals in KBr solutions after the inducement of primary nucleation. He interpreted this behaviour as evidence for a 'transitional boundary layer' of partially integrated units which could be stripped off the crystal surfaces by fluid motion.

In a series of simple experiments POWERS[129] demonstrated that the movement of a sucrose crystal in a supersaturated solution, or the movement of the solution past a stationary crystal, produced nuclei. Inert models of crystals failed to produce nuclei under the same conditions. These results tended to suggest that a fluid mechanical shearing of weak outgrowths or loosely bonded molecular units from the crystal–solution interface was responsible. More recent work by CLONTZ and McCABE[133] with $MgSO_4 \cdot 7H_2O$ has shown that, under controlled conditions at moderate levels of supersaturation, crystal contact (crystal–crystal or crystal–solid) is essential for nucleation. They gave the name 'contact nucleation' to this particular type of secondary nucleation. No evidence of fluid mechanical shearing was produced, but they stressed that it is unwise to extend their conclusions to other systems in the absence of experimental evidence.

STRICKLAND-CONSTABLE and co-workers[17, 134, 135] have made extensive studies of secondary nucleation, mainly with $MgSO_4 \cdot 7H_2O$, but also with KBr, KCl and Zn–Hg amalgams. Several possible mechanisms were suggested, such as 'initial' breeding (crystalline dust swept off a newly introduced seed crystal), 'needle' breeding (the detachment of weak outgrowths), 'polycrystalline' breeding (the fragmentation of a weak polycrystalline mass) and 'collision' breeding (a complex process resulting from the interaction of crystals with one another or with parts of the crystallisation vessel).

HEUBEL and co-workers[136–137] have also reported on a number of secondary nucleation studies with $NaNO_3$, KNO_3, $NaClO_3$ and NH_4NO_3. They distinguished two types of nuclei, viz. those arising from debris on the surfaces of the seeds introduced to the system and those arising from broken-off outgrowths from the seeds. The relative proportions of these two types of seed were recorded as functions of the initial seed surface area and supersaturation.

MELIA and MOFFITT[120] studied nucleation in aqueous solution of KCl. They established that the rate of secondary nucleation increased with supersaturation and agitation, but their results were at variance with those of other workers[134–137] in that they reported that the secondary nucleation rate was independent of the number of seeds (and, hence, the seed surface area). They confirmed that the seed surface characteristics were important. At a constant cooling rate a time-lag (induction period) was recorded before secondary nucleation commenced. In a series of papers, GINDT and KERN[138] reported their extensive investigations on nucleation in aqueous solution of alkali halides. Induction period measurements, interpreted as nucleation rates, were used to calculate other characteristics of the systems. MITSUDA, MIYAKE and NAKAI[139] measured induction times for KCl, KNO_3, $NaClO_3$ and $Na_2S_2O_3$ solutions in stagnant and stirred solutions, and correlated their results with empirical relationships.

CAYEY and ESTRIN[131] also observed an induction period with seeded solutions of $MgSO_4$ in a pilot-scale agitated batch crystalliser, and confirmed the importance of supersaturation and cooling rate. They also reported an inexplicable effect of quantity of seeds added: one seed crystal was more effect in inducing nucleation than 50 mg, but less effective than 500 mg. This anomaly was not studied in any detail. They also found that a crystal was not capable of giving rise to fresh nuclei until it had reached a critical size. A possible explanation put forward for this[140] is that, as the energy decay in homogeneous isotropic turbulence occurs at a finite eddy size, crystal seeds smaller than the eddy size will not experience shear. This explanation, of course, presupposes that the mechanism of nucleation involves fluid shearing.

BELYUSTIN and ROGACHEVA[141] studied the nucleation of $MgSO_4 \cdot 7H_2O$, which crystallises in enantiomorphic forms (see Chapter 1) at room temperature. When this salt nucleated spontaneously (unseeded), the product crystals were mostly left-handed. When the solution was seeded with right-handed crystals, the number of product crystals increased and the percentage of left-handed crystals in the total product decreased. Increases in solution velocity and supersaturation in the presence of a right-handed seed both led to decreases in the percentage of right-handed crystals in the product. Belyustin and Rogacheva concluded that secondary nucleation was not caused by fragmentation. Filtration of the solution retarded both seeded and unseeded nucleation, so they proposed that secondary nucleation was caused by foreign particles coming into contact with seed crystals, becoming activated and converted into crystal nuclei. Similar work has recently been reported by DENK and BOTSARIS[141a], who carried out careful measurements with sodium chlorate enantiomorphs, with and without the presence of impurities, at different agitation rates. Different mechanisms of secondary nucleation were shown to operate under different conditions.

MULLIN and LECI[142] studied the seeding of citric acid solutions in an agitated vessel where secondary nucleation was observed to occur in a series of pulses, mainly during the latent period. The secondary nucleation rate decreased with an increase in the seed size or in the number of seeds, of a given size. The latent period was drastically reduced by decreasing the seed size, but was relatively unaffected by the number of seeds added. Increased supersaturation increased secondary nucleation and decreased the latent period. Increased agitation increased the desupersaturation rate to a maximum and decreased the latent period to a minimum. No evidence of fluid mechanical shearing was found, and a mechanism of secondary nucleation based on molecular cluster formation in solution was proposed.

LARSON, TIMM and WOLFF[143], working with ammonium alum and ammonium sulphate in a continuously operated mixed-suspension crystalliser, found that most new crystals were generated by secondary nucleation and that the rate of secondary nucleation was related to the magma density. With ammonium perchlorate in a batch-cooled agitated-vessel crystalliser, AYERST and PHILLIPS[144] found that heterogeneous nucleation occurred to a very limited extent in clean solutions and that primary homogeneous nucleation was the predominant mechanism for determining the number of crystals produced. Some secondary nucleation was shown to occur to a limited extent.

The mechanism of secondary nucleation is still a subject of much speculation. There are undoubtedly many possible modes of nucleation, and in any given system more than one may well apply at any given time. The predominant mechanism could easily change according to the environmental conditions, e.g. level of supersaturation or intensity of agitation. It is not even possible to say with any certainty how important fluid shearing and crystal collision are in secondary nucleation or to what extent molecular cluster formation in solution contributes to the nucleation process; much more work needs to be done on a wide variety of systems before any firm conclusions can be reached.

The process of secondary nucleation can also be represented by the empirical relationship proposed for primary heterogeneous nucleation (equation 6.2). Nucleation temperatures in the presence of crystalline material can be determined by a procedure similar to that described above (*Figure 6.4*) for the measurement of unseeded nucleation data; two small crystals (~ 2 mm in size) are introduced into the flask when the solution has cooled to its pre-determined saturation temperature.

The variation of the maximum allowable undercooling, $\Delta\theta_{max}$, with the cooling rate, b, for aqueous solutions of ammonium sulphate[126] is shown in *Figure 6.5a*. The lines for seeded and unseeded solutions are not

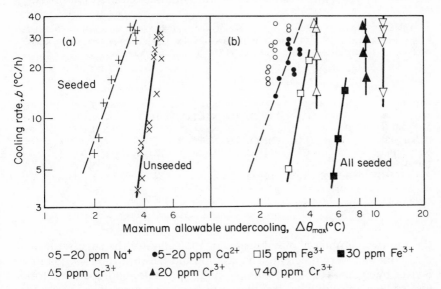

$\circ\,5\text{--}20$ ppm Na^+ $\bullet\,5\text{--}20$ ppm Ca^{2+} $\square\,15$ ppm Fe^{3+} $\blacksquare\,30$ ppm Fe^{3+}
$\triangle\,5$ ppm Cr^{3+} $\blacktriangle\,20$ ppm Cr^{3+} $\triangledown\,40$ ppm Cr^{3+}

Figure 6.5. Nucleation characteristics of ammonium sulphate solution: (a) pure solutions, seeded and unseeded; (b) effect of impurities in seeded solutions. The broken line represents data from (a)

parallel; the seeded points lie approximately $1\cdot5\text{--}2\,°C$ below the unseeded. The slopes of the lines indicate that the 'orders' of the nucleation process, m, for seeded and unseeded solutions are approximately $2\cdot6$ and $6\cdot4$, respectively, which indicates that the mechanisms of primary and secondary nucleation are different. The best straight lines through the data yield the relationships

$$b = (1\cdot28\pm0\cdot91)\times 10^{-2}\Delta\theta^{6\cdot43\pm1\cdot62} \qquad \text{seeded (secondary)}$$

and

$$b = 1\cdot38\pm0\cdot09\Delta\theta^{2\cdot64\pm0\cdot92} \qquad \text{unseeded (primary)}$$

which give a measure of the scatter of the data. The maximum allowable undercoolings for seeded and unseeded solutions are more or less inde-pendent of the saturation temperature over the range $20\text{--}40\,°C$, but do depend on the rate of cooling. At low rates of cooling ($\sim 5\,°C/h$) the values are about $1\cdot8$ and $3\cdot8\,°C$ for seeded and unseeded solutions, respectively, of ammonium sulphate compared with 3·5 and 5 °C for a cooling rate of 30 °C/h.

Undercooling data obtained from unseeded solutions have little or no industrial application. In fact it is often impossible to obtain consistent 'unseeded' values for many aqueous solutions, e.g. sodium acetate, sodium thiosulphate and citric acid. For crystalliser design purposes, the lowest 'seeded' value should be taken as the maximum allowable undercooling, and the working value of the undercooling should be kept well below this.

Some typical maximum allowable undercoolings in seeded solutions are given in *Table 6.1*. It should be noted that although the values of $\Delta\theta_{max}$

Table 6.1. MAXIMUM ALLOWABLE UNDERCOOLING*, $\Delta \theta^{max}$, FOR SOME COMMON AQUEOUS SALT SOLUTIONS AT 25 °C (MEASUREMENTS MADE IN THE PRESENCE OF CRYSTALS UNDER CONDITIONS OF SLOW COOLING ($\sim$5 °C/h) AND MODERATE AGITATION)

Substance	°C	Substance	°C	Substance	°C	Substance	°C
NH_4 alum	3·0	$MgSO_4 \cdot 7H_2O$	1·0	NaI	1·0	KBr	1·1
NH_4Cl	0·7	$NiSO_4 \cdot 7H_2O$	4·0	$NaHPO_4 \cdot 12H_2O$	0·4	KCl	1·1
NH_4NO_3	0·6	$NaBr \cdot 2H_2O$	0·9	$NaNO_3$	0·9	KI	0·6
$(NH_4)_2SO_4$	1·8	$Na_2CO_3 \cdot 10H_2O$	0·6	$NaNO_2$	0·9	KH_2PO_4	9·0
$NH_4H_2PO_4$	2·5	$Ha_2CrO_4 \cdot 10H_2O$	1·6	$Na_2SO_4 \cdot 10H_2O$	0·3	KNO_3	0·4
$CuSO_4 \cdot 5H_2O$	1·4	$NaCl$	1·0	$Na_2S_2O_3 \cdot 5H_2O$	1·0	KNO_2	0·8
$FeSO_4 \cdot 7H_2O$	0·5	$Na_2B_4O_7 \cdot 10H_2O$	3·0	K alum	4·0	K_2SO_4	6·0

* The working value for normal crystalliser operation may be 50 per cent of these values, or lower. The relation between $\Delta\theta_{max}$ and Δc_{max} is given by equation (6.5).

for any two substances may be similar, the values of the supersaturation, Δc_{max}, may be very different. The relationship between the two quantities is given by equation (6.5). For example, $\Delta\theta_{max} = 1$ °C for both sodium chloride and sodium thiosulphate, but the corresponding values of Δc_{max} are 0·25 and 18 g of crystallising substance per kg of solution, respectively.

EFFECT OF IMPURITIES

The presence of impurities in a system can affect the nucleation behaviour very considerably. It has long been known, for example, that the presence of small amounts of colloidal substances such as gelatin can suppress nucleation in aqueous solution, and certain surface-active agents also exert a strong inhibiting effect. Traces of foreign ions, especially Cr^{3+} and Fe^{3+}, can have a similar action. Recent quantitative studies include the effects of Pb^{2+} on $NaCl$[145] and KCl[146]; of $CaCl_2$, KNO_3 and $MgCl_2$ on KCl[147]; of Co^{2+}, Cr^{3+}, methylamine and docdecylamine hydrochlorides on KNO_3[148]; of sodium *p*-dodecylbenzene sulphonate and cetyldimethyl-benzene-ammonium chloride on sodium triphosphate[149]; and of Na^+, Ca^{2+}, Ni^{3+}, Fe^{3+}, Cr^{3+}, sodium carboxymethylcellulose, etc., on $(NH_4)_2SO_4$[150]. The latter work was carried out with the apparatus described earlier (*Figure 6.4*) and some typical results are shown in *Figure 6.5a, b*.

It would be unwise to attempt a general explanation of the phenomenon of nucleation suppression by added impurities with so little quantitative evidence yet available, but certain patterns of behaviour are beginning to emerge. For example, the higher the charge on the cation the more powerful the inhibiting effect, e.g. $Cr^{3+} > Fe^{3+} > Al^{3+} > Ni^{2+} > Na^+$. Furthermore there often appears to be a 'threshold' concentration of impurity above which the inhibiting effect may actually be weakened. This behaviour may well be associated with critical micelle concentrations of the ions or complexes.

The modes of action of high molecular weight substances and cations are probably quite different. The former may have their main action on the heteronuclei, rendering them inactive by adsorbing on their surfaces, whereas the latter may act as structure-breakers in solution (Chapter 2). An interesting effect is reported by BOTSARIS, DENK and CHUA[146], who studied the action of Pb^{2+} on the nucleation of KCl. If the impurity suppresses primary nucleation, secondary nucleation can occur if the uptake of impurity by the growing crystals is significant; the seed crystal creates an impurity concentration gradient about itself; the concentration of impurity near the crystal surface becomes lower than that in the bulk solution; and if it is reduced low enough, nucleation can occur.

The effects of impurities on growth and precipitation processes are discussed later.

CRYSTAL GROWTH

Despite the considerable amount of work published on crystal growth over the past 70 or more years, surprisingly little of this information is of direct use in the design and operation of crystallisers. It is only in the past decade or so that it has been recognised that for crystal growth rates to be of use in crystalliser design or assessment the independent effects of temperature, supersaturation and velocity must be known with some precision.

Many different experimental techniques have been employed to facilitate crystal growth rate measurements. These include face growth rates by direct microscopic observation[151-154], photography[155], interferometry[156]; overall growth rates of single crystals fixed in an agitated vessel[157-159], rotated in a fluid[160], from the melt[161], and from the vapour[162]; over-all growth rates measured in multiparticle systems in agitated vessels[140,144,163,164] and fluidised beds[87,165-168] and laboratory-scale[169] and pilot-scale[170,171] evaporative crystallisers.

This is by no means an exhaustive list of the numerous methods used, but it does give an idea to the diversity of techniques. NÝVLT[15] gives references to a large number of methods used, particularly by Eastern European workers in recent years.

FACE GROWTH RATES

The different faces of a crystal grow at different rates under identical environmental conditions. In general, high index faces grow faster than the low. An accurate assessment of the growth kinetics, therefore, must involve a study of the individual face growth rates.

An apparatus[153, 154] that permits precise measurement of the individual face growth rates of crystals is shown in *Figure 6.6*. This is similar to an apparatus developed earlier by CARTIER, PINDZOLA and BRUINS[151] for the

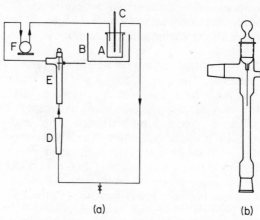

Figure 6.6. Single-crystal growth cell: (a) complete circuit, (b) the cell. A, solution reservoir; B, thermostat bath; C, thermometer; D, flow meter; E, cell; F, pump[153, 154]

growth of citric and itaconic acid crystals. Briefly the technique is as follows. A small crystal (2–5 mm) is mounted on the 1 mm tungsten wire in a chosen orientation. Solution of known temperature (± 0.05 °C), supersaturation and velocity is pumped through the cell, and the rate of advance of the chosen crystal face is observed through a travelling microscope. Several glass cells have been used ranging in internal diameter from 10 to 30 mm, permitting a wide range of solution velocities to be used.

The results in *Figure 6.7* show the effects of both solution supersaturation and velocity on the linear growth rates of the (111) faces of potash alum crystals at 32 °C. This hydrated salt $[K_2SO_4 \cdot Al_2(SO_4)_3 \cdot 24H_2O]$ grows as almost perfect octahedra, i.e. eight (111) faces.

Three interesting points may be noted. First, the growth rate is not first-order with respect to the supersaturation (concentration driving force, Δc). Second, the solution velocity has a significant effect on the growth rate. If the data are plotted on logarithmic co-ordinates (not given here) straight line correlations are obtained giving

$$v_{(111)} = K\Delta c^n \qquad (6.8)$$

where n varies from about 1·4 to 1·6. For v expressed in m/s and Δc in kg of hydrate per kg of solution K varies from about 3×10^{-5} to 2×10^{-4} as the solution velocity increases from 6 to 22 cm/s. Third, significant crystal growth does not appear to commence until a certain level of supersaturation is exceeded. This behaviour has been recorded for a number of other salts[87, 153].

The effect of solution velocity can be seen more clearly in *Figure 6.8*. The

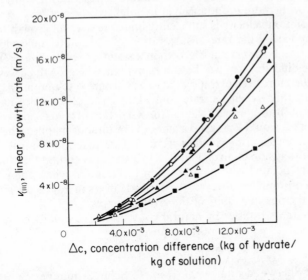

Figure 6.7. Face growth rates of single crystals of potash alum at 32 °C. Solution velocities: ● = 0·217, ○ = 0·120, ▲ = 0·064, △ = 0·022, ■ = 0·006 m/s[154]

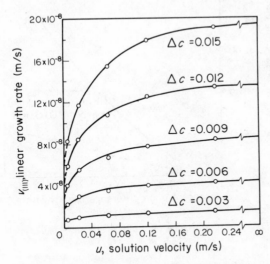

Figure 6.8. Effect of solution velocity on the (111) face growth rate of potash alum crystals at 32 °C[154]

points on this graph have been taken from the smoothed curves in *Figure 6.7*. For a given supersaturation the growth rate increases with solution velocity, the effect being more pronounced at the higher values of Δc.

If the solution velocity is sufficiently high, the over-all growth rate should be determined by the rate of integration of the solute molecules into the crystal lattice. If the crystal is grown in a stagnant solution ($u = 0$), then the rate of the diffusion step will be at a minimum. The growth curves in *Figure 6.8* have therefore been extrapolated to $u = 0$ and ∞ to obtain an estimate of the growth rates when the rate-controlling process is one of natural convection ($u = 0$) and surface reaction ($u \rightarrow \infty$). It is, of course, unlikely that the growth curves would change in a smooth continuous manner when the rate-controlling mechanism changes from surface reaction control to natural convective diffusion control, and it is by no means certain that these curves can be extrapolated with any precision to the point where the growth rate becomes constant[89]. However, the derived curves in *Figure 6.9* give an indication of the possible limits of the growth curves.

Rates of mass transfer by natural convection are usually correlated by a semi-theoretical equation of the form

$$Sh = 2 + \alpha(Gr.Sc)^{0.25} \tag{6.9}$$

where the Sherwood number $Sh = kL/D$, Schmidt number $Sc = \eta/\rho D$ and Grashof number $Gr = L^3\rho\Delta\rho g/\eta^2$. ρ = solution density, $\Delta\rho$ = difference

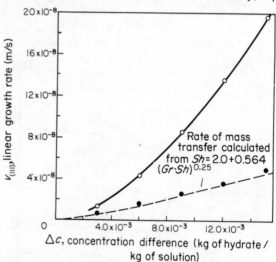

Figure 6.9. *Extrapolated growth rates of potash alum crystals at limiting velocities* ($\bigcirc = u \rightarrow \infty$, $\bullet = u \rightarrow 0$)[154]

between solution density at the interface and in the bulk solution, η = viscosity, L = crystal size, D = diffusivity and k = mass transfer coefficient. The mean value of the constant α based on the results of a number of workers is 0·56. Growth rates calculated from equation (6.9) lie very close to the experimental points and this tends to confirm that the growth process in stagnant solution is controlled by natural convection. It is of interest to note in this connection that the Grashof number contains a term $\Delta\rho$, which is directly proportional to the concentration difference, Δc, so the mass transfer rate under conditions of *natural convection* depends on $\Delta c^{1\cdot25}$. When *forced convection* is the rate-controlling process, the mass transfer rate is directly proportional to Δc.

Diffusional mass transfer rates under conditions of forced convection may be correlated by an equation of the form

$$Sh = 2 + \phi Re_p^a Sc^b \tag{6.10}$$

where Re_p is the particle Reynolds number ($\rho u L/\eta$). Equation (6.10) is frequently referred to as the Frössling equation. However, for reasonably high values of Sh (say > 100) it is common practice to ignore the constant 2 (the limiting value of Sh as $Re_p \to 0$, i.e. mass transfer in the absence of natural convection) and use the simpler expression

$$Sh = \phi Re_p^a Sc^b \tag{6.11}$$

Dissolution rate data, for example, are conveniently expressed in this way. The mass transfer coefficient, in the Sherwood number, depends on the solution velocity, u, raised to the power a. It is possible, therefore, that the effect of solution velocity on crystal growth may also be represented by an equation of this type in the region where diffusion influences the growth rate.

The effect of the two variables, Δc and u, on crystal face growth rates may thus be represented by

$$v_{hkl} = C u^a \Delta c^n \tag{6.12}$$

where C is a constant, and a and n are both functions of the solution velocity. For the growth of potash alum at 32 °C, as $u \to \infty$, $a \to 0$ and $n \to 1\cdot62$, while as $u \to 0$, $n \to 1\cdot25$.

It is of interest at this point to refer back to the theoretical work described in Chapter 5 based on the proposal that a sequence of growth steps brought about by screw dislocations should be made the basis of the theory of crystal growth[70, 71]. At low supersaturation, S, the growth rate is expected to be proportional to $(S-1)^2$, but at high supersaturation the rate tends to become a linear function of $S-1$. For growth from solution CHERNOV[71] showed that for the range $1\cdot01 < S < 1\cdot2$ (which corresponds to $0\cdot0015 < \Delta c < 0\cdot03$ kg/kg in the present case) the growth rate can be represented by

$$v_{hkl} = K(S-1)^n \tag{6.13}$$

where values of K and n are determined by the parameters in the theoretical relationship. Now $S = c/c^*$, i.e. $S-1 = \Delta c/c^*$, so equations (6.8) and (6.13) are of the same form and the value of the exponent, n, measured experimentally, is within the range predicted theoretically.

The velocity of the solution past the crystal face is thus capable of influencing the growth rate, and this velocity effect manifests itself as a crystal size effect when freely suspended crystals are grown in a crystalliser. The reason, of course, is that large crystals have higher settling velocities than small crystals, i.e. higher relative solid/liquid velocities are needed to keep the larger crystals suspended. This extremely important effect, which has not often been appreciated in the past and has rarely been measured quantitatively, can rapidly be detected in the growth cell described above.

Salts that have been established as having solution velocity dependent growth rates include ammonium and potassium alums, nickel ammonium sulphate, sodium thiosulphate and potassium sulphate, while ammonium sulphate, ammonium and potassium dihydrogen phosphates, for example, do not.

OVER-ALL GROWTH RATES

It is often much more convenient to measure crystal growth rates in terms of mass deposited per unit time per unit area of crystal surface. This may be done in agitated vessels or fluidised beds, e.g. by measuring the mass deposition on a known mass of seed crystals under carefully controlled conditions.

A laboratory-scale fluidised bed crystalliser capable of yielding reliable growth rate information is shown in *Figure 6.10*. It is constructed mainly of

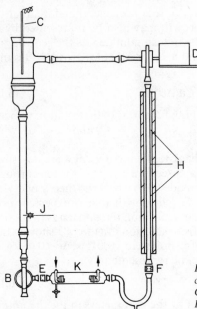

Figure 6.10. A laboratory-scale fluidised bed crystalliser: A, growth zone; B, outlet cock; C, resistance thermometer; E, F, orifice plates; H, heating tapes; J, thermometer; K, water cooler[87, 165]

glass (total capacity 10–13 l) with growth zones 5–8 cm diam. and 75 cm long. A combination of heating tapes and water cooler enables the temperature of the solution to be maintained to ± 0.03 °C. Solution concentration can be measured analytically at intervals or continuously by means of a recording density meter[172]. A typical run would consist of adding about 5 g $\pm$ 1 mg of carefully sized seed crystals and controlling the solution velocity so that the crystals are uniformly suspended in the growth zone until their weight has increased to, say, 15 g. This weight increase would allow 600 μm crystals of potassium sulphate, for example, to grow to about 800 μm. The duration of a run varies from about $\frac{1}{4}$ to 3 h, depending on the working level of supersaturation. At the end of a run the crystals are removed, dried, weighed and sieved.

A more rapid technique for measuring growth rates has recently been developed[173] in which a relatively dense bed of carefully sized seeds are grown in a fluidised bed crystalliser while the change in solution concentration is recorded continuously. With the aid of a computer program, crystal growth rates over a wide range of supersaturation can be evaluated, and the results are in good agreement with those obtained by the batch-wise technique described above.

Some typical results are shown in *Figure 6.11* for potash alum crystals

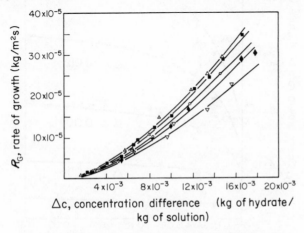

Figure 6.11. Over-all growth rates of potash alum crystals at 32 °C (mean crystal sizes: △ $= 1.96$, ■ $= 1.4$, ○ $= 0.99$, ▲ $= 0.75$, ▽ $= 0.53$ mm[165])

grown at 32 °C. These results may be compared with those shown in *Figure 6.7*. Here, again, the effect of supersaturation can be seen and so can the effect of crystal size. As explained above, solution velocity dependent growth shows up as crystal size dependent growth when freely suspended crystals are grown in a crystalliser. In this case large crystals grow faster than small.

For potash alum it has been shown that

$$R_G = K_G \Delta c^n \tag{6.14}$$

For R_G expressed as kg/m^2 s and Δc as kg hydrate/kg solution K_G varies from 0·115 to 0·218 and n from 1·54 to 1·6 for crystals ranging from 0·5 to 1·5 mm. Or, since it has already been shown that $v_{hkl} = Cu^a\Delta c^n$ for single crystals (equation 6.12), then

$$R_G = C'\bar{L}^m\Delta c^n \tag{6.15}$$

which is similar to the equation used by BRANSOM[174] as the starting point for a theoretical analysis of crystal size distributions. In the above case of potash alum

$$R_G = 16\bar{L}^{0.3}\Delta c^n \qquad (6.16)$$

Over-all growth rates for potash alum measured in the fluidised bed crystalliser coincide very well with those predicted from face growth rates measured in the single crystal cell (*Figure 6.12*). The alums grow as almost

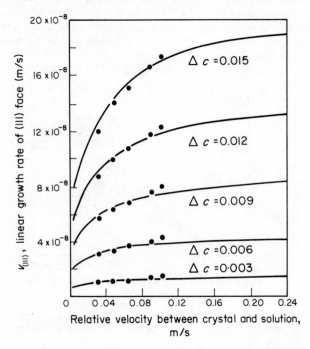

Figure 6.12. Comparison between face (smooth curve) and over-all (points) growth rates of potash alum crystals at 32 °C[165]

perfect octahedra, i.e. eight (111) faces, so it is a simple matter, using the solid density, ρ, to convert linear face velocities to over-all mass deposition rate ($R_G = \rho v_{(111)}$).

EXPRESSION OF CRYSTAL GROWTH RATES

There is no simple or generally accepted method of expressing the rate of growth of a crystal, since it has a complex dependence on temperature, supersaturation, size, habit, system turbulence, and so on. However, for carefully defined conditions crystal growth rates may be expressed as a mass deposition rate R_G (kg/m² s), a mean linear velocity $\bar{v}$ (m/s) or an over ll linear growth rate G (m/s). The relationships between these quantities are

$$R_G = K_G \Delta c^n = \frac{1}{A} \cdot \frac{dm}{dt} = \frac{3\alpha}{\beta} \cdot \rho G = \frac{3\alpha}{\beta} \rho \frac{dL}{dt} = \frac{6\alpha}{\beta} \cdot \rho \frac{dr}{dt} = \frac{6\alpha}{\beta} \cdot \rho \bar{v} \qquad (6.17)$$

where L is some characteristic size of the crystal, e.g. the equivalent sieve aperture size, and r is the radius corresponding to the equivalent sphere. The volume and surface shape factors, α and β, respectively, are defined (see Chapter 11) by $m = \alpha \rho L^3$ (i.e. $dm = 3\alpha \rho L^2 dL$) and $A = \beta L^2$, where m and A are the particle mass and area. For spheres and cubes $6\alpha/\beta = 1$. For octa-hedra $6\alpha/\beta = 0.816$.

The utility of the over-all linear growth rate, G, in the design of crystallisers is demonstrated in Chapter 10. Some typical values of the mean linear growth velocity $\bar{v}$ ($= \frac{1}{2}G$) are given in *Table 6.2*.

Table 6.2. SOME MEAN OVER-ALL CRYSTAL GROWTH RATES EXPRESSED AS A LINEAR VELOCITY

The supersaturation is expressed by $S = c/c^*$ with c and c^* as kg of crystallising substance per kg of free water. The significance of the mean linear growth velocity, $\bar{v}$, is explained by equation (6.17), and the values recorded here refer to crystals in the approximate size range 0.5–1 mm growing in the presence of other crystals. An asterisk (*) denotes that the growth rate is probably size dependent.

Crystallising substance	°C	S	$\bar{v}$ (m/s)
$(NH_4)_2SO_4 \cdot Al_2(SO_4)_3 \cdot 24H_2O$	15	1.03	1.1×10^{-8}*
	30	1.03	1.3×10^{-8}*
	30	1.09	1.0×10^{-7}*
	40	1.08	1.2×10^{-7}*
NH_4NO_3	40	1.05	8.5×10^{-7}
$(NH_4)_2SO_4$	30	1.05	2.5×10^{-7}*
	60	1.05	4.0×10^{-7}
	90	1.01	3.0×10^{-8}
$NH_4H_2PO_4$	20	1.06	6.5×10^{-8}
	30	1.02	3.0×10^{-8}
	30	1.05	1.1×10^{-7}
	40	1.02	7.0×10^{-8}
$MgSO_4 \cdot 7H_2O$	20	1.02	4.5×10^{-8}*
	30	1.01	8.0×10^{-8}*
	30	1.02	1.5×10^{-7}*
$NiSO_4 \cdot (NH_4)_2SO_4 \cdot 6H_2O$	25	1.03	5.2×10^{-9}
	25	1.09	2.6×10^{-8}
	25	1.20	4.0×10^{-8}
$K_2SO_4 \cdot Al_2(SO_4)_3 \cdot 24H_2O$	15	1.04	1.4×10^{-8}*
	30	1.04	2.8×10^{-8}*
	30	1.09	1.4×10^{-7}*
	40	1.03	5.6×10^{-8}*
KCl	20	1.02	2.0×10^{-7}
	40	1.01	6.0×10^{-7}
KNO_3	20	1.05	4.5×10^{-8}
	40	1.05	1.5×10^{-7}

(continued overleaf)

Table 6.2 *continued*

Crystallising substance	°C	S	$\bar{v}$ (m/s)
K_2SO_4	20	1·09	$2·8 \times 10^{-8}$*
	20	1·18	$1·4 \times 10^{-7}$*
	30	1·07	$4·2 \times 10^{-8}$*
	50	1·06	$7·0 \times 10^{-8}$*
	50	1·12	$3·2 \times 10^{-7}$*
KH_2PO_4	30	1·07	$3·0 \times 10^{-8}$
	30	1·21	$2·9 \times 10^{-7}$
	40	1·06	$5·0 \times 10^{-8}$
	40	1·18	$4·8 \times 10^{-7}$
NaCl	50	1·002	$2·5 \times 10^{-8}$
	50	1·003	$6·5 \times 10^{-8}$
	70	1·002	$9·0 \times 10^{-8}$
	70	1·003	$1·5 \times 10^{-7}$
$Na_2S_2O_3 \cdot 5H_2O$	30	1·02	$1·1 \times 10^{-7}$
	30	1·08	$5·0 \times 10^{-7}$
Citric acid monohydrate	25	1·05	$3·0 \times 10^{-8}$
	30	1·01	$1·0 \times 10^{-8}$
	30	1·05	$4·0 \times 10^{-8}$
Sucrose	30	1·13	$1·1 \times 10^{-8}$*
	30	1·27	$2·1 \times 10^{-8}$*
	70	1·09	$9·5 \times 10^{-8}$
	70	1·15	$1·5 \times 10^{-7}$

EFFECT OF CRYSTAL SIZE

It is probably true to say that all crystal growth rates are particle size dependent; it all depends on the size and size range under consideration. The effect of size may be quite insignificant for macro-crystals, but the situation can change dramatically for crystals of microscopic or sub-microscopic size.

The effect of crystal size on the over-all growth rates of macro-crystals has been dealt with in the previous section. Not all substances exhibit a size effect, but in cases where they do an over-all growth rate expression of the form of equation (6.15) would appear appropriate. Because of the limitations imposed by the experimental techniques, we are normally considering crystals that do not extend much outside the range 200 μm to 2 mm. In such cases any effect of size would appear to be closely linked with the effect of solution velocity, since large particles have higher terminal velocities than those of small particles.

However, another and often much more powerful effect of particle size may be exhibited at sizes smaller than a few microns. Here the effect is caused by a combination of the effects of size on growth (due to diffusion and/or surface integration) and solubility (the Gibbs–Thomson effect). These points are discussed in some detail in Chapter 5 (p. 165). It is possible to show that crystals of near-nucleic size may grow at extremely slow rates.

These growth rates, therefore, have a considerable influence on the processes of nucleation and precipitation (p. 222).

(p. 222)

EFFECT OF TEMPERATURE

The relationship between a reaction rate constant, k, and the absolute temperature, T, is given by the Arrhenius equation

$$\frac{d \ln k}{dT} = \frac{E}{RT^2} \qquad (6.18)$$

where E is the energy of activation for the particular reaction. On integration equation (6.18) gives

$$k = A \cdot \exp(-E/RT) \qquad (6.19)$$

or, taking logarithms,

$$\ln k = \ln A - \frac{E}{RT} \qquad (6.20)$$

Therefore, if the Arrhenius equation applies, a plot of $\log k$ against $1/T$ should give a straight line of slope $-E/R$ and intercept $\log k$.

Alternatively, if only two measurements of the rate constant are available, k_1 at T_1 and k_2 at T_2, the following equation may be used:

$$E = \frac{RT_1 T_2}{T_2 - T_1} \ln \frac{k_2}{k_1} \qquad (6.21)$$

Equation (6.21) is obtained by integrating equation (6.18) between the limits T_1 and T_2 assuming that E remains constant over this temperature range.

The above equations may be applied to diffusion, dissolution or crystallisation processes; k can be taken as the relevant rate constant. For example, a plot of $\log K_G$ versus $1/T$ would give a so-called activation energy for crystal growth, E_{cryst}; $\log K_D$ versus $1/T$ gives E_{diss}; $\log D$ versus $1/T$, where D = diffusivity, gives E_{diff}; and $\log \eta$ versus $1/T$, where η = viscosity, gives a value of E_{visc}; and so on.

Crystal growth processes are rarely purely surface- or diffusion-controlled. The surface reaction step may predominate at low temperature and the diffusion step at high, but over a significant intermediate range both processes may be rate-controlling. Therefore Arrhenius plots for crystal growth data usually give curves rather than straight lines, indicating that the apparent activation energy of the over-all growth process is temperature dependent.

VAN HOOK[175] calculated activation energies of growth, diffusion and viscosity for sucrose–water systems over a range of temperature in an attempt to determine the controlling step in the crystallisation process. At low temperature E_{growth} was much greater than either E_{diff} or E_{visc}, which indicates that diffusion is not an important factor in the growth process. However, above 60 °C the three activation energies were comparable, which suggests that the diffusion process becomes predominant. DEDEK (reported in reference

175) came to a similar conclusion when he compared the rate at which
sucrose crystallises from aqueous solution with the theoretical rate at which
sucrose becomes available at a growing crystal face by diffusion through the
solution. The data shown in *Table 6.3*, compiled for a supersaturation of
1.05 g/cm^3, indicate that excess sucrose is available at the growing face
at temperatures lower than about 45 °C, but above this temperature there is a
deficiency of sucrose. These results would suggest, therefore, that above
about 45 °C the crystallisation process is diffusion-controlled. Below about
45 °C the rate of diffusion is not important, and the rate of the surface
'reaction' probably controls the crystallisation rate. The recent extensive

Table 6.3. OBSERVED CRYSTALLISATION RATES OF SUCROSE COMPARED WITH
THEORETICAL DIFFUSION RATES AT DIFFERENT TEMPERATURES
(AFTER DEDEK, REPORTED BY VAN HOOK[175])

Temperature, °C	Growth rate, r_g (g/m^2 . min)	Diffusion rate, r_d (g/m^2 . min)	r_d/r_g
20	0·42	10·8	25·7
30	1·34	5·9	4·5
40	2·1	5·2	2·1
50	5·4	2·8	0·5
60	13·5	1·6	0·2
70	30	1	0·03

measurements made by SMYTHE[157] suggest that 40 °C is the critical
temperature.

In a similar manner RUMFORD and BAIN[170] showed that the crystallisation
of sodium chloride from aqueous solution is diffusion-controlled above about
50 °C and reaction rate controlled below this temperature.

EFFECT OF IMPURITIES

The presence of impurities in a system can have a profound effect on the
growth of a crystal. Some impurities can suppress growth entirely; some may
enhance growth, while others may exert a highly selective effect, acting only
on certain crystallographic faces and thus modifying the crystal habit. Some
impurities can exert an influence at very low concentrations, less than 1 part
per million, whereas others need to be present in fairly large amounts. Any
substance other than the material being crystallised can be considered an
'impurity', so even the solvent from which the crystals are grown is in the
strictest sense an impurity, and it is well known that a change of solvent
frequently results in a change of habit (see p. 207).

Impurities can influence crystal growth rates in a variety of ways. They
can change the properties of the solution (structural or otherwise) or the
equilibrium saturation concentration; they can alter the characteristics of
the adsorption layer at the crystal–solution interface and influence the
integration of growth units. They may themselves be selectively adsorbed

on to the crystal faces and exert a 'blocking effect' or, more likely, on to the growth steps and thus disrupt the flow of growth layers across the faces. They may be built into the crystal, especially if there is some degree of lattice similarity, or they may interact chemically with the crystal and selectively alter the surface energies of the different faces.

There are literally thousands of reports in the scientific literature concerning the effects of impurities on the growth of specific crystals, and it would be superfluous to attempt a summary. However, it is worth looking a little more closely at a few studies that have aimed at a more general understanding of the underlying mechanisms of growth inhibition.

In this context, reference should be made to the paper by BUNN[176] in which the properties of crystals built up by the continual adsorption and inclusion of impurity (he called them adsorption bodies) during growth were compared with those of mixed crystals. It was concluded that the two types of crystal differ only in degree; adsorption bodies may be regarded as unstable mixed crystals. It was suggested that the condition for strong adsorption is similarity of lattice structure and interatomic distances on specific planes only; the rest of the structures may be quite dissimilar. This condition is the same as that required for oriented overgrowth (epitaxy); in fact an oriented overgrowth is regarded as an adsorbed layer 'developed' to visible size. A simple explanation of habit modification is suggested, viz. that on the affected faces a mixed crystal is formed which, being unstable, tends to redissolve, thereby reducing the rate of growth.

The findings of several Russian workers have been summarised by KHAMSKII[9]. Komarova (1953) suggests that impurities affecting the crystallisation of potash alum $(KAl(SO_4)_2 \cdot 12H_2O)$ can be divided into three groups: (1) those that have common ion (KCl, KBr, KI, $(NH_4)_2SO_4$) and accelerate the growth process by reducing the solubility according to the law of mass action; (2) inorganic substances without a common ion (NaCl, NaBr, $NaNO_3$) that accelerate the growth at low concentration but reduce it at high (it is suggested that these impurities act as diffusion retarders); and (3) organic substances that retard growth by being adsorbed on to the crystal faces. Til'mans (1957) suggests that the action of impurities is caused by the electric field of the ions (governed by the ionic charge and radius and the compressibility of the ions) on the adsorption layer. The influence of cations of the same compressibility as growth inhibitors increases with increasing charge and decreasing radius. Despite their small size and charge, ions such as Cu^+, Cd^{2+} and Mn^{2+} are often useful habit modifiers because they are easily compressed. On the other hand, Kuznetzov (1953) suggests that chemical interaction between the impurity and crystal is of prime importance and that ionic fields play only a minor role.

Another possibility has been proposed by MULLIN, AMATAVIVADHANA and CHAKRABORTY[177], who studied the effects of ionic impurities, pH and supersaturation on the growth of ammonium and potassium dihydrogen phosphates (ADP and KDP) crystals. This work confirmed the findings of previous investigations on ADP and KDP, e.g. the growth-inhibiting power sequence $Cr^{3+} > Fe^{3+} > Al^{3+}$, but none of the previous explanations, such as physical blocking by adsorption or adsorption/desorption sequences, were considered satisfactory. It was suggested that the ionic influence was to some extent associated with the hydration of the ions. Metal ions exist in acid solution as aquo ions, e.g. $M(H_2O)_x^{n+}$, where $n = 3$ and $x = 6$ for Cr, Fe and Al. These ions would be attracted by phosphate groups into the vicinity of certain ADP and KDP crystal faces, but it was suggested that they do not necessarily have to be adsorbed at the surface in order to be effective. These complex ions are much larger than the 'bare' ions and would be less easily adsorbed and accommodated in the lattice. Of course, the aquo ions could be much more complex than this, since the presence of Fe^{3+} in a solution containing $H_2PO_4^-$ would most probably result in the formation of other hydrated phosphate species.

A mechanism of growth inhibition might then be considered as one in which complex aquo ions attracted into the vicinity of a crystal face exert a 'dilution' effect, retard diffusion, hinder aggregation of growth units and thus retard the growth rate. However, if adsorption did take place, the aquo ions would lose some or all of their hydration molecules, and the resulting counter-flux of water away from the interfacial region would further retard the growth rate. It is also possible that the effects of pH on growth rate are caused by similar effects with hydrated hydroxonium ions $H_3O(H_2O)_3^+$.

Some notable work has been carried out over a number of years by KERN and his co-workers[29, 178-179] into the effect of impurities adsorption on the crystal growth process. A recent review[178] summarises much of this work. Kern suggests that impurities can modify the solvation of the different crystallographic faces and thus influence their relative growth rates. Consideration is given to the possibility that adsorbed impurities interact to form adsorbed two-dimensional aggregates below a certain critical temperature.

BLITZNAKOV[180] has attempted to relate face growth rates to impurity concentration by an expression of the form

$$v = v_0 - (v_0 - v_\infty)\theta \tag{6.22}$$

where v_0, v and v_∞ are the linear growth velocities in pure solution, with a surface coverage of impurity θ, and with maximum coverage, respectively. Utilising equation (6.22) and assuming Langmuir adsorption, he obtained

$$\frac{1}{1-\mu} = \frac{B}{1-\mu_\infty} \cdot \frac{1}{C} + \frac{1}{1-\mu_\infty} \tag{6.23}$$

where B is the Langmuir constant, C is the impurity concentration in solution, $\mu = v/v_0$ and $\mu_\infty = v_\infty/v_0$. The utility of equation (6.23) is that a plot of $(1-\mu)^{-1}$ versus C^{-1} should give a straight line, and such linear correlations have been reported[180, 181].

BOTSARIS, MASON and REID[182] have studied the incorporation of Pb^{2+} in KCl crystals and showed that the impurity distribution coefficient (crystal: solution), which is usually < 1, may become > 1 at very low solution impurity concentrations. This inversion of the distribution coefficient provided an explanation for the difficulty experienced in practice in purifying certain inorganic crystals by repeated crystallisations, and led to the postulation of a model for non-equilibrium capture of impurities in growing crystals.

Crystal growth poisons do not as a rule have any influence on the dis-olution process, but SEARS[183] reports a particular type of impurity, e.g. omplex inorganic ions such as FeF_6^{3-}, which do have such an effect at oncentrations of $10^{-5}-10^{-6}$ mol/l. A mechanism based on monolayer dsorption at the growth/dissolution steps is proposed, and it is postulated hat the poison must present energetic and/or steric barriers to the addition ⋅f solute atoms so as to form an unpoisoned step by burying the adsorbed mpurities. Theoretical analyses of the effects of impurities on growth, based n the kinematic theory (p. 158), have been made by FRANK[77], CABRERA nd VERMILYEA[78] and CHERNOV[71, 184]. A surface energy approach to the ⋅roblem based on the method of STRANSKII and KAISHEV[50] has been proposed y LACKMANN and STRANSKII[185].

It is rarely possible to give a simple explanation of the mechanism by which given impurity affects the growth of a given crystal. Indeed it is unwise to ⋅elieve that only one mechanism applies to any particular case. What is not n doubt, of course, is the importance of the effect of impurities in the control ⋅f crystallisation, and this topic will be continued in the section dealing with abit modification (p. 207).

CRYSTALLISATION AND DISSOLUTION

f both crystallisation and dissolution processes were purely diffusional in ature, they should exhibit a true reciprocity; the rate of crystallisation should qual the rate of dissolution at a given temperature and under equal con-entration driving forces, i.e. at equal displacements away from the equilibrium aturation conditions. In addition, all faces of a crystal would grow and dis-olve at the same rate. These conditions rarely, if ever, occur in practice.

Crystals usually dissolve much faster than they grow; fivefold differences re not uncommon[87, 88]. Different crystallographic faces grow at different ates; they may even dissolve at different rates, but reliable measurements of his behaviour have not yet been reported. These facts have led most in-estigators to support the view that the crystallisation process can be con-idered on the basis of a simple two-step process: bulk diffusion being ⋅llowed by a surface 'reaction' at the growing crystal face. There have, how-ver, been other suggestions put forward. Some authors have suggested that rystals dissolve faster than they grow because the exposed surface is not he same in each case; etch pits rapidly form on the faces of a dissolving rystal (these occur either at random point defects or points where line lefects break the surface). Dissolution then proceeds by a pitting and layer-tripping process. It is well known that a broken or etched crystal grows nitially at a much faster rate than that when the faces are smooth, but as 'AN HOOK[19] has pointed out, even an overgenerous allowance of extra urface area due to pitting cannot possibly explain the greater rates of dis-olution compared with the rate of crystallisation of sucrose under com-⋅arable conditions. Other workers have expressed similar views, and it has ven been shown that some dissolution processes may also involve a slow 'eaction' step at the crystal surface[90, 91].

Growth and dissolution rates of crystals can be measured conveniently in the laboratory fluidised bed crystalliser described above. Some typical results for potash alum are shown in *Figure 6.13*, where it can be seen that dissolution rates are very much greater than growth rates under equal driving forces (Δc). Similar results have been reported for potassium sulphate[8] and in both cases the dissolution rates are first-order with respect to supersaturation, i.e.

$$R_D = K_D \Delta c \qquad (6.24)$$

whereas the growth processes are not, i.e. $R_G = K_G \Delta c^n$ (equation 6.14) where $n \simeq 1\cdot6$ for potash alum at 32 °C and $n \simeq 2$ for potassium sulphate at

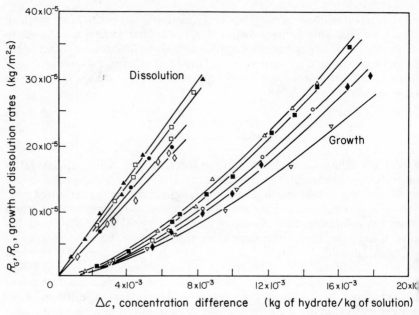

Figure 6.13. Growth and dissolution rates for potash alum crystals at 32 °C. Mean crystal sizes ▲ = 1·75, □ = 1·02, ● = 0·73, ◇ = 0·51; △ = 1·69, ■ = 1·4, ○ = 0·99, ◆ = 0·75, ▽ = 0·53 mm[172]

all temperatures from 10 to 50 °C. In equations (6.14) and (6.24) K_D and K_G are the over-all dissolution and growth mass transfer coefficients, respectively.

MASS TRANSFER

MASS TRANSFER CORRELATIONS

Dissolution rate data obtained under forced convection conditions can be correlated by means of equation (6.10) or (6.11). As described earlier, equation (6.10) is the preferred relationship on theoretical grounds, since $Sh = 2$ for mass transfer by convection in stagnant solution ($Re = 0$), whereas equation

(6.10) incorrectly predicts a zero mass transfer rate ($Sh = 0$) for this condition. However, at reasonably high values of Sh (> 100) the use of the simpler
equation (6.11) is quite justified. The exponent of the Schmidt number b is
usually taken to be $\frac{1}{3}$, since there is theoretical and experimental justification
for this assumption[186], and for mass transfer from spheres the exponent of
the Reynolds number $a = \frac{1}{2}$.

Data plotted in accordance with equation (6.11) for the dissolution of
potash alum crystals yield the relationship

$$Sh = 0.37\, Re_p^{0.62} Sc^{0.33} \tag{6.25}$$

i.e.

$$\frac{k_D \bar{L}}{\rho_s D} = 0.37 \left(\frac{\rho_s u_r \bar{L}}{\eta}\right)^{0.62} \left(\frac{\eta}{\rho_s D}\right)^{0.33} \tag{6.26}$$

where ρ_s = solution density, u_r = relative velocity between crystal and
solution, and Re_p = particle Reynolds number based on a mean crystal
size $\bar{L}$.

ROWE and CLAXTON[186] have shown that heat and mass transfer from a
single sphere in an assembly of spheres when water is the fluidising medium
can be described by

$$Sh = A + B Re_s^m Sc^{\frac{1}{3}} \tag{6.27}$$

where $A = 2[1-(1-\varepsilon)^{\frac{1}{3}}]$, $B = 2/3\varepsilon$ and $(2-3m)/(3m-1) = 4.65\, Re_s^{-0.28}$.
The solution Reynolds number, Re_s, is based on the superficial fluid velocity,
u_s, and ε = voidage.

Another correlation used for predicting rates of mass transfer in fixed and
fluidised beds is that of CHU, KALIL and WETTEROTH[187]. The j-factor for
diffusional mass transfer, j_d, given by

$$j_d = \left(\frac{K_D}{u_s \rho_s}\right) Sc^{\frac{2}{3}} \tag{6.28}$$

is plotted against the modified solution Reynolds number $Re_s'(1-\varepsilon)$, where
Re_s' contains $\bar{L}$, the diameter of a sphere with the same surface area as the
crystal under consideration. The recommended expressions[187] for calculating the mass transfer coefficients are:

$$1 < Re_s'/(1-\varepsilon) < 30: \quad j_d = 5.7 Re_s'/(1-\varepsilon)^{-0.78} \tag{6.29}$$

$$30 < Re_s'/(1-\varepsilon) < 5000: \quad j_d = 1.77 Re_s'/(1-\varepsilon)^{-0.44} \tag{6.30}$$

Dissolution rate data for potash alum are plotted in accordance with
equations (6.27) and (6.29), with $\varepsilon = 0.95$, in *Figure 6.14*, where it can be seen
that the results lie reasonably close ($\pm 20\%$) to the predicted values. However,
it should be noted that equation (6.29) is very sensitive to values of ε as $\varepsilon \to 1$,
so it cannot be applied with any reliability to very lean beds of dissolving
particles and certainly not to the dissolution of single particles[188].

Relationships of the form of equation (6.11) have also been used and

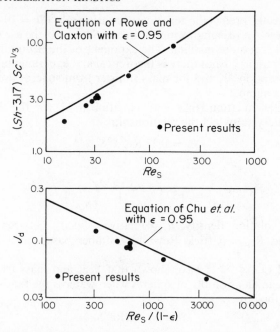

Figure 6.14. Comparison of dissolution data[172] for potash alum at 32 °C with the mass transfer correlations of ROWE and CLAXTON[186] and CHU et al.[187]

applied to growth and dissolution data for $CuSO_4 \cdot 5H_2O$ and $MgSO_4 \cdot 7H_2O$[18] and K_2SO_4[87], $b = \frac{1}{3}$ being used as the exponent of the Schmidt number.

'FILMS' AND 'BOUNDARY LAYERS'

A certain amount of confusion has been noted in recent crystallisation literature concerning the use of the term 'boundary layer' for the fluid adjacent to a crystal surface. For a complete account of the subject of boundary layers reference should be made to a standard textbook of chemical engineering, e.g. COULSON and RICHARDSON[190], but briefly it may be said that when a fluid flows past a solid surface there is a thin region near the solid-liquid interface where the velocity becomes reduced owing to the influence of the surface. This region, called the 'hydrodynamic boundary layer' δ_h, may be partially turbulent or entirely laminar in nature, but in the case of crystals suspended in their liquor the latter is most probable.

For mass transfer processes another boundary layer may be defined, viz the 'mass-transfer or diffusion boundary layer', δ_m. This is a region close to the interface across which, in the usual case of a laminar hydrodynamic boundary layer around the crystal, mass transfer proceeds by molecular diffusion. Under these conditions the relative magnitudes of the two boundary layers may be roughly estimated from

$$\frac{\delta_h}{\delta_m} \approx Sc^{\frac{1}{3}} \tag{6.31}$$

where $Sc = \eta/\rho D$ is the dimensionless Schmidt number (η = viscosity, ρ = density, D = diffusivity).

The ratio of the thicknesses of the two layers depends considerably on the solution viscosity and diffusivity. For example, for ammonium alum crystals in near-saturated solution at 25 °C, $\eta = 1\cdot2\times10^{-3}\ kg\ m^{-1}\ s^{-1}$, $D = 4\times10^{-10}\ m\ s^{-1}$, $\rho = 1\cdot06\times10^{3}\ kg\ m^{-3}$. Therefore $Sc = 2\cdot8\times10^{3}$ and $\delta_h/\delta_m \approx 15$. However, for sucrose in water at 25 °C, $\eta = 10^{-1}$, $D = 9\times10^{-11}$ and $\rho = 1\cdot5\times10^{3}$, giving $Sc = 7\cdot4\times10^{5}$ and $\delta_h/\delta_m \approx 90$.

In the description of mass transfer processes another fluid layer is frequently postulated, viz. the 'stagnant film' (see *Figure 5.17*) or, as it is sometimes called, the 'effective film for mass transfer', δ. This hypothetical film is not the same thing as the more fundamental diffusion boundary layer, but it may probably be considered to be of the same order of magnitude, i.e.

$$\delta_h > \delta_m \approx \delta \tag{6.32}$$

The stagnant film effective film for mass transfer, δ, is defined by

$$\delta = \frac{\rho D}{K} \tag{6.33}$$

where ρ = solution density, D = diffusivity and K is a mass transfer coefficient expressed as mass per unit time per unit area per unit concentration driving force, e.g. $kg\ s^{-1}m^{-2}(kg/kg)^{-1}$. As described earlier, mass transfer data are frequently correlated by relationships such as equation (6.11) in which the Sherwood number $Sh = KL/\rho D$ and particle Reynolds number $Re_p = \rho u L/\eta$. L = particle size and u = relative particle/solution velocity. Exponent b of the Schmidt number is generally taken as $\frac{1}{3}$, and in the case of a laminar boundary layer it can be shown theoretically that exponent a of the Reynolds number is $\frac{1}{2}$. However, a can vary from about 0·5 to 0·8 if the boundary layer is not truly laminar. Values of the constant ϕ for granular solids may range from about 0·3 to 0·9. So, writing a simple, arbitrary form of equation (6.11) as

$$Sh = \tfrac{2}{3}Re_p^{\frac{1}{2}}Sc^{\frac{1}{3}} \tag{6.34}$$

and expressing $Sh = L/\delta$ (equation 6.33), we get

$$\delta = \frac{3L}{2}\left(\frac{\rho u L}{\eta}\right)^{-\frac{1}{2}}\left(\frac{\eta}{\rho D}\right)^{-\frac{1}{3}} \tag{6.35}$$

and this equation may be used to give a rough estimate of the value of δ. It should be noted, however, that equation (6.35) can only be used if the mass transfer process is first-order with respect to the concentration driving force, otherwise Sh is not dimensionless and equation (6.34) is invalid.

Example

Estimate the diffusion and hydrodynamic boundary layer thicknesses, δ_m and δ_h, and the thickness of the hypothetical film, δ, for the dissolution of 2 mm crystals of ammonium alum in near-saturated solution at 25 °C using the following data: $\eta = 1.2 \times 10^{-3}$ kg m^{-1} s^{-1}, $\rho = 1.06 \times 10^3$ kg m^{-3}, $D = 4 \times 10^{-10}$ m^2 s^{-1}, $L = 2 \times 10^{-3}$ m, $u = 10^{-1}$ m s^{-1}.

$Sc = 2.8 \times 10^4$ and $Re_p = 180$, so, from equation (6.35), $\delta_m \approx \delta \approx 16 \, \mu$m. From equation (6.31), $\delta_h \approx \delta_m Sc^{\frac{1}{3}} \approx 14 \, \delta_m \approx 220 \, \mu$m.

It is thus possible to calculate a value for the thickness of the so-called stagnant film, but it is perhaps worthwhile at this point to question the meaning and utility of this quantity. The concept of a stagnant film at an interface is undoubtedly useful in providing a simple pictorial representation of the mass transfer process, but in the case of crystals growing or dissolving in multi-particle suspensions the actual existence of stable films, of the magnitude normally calculated as shown above, around each small particle is debatable.

The thickness of the stagnant liquid film, δ, can only be deduced indirectly from the mass transfer coefficient and diffusivity (equation 6.33), and it is difficult to select the appropriate value of D to use in any given situation. The question arises, therefore, as to whether or not δ is a meaningful quantity to calculate in these circumstances. In any case, the hypothetical nature of the stagnant film should be clearly appreciated, and calculated values of its thickness should be used with considerable caution.

DRIVING FORCES FOR MASS TRANSFER

There is a wide choice of possible driving forces for a mass transfer process, but provided that the driving force is clearly defined the selection is generally of little importance. However, in certain cases, e.g. under conditions of high mass flux, the choice becomes critical[191, 192].

For low mass flux mass transfer from a single sphere to an extensive fluid, the general correlation[193]

$$Sh = 2 + 0.72 \, Re^{\frac{1}{2}} Sc^{\frac{1}{3}} \qquad (6.36)$$

may be used over the range $20 < Re < 2000$.

The mass transfer coefficient in the Sherwood number may be defined by

$$R = k_c(c_0 - c_\infty) \qquad (6.37)$$

$$= k_c(\rho_0 \omega_0 - \rho_\infty \omega_\infty) \qquad (6.38)$$

$$= \rho k_c(\omega_0 - \omega_\infty) \qquad (6.39)$$

since for low mass flux $\rho_0 \simeq \rho_\infty \simeq \rho$. Other definitions of the mass transfer coefficient include

$$R = k_\omega(\omega_0 - \omega_\infty) \qquad (6.40)$$

$$= k_y(Y_0 - Y_\infty) \qquad (6.41)$$

$$= k_b B \qquad (6.42)$$

In equations (6.37) to (6.42) R = mass flux (kg m^{-2} s^{-1}), c = solution concentration (kg m^{-3}), k = mass transfer coefficient (k_c = m s^{-1}, k_ω = kg m^{-2} s^{-1} $\Delta\omega^{-1}$, k_y = kg m^{-2} s^{-1} ΔY^{-1} and k_b = kg m^{-2} s^{-1} B^{-1}), ρ = solution density (kg m^{-3}), ω = mass fraction of solute in solution (dimensionless) and Y is the mass ratio of solute to solvent in the solution (dimensionless). The subscripts 0 and ∞ refer to the interfacial and bulk solution conditions, respectively.

The dimensionless mass transfer driving force B is defined by

$$B = \frac{\omega_0 - \omega_\infty}{\omega_t - \omega_0} \qquad (6.43)$$

where ω_t is the mass fraction of the solute in the transferred solid substance, i.e. $\omega_t = 1$ for a single component. If the solute is a hydrate, then $\omega_t = 1$ only if the mass fractions are expressed as mass of hydrate per unit mass of solution.

Equation (6.36) should describe the dissolution of a solid solute into a solvent or its own solution, and either k_c or k_ω can be used, as $Sh = k_c d/D = k_\omega d/\rho D$. However, complications can arise if the solute solubility is high. First, the concentration dependence of the physical properties become significant[194] and, since $\rho_0 \neq \rho_\infty$, the Sherwood numbers based on k_c and k_ω will not be equal. Second, the mass flux from the surface of the solid alters the concentration gradient at the surface compared with that obtained under otherwise identical conditions of low mass flux.

Diffusion coefficients of electrolytes in water are greatly dependent on concentration; variations of $\pm 100\%$ from infinite dilution to near-saturation are not uncommon.[195] Moreover the change is often non-linear and accurate prediction of its effect is extremely difficult. Other physical properties, such as viscosity and density, change over this concentration range but not to such an extent.

The effects of concentration dependent physical properties on the correlation of dissolution mass transfer data have been reported in some detail[194, 196]. 'Mean' solution properties should be used for the Sherwood and Schmidt groups in equation (6.31) if the mass transfer data for moderately soluble substances are to be correlated effectively. The arithmetic mean will suffice for viscosity and density, but the integral value must be used for the diffusivity (equation 3.19). Bulk solution properties are used for the Reynolds number.

For low to moderate mass flux mass transfer studies, therefore, provided that the physical property changes are taken into account, mass transfer coefficients k_c or k_ω may be used. The dimensionless mass ratio driving force, ΔY, has been used quite successfully in crystallisation and dissolution studies[87, 88], but this has the disadvantage that each value of ω_∞ yields a different value of ϕ, even if the physical property variations are allowed for.

However, if the dimensionless driving force, B, is used, together with the appropriate physical properties, the value of ϕ in equations (6.10) and (6.11) remains substantially constant at about $0 \cdot 7-0 \cdot 8$ for a wide range of systems. There is little doubt that B is the best driving force to use for high mass flux studies[191].

MASS TRANSFER IN AGITATED VESSELS

Crystallisation and dissolution data obtained from agitated vessel studies may be analysed by the methods discussed above, but as NIENOW[195] has pointed out, a survey of the literature related to the subject of solid–liquid mass transfer in agitated vessels shows that there is an extremely wide divergence of results, correlations and theories. The difficulty is the extremely large number of variables that can affect transfer rates. the physical properties and geometry of the system and the complex liquid–solid–agitator interactions.

Relationships such as equations (6.10) and (6.11) are commonly used for correlating solid–liquid mass transfer data. However, the Reynolds number should not be based on the agitator dimension and speed, because this cannot take into account one of the most important factors, viz. the state of particle suspension. As shown by many workers, from HIXSON[198] onwards, the mass transfer coefficient increases sharply with agitator speed until the particles become fully suspended in the liquid, after which the rate of increase is reduced considerably. A maximum rate of mass transfer occurs when substantial aeration of the liquid occurs at high agitator speeds. From the 'just-suspended' to 'severe aeration' conditions the mass transfer coefficient may be enhanced by 40–50% while the power input may be increased tenfold. There is little justification, therefore, for using agitator speeds higher than those needed to suspend the particles in the system.

A much more reasonable velocity term for the Reynolds number in equations (6.10) and (6.11) is the slip velocity[199, 200], i.e. the relative velocity between particle and fluid. The slip velocity is usually assumed to be the free fall velocity of the particle, but this quantity is not easy to predict.

The critical mass transfer rate, for particles just suspended in a liquid, can be estimated from equation (6.36), the correct 'mean' solution properties being used as explained above. NIENOW[197] recommends that the terminal velocity, u_t, for use in the Reynolds number should be calculated from

$$u_t = 0 \cdot 153\, g^{0 \cdot 71} L^{1 \cdot 14} \Delta\rho^{0 \cdot 71} \rho^{-0 \cdot 29} \eta^{-0 \cdot 43} \qquad (6.44)$$

for particles smaller than 500 μm, and from

$$u_t = (4gL\Delta\rho/3\rho)^{\frac{1}{2}} \qquad (6.45)$$

for particles larger than 1 500 μm. For particles of intermediate size to be used u_t should be predicted from both relationships, the smaller value in the Re as a conservative estimate. In these relationships, u_t = cm/s,

$g = 981 \text{ cm/s}^2$, $L = \text{cm}$, $\rho = \text{g/cm}^3$ and $\eta = \text{poise (g/s cm)}$ where ρ and η refer to the bulk solution. $\Delta\rho$ is the solid–liquid density difference.

The actual expected mass transfer coefficient can be predicted from the critical value by multiplying by an enhancement factor ranging from about 1·1 for particles $\sim 200\,\mu\text{m}$ to about 1·4 for particles ~ 5 mm. Particle density also influences the rate of mass transfer. The reason for this enhancement is the increased level of turbulence at which larger and denser particles become suspended in the liquid.

Another model for mass transfer is based on the theory of homogeneous isotropic turbulence due to KOLMOGOROFF[201] developed specifically for solid–liquid systems by KOLAR[202] and others[203–206]. The energy put into the system by the agitator is considered to be transferred first to large-scale eddies and then to larger numbers of smaller isotropic eddies from which it is dissipated by viscous forces in the form of heat. KOLAR[202] suggested that for a given system the mass transfer coefficient, k, can be related to the energy input, e, to the system by $k \propto e^{0.25}$.

The Kolmogoroff theory can account for the increase in mass transfer rate with increasing system turbulence and power input, but it does not take into consideration the important effects of the system physical properties. The weakness of the slip velocity theory is the fact that the relationship between terminal velocity and the actual slip velocity in a turbulent system is really unknown. Nevertheless, on balance, the slip velocity theory appears to be the more successful for solid–liquid mass transfer in agitated vessels, and recent work on alum crystallisation tends to reinforce this view[207].

HABIT MODIFICATION

The habit of a crystal may be defined as the type of external shape which results from the different rates of growth of the various faces. Under certain conditions of crystallisation one set of faces may be induced to grow faster than others, or the growth of another set of faces may be retarded. Crystals of one given substance produced by different methods may be completely dissimilar in appearance, even though still belonging to the same crystal system. For example, one method of crystallisation may favour an acicular (needle) habit, while another may give a tabular habit (plates or flakes). A large number of factors can affect the habit of a crystal, e.g. the type of solvent, the pH of the solution, the presence of impurities, the degree of supersaturation or supercooling, the rate of cooling, the temperature of crystallisation, the degree of agitation, and so on.

Rapid cooling of a solution or melt will often cause the preferential growth of a crystal in one particular direction, leading to the formation of needles. Under these conditions there is a need for a fast rate of heat dissipation from the solid phase, and an elongated crystal is better suited for this purpose than, say, a granular or tabular crystal. Crystallisation from the vapour phase, as in sublimation processes, generally leads to the formation of needle crystals. Here again a habit which permits a rapid heat dissipation is required;

the heat of crystallisation is usually fairly high (heat of fusion plus heat of vaporisation), and the gaseous phase is a poor conductor of heat.

Dendritic growth, in which tree-like formations are produced, may also result from vapour phase crystallisation; the thin acicular branches provide the necessary exposed solid surface to dissipate the heat quickly. This type of growth is most usually associated with crystallisation from quiescent media, especially in thin layers of solution. The dendritic frosting of window panes is a typical example of this phenomenon. Many metals and organic melts crystallise initially in the form of dendrites, and several inorganic salts can be induced to crystallise in this form under the influence of traces of certain dyestuffs. Dendritic crystals of potassium chlorate are readily deposited from aqueous solution in the presence of Congo Red.

A change of solvent will often result in a change of the habit of the crystallising solute. Naphthalene, for instance, crystallises in the form of needles from cyclohexane, as thin plates from methanol. Iodoform crystallises as hexagonal prisms from cyclohexane, as hexagonal pyramids from aniline. Pentaerythritol crystallises in the form of tetragonal bipyramids from water, of tetragonal plates from acetone. The possibility of a habit change is an important factor to be borne in mind when a solvent for crystallisation is being chosen.

Occasionally the pH of an aqueous solution influences the type of crystal which grows. Copper sulphate, for example, which normally crystallises in the form of large granules, can be crystallised in the form of thin plates from acid solution. Ammonium and potassium dihydrogen phosphates crystallise as needles from pure solution (pH ~ 4) and squat prisms at pH ~ 5. In general, however, pH is not regarded as one of the major factors affecting crystal habit.

Supersaturation can influence the crystal habit. The individual crystal face growth kinetics usually depend to a different extent on supersaturation, so by raising or lowering the operating level it is sometimes possible to effect a considerable control over the crystal habit. It may be, of course, that the desired habit can only be grown at a high supersaturation, above the metastable limit, and in such cases a nucleation inhibitor may have to be added to allow growth to proceed as planned.

One of the most common causes of habit modification is the presence of impurities in the crystallising solution. These impurities may already be present, as in the crystallisation of beet sugar, where the presence of raffinose induces a characteristic flat crystal. Sometimes they are added deliberately, e.g. Pb^{2+} ions, which allow large strong crystals of alkali halides to be grown from aqueous solution. In some cases minute traces (< 1 p.p.m.) of an impurity can cause startling changes in the crystal habit, but this is not always the case; sometimes habit modification only occurs in the presence of large quantities of impurity in the crystallising system, e.g. > 5 per cent of biuret is needed to change the habit of urea from needles to 'bricks'. Tervalent ions such as Cr^{3+}, Fe^{3+} and Al^{3+} are particularly active impurities and these are frequently used at levels of ~ 100 p.p.m. for habit modification purposes.

Surface-active agents (surfactants) and other organic substances are being increasingly used industrially to change crystal habits. Common anionic surfactants include the alkyl sulphates, alkane sulphonates and aryl alkyl sulphonates. Quaternary ammonium salts are frequently used as cationic agents. Occasionally non-ionic surfactants find application as habit modifiers. Some polyelectrolytes, e.g. polyacrylamides and sodium carboxymethyl cellulose, can exert a habit-modifying effect at very low concentrations.

Many dyestuffs act as habit modifiers for inorganic salts, and BUCKLEY[1] has summarised a large number of case histories giving an indication of the concentrations necessary to induce the required change. From about 400 known active habit-modifying dyes, Buckley selects about 30 which are particularly effective for the four salts $KClO_3$, K_2SO_4, K_2CrO_4 and NH_4ClO_4 and from these again five which he considers to be the best all-rounders: Brilliant Azurine B (511), Tryptan Red (438), Naphthol Black B (315), Brilliant Congo R (456) and Sky Blue FF (518). The numbers in parentheses are the Colour Index numbers.

There are literally thousands of references in the technical literature to the effects of impurities on the habit of crystals, but many of these are qualitative only and the information should be treated with great caution. In recent years, however, a number of careful quantitative studies[176-181] have been made and these have helped towards a better understanding of the possible mechanisms of habit modification. Some of these have already been mentioned above (see 'Effect of Impurities', p. 196). The paper by MICHAELS and

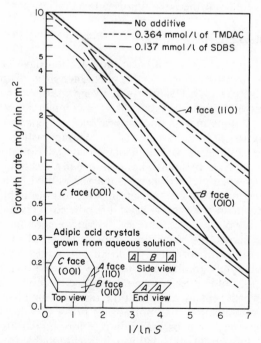

Figure 6.15. Specific growth rates of the faces of adipic acid crystals as a function of supersaturation, S, with and without the presence of surface-active agents[152]

COLVILLE[208] is a good example of the use of face growth rate measurements in the prediction of crystal habit. They studied the growth of adipic acid crystals from aqueous solution in the presence of surfactants. The two agents, used in concentrations of 50–100 p.p.m., were sodium dodecylbenzenesulphonate (SDBS) (anionic) and trimethyl dodecylammonium chloride (TMDAC)(cationic). Some of the results of this work are shown in *Figure 6.15*.

The growth rates of the normal faces of adipic acid crystals were found to increase in the order (110)>(010)>(001), and the rate constants were correlated with the hydroxyl densities on the faces. Trace quantities of the anionic agent caused a much greater reduction in the growth rate of the (010) and (110) faces than of the (001) face, leading to the formation of prisms or needles. The cationic agent had the opposite effect, favouring the formation of plates or flakes. Anionic agents appeared to be physically adsorbed on to the crystal faces, whereas the cationic agents appeared to be chemisorbed. It was also observed that surfactants in general exhibited a far greater retarding influence on the growth of very small crystals than of large ones, and it was suggested that growth from dislocations is less sensitive to the influence of adsorbed contaminants than is the growth by two-dimensional nucleation.

A similar study was made by MULLIN *et al.*[177] on the face growth rates of ammonium and potassium dihydrogenphosphates in the presence of Cr^{3+}, Fe^{3+} and Al^{3+}. The effect of pH was also studied and the results of the single-crystal studies were confirmed in a fluidised bed crystalliser[209].

INDUSTRIAL IMPORTANCE

The majority of reported cases of habit modification have been concerned with laboratory investigations, but the phenomenon is of the utmost importance in industrial crystallisation and by no means a mere laboratory curiosity. Certain crystal habits are disliked in commercial crystals because they give the crystalline mass a poor appearance; others make the product prone to caking (p. 293), induce poor flow characteristics or give rise to difficulties in the handling or packaging of the material. For most commercial purposes a granular or prismatic habit is usually desired, but there are specific occasions when plates or needles may be wanted.

In nearly every industrial crystallisation some form of habit modification is necessary to control the type of crystal produced. This may be done by controlling the rate of crystallisation, e.g. the rate of cooling or evaporation, the degree of supersaturation or the temperature at which crystallisation occurs, by choosing the correct type of solvent, adjusting the pH of the solution, or deliberately adding some impurity that acts as a habit modifier to the system. A combination of several of these methods may have to be used. It is also worth remembering that the results of small-scale laboratory trials on habit modification may not always prove of value for large-scale application; they may even be misleading. Pilot-plant trials, however, on batches greater than about 100 l will usually yield reliable information.

The suppression of $CaSO_4$ scaling in boilers by the use of phosphates owes as much to the habit modification of the deposited crystal as it does to nucleation inhibition. The same is true for the suppression of ice nucleation and growth in ice-cream by the addition of sodium carboxymethylcellulose. Lecithin (a triglyceride in which one fatty acid radical has been replaced with phosphoric acid) is widely used as an oil-soluble wetting agent in the manufacture of chocolate confections; it lowers the viscosity and reduces 'bloom' on the chocolate surfaces by its dispersive effect on the chocolate fat crystals. Lecithin is also used as a crystallisation inhibitor in cotton seed oil.

GARRETT[210] has reviewed many reported cases of the industrial application of habit modification, and also discussed the factors that must be considered when selecting a suitable habit modifier. A few of his examples may be quoted. Borax crystals can be modified to a tabular or flaky habit by the addition of gelatin or casein to the crystallising solution; these substances are selectively adsorbed on to the basal pinacoid faces and reduce their growth rate. Large crystals of sodium, potassium and ammonium chlorides, normally difficult to produce, can be grown when Pb^{2+} ions are present in the solution. Large granular crystals of sodium sulphate decahydrate (Glauber's salt) can be produced from by-product rayon liquors when certain surfactants, e.g. alkyl aryl sulphonates, are added; otherwise the salt usually tends to crystallise in the form of small needles from these liquors. SVANOE[211] discusses many of the factors affecting the habit of industrial crystals, i.e. the effects of supersaturation, the type of solvent, pH, the presence of inorganic and organic impurities and surface-active agents on the type of crystal produced. Several interesting photographs of the 'before and after' type are given showing crystals of K_2CO_3, KCl, $CuSO_4 \cdot 5H_2O$, $Na_2CO_3 \cdot 10H_2O$ and $(NH_4)_2SO_4$, produced under different conditions.

PHOENIX[212] has reported on the effects of a wide variety of inorganic and organic additives on NaCl, NaBr, KCl, KCN, K_2SO_4, NH_4Cl, NH_4NO_3 and $(NH_4)_2SO_4$. A considerable amount of valuable quantitative information is given concerning the effects of the different additives on crystal habit, growth and dissolution rates, and anti-caking effectiveness (see Chapter 8). The remarkable influence of ferrocyanide ions, which are effective in concentrations as low as 0·1 p.p.m., in producing dendritic crystals of NaCl is discussed in some detail. The mechanism and commercial importance of this discovery are also discussed by COOKE[145] and VAN DAMME[213]. If the habit modification is influenced by more than one variable, e.g. supersaturation and impurities, the combined effects can be expressed on a morphogram[145] (see *Figure 6.16*), which indicates the most likely behaviour in the different regions of the diagram. KERN and co-workers[179] have made extensive use of these diagrams, which they call morphodromes. Some habit modifications of industrial interest are listed in *Table 6.4*.

The use of temperature and concentration control in habit modification is illustrated by work carried out by EDWARDS[214] on paraffin wax, where the type of habit formed is a very important factor in its efficient processing. When crystallisation occurs at temperatures near the melting point of paraffin wax, or from concentrated solutions, needle crystals are usually formed. Thin plates grow when the crystallising temperature is much lower than the melting point, or on crystallisation from dilute solution. The cooling rate was also shown to exert a considerable influence on the crystal habit. CHICHAKLI and

Table 6.4. SOME HABIT MODIFICATIONS OF INDUSTRIAL INTEREST

Substance	Normal habit	Habit modifier	Changed habit
NH_4 alum	octahedra	borax	cubes
NH_4Cl	dendrites	Cd^{2+}, Ni^{2+}	cubes
		$MnCl_2 + HCl$	granules
$NH_4H_2PO_4$	needles	Al^{3+}, Fe^{3+}, Cr^{3+}	tapered prisms
NH_4NO_3	squat crystals	Acid Magenta	needles
$(NH_4)_2SO_4$	prisms	$FeCl_3$	irregular crystals
H_3BO_3	needles	gelatin, casein	flaky crystals
$CaCO_3$		alkyl aryl	
		sulphonates	granular precipitate
$CaSO_4 \cdot 2H_2O$	needles	sodium citrate	prisms
		alkyl aryl	
		sulphonates	prisms
$MgSO_4 \cdot 7H_2O$	needles	borax	prisms
$AgNO_3$	plates	sodium oleate	dendrites
K alum	octahedra	borax	cubes
KBr	cubes	phenol	octahedra
KCN	cubes	Fe^{3+}	dendrites
KCl	cubes	$Fe(CN)_6^{4+}$	dendrites
K_2SO_4	rhombic prisms	$FeCl_3$	irregular crystals
NaBr	cubes	$Fe(CN)_6^{4+}$	dendrites
NaCN	cubes	Fe^{3+}	dendrites
NaCl	cubes	$Fe(CN)_6^{4+}$	dendrites
		formamide	octahedra
		Pb^{2+}, Cd^{2+}	large crystals
		polyvinylalcohol	needles
		$Na_6P_4O_{13}$	octahedra
$Na_2B_4O_7 \cdot 10H_2O$	elongated	NaOH, Na_2CO_3	squat crystals

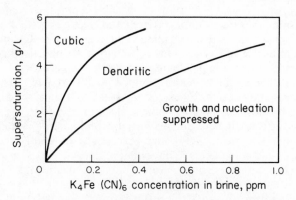

Figure 6.16. Morphogram for the crystallisation of sodium chloride from the system $K_4Fe(CN)_6$–NaCl–H_2O at 42 °C[145]

JESSEN[215] made a systematic investigation of the structure, mechanism of crystallisation and orientation of paraffin wax crystals. Crystallisation from solutions and melts and the use of habit modifiers were considered. Problems associated with the production of thin flat crystals are discussed by BLATCHLY and HARTSHORNE[216], who show that curling of the crystals during growth can be caused by differential adsorption of solvent on opposite faces of a platelet.

INCLUSIONS

Crystals generally contain foreign impurities, which may be solid, liquid or gas, and the pockets of impurity are called 'inclusions'. The term 'occlusion', which has also been used in this connection, is more generally applied to the fluid adhering to the surfaces of crystals, or trapped between agglomerated crystals, e.g. after filtration. The term 'inclusion' is preferred because it emphasises the entrapment of impurity *inside* a crystal.

It goes without saying that inclusions are undesirable; they are a frequent source of trouble in industrial crystallisation[217]. Crystals grown from aqueous solution can contain up to 0·5% by weight of liquid inclusions, and their presence can significantly affect the purity of analytical reagents, pharmaceutical chemicals and foodstuffs such as sugar. Inclusions can cause caking (Chapter 7) of stored crystals by the seepage of liquid if the crystals become broken.

The subject of crystal inclusions has been studied extensively by geologists for a century or more. In fact it is within this branch of science that most work is still being done; the reason for this geological interest is that fluid inclusions preserve a record of the conditions under which the crystalline minerals were formed. Fluid inclusions may often be observed with the aid of a simple magnifying glass, although a more detailed picture is revealed under a low-power microscope. A technique suggested for this purpose[218] consists of immersing the crystal in an inert liquid of similar refractive index; alternatively, as second-best, a saturated solution of the crystalline material may be used. The use of a solution as the medium can aid identification of the inclusion under the microscope as follows. The temperature is raised slightly to dissolve the crystal: if the inclusion is a liquid, concentration streamlines will be seen as the two fluids meet; if it is a vapour, the bubble will bob to the surface.

A number of terms are used to describe inclusions, some of which are self-explanatory, such as bubbles, fjords (parallel channels), veils (thin sheets of small bubbles), clouds (random clusters of small bubbles), negative crystals, and so on. Most frequently inclusions appear in random array, but sometimes they show a remarkable regularity, as illustrated by photographs of hexamine crystals[219]. On the other hand, over-all patterns may appear in the crystals such as 'hour-glasses' or Maltese crosses, as shown by photographs of sucrose crystals[218].

Inclusions may be classified as primary (formed during growth) or secondary (formed later). Primary fluid inclusions constitute samples of fluid in which the crystals grew, whereas secondary inclusions give evidence of later environments. Many inclusions, particularly those found in minerals, show a remarkable resistance to supercooling. Some water inclusions have been known to withstand supercoolings of 100 °C for prolonged periods before solidifying. One explanation frequently noted is that this is due to the absence of heteronuclei in the very small liquid sample. This may well be true, but it is also interesting to speculate on the 'structure' of these small pockets of liquid that have been entrapped for aeons.

Inclusions may be formed in a variety of ways. One common mechanism, especially in melts, is dendritic growth followed later by a filling-in process that traps fluid in small pockets. Another arises from growth instabilities; inclusions often begin as depressions on the crystal surface. This might be caused, for example, by the presence of impurities at the growing crystal faces which impede the diffusion of solute. On the other hand, the level of supersaturation may vary across the crystal surface, the edges growing faster than the centre, for example, giving rise to a centre depression. Inclusions readily form in a crystal that has been subjected to dissolution — rapid growth occurs on the partially rounded surfaces and entraps mother liquor. The rapid healing of etch pits will do the same. These 'regeneration' inclusions[220], which usually lie in lines, i.e. along the former crystal edges, are characteristic of crystals grown from seeds.

Fast growth leads to inclusions and so does interrupted growth. BROOKS, HORTON and TORGESEN[221] correlated the formation of inclusions in ADP and NaClO$_3$ crystals to the introduction of sudden step changes in supersaturation. They showed that inclusions are associated with certain faces whose growth rates, relative to the rates of the contiguous faces, are affected by the sudden changes in conditions. The suggested sequence of events leading to inclusion is: formation of the special faces when the growth rate was slow; development of imperfections on the faces, which became fast-growing when the supersaturation increased; overgrowth of the imperfections by the enlargement of the adjacent slow-growing faces. A study by BELYUSTIN and FRIDMAN[222] with KDP and NaClO$_3$ suggested that layer growth by the advancement of steps across the surface could be used to explain inclusions. They concluded that the production of inclusions at any point is governed entirely by local conditions, and more specifically by the concentration gradient along the height of a step. A critical step height is postulated beyond which a layer of solution is trapped. These two models are not inconsistent with each other.

The adsorption of impurities, including the solvent, on the crystal surface can also lead to impeded growth and, hence, to inclusions. There is evidence that crystals reject impurity during growth, and for this reason liquid inclusions may be richer than the mother liquor from which the crystals grew, since the rejected impurity is concentrated in the entrapped fluid. Solid particles may be included into the crystal at the nucleation stage, i.e. foreign bodies acting as heteronuclei. Secondary inclusions are formed by crystals cracking during growth and resealing later. This mechanism occurs frequently in minerals, but it is not known how important it is in industrial crystallisation.

Large crystals and/or fast growth increase the likelihood of inclusions. The study of DENBIGH and WHITE[219] affords an excellent confirmation of these simple 'rules'. One conclusion of their work is that crystals of hexamine grow regularly when they are quite small, but when a critical size is reached cavities begin to form at the centre of the faces and these are eventually sealed off to form a regular pattern of inclusions. Cavities appeared to form only if the growth rate exceeded its critical value when the crystal had reached its critical size. It should be possible, therefore, to minimise inclusion formation by ensuring that the growth rate is below the critical once the largest crystals pass the critical size. In practice this could be achieved by permitting only a low rate of crystallisation, especially in the early stages, or by promoting an initial high rate of nucleation.

The questions of why a cavity forms at a face centre and why it subsequently seals over have not yet been answered with any conviction. The work of BUNN[92], BERG[93] and HUMPHREYS-OWEN[156], referred to earlier, showed that the supersaturation is generally lower at the centre of the face than at the edges of a crystal, but small crystals tend to grow layerwise away from the centres. Bunn's explanation for this unexpected finding was that the diffusion field around a small crystal would tend to develop spherical symmetry and this results in the component of the concentration gradient normal to the face being greater near the centre than near the edges, thus causing more solute to be transported to the centre. However, when the crystal is large enough, the above situation is reversed[219] and the corners and edges grow more rapidly than the face centres, and cavities form. Later, when the face grows beyond a certain size, a 'healing process' occurs and an inclusion is formed at its centre.

CHERNOV[223] put forward a somewhat different interpretation, but his view that an instability develops at a face centre is consistent with the above reasoning. A simple criterion for crystal face stability, i.e. for the avoidance of inclusions, arises from Chernov's mathematical theory, viz. $L \ll D/K$, where L = crystal size, D = diffusivity and K is the constant of proportionality in an assumed linear dependence of growth rate on supersaturation.

Inclusions may sometimes be prevented if the crystals are grown in the presence of ionic impurities, e.g. Pb^{2+} and Ni^{2+}, traces of which allow near-perfect crystals of ADP to be grown for piezo-electric use. Traces of Pb^{2+} help good crystals of NaCl to be grown. A change of solvent may have a significant effect; hexamine grown from methanol or ethanol, for example, contains no inclusions[219]. An increase in the viscosity of the liquor may also help; small amounts of carboxymethyl cellulose added to the liquor have been known to have a beneficial effect. Ultrasonic vibrations have also been tried with moderate success.

Under isothermal conditions inclusions may change shape or coalesce as the system adjusts itself towards the condition of minimum surface energy. If the temperature is raised, negative crystals (faceted inclusion cavities) may be formed by a process of recrystallisation. Fluid inclusions cannot be removed by heating alone. In fact heating to decrepitation frequently fails to destroy all the inclusions. However, liquid inclusions can actually move under the influence of a temperature gradient[224]. Since solubility is temperature dependent, crystalline material dissolves on the high solubility side of the inclusion, diffuses across the liquid and crystallises out on the low solubility side. It is feasible, therefore, to remove inclusions from large single crystals by this method.

For a detailed account of the subject of inclusions reference should be made to the comprehensive reviews of DEICHA[225], POWERS[218] and WILCOX[224].

In addition, a world-wide coverage of research on inclusions, although mainly of geological interest, is provided by the annual COFFI[226] reviews (Commission on Ore-Forming Fluids in Inclusions).

PRECIPITATION

Precipitation is a widely used industrial process. It is also a very popular laboratory technique, especially in analytical chemistry, and the literature on this aspect of the subject is voluminous[20, 227]. Precipitation plays an important role not only in chemistry but also in metallurgy, geology, physiology, and other sciences. In the industrial field, the manufacture of photographic chemicals, pharmaceuticals, paints and pigments, polymers and plastics utilises the principles of precipitation. For the production of microcrystalline powders precipitation may be considered a useful alternative to mechanical comminution.

PRECIPITATION FROM HOMOGENEOUS SOLUTION

For the purpose of gravimetric analysis, where it is necessary to effect an efficient separation of solid from liquid, it is generally accepted that precipitation should be carried out slowly from dilute solution. However, some substances, such as the hydroxides and basic salts of aluminium, iron and tin, demand extremely high dilutions and excessively long times for dense particles to be produced. But the method known as 'precipitation from homogenous solution' (PFHS) originally developed in 1937 by WILLARD and TANG[227] allows coarse precipitates to be produced in relatively short times.

Briefly, the technique of PFHS consists of slowly generating the precipitating agent homogeneously within the solution by means of a chemical reaction. Undesirable concentration effects are eliminated, a dense granular precipitate is formed and co-precipitation is minimised. For example, silver chloride crystals can be produced in aqueous solution, by using allyl chloride as the chloride-generating reagent:

$$C_3H_5Cl + H_2O \rightarrow Cl^- + C_3H_5OH + H^+$$

$$Cl^- + Ag^+ \rightarrow AgCl\downarrow$$

The growth kinetics of this process are reported to be second-order and surface reaction controlled[20]. The precipitation of silver iodide in ethanol by the reaction

$$2C_2H_5I + 2AgNO_3 + C_2H_5OH \rightarrow 2AgI\downarrow + (C_2H_5)_2O + HNO_3 + C_2H_5NO_3$$

is reported to be first-order and diffusion-controlled. This is one of the few PFHS reactions that have been studied in non-aqueous solution.

Also first-order and diffusion-controlled are the precipitation of barium sulphate by a persulphate–thiosulphate reaction in the presence of Ba^{2+}:

$$S_2O_8^{2-} + 2S_2O_3^{2-} \rightarrow 2SO_4^{2-} + S_2O_6^{2-}$$

and the production of sulphur (as a spherical monodisperse sol) by the decomposition of thiosulphate in acid solution:

$$4S_2O_3^{2-} \rightarrow 3SO_4^{2-} + 5S \downarrow$$

Other PFHS methods that have been used for the production of crystalline precipitates by the controlled generation of the required anions in an appropriate aqueous solution include the hydrolysis of allyl chloride (Cl^-), dimethyl oxalate ($C_2O_4^{2-}$), triethylphosphate (PO_4^{3-}), dimethylsulphate (SO_4^{2-}) and thioacetamide (S^{2-}).

PFHS plays a very important role in modern analytical chemistry. It is also being used in studies of co-precipitation and nucleation phenomena because the slow controlled precipitation allows a close approach to equilibrium between the solid and solution. The applications of PFHS on the industrial scale have so far been limited, but it does appear to be a promising technique. Fractional precipitation methods are usually improved by PFHS and it has been applied to the difficult separation of radium and barium[227]. It has also been used for the production of carriers for radioactive materials and for the preparation of monodisperse suspensions of pigments and polishing agents.

PRECIPITATION BY DIRECT MIXING

Another common method for producing a precipitate is to mix two reacting solutions together as quickly as possible, but the kinetic analysis of this apparently simple operation is exceedingly complex. Primary nucleation does not necessarily commence as soon as the reactants are mixed, unless the level of supersaturation is very high. The mixing stage may be followed by an appreciable time-lag (induction period), which depends on the temperature, supersaturation, efficiency of mixing, state of agitation, presence of impurities, and so on, before nuclei appear. Some time after the induction period a rapid desupersaturation ensues, during which both primary (homogeneous and heterogeneous) and secondary nucleation may occur together, but the predominant process here is growth of the nuclei. Later on, if sufficient time is allowed to elapse, an ageing or particle-coarsening process may occur. Particle agglomeration, another complicating factor, can occur at almost any stage of the process.

INDUCTION AND LATENT PERIODS

A period of time usually elapses between the achievement of supersaturation, or supercooling, and the appearance of crystals in a given system. This time-lag is known as the 'induction period'. The term 'latent period' has also been used in this context by some authors but, as will be described later, it is preferable to restrict the use of this latter term to denote a more general condition of a system, not necessarily associated with nucleation (see *Figure 6.18*).

The duration of the induction period can be affected by the level of supersaturation, the temperature, the state of agitation, the viscosity of the system, the presence of impurities, and so on. Induction periods may be measured visually, i.e. by recording the time after the creation of supersaturation when the first nuclei appear, e.g. as bright twinkling particles in a beam of light. On the other hand, concentration changes may be recorded over a period of time by turbidity, conductivity or refractive index measurements, although these concentration-based methods are generally more appropriate for the measurement of 'latent' rather than induction periods.

The existence of an induction period in a supersaturated system is contrary to expectations from the classical, Volmer–Becker–Döring (VBD) theory of homogeneous nucleation, which assumes ideal steady-state conditions and predicts immediate nucleation once supersaturation is achieved. The induction period may therefore be considered as being made up of several parts. For example, a certain 'relaxation time', t_r, is required for the system to achieve a quasi-steady-state distribution of molecular clusters. Time is also required for the formation of a stable nucleus, t_n, and for the nucleus to grow to a detectable size, t_g. So the induction period, τ, may be written

$$\tau = t_r + t_n + t_g \tag{6.46}$$

It is difficult, if not impossible, to isolate and measure these separate quantities, but the nucleus formation time, t_n, is often assumed to be the largest, although this may not always be so. The nucleus growth time, t_g, may be quite small, especially in aqueous solutions of inorganic salts. For example, a nucleus with a linear growth rate of 10^{-7} m/s would reach 1 μm in about 5 s. Such a calculation is quite speculative, of course, because it is by no means certain that the rate of growth of a nucleus has the same order of magnitude as that of a macrocrystal; the mechanism and rate may well be quite different (see pp. 165 ff.). The relaxation time, t_r, depends to a great extent on the system viscosity and, hence, diffusivity. NIELSEN[14] has suggested that $t_r \sim 10^{-13} D^{-1}$, where D = diffusivity, so in an aqueous solution of an electrolyte with $D \sim 10^{-5}$ cm^2/s, the relaxation time would be about 10^{-8} s. However, in highly viscous systems values of D can be extremely low and t_r very high. Indeed, some systems set to a glass before nucleation occurs.

The induction time is undoubtedly related in some way to the size and complexity of the critical nucleus. It should be emphasised, however, that, as it can be affected profoundly by external influences, it cannot be regarded as a fundamental property of a system. Nor can it be relied upon to yield basic information on the process of nucleation. Nevertheless, despite its complexity and uncertain composition, the induction period has frequently been used as a measure of the nucleation rate. A number of authors[6,9,96,228–231] have made the simplifying assumption that the induction period is essentially devoted to nucleus formation and can therefore be considered inversely proportional to the rate of nucleation:

$$\tau \propto J^{-1} \tag{6.47}$$

which from the classical nucleation relationship (equation 5.17) gives

$$\log\left(\frac{1}{\tau}\right) \propto \left(\frac{\sigma^3}{T^3(\log S)^2}\right) \tag{6.48}$$

Thus for a given temperature a plot of $\log \tau$ versus $(\log S)^{-2}$ should yield a straight line and allow a value of σ, the surface energy, to be calculated. For precipitations of barium and lead salts at 22 °C NIELSEN[95] reports values of σ in the range 90–150 erg/cm^2 (0·09–0·15 J/m^2). Much lower values than these are reported for the surface energy of more soluble salts such as the alkali halides[138] (1–17 erg/cm^2), sodium bicarbonate[230] (11–17 erg/cm^2) and nickel ammonium sulphate[231] (4–5 erg/cm^2).

A plot of $\log \tau$ versus $(\log S)^{-2}$ does not always result in a straight line, as shown by KHAMSKII[9], who points out that the assumption of equation (6.44) ignores any possible growth contribution to the nucleation process. Furthermore there is no guarantee that the classical VBD theory remains valid for all values of S. The work of GINDT and KERN[138] on the nucleation of alkali halides lends support for the Frenkel–Zeldovich theory of transition nuclei, leading on from the VBD theory, according to which the relaxation time of the transition phenomenon is due to the fact that the medium is a condensed phase and its strong dependence on the degree of supersaturation is connected with two-dimensional nucleation. MITSUDA et al.[139] used rather complex empirical relationships to correlate induction periods for KNO_3, KCl and other inorganic salts.

Other empirical τ–S relationships have been suggested such as

$$\log \tau = \alpha + \beta \log S \tag{6.49}$$

which does allow for the possibility of a growth contribution to the induction period. From equation (6.49) a linear relationship may be expected between $\log \tau$ and $\log S$, and data for the nucleation of ammonium titanyl sulphate[9] from aqueous solution support this.

Another empirical approach is that of CHRISTIANSEN and NIELSEN[51], who proposed an ion-clustering mechanism of nucleation and suggested that the induction time, τ, could be related to the primary ion concentration, c, of the solution by

$$\frac{dc}{d\tau} \propto c^p$$

which on integration gives

$$\tau = kc^{1-p} \tag{6.50}$$

where k is a constant and p is the number of ions required to form a critical cluster. For many sparingly soluble salts p is reported to be <10.

The effect of agitation, temperature and supersaturation on the induction period for the precipitation of nickel ammonium sulphate by the controlled mixing of aqueous solutions of nickel and ammonium sulphates can be seen in *Figure 6.17a* and *b*. At a given temperature τ decreases rapidly with an increase in supersaturation in non-agitated systems. At 25 °C, for example, at the lowest supersaturation used ($S = 1·24$) no nuclei appeared 2 h after mixing the reactant solutions, but at $S > 4$ the precipitation was more or less instantaneous. At 25 °C a 'critical' supersaturation could be considered to exist at $S \sim 1.6$ (*Figure 6.17a*); above this value induction times could be measured in seconds rather than minutes or hours. For a given level of supersaturation the induction time decreases with an increase in temperature. Agitation reduces the induction time considerably and induces nucleation at much lower supersaturations (*Figure 6.17b*).

These data cannot be correlated satisfactorily by any of the methods referred to above, e.g. through plots of log τ versus $(\log S)^{-2}$ or log τ versus log S. A plot of log τ versus T^{-3} $(\log S)^{-2}$ for the agitated data gives a

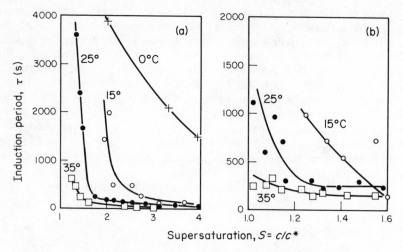

Figure 6.17. Effect of supersaturation, S, on the induction period, at different temperatures in (a) static and (b) stirred solutions of nickel ammonium sulphate[231]

family of curves at all temperatures. The same plot for the non-agitated data gives a family of curves above 20 °C, and this behaviour leads to the postulation of a change-over in the predominant mode of nucleation from homogeneous at low temperature to heterogeneous at high. On the other hand, CHEPELEVETSKI[228], who also found that his experimental data did not support the expected temperature dependence, suggested that solution viscosity, η, should be taken into account such that the slope of the line log τ versus $(\log S)^{-2}$ is given by $K\eta^2 T^{-3}$, where K is a constant for the system.

The presence of impurities can also affect the induction time, but it is virtually impossible to predict the effect. Ionic impurities, especially Fe^{3+} and Cr^{3+}, may reduce the induction period considerably in aqueous solutions of inorganic salts. Others, such as sodium carboxymethylcellulose or polyacrylamide, can increase τ, whereas others may have no effect at all. The effects of soluble impurities may be caused by changing the equilibrium solubility or the solution structure, by adsorption or chemisorption on nuclei or heteronuclei, by chemical reaction or complex formation in the solution, and so on. The effects of insoluble impurities are also unpredictable.

Nucleation rates of solutions and melts can often be affected considerably by traces of inert impurities although, as TAMMAN[48] observed, the temperature at which the maximum rate of nucleation occurs may not be appreciably altered. However, an impurity that acts as a nucleation inhibitor in one case may act as an accelerator in another. No general rule applies; each case must be treated separately.

Figure 6.18 indicates diagrammatically a typical desupersaturation curve. Supersaturation is created at *A* ($t = 0$) and a certain induction time ($t = \tau$) elapses before nuclei are first detected visually (*B*). However, little change may occur in the solution concentration for some considerable time after this, i.e. until the end of the latent period, *C* ($t = \phi$), after which a rapid desupersaturation ensues (*D*). Crystal growth predominates in this region. Towards the end of the gradual approach to equilibrium, *E*, which may take hours or days, an ageing process may occur (see p. 222).

The presence of seeds generally reduces the induction period, but does not necessarily eliminate it. Even if the system is seeded at time $t = 0$, a measurable induction period, τ, may elapse before nuclei appear. By definition, these are 'secondary' nuclei (see p. 137) and they may appear in several bursts throughout the latent period, making it difficult to attach any real significance to the induction time itself. For these reasons it may be preferable to record the latent period as the more reliable characteristic of the system. The rate of desupersaturation (region *D*) may also be used to characterise the system[142]. Factors that can influence the induction and latent

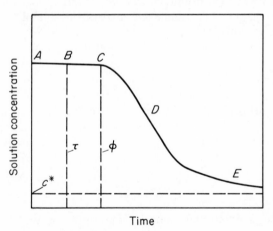

Figure 6.18. A typical desupersaturation curve: c^* = equilibrium saturation, τ = induction period, ϕ = latent period

periods and the rate of desupersaturation are temperature, agitation, heat effects during crystallisation, seed size, seed area and the presence of impurities.

Some inorganic systems show three distinct induction regions[20]. At low supersaturation, heterogeneous nucleation and discrete growth occur; but as the supersaturation is increased, dendritic growth may develop. At high supersaturation homogeneous nucleation predominates and the particle size of the precipitate is greatly reduced.

When solid particles are dispersed in their own saturated or supersaturated solution, there is a tendency for the smaller to dissolve and the solute to be deposited later on the larger particles. Thus the small particles disappear, the large grow larger and the particle size distribution gradually changes towards that of a monosized dispersion. The reason for this behaviour lies in the tendency of the solid material in the system to adjust itself to a minimum total surface free energy. The process of particle coarsening, first discovered by Ostwald (1900), is called 'Ostwald ripening' or 'ageing'.

The driving force for ripening is the difference in solubility between small and large particles, as given by the Gibbs–Thomson (also known as the Gibbs–Kelvin or Ostwald–Freundlich) relationship (equations 2.18 and 5.6), which for the present purpose may be written as

$$\ln\left(\frac{c(r)}{c^*}\right) = \frac{2\sigma v}{RTr} \tag{6.51}$$

where v = molecular volume of the solute, $c(r)$ = solubility of small particles of size r and c^* = equilibrium saturation for large particles ($r \to \infty$). As discussed in Chapter 2, a significant increase in solubility occurs when $r < 1\,\mu m$. From equation (6.51), expanding the logarithm for $c(r)/c^* \sim 1$ (ripening takes place at very low supersaturations), we get

$$c(r) - c^* \approx \frac{2\sigma v c^*}{RTr} \tag{6.52}$$

If mass transport is possible between the particles in a polydisperse precipitate, the large will grow at the expense of the small. If the growth kinetics are first-order, diffusion-controlled, then for a change in particle radius with time (see equation 5.72)

$$\frac{dr}{dt} = \frac{Dv[c - c(r)]}{r} \tag{6.53}$$

where c is the average bulk solution concentration, and from equation (6.52)

$$\frac{dr}{dt} = \frac{Dv}{r}\left[(c - c^*) - \frac{2\sigma v c^*}{RTr}\right] \tag{6.54}$$

where $c - c^*$ is always positive during precipitation. Setting equation (6.54) equal to zero, it follows that all particles of size

$$r = \frac{2\sigma v c^*}{RT(c - c^*)} \tag{6.55}$$

are in equilibrium with the bulk solution ($dr/dt = 0$). All particles smaller than this will dissolve ($dr/dt < 0$), and all particles larger will grow ($dr/dt > 0$).

The speed with which ripening occurs depends to a large extent on the particle size and solubility. For example, small crystals of a moderately soluble salt such as K_2SO_4 with a solubility $c_1^* \sim 10^3$ mol/m^3 will age much faster than those of the almost insoluble $BaSO_4$ ($c_2^* \sim 10^{-2}$ mol/m^3) if kept in solutions of the same supersaturation, say $S = c/c^* = 1 \cdot 1$. The driving force for mass transfer $\Delta c_1 = (S-1)c_1^* = 10^2$ and $\Delta c_2 = 10^{-3}$ mol/m^3, and if a growth law $R_G = K_G \Delta c^n$ is assumed (equation 6.14), with the growth coefficients not differing by more than an order of magnitude, the relative growth rates of the two salts are $\sim 10^5:1$ for first-order growth kinetics rising to $\sim 10^{12}:1$ for second-order.

For diffusion-controlled growth kinetics it can be shown[14, 20, 232] that the linear growth velocity approximates to

$$\frac{dr}{dt} \approx \frac{\sigma v^2 D c^*}{3RTr^2} \tag{6.56}$$

which on integration gives

$$t \approx \frac{RTr^3}{\sigma v^2 D c^*} \tag{6.57}$$

thus the smaller the particle size, r, or the higher the solubility, c^*, the faster the ripening process.

Of course, the ripening process, which occurs at very low supersaturation, is probably more likely to be controlled by a surface reaction than by a diffusion process, and under these circumstances ripening could be considerably retarded. For surface reaction growth kinetics the linear growth velocity obeys a relationship such as

$$\frac{dr}{dt} = k(c - c^*)^n \tag{6.58}$$

where k is a growth rate constant and if an arbitrary value of $n = 2$ is assumed, then[232]

$$t = \left(\frac{RT}{\sigma v c^*}\right)^2 \frac{r^3}{k} \tag{6.59}$$

Therefore, from equations (6.57) ($t = t_D$) and (6.59) ($t = t_R$)

$$t_R \approx \left(\frac{RTD}{\sigma k c^*}\right) t_D \tag{6.60}$$

Substituting typical values in equation (6.60) for a moderately insoluble salt such as CaF_2 ($T = 300$ K, $D = 10^{-9}$ m^2/s, $\sigma = 5 \times 10^{-2}$ J/m^2, $R = 8 \cdot 3$ J/mol K, $c^* = 1$ mol/m^3, $k = 10^{-6}$ m^7/mol^2 s), we get $t_R \simeq 50 \, t_D$, which indicates that the interfacial reaction can exert a dominant effect on the ripening process. Any additive to the system which slows down the surface reaction step will automatically retard the ripening process and thus stabilise the suspension. Substances such as gelatin and gum acacia are commonly used for this purpose.

The ripening process will change the particle size distribution of a precipitate over a period of time, even in an isothermal system. However, the change can be accelerated by making the temperature fluctuate in a controlled manner. This process, known as 'temperature cycling', has been used by CARLESS and FOSTER[233] to alter the physical characteristics of pharmaceutical preparations. SUTHERLAND[234] has shown that temperature cycling can change the habit of sucrose crystals, but this effect arises from the non-reciprocal nature of the growth and dissolution processes.

It has often been suggested that the ideal size–frequency distribution of a precipitate should be the log-normal distribution (p. 391). Such a development can be demonstrated by assuming a normal (Gaussian) distribution of nucleation times accompanied with a growth rate directly proportional to the crystal size. The normal distribution of times is transformed into a log-normal distribution of sizes[235].

Much of this theoretical work has been reviewed critically by BERRY and SKILLMAN[236], who also reported an experimental study of the growth of silver bromide micro-crystals. They could not detect any normal–log-normal transformation, and they concluded that the log-normal distribution of crystal sizes cannot be considered to have a fundamental basis, at least when applied to the nucleation and growth of silver halide crystals.

AGGLOMERATION AND PRECIPITATE MORPHOLOGY

Precipitates do not usually consist entirely of discrete crystalline particles. Agglomerated crystals are frequently present, and may even predominate, and characterisation of the particle size distribution under these circumstances becomes difficult. Crystalline particles can adhere and grow together in many different ways and for a wide variety of reasons.

Interparticle collision may result in permanent attachment if the particles are small enough for the van der Waals' forces to exceed the gravitational forces; this condition does not generally occur for sizes >1 µm. From the work of SMOLUCHOWSKI[237] it can be deduced that the half-time, t^*, the time needed to halve the number of particles in a monodisperse system, may be expressed as

$$t^* = \left(\frac{n_t}{n_0 - n_t} \right) t \qquad (6.61)$$

where n_t and n_0 are the number of particles present at time t and in the original monodispersion ($t = 0$), respectively.

For a binary collision process in a water-dispersed phase, in which all collisions are effective, WALTON[20] suggests that $n_0 t^* \sim 2 \times 10^{11}$. If an agglomerated system were arbitrarily defined as one in which $>10\%$ of the particles have agglomerated in <1000 s, then aqueous systems containing $<10^7$ particles/cm³ can be said to be non-agglomerating. The rather interesting conclusion to be drawn from this is that agglomeration may not be expected in systems where heteronucleation occurs (the number of hetero-nuclei is generally $< 10^7 \, cm^{-3}$), but for homogeneous nucleation, where n can greatly exceed $10^7 \, cm^{-3}$, agglomeration may be expected. Of course, not all interparticle collisions result in permanent contact, and in lyophobic systems charge stabilisation greatly decreases the rate of agglomeration

Probably the most important factor influencing precipitate morphology is supersaturation. Well-formed crystals are generally produced only at low supersaturation. Dendritic growth is often favoured at high supersaturation, while at very high supersaturations homogeneous nucleation results in the production of large numbers of particles and, hence, a very fine precipitate. Agglomeration may occur under these conditions, but the particles are often so small and numerous that the suspension can approach that of a colloidal dispersion.

If often appears that, for a given system, there is a certain level of super-saturation for maximum agglomeration. It is as if under these conditions the nuclei become 'sticky'. It is interesting to speculate on the link between this behaviour and the character of the adsorption layer surrounding a growing crystal. At very low supersaturations the adsorption layer or 'third phase' will be thin and strongly attracted to the crystal face, and the growth units will be readily incorporated at the growth sites. At very high supersaturations the adsorption layer will be thick and the outer layers of growth units will only be loosely attracted to the parent crystal. Under these conditions fluid shearing, as the crystal moves through the solution, can sweep portions of the loosely bound growth units into the solution, where, because they are already partially integrated, they become nuclei. Indeed this is one of the possible mechanisms of secondary nucleation (p. 180). However, at some inter-mediate level of supersaturation the adsorption layer may be neither 'thick' nor 'thin' and contact between two crystals in the same condition could rapidly result in a permanent bond through their integrated adsorption layers.

The influence of supersaturation on the particle size or characteristics of a precipitate has been summed up in one of the so-called laws of precipitation proposed by WEIMARN[236]: as the supersaturation is increased, the mean size of the precipitate (measured at a given time after the mixing of the reactants so long as this time is less than that required for complete precipitation) increases to a maximum and then decreases.

REFERENCES FOR CHAPTERS 5 AND 6

1. BUCKLEY, H.E., *Crystal Growth*, 1952. London; Chapman and Hall
2. BLACKADDER, D. A., *Report on Nucleation and Crystal Growth*, 1964. London; Instn chem. Engrs
3. BRICE, J. C., *Growth of Crystals from the Melt*, 1965. Amsterdam; North-Holland
4. FINE, M. E., *Introduction to Transformations in Condensed Systems*, 1964. New York; Macmillan
5. CHALMERS, B., *Principles of Solidification*, 1964. New York; Wiley
6. DUNNING, W. J., *General and Theoretical Introduction to Nucleation*. pp. 1–67 in reference 21

7. HIRTH, J. P. and POUND, G. M., *Condensation and Evaporation*, 1963. Oxford; Pergamon
8. JACKSON, K. A., *Solidification and Crystal Growth*, 1968. New York; Macmillan
9. KHAMSKII, E. V., *Crystallisation from Solutions*, 1969. New York; Consultants Bureau
10. KNIGHT, C. A., *The Freezing of Supercooled Liquids*, 1967. Princeton; Van Nostrand
11. MATUSEVICH, L. N., *Crystallisation from Solutions* (in Russian), 1968. Moscow
12. MATZ, G., *Kristallization*, 2nd Ed., 1969. Berlin; Springer-Verlag
13. NANCOLLAS, G. H. and PURDIE, N., 'Review of nucleation phenomena', *Q. Rev. chem. Soc.*, **18** (1964), 1
14. NIELSEN, A. E., *Kinetics of Precipitation*, 1964. Oxford; Pergamon
15. NÝVLT, J., *Industrial Crystallisation from Solutions*, 1971. London; Butterworths
16. SARATOVKIN, D. D., *Dendritic Crystallisation*, 1959. New York; Consultants Bureau
17. STRICKLAND-CONSTABLE, R. F., *Kinetics and Mechanism of Crystallisation*, 1968. London; Academic Press
18. UBBELOHDE, A. R., *Melting and Crystal Structure*, 1965. Oxford University Press
19. VAN HOOK, A., *Crystallisation: Theory and Practice*, 1961. New York; Reinhold
20. WALTON, A. G., *The Formation and Properties of Precipitates*, 1967. New York, Interscience
21. ZETTLEMOYER, A. C. (Ed.), *Nucleation*, 1969. New York; Dekker
22. ZIEF, M. and WILCOX, W. R. (Eds.), *Fractional Solidification*, Vol. I, 1967. New York; Dekker
23. MULLIN, J. W., 'Recent advances in crystallisation', *Br. chem. Engng*, **7** (1962), 12; **9** (1964), 438
24. 'Crystallisation (Annual Reviews)' (various authors), *Ind. Engng Chem.*, **56** (1964) (10), 38; **57** (1965) (11), 69; **58** (1966) (11), 67; **60** (1968) (4), 65; **61** (1969) (10), 86; **61** (11), 93; **61** (12), 65; **62** (1970) (11), 52; **62** (12), 148
25. 'Symposium on Crystal Growth', *Discuss. Faraday Soc.*, No. 5 (1949)
26. 'Symposium on Nucleation Phenomena', *Ind. Engng Chem.*, **44** (1952), 1269
27. R. H. DOREMUS, B. W. ROBERTS and D. TURNBULL (Eds.), *Growth and Perfection of Crystals* (Symposium proceedings), 1958. New York; Wiley
28. J. J. GILMAN (Ed.), *The Art and Science of Growing Crystals* (Collection of papers), 1963. New York; Wiley
29. R. KERN (Ed.), *Adsorption et Croissance Cristalline* (Symposium proceedings), Colloq. No. 152, 1965. Paris; C.N.R.S.
30. A. V. SHUBNIKOV and N. N. SHEFTAL (Eds.), *Growth of Crystals* (Symposium Proceedings and Collections of Papers), Vol. 1 (1958); Vol. 2 (1959); Vol. 3 (1962); Vol. 4 (1966); Vol. 5 (1968); Vol. 6 (1968); Vol. 7 (1969); Vol. 8 (1969)
31. N. N. SIROTTA, F. K. GORSKII and V. M. VARIKASH (Eds.), *Crystallisation Processes* Symposium Proceedings and Collections of Papers), 1966. New York; Consultants Bureau
32. H. S. PEISER (Ed.), *Crystal Growth* (1st I.C.C.G., Boston), 1967. New York; Pergamon. Also issued as *J. Phys. Chem. Solids*, Suppl. 1 (1967)
33. F. C. FRANK, J. B. MULLIN and H. S. PEISER (Eds.), *Crystal Growth* 2nd I.C.C.G., Birmingham), 1968. Amsterdam, North-Holland. Also issued as *J. Crystal Growth*, **3/4** (1968)
34. J. B. MULLIN, B. MUTAFTSCHIEV, H. PEISER (Eds.), *Crystal Growth* (3rd I.C.C.G., Marseilles), 1971. Amsterdam, North-Holland. Also issued as *J. Crystal Growth*, (1972)
35. J. A. PALERMO and M. A. LARSON (Eds.), 'Crystallisation from Solutions and Melts', *Chem. Engng Prog. Symp. Ser.*, **65**, No. 95 (1969)
36. *Industrial Crystallisation* (Symposium Proceedings), 1969. London; Instn. chem. Engrs
37. 'Crystallisation from solution: factors influencing distribution size', *Chem. Engng Prog. Symp. Ser.*, **67**, No. 110 (1971)
38. 'Nucleation Phenomena in Systems of Growing Crystals' (A.I.Ch.E. Symposium, Denver, 1970). *Chem. Engng Prog. Symp. Ser.*, (to be published)
39. GIBBS, J. W., *Collected Works*, 1928, London; Longmans Green
40. VOLMER, M., *Kinetik der Phasenbildung*, 1939. Dresden and Leipzig; Steinkopff
41. BECKER, R. VON and DÖRING, W., 'Kinetische Behandlung der Keimbildung in übersättigen Dämpfen', *Annln Phys.*, **24** (1935), 719
42. LA MER, V. K., 'Nucleation in phase transitions', p. 1270 in reference 26
43. TURNBULL, D. and FISHER, J. C., 'Rate of nucleation in condensed systems', *J. chem. Phys.*, **17** (1949), 71
44. TURNBULL, D., 'Phase changes', *Solid St. Phys.*, **3** (1956), 225

45. JACKSON, K. A., 'Nucleation from the melt', *Ind. Engng Chem.*, **57** (12) (1965), 28
46. TURNBULL, D. and CORMIA, R. L., 'Kinetics of nucleation in some normal alkane liquids', *J. chem. Phys.*, **34** (1961), 820
47. PHIPPS, L. W., 'Heterogeneous and homogeneous nucleation in supercooled triglycerides and n-paraffins', *Trans. Faraday Soc.*, **60** (1964), 1873
48. TAMMAN. G., *States of Aggregation* (transl. R. F. Mehl), 1925. New York; van Nostrand. (1926. London; Constable)
49. MULLIN, J. W. and LECI, C. L., 'Some nucleation characteristics of aqueous citric acid solutions', *J. Crystal Growth*, **5** (1969), 75
50. STRANSKII, I. N. and KAISHEV, R., *Usp. Fiz. Nauk*, **21** (1939), 408
51. CHRISTIANSEN, J. A. C. and NIELSEN, A. E., *Acta. chem. scand.*, **5** (1951), 673
52. MASON, B. J., *The Physics of Clouds*, 1957. Oxford; Clarendon
53. EDWARDS, G. R. and EVANS, L. F., 'Effective surface charge on ice nucleation by AgI', *Trans. Faraday Soc.*, **58** (1962), 1649
54. TURNBULL, D., 'Kinetics of heterogeneous nucleation', *J. chem. Phys.*, **18** (1950), 198
55. TURNBULL, D. and VONNEGUT, B., 'Nucleation catalysts', p. 1292 in reference 26
56. FLETCHER, N. H., 'Nucleation by crystalline particles', *J. chem. Phys.*, **38** (1963), 237
57. CAHN, J. W. and HILLIARD, J. E., 'Nucleation in a two-component incompressible fluid', *J. chem. Phys.*, **31** (1959), 688
58. WELLS, A. F., 'Crystal growth', *Ann. Rep. chem. Soc.*, **43** (1946), 62
59. CURIE, P., *Bull. Soc. fr. Minér.*, **8** (1885), 145
60. WULFF, G., *Z. Kristallogr.*, **34** (1901), 449
61. LAUE, M. VON, *Z. Kristallogr.*, **105** (1943), 124
62. HERRING, C., 'Some theorems on the free energy of crystal surfaces', *Phys. Rev.*, **82** (1951), 87
63. BRANDES, H., 'Zur Theorie des Kristallwachstums', *Z. phys. Chem.*, **126** (1927), 196
64. STRANSKII, I. N., 'Zur Theorie des Kristallwachstums', *Z. phys. Chem.*, **136** (1928), 259
65. KOSSEL, W., 'Zur Energetik von Oberflächenvorgängen', *Annln Phys.*, **21** (1934), 457
66. VOLMER, M. and SCHULTZ, W., 'Kondensation an Kristallen', *Z. phys. Chem.*, **156** (1931), 1
67. FRANK, F. C., 'The influence of dislocations on crystal growth', p. 48 in reference 25
68. VERMA, A. R., *Crystal Growth and Dislocations*, 1953. London; Butterworths
69. READ, W. T., *Dislocations in Crystals*, 1953. New York; McGraw-Hill
70. BURTON, W. K., CABRERA, N., and FRANK, F. C., 'The growth of crystals and the equilibrium structure of their surfaces', *Phil. Trans.*, **A243** (1951), 299
71. CHERNOV, A. A., 'The spiral growth of crystals', *Soviet Phys. Usp.*, **4** (1961), 116
72. BENNEMA, P. and HANEVELD, H. B. K., *J. Crystal Growth*, **1** (1967), 225, 232
73. BENNEMA, P., *J. Crystal Growth*, **1** (1967), 278, 287; **3/4** (1968), 331; **5** (1969), 29
74. GILMER, G. H., GHEZ, R., and CABRERA, N., 'An analysis of combined surface and volume diffusion processes in crystal growth', *J. Crystal Growth*, **8** (1971), 79
75. ALBON, N., 'The origin of dislocations during crystal growth', *Phil. Mag.*, **8** (1963), 1335
76. ALBON, N. and DUNNING, W. J., 'Studies on the behaviour of growth steps on sucrose crystals', *Acta. Cryst.*, **13** (1960), 495
77. FRANK, F. C., 'Kinematic theory of crystal growth and dissolution processes', p. 411 in reference 27
78. CABRERA, N. and VERMILYEA, D. A., 'Growth of crystals from solution', p. 393 in reference 27
79. CABRERA, N. and COLEMAN, R. V., 'Theory of crystal growth from the vapour', p. 3 in reference 28
80. SCHWOEBEL, E. L. and SHIPSEY, E. J., 'Step motions on crystal surfaces', *J. appl. Phys.*, **37** (1966), 3682
81. NOYES, A. A., and WHITNEY, W. R., 'Rate of solution of solid substances in their own solution', *J. Am. chem. Soc.*, **19** (1897), 930; *Z. phys. Chem.*, **23** (1897), 689
82. NERNST, W., 'Theorie der Reaktionsgeschwindigkeit in heterogenen Systemen', *Z. phys. Chem.*, **47** (1904), 52
83. MARC, R., ' Über die Kristallisation aus wässerigen Lösungen', *Z. phys. Chem.*, **61** (1908), 385; **67** (1909), 470; **68** (1909), 104; **73** (1910), 685
84. MIERS, H. A., 'The concentration of the solution in contact with a growing crystal', *Phil. Trans.*, **A202** (1904), 492
85. BERTHOUD, A., 'Theorie de la formation des faces d'un crystal', *J. chim. Phys.*, **10** (1912), 624

86. VALETON, J. J. P., 'Wachstum und Auflösung der Kristalle', Z. Kristallogr., 59 (1923), 135, 335; 60 (1924), 1

87. MULLIN, J. W. and GASKA, C., 'Growth and dissolution of potassium sulphate crystals in a fluidised bed', Can. J. chem. Engng, 47 (1969), 483

88. GARSIDE, J. and MULLIN, J. W., 'Growth and dissolution rates of potash alum crystals', Trans. Instn chem Engrs, 46 (1968), 11

89. GARSIDE, J., 'The concept of effectiveness factors in crystal growth', Chem. Engng Sci., 26 (1971) 1425

90. BENNETT, J. A. R. and LEWIS, J. B., 'Dissolution of solids in mercury and aqueous liquids; development of a new type of rotating dissolution cell', A.I.Ch.E.Jl., 4 (1958), 418

91. BOVINGTON, C. H. and JONES, A. L., 'Tracer studies of the kinetics of dissolution of lead sulphate', Trans. Faraday Soc., 66 (1970), 2088

92. BUNN, C. W., 'Concentration gradients and the rates of growth of crystals', p. 132 in reference 25

93. BERG, W. F., 'Crystal growth from solutions', Proc. R. Soc., A164 (1938), 79

94. BRICE, J. C., 'Kinetics of crystal growth from solution', J. Crystal Growth, 1 (1967) 218

95. NIELSEN, A. E., 'Nucleation in aqueous solutions', p. 419 in reference 32. See also Krist. Tech., 4 (1969), 17

96. TURNBULL, D., 'Kinetics of precipitation of $BaSO_4$', Acta. Met., 1 (1953), 684

97. MATZ, G., 'Mittlere Kristallwachstumsgeschwindigkeit des Einzelkorns', Chemie-Ingr-Tech., 42 (18) (1970), 1134

98. PALERMO, J. A., 'Studies of crystallisation and melting processes in small equipment', Chem. Engng Prog. Symp. Ser., 63, No. 70 (1967), 21

99. KIRWAN, D. J. and PIGFORD, R. L., 'Crystallisation kinetics of pure and binary melts', A.I.Ch.E.Jl., 15 (1969), 442

100. HARTMAN, P. and PERDOK, W. G., Acta. Cryst., 8 (1955), 49

101. HARTMAN, P., 'The dependence of crystal morphology on crystal structure', p. 3 in Vol. 7 of reference 30. See also Z. Kristallogr., 119 (1963), 65

102. BENNEMA, P., 'The importance of surface diffusion for crystal growth from solution', J. Crystal Growth, 5 (1969), 29

103. TROOST, S., 'Crystallisation of sodium triphosphate hexahydrate', Doctorate Thesis, University of Groningen, 1969

104. SIMON, B. and BIENFAIT, M., 'Structure et mécanisme de croissance du gypse', Acta. Cryst., 19 (1965), 750

105. YOUNG, S. W., 'Mechanical stimulus to crystallisation in supercooled liquids', J. Am. chem. Soc., 33 (1911), 148

106. BERKELEY, The Earl of, 'Solubility and supersolubility from the osmotic standpoint', Phil. Mag., 24 (1912), 254

107. TIPSON, R. S., 'Crystallisation and recrystallisation', in Techniques of Organic Chemistry, Vol. 3 (ed. A. WEISSBERGER), 1950. New York; Interscience

108. KAPUSTIN, A. P., The Effects of Ultrasound on the Kinetics of Crystallisation, 1963. New York, Consultants Bureau

109. HUNT, J. D. and JACKSON, K. A., 'Nucleation of solid in an undercooled liquid by cavitation', J. appl. Phys., 37 (1966), 254

110. FISHER, J. C., HOLLOMAN, J. H. and TURNBULL, D., 'Rate of nucleation of solid particles in a sub-cooled liquid', Science, N. Y., 109 (1949), 168

111. MULLIN, J. W. and RAVEN, K. D., 'Influence of mechanical agitation on the nucleation of aqueous salt solutions', Nature, Lond., 190 (1961), 251; 195 (1962), 35

112. MULLIN, J. W. and LECI, C. L., 'Evidence for molecular cluster formation in supersaturated solutions of citric acid', Phil. Mag., 19 (1969), 1075

113. TIKHOMIROFF, N., 'Association moléculaire au cours de la periode de precristallisation des solutions aqueuses sursaturées de saccharose', Industr. Aliment. Agric., 82 (1965), 755

114. SVORONOS, D. R., 'Recherches sur la stabilité des solutions salines sursaturée', Revue der Chimie minérale, 5 (1968), 59

115. VONNEGUT, B., 'Variation with temperature of the nucleation rate of supercooled liquid tin and water drops', J. Colloid Sci., 3 (1948), 563

116. POUND, G. M. and LA MER, V. K., 'Kinetics of crystalline nucleus formation in supercooled liquid tin', J. Am. chem Soc., 74 (1952), 2323

17. TURNBULL, D., 'Kinetics of solidification of supercooled mercury droplets', *J. chem. Phys.*, **20** (1952), 411

18. BIGG, E. K., 'The supercooling of water', *Proc. phys. Soc.*, **66B** (1953), 688

19. WHITE, M. L. and FROST, A. A., 'The rate of nucleation of supersaturated KNO_3 solutions', *J. Colloid Sci.*, **14** (1959), 247

20. MELIA, T. P., and MOFFITT, W. P., 'Secondary nucleation from aqueous solution', *Ind. Engng Chem. Fundam.*, **3** (1964), 313

21. PRICE, F. P. and GORNICK, F., 'Estimation of nucleation parameters from continuous cooling droplet experiments', *J. appl. Phys.*, **38** (1967), 3883

22. GARTEN, V. A. and HEAD, R. B., 'Crystalloluminescence and the nature of the critical nucleus', *Phil. Mag.*, **8** (1963), 1793; **14** (1966), 1243

23. OTPUSHCHENNIKOV, N. F., *Soviet Phys. Crystallogr.*, **7** (1962), 237

24. ADAMSKI, T., 'Commination of crystal nucleation by a precipitation method', *Nature, Lond.*, **197** (1963), 894

25. NÝVLT, J., 'Kinetics of nucleation in solutions', *J. Crystal Growth*, **3/4** (1968), 377

26. MULLIN, J. W., CHAKRABORTY, M. and MEHTA, K., 'Nucleation and growth of ammonium sulphate crystals', *J. appl. Chem.*, **20** (1970), 367

27. TING, H. H. and MCCABE, W. L., 'Supersaturation and crystal formation in seeded solutions of $MgSO_4$', *Ind. Engng Chem.*, **26** (1934), 1201, 1207

28. GYULAI, Z., 'Betrage zur Kenntnis der Kristallwachstumsvorgänge', *Z. Phys.*, **125** (1948), 1

29. POWERS, H. E. C., 'Nucleation and early crystal growth', *Ind. Chemist*, **39** (1963), 351

30. WALTON, A. G., 'Nucleation of crystals from solution', *Science, N. Y.*, **148** (1965), 601

31. CAYEY, N. W. and ESTRIN, J., 'Secondary nucleation in agitated $MgSO_4$ solutions', *Ind. Engng Chem., Fundam.*, **6** (1967), 13

132. MCCABE, W. L. and STEVENS, R. P., 'Rate of growth of crystals in aqueous solution', *Chem. Engng Prog.*, **47** (1951), 168

133. CLONTZ, N. A. and MCCABE, W. L., 'Contact nucleation of $MgSO_4 \cdot 7H_2O$', p. 6 in reference 37

134. MASON, R. E. A. and STRICKLAND-CONSTABLE, R. F., 'Breeding of crystal nuclei', *Trans. Faraday Soc.*, **62** (1966), 455

135. LAL, D. P., MASON, R. E. A. and STRICKLAND-CONSTABLE, R. F., 'Collision breeding of nuclei', *J. Crystal Growth*, **5** (1969), 1

136. CHRETIEN, A. and HEUBEL, J., *C. r. hebd. Séanc. Acad. Sci., Paris*, **239** (1954), 814; **242** (1956), 2837

137. HEUBEL, J. and DEVRAINE, P., *C. r. hebd. Séanc. Acad. Sci., Paris*, **252** (1961), 1158; **254** (1962), 290; **256** (1963), 2393

138. GINDT, R. and KERN, R., 'Temps de latence, germination et croissance dans solutions des halogenures alcalins', *C. r. hebd. Séanc. Acad. Sci., Paris*, **256** (1963), 4186, 4400; **258** (1964), 5011. 'Germination en milieu condensé', *Ber. Bunsenges phys. Chem.*, **69** (1965), 124; **72** (1968), 460

139. MITSUDA, H., MIYAKE, K. and NAKAI, T., 'Experimental study of the formation of seed crystals', *Int. chem. Engng*, **8** (1968), 733

140. BRANSOM, S. H., BROWN, D. E., and HEELEY, G. P., 'Crystallisation studies in continuous flow stirred tank crystallisers', pp. 24 and 43 in reference 36

141. BELYUSTIN, A. V. and ROGACHEVA, E. D., 'Production of nuclei in the presence of seed crystals', p. 3 in Vol. 4 of reference 30

141a. DENK, E. G. and BOTSARIS, G. D., 'Fundamental studies in secondary nucleation', in reference 34

142. MULLIN, J. W. and LECI, C. L., 'Desupersaturation of seeded citric acid situations in a stirred vessel', in reference 38

143. LARSON, M. A., TIMM, D. C. and WOLFF, P. R., 'Effect of suspension density on crystal size distribution', *A.I.Ch.E.Jl.*, **14** (1968), 448

144. AYERST, R. P. and PHILLIPS, M. I., 'Crystallisation from agitated ammonium perchlorate solutions', p. 56 in reference 36

145. COOKE, E. G., 'Influence of impurities on the growth and nucleation of NaCl', *Krist. Tech.*, **1** (1966), 119

146. BOTSARIS, G. D., DENK, E. G., and CHUA, J. O., 'Secondary nucleation in an impurity gradient', in reference 38

147. SVORONOS, D. R., 'Sur la cristallisation du KCl en solution aqueuse', *C. r. hebd. Séanc. Acad. Sci., Paris*, **269C** (1969), 133

148. SHOR, S. M. and LARSON, M. A., 'Effect of additives on crystallisation kinetics', p. 32 in reference 37

149. TROOST, S., 'Influence of surfactants on the crystal growth of sodium triphosphate', p. 340 in reference 33

150. MULLIN, J. W., 'Effect of impurities on the nucleation of $(NH_4)_2SO_4$', to be published

151. CARTIER, R., PINDZOLA, D. and BRUINS, P. F., 'Particle integration rate in crystal growth', Ind. Engng Chem., 51 (1959), 1409

152. MICHAELS, A. S. and COLVILLE, A. R., 'The effect of surface active agents on crystal growth rate and crystal habit', J. phys. Chem., 64 (1960), 13

153. MULLIN, J. W. and AMATAVIVADHANA, A., 'Growth kinetics of ammonium and potassium dihydrogenphosphate crystals', J. appl. Chem., 17 (1967), 151

154. MULLIN, J. W. and GARSIDE, J., 'Crystallisation of aluminium potassium sulphate. I–Single crystals', Trans. Instn chem. Engrs, 45 (1967), 285

155. KREUGER, G. C. and MILLER, C. W., 'A study of the mechanism of crystal growth from a supersaturated situation', J. chem. Phys., 21 (1953), 2018

156. HUMPHREYS-OWEN, S. P. F., 'Crystal growth from solution', Proc. R. Soc., A197 (1949), 218

157. SMYTHE, B. M., 'Sucrose crystal growth', Aust. J. Chem., 20 (1967), 1087, 1097, 1115

158. BENNEMA, P., 'Techniques for measuring the rate of growth of crystals from solution', Phys. Status Solidi, 17 (1966), 555

159. BOTSARIS, G. D. and DENK, E. G., 'Growth rates of aluminium potassium sulphate crystals in aqueous solution', Ind. Engng Chem., Fundam., 9 (1970), 276; see also J. Crystal Growth, 6 (1970), 241

160. TORGESEN, J. L., HORTON, A. T. and SAYLOR, C. P., 'Equipment for single crystal growth from aqueous solution', J. Res. natn Bur. Stand., 67C (1963), 25

161. HORTON, A. T. and GLASGOW, A. R., 'Equipment for single crystal growth from the melt available for substances with a low melting point', J. Res. natn Bur. Stand., 69C (1965), 195

162. KITCHENER, S. A. and STRICKLAND-CONSTABLE, R. F., 'Growth and evaporation of crystals from the vapour phase', Proc. R. Soc., A245 (1958), 93

163. LARSON, M. A., 'Apparatus for the study of crystallisation kinetics', Chem. Engng Prog. Symp. Ser., 63, No. 70 (1967), 58

164. RANDOLPH, A. D. and RAJAGOPAL, K., 'Direct measurement of K_2SO_4 nucleation and growth kinetics in a backmixed crystal slurry', Ind. Engng Chem., Fundam., 9 (1970), 165

165. MULLIN, J. W. and GARSIDE, J., 'Crystallisation of aluminium potassium sulphate. II–In a fluidised bed', Trans. Instn chem. Engrs, 45 (1967), 291

166. BRANSOM, S. H. and PALMER, A. G. C., 'An experimental Oslo crystalliser', Br. chem. Engng, 9 (1964), 672

167. ROSEN, H. N. and HULBURT, H. M., 'Growth rate of K_2SO_4 in a fluidised bed crystalliser', p. 27 in reference 37

168. GLASBY, J. and RIDGWAY, K., 'Crystallisation of aspirin from ethanol', J. Pharm. Pharmac., Suppl., 20 (1968), 94S

169. BENNETT, M. C. and FENTIMAN, Y. L., 'Superheating effects on sucrose crystallisation under ebullient conditions', Int. Sugar J., 70 (1968), 9, 36. See also p. 217 in reference 36

170. RUMFORD, F. and BAIN, J., 'The controlled crystallisation of sodium chloride', Trans. Instn chem. Engrs, 38 (1960), 10

171. AYERST, R. P. and PHILLIPS, M. I., 'A pilot-scale forced-circulation evaporating crystalliser for ammonium perchlorate', p. 177 in reference 36

172. GARSIDE, J. and MULLIN, J. W., 'Continuous measurement of solution concentration in a crystalliser', Chemy Ind., 1966, 2007

173. BUJAC, P. D. B. and MULLIN, J. W., 'A rapid method for the measurement of crystal growth rates in a fluidised bed crystalliser', p. 121 in reference 36

174. BRANSOM, S. H., 'Factors in the design of continuous crystallisers', Br. chem. Engng, 5 (1960), 838

175. VAN HOOK, A., 'Nucleation in supersaturated sucrose solutions', in Principles of Sugar Technology, Vol. 2 (ed. P. HONIG), 1959. Amsterdam; Elsevier

176. BUNN, C. W., 'Adsorption, oriented overgrowth and mixed crystal formation', Proc. R. Soc., 141 (1933), 567

177. MULLIN, J. W., AMATAVIVADHANA, A. and CHAKRABORTY, M., 'Crystal habit modification studies with ADP and KDP', *J. appl. Chem.*, **20** (1970), 153
178. KERN, R., 'Croissance cristalline et adsorption', *Bull. Soc. fr. Minér. Cristallogr.*, **91** (1968), 247. See also p. 3 in Vol. 8 of reference 30 (in English)
179. BIENFAIT, M., BOISTELLE, R. and KERN, R., 'Les morphodromes de NaCl en solution et l'adsorption d'ions etrangers', p. 577 in reference 29
180. BLITZNAKOV, G., 'Le mechanism de l'action des additives adsorbants dans la croissance cristalline', p. 283 in reference 29
181. BLITZNAKOV, G. and KIRKOVA, E., 'Effect of SO_4^{2+} on the growth of $NaClO_3$', *Krist. Tech.*, **4** (1969), 331
182. BOTSARIS, G. D., MASON, E. A. and REID, R. C., 'Incorporation of ionic impurities in crystals growing from solution (Pb^{2+} in KCl)', *A.I.Ch.E.Jl.*, **13** (1967), 764
183. SEARS, G. W., 'The effect of poisons on crystal growth', p. 441 in reference 27
184. CHERNOV, A. A., 'Some aspects of the theory on crystal growth forms in the presence of impurities', p. 265 in reference 29
185. LACMANN, R. and STRANSKII, I. N., 'The effect of adsorption of impurities on the equilibrium and growth forms of crystals', p. 427 in reference 27
186. ROWE, P. N. and CLAXTON, K. T., 'Heat and mass transfer from a single sphere to fluid flowing through an array', *Trans. Instn chem. Engrs*, **43** (1965), 321
187. CHU, J. C., KALIL, J. and WETTEROTH, W. A., 'Mass transfer in a fluidised bed', *Chem. Engng Prog.*, **49** (1953), 141
188. MULLIN, J. W. and TRELEAVEN, C. R., 'Solid–liquid mass transfer in multiparticulate systems', in *Instn Chem. Engrs Symposium (Interaction between Fluids and Particles)*, (1962), 203
189. HIXSON, A. W. and KNOX, K. L., 'Effect of agitation on the rate of growth of single crystals', *Ind. Engng Chem.*, **43** (1951), 2144
190. COULSON, J. M. and RICHARDSON, J. F., *Chemical Engineering*, Vol. 1. 2nd Ed., 1964. Oxford; Pergamon. New York; Macmillan
191. NIENOW, A. W., UNAHABHOKHA, R. and MULLIN, J. W., 'The mass transfer driving force for high mass flux', *Chem. Engng Sci.*, **24** (1969), 1655
192. ROSNER, D. E., 'Lifetime of a highly soluble dense spherical particle', *J. chem. Phys.*, **73** (1969), 382
193. ROWE, P. N., CLAXTON, K. T. and LEWIS, J. B., 'Heat and mass transfer from a single sphere in an extension flowing fluid', *Trans. Instn Chem. Engrs*, **43** (1965), 14
194. NIENOW, A. W., UNAHABHOKHA, R. and MULLIN, J. W., 'Concentration-dependent physical properties and rates of mass transport', *Ind. Engng Chem., Fundam.*, **5** (1966), 578
195. NIENOW, A. W., 'Diffusivity in the liquid phase', *Br. chem. Engng*, **10** (1965), 827
196. NIENOW, A. W., UNAHABHOKHA, R. and MULLIN, J. W., 'Diffusion and mass transfer of NH_4Cl and KCl in aqueous solution', *J. appl. Chem.*, **18** (1968), 154
197. NIENOW, A. W., 'Dissolution mass transfer in a turbine agitated baffled vessel', *Can. J. chem. Engng*, **47** (1969), 248
198. HIXSON, A. W. and BAUM, S. J., 'Agitation: heat and mass transfer coefficients in liquid–solid systems', *Ind. Engng Chem.*, **33** (1941), 478, 1433
199. CALDERBANK, P. and MOO-YOUNG, M. B., 'The continuous phase heat and mass transfer properties of dispersions', *Chem. Engng Sci.*, **16** (1961), 39
200. HARRIOTT, P., 'Mass transfer to particles in agitated tanks', *A.I.Ch.E.Jl.*, **8** (1962), 93
201. KOLMOGOROFF, A. N., *C. r. Acad. Sci. U.R.S.S.*, **30** (1941), 301
202. KOLAR, V., *Colln Czech. Chem. Commun.*, **23** (1958), 1680; **24** (1959), 301
203. SHINNAR, R. and CHURCH, J. M., 'Predicting particle size in agitated dispersions', *Ind. Engng Chem.*, **52** (1960), 253
204. LEVICH, V., *Physico-Chemical Hydrodynamics*, 1962. New York; Prentice-Hall
205. MIDDLEMAN, S., 'Mass transfer from particles in agitated systems: applications of Kolmogoroff theory', *A.I.Ch.E.Jl.*, **11** (1965), 750
206. HUGHMARK, G. A., 'Mass transfer for suspended solid particles in agitated liquids', *Chem. Engng Sci.*, **24** (1969), 291
207. NIENOW, A. W., BUJAC, P. D. B. and MULLIN, J. W., 'Slip velocities in agitated vessel crystallisers', p. 488 in reference 34
208. MICHAELS, A. S. and COLVILLE, A. R., 'The effect of surface active agents on crystal growth rate and crystal habit', *J. phys. Chem.*, **64** (1960), 13

209. MULLIN, J. W., 'The role of laboratory studies in the design of industrial crystallisers', in *Instn. Chem. Engrs, Pilot Plant Symposium*, Manchester (1969)
210. GARRETT, D. E., 'Industrial crystallisation: influence of chemical environment', *Br. chem. Engng,* 4 (1959), 673
211. SVANOE, H., 'Solids recovery by crystallisation', *Chem. Engng Prog.,* 55 (5) (1959), 47
212. PHOENIX, L., 'How trace additives inhibit the caking of inorganic salts', *Br. chem. Engng,* 11 (1966), 34
213. VAN DAMME-VAN WEELE, M. A., 'Influence of ferrocyanide complexes on the dissolution of NaCl', p. 433 in reference 29
214. EDWARDS, R. T., 'Crystal habit of paraffin wax', *Ind. Engng Chem.,* 49 (1957), 750
215. CHICHAKALI, M. and JESSEN, F. W., 'Crystal morphology in hydrocarbon systems', *Ind. Engng Chem.,* 59 (5) (1967), 86
216. BLATCHLY, J. M. and HARTSHORNE, N. H., 'Curling of 2·5 di·t·butyl 1·4 dimethoxybenzene crystals during growth', *Trans Faraday Soc.,* 62 (1966), 512
217. GARRETT, D. E., 'Overgrowth during industrial crystallisation', *Chem. Engng Prog.,* 62 (12) (1966), 74
218. POWERS, H. E. C., 'Sucrose crystals: inclusions and structure', *Sugar Technol. Rev.,* 1 (1969/70), 85
219. DENBIGH, K. G. and WHITE, E. T., 'Studies on liquid inclusions in crystals', *Chem. Engng Sci.,* 21 (1966), 739
220. PETROV, T. G., 'Growth of KNO_3 crystals and the formation of inclusions', p. 107 in Vol. 3 of reference 30
221. BROOKS, R., HORTON, A. T. and TORGESEN, J. L., 'Occlusion of mother liquor in solution-growth crystals', *J. Crystal Growth,* 2 (1968), 279
222. BELYUSTIN, A. V. and FRIDMAN, S. S., 'Trapping of solution by a growing crystal', *Soviet Phys. Crystallogr.,* 13 (1968), 298
223. CHERNOV, A. A., 'Crystal growth forms and their kinetic stability', *Soviet Phys. Crystallogr.,* 7 (1963), 728; 8 (1964), 63, 401
224. WILCOX, W. R., 'Removing inclusions from crystals by gradient techniques', *Ind. Engng Chem.,* 60 (3) (1968), 13
225. DEICHA, G., *Lacunes des Cristeaux et leurs Inclusions Fluides,* 1955. Paris; Masson
226. *COFFI Proceedings* (ed. E. ROEDDER), 1968ff. U.S. Geol. Survey, Washington
227. GORDON, L., SALUTSKY, M. L. and WILLARD, H. H., *Precipitation From Homogeneous Solution,* 1959. New York; Wiley
228. CHEPELEVETSKII, M. L., p. 112 in reference 25
229. VAN HOOK, A., and BRUNO, A. J., 'Nucleation and growth in sucrose solutions', p. 112 in reference 25
230. FELBINGER, A. and NEELS, H., 'Die natriumbikarbonat kristallisation in ammoniak-soda-prozess', *Krist Tech.,* 1 (1966), 137
231. MULLIN, J. W. and OSMAN, M. M., 'Nucleation and precipitation of nickel ammonium sulphate from aqueous solution' (in press 1972)
232. HANITZSCH, E. and KAHLWEIT, M., 'Ageing of precipitates', p. 130 in reference 36
233. CARLESS, J. E. and FOSTER, A. A., 'Accelerated crystal growth of sulphathiazole by temperature cycling', *J. Pharm. Pharmac.,* 18 (1966), 697, 815
234. SUTHERLAND, D. N., 'Crystallisation-dissolution cycling of sucrose', *Chem. Engng Sci.,* 24 (1969), 192
235. KOTTLER, F., 'The distribution of particle sizes', *J. Franklin Inst.,* 250 (1950), 339 and 419
236. BERRY, C. R. and SKILLMAN, D. C., 'Ideal size distribution and growth rate in microcrystalline precipitates', *J. Chem. Phys.,* 67 (1963), 1827
237. SMOLUCHOWSKI, M. VON, 'Versuch einer mathematischen Theorie der Koagulations-kinetic kolloider Lösungen', *Z. Phys. Chem.,* 92 (1918), 129
238. WEIMARN, P. P. VON, 'The precipitation laws', *Chem. Rev.,* 2 (1926), 217

7

Crystallisation Techniques

A SINGLE crystallisation operation performed on a solution or melt often fails to produce a pure crystalline product. It may be that the impurity has solubility characteristics similar to those of the desired pure component, and both substances are co-precipitated. On the other hand, the impurity may be present in large amounts and the deposited crystals become contaminated. A pure substance cannot be produced from a melt in one crystallisation stage if the impurity and the required substance form a solid solution. The word 'pure' is used rather indiscriminately; no substance can be prepared absolutely pure, so the meaning of the word is somewhat arbitrary. For example, a chemist who requires a sample of a compound for the determination of its physical characteristics may demand a purity of 99·999 per cent, while an industrial chemist may reckon that he has produced a pure chemical, i.e. pure enough for his purpose, if a purity of 99 per cent has been achieved.

Recrystallisation, i.e. repeated crystallisation steps, from solutions or melts is an extremely useful and widely employed separation technique; if the operation is repeated often enough, very high degrees of purity can be attained. Recrystallisation often offers several advantages over the processes of extraction and distillation: a less complicated apparatus is usually required, and one crystallisation step can, in many cases, effect a much larger removal of impurity than one extraction or distillation step. Unfortunately, the inclusion of impure mother liquor by a crystalline mass can reduce the efficiency of a crystallisation process quite considerably. Many systems, such as azeotropes, close-boiling mixtures and heat-sensitive products which cannot be separated by distillation, can often be processed with ease by crystallisation. Another point to be borne in mind is that less heat energy is required for crystallisation than for distillation, because latent heats of fusion, solution and crystallisation are very much lower than latent heats of vaporisation, as indicated in *Table 7.1*.

RECRYSTALLISATION FROM SOLUTIONS

It is often possible to remove the impurities from a crystalline mass by dissolving the crystals in a small amount of fresh hot solvent and cooling the solution to produce a fresh crop of purer crystals, provided that the impurities

Table 7.1. ENERGIES OF CRYSTALLISATION AND DISTILLATION

Substance	Crystallisation		Distillation	
	Melting point, °C	Enthalpy of crystallisation, kJ/kg	Boiling point, °C	Enthalpy of vaporisation, kJ/kg
o-Cresol	31	115	191	410
m-Cresol	12	117	203	423
p-Cresol	35	110	202	435
o-Xylene	−25	128	141	347
m-Xylene	−48	109	139	343
p-Xylene	13	161	138	340
o-Nitrotoluene	−4·1	120	222	344
m-Nitrotoluene	15·5	109	233	364
p-Nitrotoluene	51·9	113	238	366
Benzene	5·4	126	80	394
Water	0	334	100	2260

are more soluble in the solvent than is the main product. This step may have to be repeated many times before a yield of crystals of the desired purity is obtained. Such an operation is called a simple recrystallisation. A typical scheme is shown in *Figure 7.1*. An impure crystalline mass *AB*

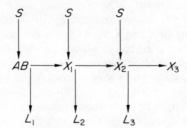

Figure 7.1. Simple recrystallisation

(*A* is the less soluble pure component, *B* the more soluble impurity) is dissolved in the minimum amount of hot solvent *S* and then cooled. The crop of crystals X_1 will contain less impurity *B* than the original mixture; but if the desired degree of purity has not been achieved, the procedure can be repeated: crystals X_1 are dissolved in more fresh solvent *S* and recrystallised to give a crop X_2, and so on.

In a sequence of operations of the above kind the losses of the desired component *A* can be considerable, and the final amount of 'pure' crystals may easily be a minute fraction of the starting mixture *AB*. This question of yield from recrystallisation processes is of paramount importance, and many schemes have been designed with the object of increasing both yield and separation efficiency. The choice of solvent depends on the nature of the required substance *A* and the impurity *B*. Ideally, *B* should be very soluble in the solvent at the lowest temperature employed, and *A* should have a high temperature coefficient of solubility so that high yields of *A* can be obtained from operation within a small temperature range. Some of the factors affecting the choice of a solvent are discussed in Chapter 2.

Figure 7.2 indicates a modification of the simple recrystallisation scheme; additions of the original impure mixture AB are made to the system at certain intervals. The first two stages are identical with those illustrated in *Figure 7.1*, but the crystals X_2 are set aside and fresh feedstock AB is dissolved by heating in mother liquor L_2. On cooling, a crop of crystals X_3 is obtained and a mother liquor L_3, which is discarded. Crystals X_3 are recrystallised

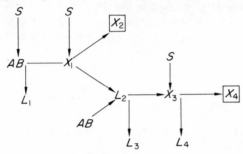

Figure 7.2. Simple recrystallisation with further additions of feedstock

from fresh solvent to give a crop X_4, which is set aside; the mother liquor L_4 can, if required, be used as a solvent from which more feedstock can be crystallised. If the crops X_2, X_4, X_6, etc., are still not pure enough, they can be bulked together and recrystallised from fresh solvent.

The triangular fractional recrystallisation scheme shown in *Figure 7.3* makes better use of the successive mother liquor fractions than does that of *Figure 7.2*. Again, A and B are taken to be the less and more soluble constituents, respectively. The mixture AB is dissolved in the minimum amount of hot solvent S, and then cooled. The crop of crystals X_1 which is deposited is separated from the mother liquor L_1 and then dissolved in fresh solvent.

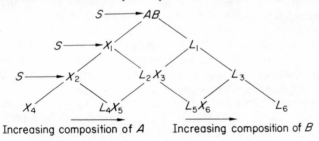

Increasing composition of A Increasing composition of B

Figure 7.3. Fractional recrystallisation of a solution

The cooling and separating operations are repeated, giving a further crop of crystals X_2 and a mother liquor L_2. The first mother liquor L_1 is concentrated to yield a crop of crystals X_3 and a mother liquor L_3. Crystals X_3 are dissolved by warming in mother liquor L_2 and then cooled to yield crystals X_5 and liquor L_5. Crystals X_2 are dissolved in fresh hot solvent and cooled to yield crystals X_4 and liquor L_4. Liquor L_3 is concentrated to give crystals X_6 and liquor L_6. The scheme can be continued until the required degree of separation is effected. The less soluble substance A is concentrated in the fractions on the left-hand side of the diagram, the more soluble constituent B on the right-hand side. If any other substances with intermediate solubilities were present, they would be concentrated in the fractions in the centre of the diagram.

If the starting material contained a unit quantity of component A, and each crystallisation step resulted in the deposition of a proportion P of this component, the proportions of A which would appear at any given point in the triangular scheme (see *Figure 7.4*) would be given by a term in the binomial expansion

$$[P+(1-P)]^n = 1 \tag{7.1}$$

$$p_{r,n} = \frac{n!}{r!(n-r)!} \cdot P^{n-r}(1-P)^r \tag{7.2}$$

where $p_{r,n}$ = proportion of A at a point represented by row r and stage n Thus, for example,

$$p_{2,3} = \frac{3!}{2!} \cdot P^{(3-2)}(1-P)^2$$

$$= 3P(1-P)^2$$

A modification of the triangular scheme is shown in *Figure 7.5*. In this case further quantities of the feedstock AB are added to the system by dissolving it in successive mother liquors on the right-hand side of the diagram This scheme is particularly useful if component A has a high temperature coefficient of solubility.

Several other much more complex schemes for fractional recrystallisation can be used, their aim being to increase the yield of the desired constituent by further re-use of the mother liquors. A detailed account of these methods has been given by TIPSON[1, 2]; *Figure 7.6* illustrates two of them. In the 'diamond' scheme (*Figure 7.6a*) the outermost fractions are set aside when they have reached a predetermined degree of purity, and fractionation is continued until all the material is obtained either in a crystalline form or in solution in a final mother liquor. If necessary, various crystal fractions can be bulked together and recrystallised, and in a similar manner the mother liquors can receive further treatment.

Unfortunately, the various fractions obtained by the above method will differ in composition, and relatively pure crystals will be mixed with relatively impure ones. This difficulty can be overcome by the use of the 'double withdrawal' scheme shown in *Figure 7.6b*. The procedure is the same as that used in the diamond scheme up to the point where no further fresh solvent is used (line 5 in *Figure 7.6a*). At this point it is assumed that crystals 5a and mother liquor 5f have reached the desired degree of purity, and they are both set aside from the system. Fresh solvent is then added to the crystal crop 6b to yield a purer crop 7b, which is arranged to have a purity similar to that of crop 5a. Crop 7b is set aside, crop 8c crystallised from pure solvent and so on.

Theoretical analyses and surveys of the factors affecting the choice of many

237

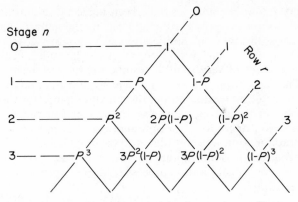

Figure 7.4. Analysis of the triangular fractional crystallisation scheme

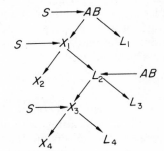

Figure 7.5. Simple recrystallisation with further additions of feedstock

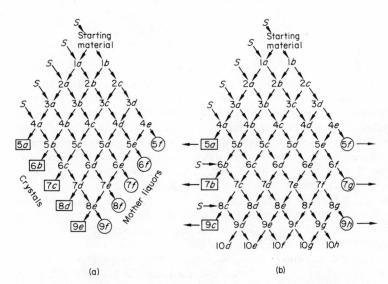

Figure 7.6. Fractional recrystallisation schemes: (a) diamond; (b) double withdrawal. (After R. S. TIPSON[1, 2])

fractional crystallisation schemes have been made by several authors[1, 3, 4]. GARBER and GOODMAN[5] presented a rigorous mathematical analysis of the fractional crystallisation of multicomponent systems. JOY and PAYNE[6] investigated the fractional crystallisation of similar substances when one substance is present in very small amounts, e.g. radium–barium chromate mixtures. The separation efficiency for a particular fractionation step was considered to be given by the ratio of the entropy change for the step to the maximum obtainable entropy change for the system. For many precipitation and crystallisation operations the optimum cut at each step for maximum entropy of fractionation was considered to be about 0·5.

The distribution of isomorphous impurities between solid and solution phases in fractional crystallisation processes may be expressed in terms of the CHLOPIN[7] relation

$$\frac{x}{y} = D\left(\frac{a-x}{b-y}\right)$$

(7·3)

where a and b are the amounts of components A and B in the original solid, x and y are the amounts of A and B in the crystallised solid, and $a-x$ and $b-y$ are the amounts of A and B retained in the solution. D is a distribution constant. Alternatively, the logarithmic relation due to DOERNER and HOSKINS[3] may be used:

$$\ln\left(\frac{a}{x}\right) = \lambda \ln\left(\frac{b}{y}\right)$$

(7.4)

where λ is a constant. If component A is the impurity, $\lambda > 1$ if crystallisation is to result in purification. Both of these relationships have been widely used to correlate the results of fractional crystallisation processes, although neither is entirely satisfactory from a theoretical viewpoint.

Fractional Crystallisation

The process of fractional crystallisation from solution, as pointed out by MATZ[8], can be analysed by the well-known McCabe–Thiele and Ponchon–Savarit graphical methods commonly used for fractional distillation. In the Ponchon–Savarit diagram (upper section of *Figure 7.7*) the abscissa records crystal compositions, x, or mother liquor concentrations, y. The ordinate represents the solvent–solute mass ratio, N. The system used here as an example is lead and barium nitrates in water, which form a continuous series of solid solutions with no hydrate formation. The $N–x$ curve, therefore, is the abscissa of the diagram ($N = 0$) and the $N–y$ curve has no discontinuities. The region between the $N–x$ and $N–y$ curves represents solid–liquid mixtures. As in similar diagrams for liquid–liquid and vapour–liquid systems, this area is interlaced with tie-lines whose endpoints correspond to the solid (x) and liquid (y) phase compositions in equilibrium.

Equilibrium values from the upper Ponchon–Savarit diagram are used to construct the equilibrium curve in the lower McCabe–Thiele diagram as follows. For a given tie-line, a vertical line is drawn down from the end-point on the $N–y$ curve on to the diagonal ($x = y$) in the lower diagram. A

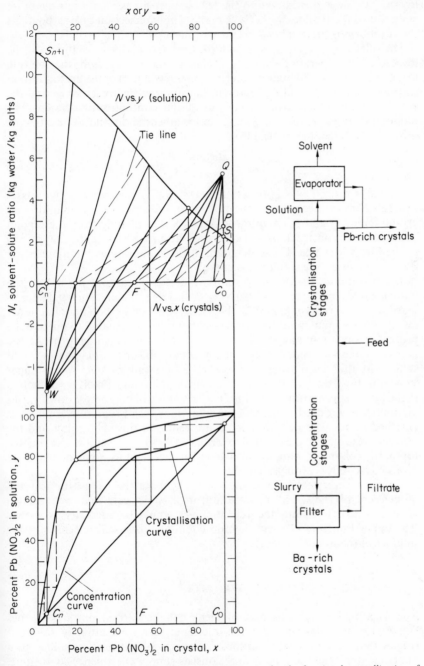

Figure 7.7. *Ponchon–Savarit and McCabe–Thiele diagrams for the fractional crystallisation of lead and barium nitrates from aqueous solution*

horizontal line is then drawn to the left to meet the vertical from the other endpoint of the tie-line on the $N-x$ curve. The intersection gives a point on the equilibrium curve. This procedure is repeated.

The inlet and exit streams (F = feed, C = crystals, S = solution) can be located on the operating diagram. Points S_1 and C_0 represent the solution leaving and the crystal 'reflux' entering the top (stage 1) of the crystallisation section. Points S_{n+1} and C_n represent the solution 'reflux' entering and the crystals leaving the bottom (stage n) of the concentration section. The minimum crystal reflux ratio, R_{min}, is obtained by extending the tie-line through F to meet the vertical from C_0 at P. Then

$$R_{min} = \frac{\text{distance } PS_1}{\text{distance } S_1 C_0} \qquad (7.5)$$

For a reflux ratio $R > R_{min}$ (where $R = QS_1/S_1C_0$) the operating line passes through F more steeply than the tie-line and determines the solution and crystal 'poles' Q and W. Arbitrarily drawn lines radiating from these poles are used in the construction of the 'crystallisation' and 'concentration' curves in the lower McCabe–Thiele diagram, in the same way as the tie-lines are used to construct the equilibrium curve.

The continuous fractional crystallisation of a mixture of $Pb(NO_3)_2$ (the more soluble component) and $Ba(NO_3)_2$ (the less soluble) is shown diagrammatically at the right of *Figure 7.7*. This scheme is similar to that for fractional distillation or countercurrent extraction, but it is not essential to operate the process in a column; any arrangement of contact vessels which provides the necessary stage-wise countercurrent contact of solid and liquid phases would suffice. In the scheme depicted $Ba(NO_3)_2$ crystallises out in the upper 'crystallisation' section, above the feed entry point, and $Pb(NO_3)_2$ dissolves in the lower 'concentration' section. The crystals, which progress downwards, are therefore enriched in Ba, and the solution, which progresses upwards, is enriched in Pb. The Ba-rich crystals are removed through a filter and the filtrate is returned to the column as 'reflux'. The Pb-rich solution leaving the top of the column is evaporated to yield Pb-rich crystals, some of which are returned as 'reflux' at the top of the column.

To produce crystals containing 95 per cent Ba $(NO_3)_2$ at the top and 95 per cent Pb $(NO_3)_2$ at the bottom from a 50 per cent feedstock, using a reflux ratio of 1·36, four theoretical stages would be adequate as shown. The upper Ponchon–Savarit diagram could also be used to determine the number of theoretical stages.

RECRYSTALLISATION FROM MELTS

The fractional recrystallisation of melts can be carried out by schemes similar to those described for the fractional recrystallisation of solutions (*Figures 7.1–7.6*). Of course, in most cases no solvent need be added to a molten system; the usual procedure consists simply of sequences of melting, partial freezing and separation. Selected fractions may be mixed at intervals according to the type of scheme employed, and fresh additions of feedstock may be introduced at certain stages if necessary. Several schemes for the purification of fats and waxes have been described by BAILEY[9].

Solid Solutions

When two substances constitute a system that forms a simple series of solid solutions, it should be possible to separate them almost completely by repeated melting, cooling and separating. The process is similar in many respects to fractional distillation. *Figure 7.8* shows the phase diagram for a solid solution of two components, A and B, which do not form a eutectic, or maximum or minimum melting point mixtures. In this particular example the melting point of A is lower than that of B. The region above the liquidus represents the homogeneous, that below the solidus the solid phase. Liquid and solid coexist in equilibrium in the region between the liquidus and

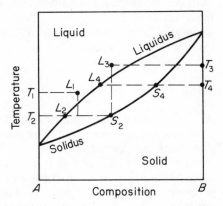

Figure 7.8. Fractional recrystallisation of a solid solution

solidus, and it is within this region that phase separation is effected. The recrystallisation process can be illustrated as follows. If the solid solution is first melted by heating to some temperature T_1, to give a liquid of composition L_1, then cooled to some temperature T_2 within the region bounded by the liquidus and solidus curves, it will partially solidify to give a liquid of composition L_2 and a solid (solid solution) of composition S_2. If this solid S_2 is separated from the liquid, melted by heating to some temperature T_3 and then cooled to some temperature T_4, a liquid L_4 and solid S_4 will be obtained. It can be seen, therefore, that the solid fractions become richer and richer in component B, and a repetition of this process will result in an almost complete separation of the two components.

This simple recrystallisation method is also applicable to solid solutions containing more than two components, but many recrystallisation steps will be required to effect any appreciable degree of separation. When two substances form a solid solution with a maximum or minimum melting point (see Chapter 4), a complete separation of both components by fractional recrystallisation will be impossible; these cases are analogous to the formation of azeotropes in distillation processes.

FORSYTH and WOOD[10] investigated the possibilities of the separation of organic mixtures by crystallisation from the melt. A molten binary organic system, e.g. naphthalene-β-naphthol, was pumped in turbulent flow down the inside of a cooled tube. The solid that was deposited on the wall differed in composition from the melt that continued down the tube, and such an arrangement acted as a combined crystallisation and phase separation unit in a stagewise countercurrent cascade. For systems that formed solid solutions, stage efficiencies of over 50 per cent at a rate of deposition of solid of 10 lb/h ft^2 were reported.

Impure systems that do not form solid solutions can be purified by partial melting or partial freezing, followed by separation of the liquid and solid phases. Liquids, too, can be submitted to these operations. Commercial grade benzene, for example, can be partially frozen or completely frozen and then partially melted, depending on the nature of the impurity that has to be removed. ASTON and MASTRANGELO[11] discussed the method of fractional melting as an alternative to fractional freezing, or crystallisation from solution, for certain organic compounds. They later described an apparatus for the stepwise melting of many impure hydrocarbons, including cyclohexane, iso-octane, n-heptane and *cis*-2-butene.

Eutectic Systems

As in the case of solid solutions that form maximum and minimum melting mixtures, the components of a eutectic system cannot be isolated in the pure state by normal fractional crystallisation techniques. The best that can be done in the case of a binary eutectic system, for example, is the isolation of one pure component and a eutectic mixture containing a fixed proportion of both components. The phase reactions that can occur in eutectic systems have been described in Chapter 4. *Figure 7.9* shows the phase equilibria for the ternary system, *ortho-, meta-* and *para*-nitrotoluene; the three pure components are represented by the letters O, M and P, respectively, at the apexes of the triangle. Four different eutectics can exist in this system, three binaries and one ternary:

| Eutectic points | | Per cent by weight | | | Temperature, |
Symbol	Components	O	M	P	°C
A	O–M	52	48	—	−31·7
B	O–P	76	—	24	−16·3
C	M–P	—	67	33	− 2·8
D	O–M–P	42	44	14	−40·0

COULSON and WARNER[12] used a diagram similar to *Figure 7.9* to demonstrate the analysis of a crystallisation problem encountered in the commercial manufacture of mononitrotoluene; the following example is taken from this publication.

Example

The bottom product from a continuous distillation column has the composition 3·0 per cent *ortho*-, 8·5 per cent *meta*- and 88·5 per cent *para*-nitrotoluene. Suggest the operation conditions of a cooling crystalliser to recover pure *para*-nitrotoluene from this material, and estimate the yield.

Solution

Point *X* in *Figure 7.9* represents the composition of feedstock; it is located between the 40 and 50 °C isotherms. By interpolation, the temperature at

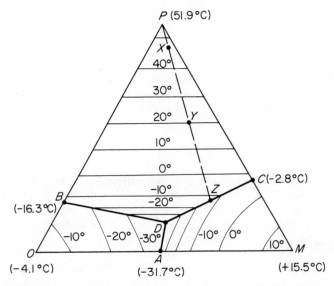

Figure 7.9. Phase diagram for the ternary system o-, m- *and* p-*nitrotoluene (point D = −40·0 °C)*

which this system starts to freeze can be estimated as about 46 °C. As point *X* lies in the region *PBDC*, pure *para*- will crystallise out once the temperature falls below 46 °C, and the composition of the mother liquor will follow line *XYZ* (i.e. away from point *P*) as cooling proceeds. At point *Z* (about −15 °C) on curve *DC*, *meta*- also starts crystallising out. It is not necessary, however, to cool to near −15 °C in order to get a high yield of *para*-, as *Table 7.2*, based on 100 kg of feedstock, shows.

The above mother liquor compositions are read off *Figure 7.9* at the point at which the line *P* → *Z* cuts the particular isotherm. The total weight of *para*- crystallised out is calculated by the mixture rule. For example, for 100 kg of original mixture *X* at 20 °C (point *Y*)

$$para\text{- deposited} = 100\left(\frac{\text{distance } XY}{\text{distance } PY}\right) = 75 \text{ kg}$$

Table 7.2. YIELD OF *para*-NITROTOLUENE FROM AN $O-P-M$ MIXTURE BY
COOLING CRYSTALLISATION
(AFTER COULSON AND WARNER[12])

Temperature, °C	Para-deposited, kg	Mother liquor, kg	Composition of mother liquor (per cent by weight)		
			O	M	P
46	0	100	3·0	8·5	88·5
40	39·6	60·4	5·0	14·0	81·0
30	66·7	33·3	8·7	24·8	66·5
20	75·0	25·0	12·0	34·0	54·0
10	79·6	20·4	14·7	41·6	43·7
0	82·3	17·7	16·7	48·0	35·3
−10	84·8	15·2	17·0	53·9	27·1

In commercial practice 20 °C is a reasonable temperature which can be attained by the use of normal cooling water. If this is adopted as the operating temperature, the yield of pure *para-* crystals per 100 kg of feedstock is 75 kg, equivalent to a recovery of about 85 per cent.

ZONE REFINING

For the removal of the last traces of impurity from a substance, fractional crystallisation from the melt cannot be applied with any degree of success. Apart from the fact that an almost infinite number of recrystallisation steps would be necessary, there would only be a minute quantity of the pure substance left at the end of the process. High degrees of purification combined with a high yield of purified material can, however, be obtained by the technique known as zone melting or zone refining, originally developed by PFANN[13, 14] for the purification of germanium for use in transistors. Purification by zone refining can be effected when a concentration difference exists between the liquid and solid phases that are in contact during the melting or solidification of a solid solution. The method is best explained by means of a phase diagram.

Figure 7.10a shows a phase diagram for two substances A and B that form a simple solid solution; in this case the melting point of A is higher than that of B. When a homogeneous melt M of composition x is cooled, solid material will first be deposited at some temperature T. The composition of this solid, which is in equilibrium with the melt of composition x (point L on the liquidus), is given by point S^* on the solidus. As cooling proceeds, more solid is deposited, and the concentration of B in the solid increases towards S. A similar reasoning may be applied to the reversed melting procedure starting from temperature T', where a liquid of composition L^* is in equilibrium with a solid S of composition x. The solidification of a homogeneous molten mixture M yields a solid with an over-all composition x, but owing to segregation the solid mixture is not homogeneous. Adjustments in the composition of the successive depositions of solid matter will not take place because of the very slow rate of diffusion in the solid state. The further apart the liquidus and solidus lines are, the greater will be the difference in concentration between the deposited solid and residual melt.

A measure of the expected efficiency of separation is given by a factor known as the segregation coefficient k. The significance of this coefficient can be seen in *Figure 7.10b*, where the liquidus and solidus for a binary solid solution in the regions of low B and low A concentrations are represented by straight lines. As zone melting is only useful for the refining of substances

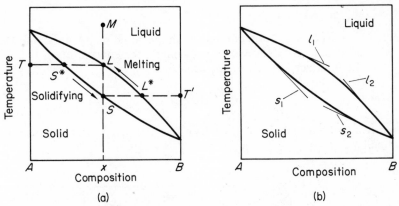

Figure 7.10. *Phase diagram for simple solid solution: (a) solidification of a mixture; (b) liquidus and solidus drawn as straight lines in regions of near-pure A and B*

with low impurity contents, these are the regions that are of interest. For dilute solutions the segregation coefficient k is defined by

$$k = \frac{\text{concentration of impurity in solid}}{\text{concentration of impurity in liquid}} \text{ (at equilibrium)}$$

$$= \frac{\text{slope of liquidus}}{\text{slope of solidus}} = \frac{l}{s} \tag{7.6}$$

In *Figure 7.10b* it can be seen that $k_1 = l_1/s_1$ and $k_2 = l_2/s_2$, and also that $s_1 > l_1$ and $s_2 < l_2$. In general, therefore, it may be said that

$$k < 1 \text{ when the impurity lowers the melting point}$$

and

$$k > 1 \text{ when the impurity raises the melting point}$$

When $k < 1$, the impurity will concentrate in the melt, when $k > 1$ in the solidifying mass. The nearer the value of k approaches unity, i.e. the closer the liquidus and solidus approach, the more difficult the segregation becomes. When $k = 1$, no zone refining is possible.

The impurity concentration at any point along a solid bar or ingot that was originally molten and then progressively cooled and solidified along its length (*Figure 7.11*) can be expressed in terms of the segregation coefficient.

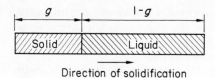

Direction of solidification

Figure 7.11. Molten bar of unit length under-going progressive directional solidification

The concentration C of impurity (solute) in the solid at the solid–liquid interface is given by

$$C = kC_l \tag{7.7}$$

where C_l is the concentration of impurity in the liquid phase at the interface. At any distance g along the bar of unit length, the concentration C at the solid–liquid interface is given by the equation

$$C = kC_0(1-g)^{k-1} \tag{7.8}$$

where C_0 is the initial concentration of the impurity in the homogeneous molten bar.

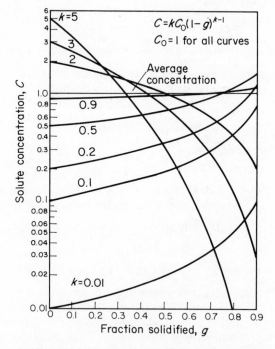

Figure 7.12. Pfann's curve for normal freezing. (From w. g. pfann[13], *by courtesy of* American Institute of Mining and Metallurgical Engineers)

The distribution of impurity along a bar subjected to this process of *directional solidification* can be calculated from equation (7.8), and *Figure 7.12* indicates the various distributions that would be expected for different values of k. The greater the deviation of k from unity, the greater the concentration gradient along the bar.

When a crystal grows in a melt or solution, impurity is rejected at the solid–liquid interface. In a stagnant fluid, e.g. with the crystalline phase gradually advancing into an unstirred melt, the impurity may not be able to diffuse away fast enough, in which case it will accumulate near the crystal face and lower the melting point in that region. Thus the effective temperature driving force, $\Delta T = T^* - T$, is decreased and the crystal growth rate is retarded. However, the melting point in the main bulk of the melt remains unaffected, so ΔT increases away from the interface.

This condition, known as *constitutional supercooling*, represents an instability and the advancing face usually breaks up into finger-like cells, which progress in a more or less regular bunched array. In this manner heat of crystallisation is more readily dissipated and the tips of the projections advance clear of the concentrated impurity and grow under conditions of near-maximum driving force. Impure liquid may be entrapped in the regions between the fingers, and a succession of regular inclusions may be left behind. At low constitutional supercooling, i.e. in a relatively pure system, the advancing interface is generally cellular. At high constitutional supercooling (high impurity) dendritic branching may occur. This sort of behaviour is common in, for example, the casting of metals.

Zone Purification

Directional solidification is not readily adaptable for purification purposes, because the impurity content varies considerably along the bar at the end of the operation. The end portion of the bar, where the impurity concentration is highest, could be rejected and the process repeated after a remelting operation, but this would be extremely wasteful. Nevertheless, progressive freezing techniques have proved useful in a number of areas, as described by RICHMAN, WYNNE and ROSI[15].

The technique known as *zone melting* does lend itself to repetitive purification without undue wastage. In this process a short molten zone is passed along the solid bar of material to be purified. If $k < 1$, the impurities pass into the melt and concentrate at the end of the bar. If $k > 1$, the impurities concentrate in the solid; thus in this case a cooled solid zone could be passed through the melt. The sequence of operations for molten zone refining is shown in *Figure 7.13a, b*. Although the impurity concentration in the bar, after the rejection of the high-impurity end portion, is much lower than initially, there is a considerable concentration gradient along the bar. In order to make the impurity concentration uniform along the bar, the process known as *zone levelling* must be employed: further zoning is carried out in alternate

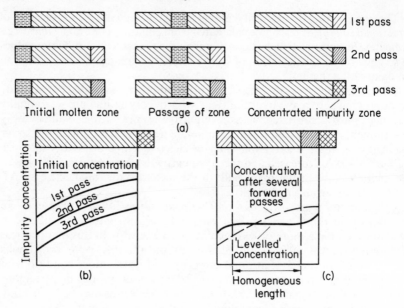

Figure 7.13. (a) and (b): A 3-pass zone refining operation; (c) zone levelling

directions until a homogeneous mid-section of the bar is obtained (*Figure 7.13c*). Both the high- and low-impurity ends of the bar are discarded.

Zone Refining Methods

A few of the basic arrangements used in zone refining[16] are shown in *Figure 7.14*. *Figure 7.14a* shows the material contained in a horizontal crucible along which a heater is passed; or the crucible may be pulled through a stationary heater, and the molten zone travels through the solid. Several zones may be passed simultaneously at fixed intervals along the bar (*Figure 7.14b*) in order to reduce purification time. It is essential that the heaters be spaced at a sufficient distance to prevent any spread of the molten zones; for materials of high thermal conductivity and substances that exhibit high degrees of supercooling, alternate cooling arrangements may have to be fitted between the heaters. The ring method (*Figure 7.14c*) permits a simple multi-pass arrangement. The vertical tube method (*Figure 7.14d*) is useful when impurities that can sink or float in the melt are present. A downward zone pass can be used in the former case, an upward pass in the latter.

Different methods of heating may be used to produce molten zones; the choice of a particular method is usually governed by the physical characteristics of the material undergoing purification. Resistance heating is widely employed, and direct-flame and focused-radiation heating are quite common. Heat can be generated inside an ingot by induction heating, and this method is often used for metals and semiconductors. Heating by electron bombardment or by electrical discharge also have their specific uses. Solar radiation, focused by lenses or reflectors, affords an automatic method for zone movement, on account of the sun's motion. Details of the various heating techniques are discussed by a number of authors[16-19].

The speed of zoning is a factor that can have a considerable effect on the efficiency of zone refining. The correct speed is that which gives a uniform zone passage and at the same time permits impurities to diffuse away from the solidifying face into the melt. If the zone speed is too fast, irregular crystallisation will occur at the solidifying face, and impure melt will be trapped before it can diffuse into the bulk of the moving zone (*Figure 7.15a*).

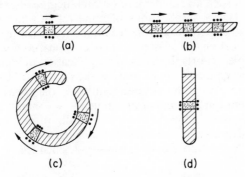

(a)

(b)

(c)

(d)

Figure 7.14. Simple methods of zone refining (after N. L. PARR[16]): *(a) horizontal, single-pass; (b) horizontal concurrent passes; (c) continuous concurrent passes in 'broken ring' crucible; (b) vertical tube, upward or downward passes*

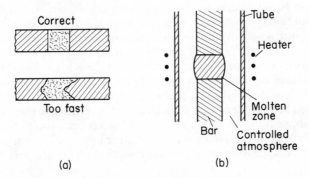

Correct

Too fast

(a)

Tube

Heater

Molten zone

Bar   Controlled atmosphere

(b)

Figure 7.15. (a) Influence of zone speed; (b) floating-zone technique

For the purification stages, the speed of zoning can vary from about 50 to 200 mm/h and speeds of about 5 to 15 mm/h are common for zone levelling.

The tube or crucible that contains the material undergoing purification should not provide a source of contamination and must be capable of withstanding thermal and mechanical stresses. The purified solid must be easily removable from its container, so the melt should not wet the container walls. The choice of the container material depends on the substance to be purified. Glass and silica are commonly used for organic substances, and silica for many sulphides, selenides, arsenides and antimonides; graphite-lined silica is often used for metals.

External contamination can be prevented by dispensing with the use of a conventional container. For example, a vertical zone refining technique that uses a 'floating zone' is shown in *Figure 7.15b*. The bar of material is fixed inside a container tube without touching the walls and the annulus between the bar and tube may contain a controlled atmosphere. Surface tension plays a large part in preventing the collapse of the molten zone, but the control of such a zone demands a high degree of experimental skill.

Although the largest applications of zone refining are in the fields of semiconductors and metallurgy, the technique has been used very success-fully for the purification of a large number of chemical compounds. HERRING-TON[18], for example, reports the particular application of zone refining to the preparation of very pure organic compounds. NICOLAU[20] has described the purification of inorganic salts by zone refining in the presence of a solvent, which allows operation at temperatures much lower than the melting point. He calls the technique zone-dissolution-crystallisation, and describes the production of reagent-grade di-sodium phosphate, ammonium alum and copper sulphate from impure commercial-grade chemicals.

Zone precipitation

A modification of the zone melting principle, in which the material being processed is originally dissolved in a solvent, has been called 'zone precipita-tion' by ELDIB[21]. This technique is particularly suitable for materials that do not readily crystallise out of solution, but form gel-like or amorphous masses. The material to be purified is dissolved in the requisite amount of hot solvent and poured into a vertical tube in which separation is to take place. On cool-ing, the mixture sets to a gel. A heater is moved down the vertical column, which allows a molten zone to progress through the gel. The liquid behind the heater solidifies rejecting the more soluble components, which are con-centrated in the direction of movement. The less soluble components are concentrated behind the heater. Repeated passes of the zone may be made, as described above for zone melting. The process has been successfully applied to the purification of petroleum waxes and other hydrocarbon fractions, poly-propylene resins, heat-sensitive protein materials, dyestuffs, fats and gums.

SINGLE CRYSTALS

For many centuries the growing of large crystals has been the pastime of devoted scientists, often just for curiosity's sake. To-day, however, there is an increasing demand for perfect crystals of innumerable substances for research work on the chemical and physical properties of pure solids and for use in the electrical industries, where crystals of certain substances are required for their dielectric, piezo-electric, paramagnetic and semiconductor properties. More recently single crystals have been in demand for masers and lasers. The production of large single crystals demands specialised and exacting techniques, but broadly speaking there are three general methods available, viz. growth from solutions, from melts, and from vapours. Several comprehensive reviews have been made of this rather special aspect of crystallisation practice[22-29], and only a brief summary will be attempted here.

Growth from Solutions

Slow crystallisation from solution in water or organic solvents has long been a standard method for growing large pure crystals of inorganic and organic substances. Basically, a small crystal seed is immersed in a super-saturated solution of the given substance and its growth is regulated by a careful control of the temperature, concentration and degree of agitation of the system. For instance, a tiny selected crystal may be mounted on a suit-able support and suspended in a vessel containing a solution of the substance, maintained at a fixed temperature. Slow rotation of the vessel will give an adequate movement of the solution around the crystal, and slow, controlled evaporation of the solvent will produce the degree of supersaturation neces-sary for crystal growth. Alternatively, the vessel may remain stationary and the growing crystal, or several suitably mounted crystals, may be gently rotated in the solution.

The actual operating conditions vary according to the nature of the crystallising substance and solvent; the optimum supersaturation and solution movement past the crystal must be found by trial and error. Generally speaking, the degree of supersaturation must not be high, and in any case the solution must never be allowed to approach the labile condition. The degree of supersaturation should be kept as constant as possible to ensure a constant rate of deposition of solute on the crystal seed. HOLDEN[28] states that, for salts with solubilities in the approximate range 20–50 per cent by weight, the maximum linear growth rate that can be tolerated by the fastest-growing faces of a crystal is about 1–3 mm per day; faster rates tend to give imperfections. Single crystals are better grown by a cooling than by an evapo-ration process, as supersaturation can then be much more closely controlled.

A typical example of a cooling-type growth unit[29] is shown in *Figure 7.16a*. The carefully selected seed crystals are supported on a rotating arm, which is submerged in the gently agitated system, and grow under carefully con-trolled conditions. A very slow cooling rate is used, often less than 0·1 °C/h. TORGESEN, HORTON and SAYLOR[30] have given a detailed description of an apparatus and procedure for growing large single crystals by this method. Single crystals can also be grown by holding them in a fixed position in a growth cell of the type described in Chapter 6 (see *Figure 6.6*) through which solution of the required level of supersaturation flows at a controlled steady rate.

Hydrothermal Crystallisation

Many substances normally considered insoluble in water have an appreciable solubility at elevated temperatures and pressures. This property is utilised in the technique called 'hydrothermal crystallisation', which is basically crystallisation from aqueous solution at high temperature (350–550 °C) and pressure (1–3 kbar). The operation is carried out in a steel autoclave (*Figure 7.16b*), which can be provided with a silver or platinum liner for protection. The technique has proved satisfactory for the growth of silica and aluminosilicates from aqueous alkaline solutions. Quartz crystal, the

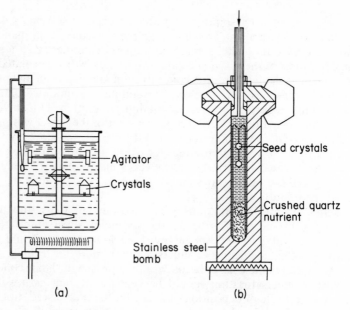

Figure 7.16. Crystallisation from solution: (a) temperature-lowering methods, (b) hydrothermal growth[29]

ideal piezoelectric material, is now grown in this manner on a commercial scale[31]. Hydrothermal growth and its industrial applications has been the subject of a comprehensive review by BARRER[32].

Growth in Gels

A room-temperature method, particularly useful for the production of single crystals of substances that are thermally unstable or have a very low solubility in water, is the technique of growth in a gel. The gel provides a kind of protective barrier for the growing crystals; it eliminates turbulence, yet it imposes no strain on the crystals, and permits a steady diffusion of the growth components. The diffusion and reaction processes are retarded and very much larger crystals are grown than could be produced by normal reaction and precipitation techniques. The first reports on crystal growth in gels date back to the work of Hatschek (1911), Liesegang (1914), Holmes

(1918) and Lord Rayleigh (1919), but more recent investigations have revived interest in the method for the growth of large single crystals[33-35].

One technique for growth by controlled diffusion in a gel may be illustrated by the growth of calcium tartrate tetrahydrate crystals by reacting calcium chloride with D-tartaric acid in a silica gel[35]. A freshly made solution of sodium metasilicate (21·6 g in 250 cm^3 of water, density 1·034, acidified to pH 3·5 with normal tartaric acid solution) is poured into test-tubes (6 × 1 in, two-thirds full) sealed to prevent loss of water and allowed to set over a period of 3 days at 40 °C. A normal solution of calcium chloride is then carefully pipetted on to the gel and nucleation immediately takes place at the gel–nutrient interface. After several hours single crystals appear below the interface and grow until they can no longer be supported by the gel. Crystals from 2 to 10 mm have been grown by this method over a period of a few weeks. Doped crystals can also be grown by doping the nutrient solution.

Growth from the Melt

A single crystal can be grown in a pure melt in a manner similar to that described earlier for growth in a solution. For example, a small crystal seed could be rotated in a supercooled melt, or the melt could flow past a fixed seed in a growth cell. However, these methods do not find any significant application. The most widely used techniques for the production of single crystals from the melt can be grouped into four categories: the withdrawal or pulling techniques, the crucible methods, flame fusion and zone refining. The latter method has been described in some detail above.

The pulling techniques are typified by the Kyropoulos (1926) and Czochralski (1918) methods. In both cases a small, carefully selected seed crystal is partially immersed in a melt, kept just above its melting point, and the growing crystal is maintained just below its melting point by holding it on a water-cooled rod or tube (see Figure 7.17). The apparatus is usually kept under reduced pressure or supplied with an inert atmosphere. Silicon, germanium, bismuth, tin, aluminium, cadmium, zinc, intermetallic compounds, and many organic substances have been grown as large strainfree single crystals in this manner.

The main difference between the methods of Kyropoulos (see Figure 7.17a) and Czochralski (see Figure 7.17b) is that in the former the seed is permitted to grow into the melt, while in the latter the seed is withdrawn at a rate that keeps the solid–liquid interface more or less in a constant position. Pull rates depend on the temperature gradient at the crystal–melt interface and can vary from 1 to 40 mm/h. The steeper the gradient the faster the growth rate and, hence, the faster the permissible rate of withdrawal.

Compounds that dissociate on melting cannot be grown by the vertical pulling technique unless means are provided for suppressing dissociation. A simple elegant means is provided by the *liquid encapsulation* technique[36], in which the melt surface is covered by a floating layer of transparent liquid. The encapsulant, e.g. boric oxide, B_2O_3, a low-melting-point glass, acts as a liquid seal provided that the inert gas pressure on its surface is greater than

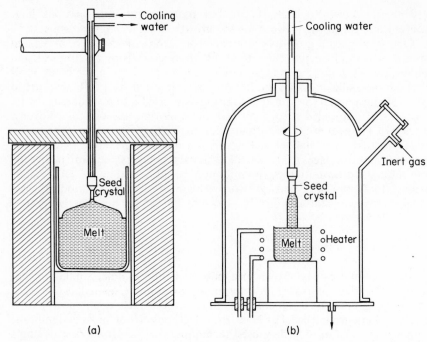

Figure 7.17. Crystal pulling and withdrawal techniques[29] : (a) Kyropoulos, (b) Czochralski

the dissociation pressure of the compound. As the crystal is pulled from the melt, a thin film of encapsulant adheres to its surface and suppresses dissociation. The technique is particularly useful for the growth of semiconductor and metal crystals; even highly dissociable compounds such as GaP (vapour pressure 35 bar at 1470 °C) can be grown successfully on a commercial scale by this technique[37, 38].

Table 7.3. SOME USES OF SINGLE CRYSTALS

Uses	*Substances.*
Piezo-electric	quartz, Rochelle salt, ammonium and potassium dihydrogen phosphates, ethylenediamine tartrate
Optical materials	CaF_2, α-SiO_2, $SrTiO_2$
Masers and lasers	$CaWO_4$ impregnated with rare earths
Paramagnetic studies	α-Al_2O_3, TiO_2, CaF_2
Semiconductors	Ge, Si, GaAs
Luminescence and fluorescence	ZnS and CsS activated with Ag, Tl, Cr, Mn and Cu salts; various organic crystals
Gemstones	alumina (sapphires and ruby)

An example of the crucible method is that due to Bridgman (1925) and Stockbarger (1949) (see *Figure 7.18a*). The melt is contained in a crucible with a conical bottom which is lowered slowly from a hot to a cold zone. The two sections of the furnace, which are kept at about 50 °C above and

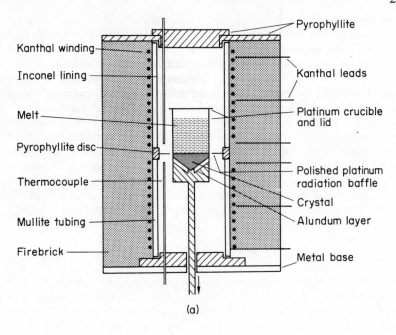

Kanthal winding

Inconel lining

Melt

Pyrophyllite disc

Thermocouple

Mullite tubing

Firebrick

Pyrophyllite

Kanthal leads

Platinum crucible
and lid

Polished platinum
radiation baffle

Crystal

Alundum layer

Metal base

(a)

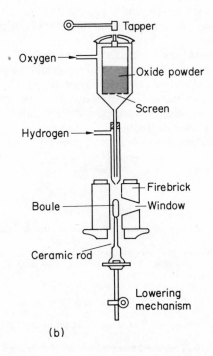

Tapper

Oxygen

Oxide powder

Screen

Hydrogen

Firebrick

Boule

Window

Ceramic rod

Lowering
mechanism

(b)

Figure 7.18. (a) The Bridgman–Stockbarger technique; (b) the Verneuil boule furnace[39]

below the melting point of the substance, respectively, are isolated by a thermal shield. The bulk of the solidified material, apart from the tip of the cone where the nuclei gather, is a single crystal. This method has proved successful for many semiconductor materials and alkali halides.

The flame fusion technique (see *Figure 7.18b*) was originally devised by Verneuil (1904) for the manufacture of artificial gemstones, such as corundum (white sapphire) and ruby. This method is now used for the mass production of jewels for watches and scientific instruments. A trickle of fine alumina powder plus traces of colouring oxides is fed at a controlled rate into an oxyhydrogen flame. Fusion occurs and the molten droplets fall on to a ceramic collecting rod. A seed crystal cemented to the rod is fused in the flame and the rod is lowered at a rate that allows the top of the growing crystal (known as a *boule*) to remain just molten. Renewed interest has recently been shown in this method for the production of rubies for lasers.

Growth from Vapour

Crystal growth from a vapour by direct condensation, without the intervention of the liquid phase, can be used to produce small strain-free single crystals of substances that sublime readily. Large single crystals cannot be grown by this method. A number of techniques are available[23, 25]. For example, a gas stream, such as N_2 or H_2S, may be passed over a sublimable substance, such as cadmium sulphide, in a heated container. The vapours then pass to another part of the apparatus, where they condense, in crystalline form, on a cold surface.

Sublimation in sealed tubes is also used for the preparation of single crystals of metals, such as zinc and cadmium, and non-metallic sulphides. A quantity of the material is placed at one end of the tube, along which a temperature gradient is maintained, so that sublimation occurs at the hot end and crystallisation at the other. An electric furnace with a number of independently controlled windings is used to maintain the temperature gradient to give a rate of sublimation sufficiently slow for the grown crystals to be single and not polycrystalline.

REFERENCES

1. TIPSON, R. S., 'Theory and scope of methods of recrystallisation', *Analyt. Chem.*, **22** (1950), 628
2. TIPSON, R. S., 'Crystallisation and recrystallisation', in *Techniques of Organic Chemistry*, Vol. III, Part I (ed. A. WEISSBERGER) 1956. New York; Interscience
3. DOERNER, H. A. and HOSKINS, W. M., 'The co-precipitation of radium and barium sulphates', *J. Am. chem. Soc.*, **47** (1925), 662
4. SUNIER, A. A., 'A critical consideration of some schemes of fractionalism', *J. phys. Chem.*, **33** (1929), 577
5. GARBER, H. J. and GOODMAN, A. W., 'Fractional crystallisation', *J. phys. Chem.*, **45** (1941), 573
6. JOY, E. F. and PAYNE, J. H., 'Fractional precipitation or crystallisation systems', *Ind. Engng Chem.*, **47** (1955), 2157

7. CHLOPIN, V., *Z. anorg. allg. Chem.*, **143** (1925), 97
8. MATZ, G., *Kristallization*, 2nd Ed., 1969. Berlin; Springer-Verlag
9. BAILEY, A. E., *Melting and Solidification of Fats and Waxes*, 1950. New York; Interscience.
10. FORSYTH, J. S. and WOOD, J. T., 'Separation of organic mixtures by crystallisation from the melt', *Trans. Instn chem. Engrs*, **33** (1955), 122
11. ASTON, J. G. and MASTRANGELO, S. V. R., 'Purification by fractional melting', *Analyt. Chem.*, **22** (1950), 636; **26** (1956), 764
12. COULSON, J. M. and WARNER, F. E., *A problem in chemical engineering design; the manufacture of mono-nitrotoluene*, Instn. chem. Engrs London, 1949
13. PFANN, W. G., 'Principles of zone melting', *J. Metals, N.Y.*, **4** (1952), 747
14. PFANN, W. G., *Zone Melting*, 1958. New York; Wiley
15. RICHMAN, D., WYNNE, E. A. and ROSI, F. D., 'Progressive freezing', p. 257 in reference 19
16. PARR, N. L., *Zone Refining and Allied Techniques*, 1960. London; Newnes
17. SCHILDKNECHT, H., *Zone Melting*, 1966. London; Academic Press
18. HERRINGTON, E. F. G., *Zone Melting of Organic Compounds*, 1963. New York; Wiley
19. ZIEF, M. and WILCOX, W. R., *Fractional Solidification*, 1967. New York; Dekker
20. NICOLAU, I. F., 'Purification process by solution zone passages', *J. Mater. Sci.*, **5** (1970), 623; **6** (1971), 1049
21. ELDIB, A., 'Zone precipitation and allied techniques', p. 369 in reference 19
22. WALKER, A. C., 'Growing piezoelectric crystals'. *J. Franklin Inst.*, **250** (1950), 481
23. LAWSON, W. D. and NIELSEN, S., *Preparation of Single Crystals*, 1958, London; Butterworths
24. WHITE, E. A. D., 'Synthetic gemstones', *Q. Rev. chem. Soc.*, **15** (1961), 1
25. SMAKULA, A., *Einkristalle*, 1962. Berlin; Springer
26. R. H. DOREMUS, B. W. ROBERTS, D. TURNBULL (Eds.), *Growth and Perfection of Crystals*, 1958. New York; Wiley
27. J. J. GILMAN (Ed.), *The Art and Science of Growing Crystals*, 1963 New York; Wiley
28. HOLDEN, A. N., 'Growing single crystals from solution', *Discuss Faraday Soc.*, **5** (1949), 312
29. WHITE, E. A. D., 'Crystal growth techniques', *G.E.C. Jl.*, **31** (1) (1964), 43
30. TORGESEN, J. L., HORTON, A. T. and SAYLOR, C. P., 'Equipment for single crystal growth from aqueous solution', *J. Res. natn. Bur. Stand.*, **67C** (1963), 25
31. LAUDISE, R. A. and SULLIVAN, R. A., 'Pilot plant production of synthetic quartz', *Chem. Engng Prog.*, **55** (5) (1959), 55
32. BARRER, R. M., 'Mineral synthesis by the hydrothermal technique', *Chem. in Br.*, **2** (1966), 380
33. DENNIS, J. and HENISCH, H. K., 'Nucleation and growth of crystals in gels', *J. electrochem. Soc.*, **114** (1967), 263
34. HENISCH, H. K., *Crystal Growth in Gels*, 1970. Pennsylvania State Univ. Press.
35. RUBIN, B., 'Growth of single crystals by controlled diffusion in silica gel', *A.I.Ch.E.Jl.*, **15** (1969), 206
36. MULLIN, J. B., STRAUGHAN, B. W. and BRICKNELL, W. S., 'Liquid encapsulation techniques', *Physics Chem. Solids*, **26** (1965), 782
37. MULLIN, J. B., HERITAGE, R. J., HOLLIDAY, C. H. and STRAUGHAN, B. W., 'Liquid encapsulation pulling at high pressures', *J. Crystal Growth*, **3/4** (1968), 281
38. BASS, S. J. and OLIVER, P. E., 'Pulling of gallium phosphide crystals by liquid encapsulation', *J. Crystal Growth*, **3/4** (1968), 286
39. SHORT, M. A., 'Methods of growing crystals', *Ind. Chemist*, **33** (1957), 3

8

Industrial Crystallisation Processes

CRYSTALLISATION ranks high in the list of industrial processes devoted to the production of pure chemicals. Apart from the fact that its final product has an attractive appearance, crystallisation frequently proves to be the cheapest and sometimes the easiest way in which a pure substance can be produced from an impure solution. Conventional distillation techniques cannot separate efficiently close-boiling liquids or those that form azeotropes, yet crystallisation may often lead to their complete separation. There is evidence that the petroleum industry is now turning its attention to crystallisation techniques to deal with difficult separations.

The methods available for crystallisation are many and varied. Crystals can be grown from the liquid or the vapour phase, but in all cases the state of supersaturation has first to be achieved. The way in which supersaturation is produced depends on the characteristics of the crystallising system; some solutes are readily deposited from their solutions merely by cooling, while others have to be evaporated to a more concentrated form. In cases of very high solubility, or for heat-labile solutions, another substance may have to be added to the system to reduce the solubility of the solute in the solvent. Again, supersaturation of the liquid or gaseous phase may be caused by the chemical reaction of two substances; one of the reaction products is then precipitated.

In this chapter an account is given of the ways in which crystallisation can be performed. Little attention will be devoted to the actual types of processing equipment employed, because these are discussed in greater detail in Chapter 9. Problems in the crystal-producing industries do not vanish once the solid phase has been deposited out of solution; indeed, many technologists would assert that they just begin there. Brief accounts, therefore, are given of some aspects of crystal washing and the caking of crystals on storage.

COOLING AND EVAPORATION

One of the most common ways in which the supersaturation of a liquid can be achieved is by means of a cooling process. If the solubility of the solute in the solvent decreases with a decrease in temperature, some of the solute will be deposited on cooling; a slow controlled rate of cooling in an agitated system can result in the production of crystals of regular size. The crystal yield may be slightly increased if some of the solvent evaporates during the cooling process.

If the solubility characteristics of the solute in the solvent are such that there is little change with a reduction in temperature, some of the solvent may have to be deliberately evaporated from the system in order to effect the necessary supersaturation and crystal deposition. Cooling and evaporative techniques are widely used in industrial crystallisation; the majority of the solute–solvent systems of commercial importance can be processed by one or other of these methods. Descriptions of many of the cooling and evaporating crystallisers commonly encountered are given in Chapter 9.

The yield from a cooler or evaporator can be calculated, as explained in Chapter 2, from the general equation

$$Y = \frac{WR[C_1 - C_2(1 - V)]}{1 - C_2(R - 1)} \tag{2.10}$$

where Y = crystal yield (kg); W = weight of solvent present initially (kg); V = weight of solvent lost, either deliberately or unavoidably, by evaporation (kg per kg of original solvent); R = ratio of the molecular weights of solvate (e.g. hydrate) and unsolvated (e.g. anhydrous) solute; and C_1, C_2 = initial and final solution concentrations, respectively (kg of unsolvated solute per kg of solvent). The yield calculated from equation (2.10) is the theoretical maximum, on the assumptions (a) that C_2 refers to the equilibrium saturation at the final temperature, and (b) that no solute is lost when the crystals are washed after being separated from the mother liquor.

VACUUM COOLING

If a hot saturated solution of a substance is fed into a vessel maintained at low pressure, some of the solvent will 'flash off' and the liquor will cool adiabatically. Supersaturation is thus achieved by both cooling and evaporation. Industrial crystallisers based on these principles are usually referred to as vacuum crystallisers; a further discussion of this method of operation is given in Chapter 9.

The yield from a vacuum crystalliser can be calculated from equation (2.10); before it can be applied, however, the quantity V, the amount of evaporation, must be known. It is possible to predict the evaporation capacity of a vacuum crystalliser, given certain information, as follows.

The amount of solvent evaporated depends upon the heat made available during the operation of the crystalliser. This is the sum of the sensible heat drop of the solution, which cools from the feed temperature to the equilibrium temperature in the vessel, and the heat of crystallisation liberated. The heat balance, therefore, will be

$$\begin{pmatrix} \text{solvent} \\ \text{evaporated} \end{pmatrix} \times \begin{pmatrix} \text{latent} \\ \text{heat} \end{pmatrix} = \text{sensible heat drop} + \text{heat of crystallisation}$$

Let l_v = latent heat of evaporation of solvent (kJ/kg)
   q = heat of crystallisation of product (kJ/kg)
   t_1 = initial temperature of solution (°C)
   t_2 = final temperature of solution (°C)
   c = heat capacity of solution–assumed constant (kJ/kg K)

In addition, let the symbols used in equation (2.10) be used, viz. C_1 and C_2 = initial and final solution concentrations (kg/kg of solvent), W = initial weight of solvent (kg), V = loss of solvent by evaporation (kg/kg of original solvent), R = ratio of the molecular weights of the hydrate and anhydrous salt, and Y = crystal yield (kg). Then

$$VWl_v = c(t_1 - t_2)W(1 + C_1) + qY$$

Substituting for the value of Y from equation (2.10) and simplifying,

$$V = \frac{qR(C_1 - C_2) + c(t_1 - t_2)(1 + C_1)[1 - C_2(R - 1)]}{l_v[1 - C_2(R - 1)] - qRC_2} \tag{8.1}$$

Once the value of the evaporation capacity V is determined from equation (8.1), the yield from the crystalliser can be calculated from equation (2.10).

Example

Determine the yield of sodium acetate crystals ($CH_3COONa \cdot 3H_2O$) obtainable from a vacuum crystalliser operating with an internal pressure of 10 mm Hg (1.33×10^3 N/m²) when it is supplied with 2000 kg/h of a 40 per cent aqueous solution of sodium acetate at 80 °C. The boiling point elevation of the solution may be taken as 11.5 °C.

Data

Heat of crystallisation, q = 144 kJ/kg of trihydrate
Heat capacity of the solution, c = 3.5 kJ/kg K
Latent heat of water at 1.33 kN/m², l_v = 2.46 MJ/kg
Boiling point of water at 1.33 kN/m² = 17.5 °C
Solubilities of sodium acetate in water are given in the Appendix, *Table A4*

Solution

Equilibrium temperature of liquor = 17.5 + 11.5 = 29 °C
Initial concentration, C_1 = 0.4/0.6 = 0.667 kg/kg of solvent
Final concentration at 29 °C, C_2 = 0.539 kg/kg of solvent
Initial weight of water = 0.6 × 2000 = 1200 kg
Ratio of molecular weights, R = 136/82 = 1.66
The vaporisation, V, is calculated from equation (8.1)
   V = 0.153 kg/kg of water present originally
This value of V substituted in equation (2.10) gives the crystal yield
   Y = 660 kg of sodium acetate trihydrate

For crystallisation from complex mixtures it is necessary to find the optimum solubility relationships which lead to maximum yield at specific feedstock and cooling water temperatures. Problems of this sort are usually solved graphically, by laborious trial and error methods, but ENYEDY[1] has described a computer program for the continuous vacuum crystallisation of a double salt from a ternary system [$FeSO_4 \cdot (NH_4)_2SO_4 \cdot 6H_2O$ from $FeSO_4-(NH_4)_2SO_4-H_2O$] which could be used with appropriate modification in similar situations.

CONTROLLED SEEDING

During a crystallisation operation the accidental production of nuclei ('false grain') must be avoided at all costs; the solution must never be allowed to become labile. The deliberate addition of carefully selected seeds, however, is permitted so long as the deposition of crystalline matter takes place on these nuclei only. The seeds should be dispersed uniformly throughout the solution by means of gentle agitation; and if the temperature is carefully regulated, considerable control is possible over the final product size. Deliberate seeding is frequently employed in industrial crystallisations; the actual weight of seed material to be added depends on the solute deposition, the size of the seeds and the product:

$$W_s = W_p(L_s^3/L_p^3) \qquad (8.2)$$

where W_s and W_p are the weights and L_s and L_p are the mean particle sizes of the seeds and product, respectively. The product weight, W_p, is the crystal yield, Y, e.g. as calculated from solubility data using equation (2.10), plus the weight of added seeds, i.e. $W_p = Y + W_s$, and

$$W_s = Y\left(\frac{L_s^3}{L_p^3 - L_s^3}\right) \qquad (8.2a)$$

Seeds as small as 5 μm have been used in sugar boiling practice; these tiny particles are produced by prolonged ball-milling in an inert medium, e.g. isopropyl alcohol or mineral oil, and 500 g of such seeds may be quite sufficient for 50 m^3 of massecuite.

The seeds do not necessarily have to consist of the material being crystallised, unless absolute purity of the final product is required. A few tiny crystals of some isomorphous substance may be used to induce crystallisation. For example, phosphates will often nucleate solutions of arsenates. Small quantities of sodium tetraborate decahydrate can induce the crystallisation of sodium sulphate decahydrate. Crystalline organic homologues, derivatives and isomers are frequently used for inducing crystallisation; phenol can nucleate m-cresol, and ethyl acetanilide can nucleate methyl acetanilide.

GRIFFITHS[2] investigated the crystallisation of solutions by cooling, with and without the presence of seeding crystals, and his results are shown on the solubility–supersolubility diagrams in *Figure 8.1*.

The rapid cooling of an unseeded solution is represented by *Figure 8.1a*. The metastable region is soon penetrated, and spontaneous crystallisation occurs when conditions corresponding to some point on the supersolubility curve are reached; a shower of tiny crystals is suddenly deposited from the solution. The slow cooling of an unseeded solution is shown in *Figure 8.1b*; again, crystallisation cannot occur until the supersolubility curve is reached, but the rate of crystallisation is slower than that in case (a) because the rate of heat removal is reduced. Control over the growth of crystals by this method is strictly limited and the crystals vary considerably in size. *Figure 8.1c* shows the rapid cooling of a seeded solution. As soon as the solution becomes supersaturated, growth begins on the crystal seeds and the concentration of the solution decreases as more and more substance is deposited. However, because of the rapid rate of cooling, the labile condition is soon

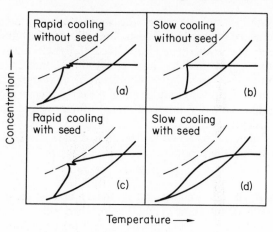

Figure 8.1. The effect of cooling rate and seeding on a crystallisation operation[2]

reached, and the eventual spontaneous deposition of fine crystals cannot be avoided.

The effect of a slow rate of cooling on a seeded solution is seen in *Figure 8.1d*. The temperature is controlled so that the system is kept in the metastable state throughout the operation, and the rate of growth of the small crystal seeds is governed solely by the rate of cooling. There is no sudden deposition of fine crystals, because the system does not enter the labile zone. This method of operation is usually referred to as 'controlled crystallisation'; crystals of a regular and predetermined size can be grown. Many large-scale crystallisation operations are carried out in this manner.

The mass of crystal seeds accidentally produced in large crystallisers may be very small indeed, yet on account of their minute size the number of seeds can be exceptionally large. If, for simplicity, spheres of diameter d and density ρ are considered, the mass of one seed is $\frac{\pi}{6} \rho d^3$. Thus 100 g of 0·1 mm ($\sim$150-mesh) seeds of a substance of density 2 g/cm^3 will contain about 100 million separate particles, and every seed is a potential crystal. *Figure 8.2* indicates the mass deposition of solute necessary for 100 g of 0·1 mm seeds to grow into crystals of various sizes ranging up to nearly 3 mm. It can be seen, for example, that 1 mm crystals can be grown by the deposition of only 100 kg of solute, but 800 kg has to be deposited to produce 2 mm crystals, and for $2\frac{1}{2}$ mm crystals the required mass deposition increases to more than 1500 kg.

False grain smaller than 0·1 mm can exist in commercial crystallisers where high degrees of supersaturation may have been produced, and these seeds cannot all be allowed to grow; 1 g of 0·01 mm seeds, for example, represents about 1000 million particles. For controlled growth, therefore,

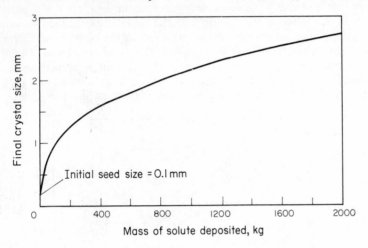

Figure 8.2. Crystal sizes and mass depositions on 100 g of 0.1 mm seeds

the liquor in the crystalliser must not be allowed to nucleate. Vigorous agitation and mechanical and thermal shock should be avoided, and the supersaturation should be kept to the absolute working minimum; high magma densities, up to 20 or 30 per cent suspended solids, help in this respect. If, despite these precautions, unwanted nucleation still occurs, some system of false grain removal must be operated; continuous circulation of the spent liquor through a fines trap is one such method.

SALTING-OUT CRYSTALLISATION

Another way in which the supersaturation of a solution can be effected is by the addition to the system of some substance that reduces the solubility of the solute in the solvent. The added substance, which may be a liquid, solid or gas, is often referred to as a diluent or precipitant. Liquid diluents are most frequently used. Such a process is known as salting-out, precipitation or crystallisation by dilution. The properties required of the diluent are that it be miscible with the solvent of the original solution, at least over the ranges of concentration encountered, that the solute be relatively insoluble in it, and also that the final solvent–diluent mixture be capable of easy separation, e.g. by distillation.

Although salting-out is widely employed industrially, relatively few published data are available regarding its use in crystallisation operations. The process is commonly encountered, for instance, in the crystallisation of organic substances from water-miscible organic solvents by the controlled addition of water to the solution; the term 'watering-out' is used in this connection. Some of the advantages of salting-out or dilution crystallisation are as follows. Very concentrated initial solutions can be prepared, often with great ease, by dissolving the impure crystalline mass in a suitable solvent. A high solute recovery can be made by cooling the solution as well as salting it out. If the solute is very soluble in the initial solvent, high dissolution temperatures are not necessary, and the temperature of the batch during the crystallisation operation can be kept low; this is advantageous when heat-labile substances are being processed. Purification is sometimes greatly simplified when the mother liquor retains undesirable impurities owing to their greater solubility in the solvent–diluent mixture. Probably the biggest

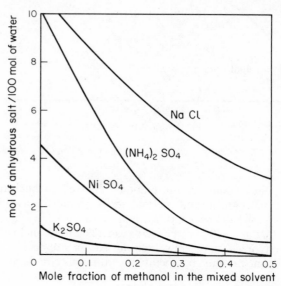

Figure 8.3. Solubility of some common salts in methanol–water mixtures at 20 °C

disadvantage of dilution crystallisation is the need for a recovery unit to handle fairly large quantities of the mother liquor in order to separate solvent and diluent, one or both of which may be valuable.

Salting-out crystallisation induced by an added organic solvent may be useful for the preparation of pure inorganic substances, despite the fact that the organic diluents are usually much more expensive than the required inorganic products. The solubility of an inorganic salt in water often falls rapidly when an organic solvent is added (see *Figure 8.3*). An example of this use of salting-out crystallisation was demonstrated by THOMPSON and MOLSTAD[3], who determined solubility and density isotherms for potassium and ammonium nitrates in aqueous isopropanol solutions over the temperature range 25–70 °C. Both salts are very soluble in water, and their solubilities increase greatly with increasing temperature. The work was carried out with a view to investigating the possibility of increasing the purity, as well as the crystal yield, of these substances. It was shown, for example, that 15 kg of isopropanol added to 100 kg of a saturated aqueous solution of KNO_3 at 40 °C would result in the precipitation of 44 per cent of the dissolved salt. The salt recovery would be increased to 68 per cent if 50 kg of isopropanol were added.

GEE, CUNNINGHAM and HEINDL[4] presented a detailed account of a pilot plant (2 ton/day) investigation into the commercial possibilities of the production of iron-free alum by the use of ethanol as a precipitant for aqueous solutions of crude alum. Under optimum conditions, alcohol losses from the system of less than $\frac{1}{2}$ per cent were reported. A detailed cost evaluation showed that the large-scale production of pure alum (20 ton/day) by this method is economically feasible. The use of ethanol has also been proposed by THOMPSON and BLECHARCZYK[5] for the commercial fractional crystallisation of natural brines.

In the so-called Somet process[6] for the commercial production of pure inorganic salts from impure mixtures, selective precipitation is effected by a suitable choice of solvent and operating procedure. Recommended solvents include methanol, ethanol, acetone and isopropanol. The dehydration of hydrated salts is a special case of salting-out pure salts from single-component salt solutions. The Somet process has been applied for this purpose to salts that dissolve in their own water of crystallisation when heated (e.g. $Na_2SO_4 \cdot 10H_2O$) and other highly soluble salts. It is claimed that the coarse-grained anhydrous precipitates are superior to those produced by conventional calcining techniques.

The crystal yield, Y, from a salting-out operation can be calculated, if no solvate is formed, from the simple equation

$$Y = W(C_1 - C_2') \tag{8.3}$$

where W = weight of initial solvent, and C_1 and C_2' = initial and final concentrations of the solute (kg/kg of original solvent). The presence of the diluent affects the value of C_2'; this will be very different from the equilibrium solubility of the solute in the original solvent at the final temperature. Solubility data for mixed solvents are rarely found in the literature, but in any case it is always wise to determine the solubility characteristics of the solute with the actual liquors to be encountered in the crystallisation process; small traces of impurity can affect solubility to a marked degree. If evaporation and solute formation also occur in a salting-out crystallisation, the crystal yield can be calculated by means of a suitably modified form of equation.

As previously mentioned, the salting-out substance need not be a liquid. A suitable solid can be added to the solution, provided that it meets the above requirement, i.e. it must be highly soluble in the original solvent and must not react with the required solute. To avoid crystal deposition occurring on the added solid, the added substance could be introduced in the form of a concentrated solution in the main solvent. The use of solid substances as salting-out aids is not very common, but the addition of solid salts can be used to precipitate other salts from solution as a result of the formation of a stable salt pair. This behaviour is encountered when two solutes, AX and BY, usually without a common ion, react in solution and undergo a double decomposition

$$AX + BY \rightleftharpoons AY + BX$$

The four salts AX, BY, AY and BX constitute a reciprocal salt pair. One of these pairs AX, BY or AY, BX is a stable salt pair (compatible salts), the other an unstable salt pair (incompatible salts); the stable pair comprises the two salts that can coexist in solution in stable equilibrium. Reciprocal salt pairs are discussed in Chapter 4, where the graphical representation of these systems is also described. PURDON and SLATER[17] give several examples of the commercial importance of salt pair formation in the recovery of pure crystalline solids from complex solutions. GARRETT[18] described the large-scale production of potassium sulphate from glaserite ($3K_2SO_4 \cdot Na_2SO_4$) and potassium chloride; glaserite is added to a concentrated solution of KCl, conversion to a stable salt pair occurs, and the Na_2SO_4 and KCl remain in solution while the K_2SO_4 is deposited.

REACTION CRYSTALLISATION

The precipitation of a solid product as the result of the chemical reaction between gases and/or liquids is a standard method for the preparation of many industrial chemicals. Indeed, in the pharmaceutical industry, for example, precipitation is frequently considered as an alternative method to comminution for the production of finely dispersed substances. Precipitation occurs because the gaseous or liquid phase becomes supersaturated with respect to the solid component. A crude precipitation operation, therefore, can be transformed into a crystallisation process by careful control of the degree of supersaturation. Reaction crystallisation is practised widely, especially in industries where valuable waste gases are produced. For instance, ammonia can be recovered from coke oven gases by converting it into ammonium sulphate by reaction with sulphuric acid.

A typical saturator used for the production of ammonium sulphate in coke-oven plants[9] is shown in *Figure 8.4*. Agitation of the crystals within the saturator is effected by a combination of the vigorous nature of the exothermic reaction and air sparging. The heat of reaction is removed by the evaporation of water added to the reaction zone. BAMFORTH[9] gives a detailed heat and materials balance for a commercial installation producing 32 tonnes/h of $(NH_4)_2SO_4$.

Another example of a reaction crystalliser is the carbonation tower (*Figure 8.5*) used for the production of sodium bicarbonate by the interaction between brine and flue gases containing about 10–20 per cent of carbon dioxide. The 15 m high tower is kept full of brine, and the flue gas, which enters at the bottom of the tower, flows upwards countercurrently to the

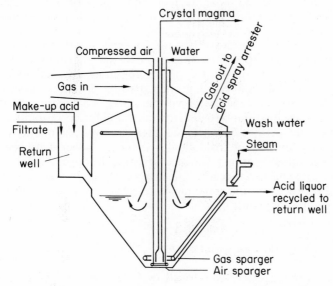

Figure 8.4. A reaction crystalliser for the production of ammonium sulphate. (After BAMFORTH[9])

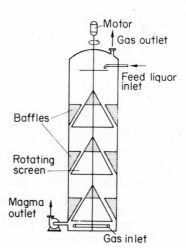

Figure 8.5. Reaction crystalliser—a carbonation tower for the production of sodium bicarbonate from brine and flue gas. (After D. E. GARRETT[18])

brine flow. Carbonated brine is pumped continuously out of the bottom of the tower. To effect efficient absorption of CO_2, three large rotating screens continually redisperse the gas stream in the form of tiny bubbles in the liquor. The operating temperature is controlled at about 38 °C, which has been found from experience to give both good absorption and crystal growth.

STEVENSON[10] has discussed the design and operation of continuous precipitators, and reported pilot plant data for the precipitation of ammonium diuranate from uranyl nitrate solutions with ammonia. Among the many variables that affect particle growth, the mean solids residence time in the precipitator is a vital parameter. An examination of the particle mass frequency distribution indicated that equilibrium is not attained until 8–10 residence periods have elapsed after start-up. The quality of the precipitate is a function of the solubility, which is affected by common ion effects or complexing agents. The effect of temperature, the influence of surface potentials, ageing and annealing are also considered. It is shown that a compromise between several conflicting requirements may be achieved by multistage operation with feed-back.

EMULSION CRYSTALLISATION

Organic substances may be purified by fractional crystallisation from the melt or from organic solvents, but these operations frequently present

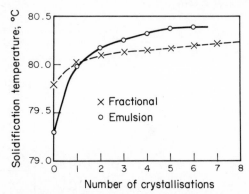

Figure 8.6. Comparison of fractional and emulsion crystallisation cycles for the purification of naphthalene. (After HOLECI[11])

difficulties in large-scale production. In the latter case, apart from expensive solvent losses and the potential fire and explosion hazards, yields are often low on account of the high solubility of the crystals. The method known as emulsion crystallisation, described by HOLECI[11], is generally free of these shortcomings.

Briefly, crystallisation is carried out by cooling from an aqueous emulsion. Impurities, in the form of eutectic mixtures, remain in the emulsion, from which they may be recovered by further cooling. The organic substance should be (a) practically insoluble in water and (b) able to melt and solidify in a heterogeneous aqueous medium, and remain stable.

The organic melt is emulsified in water with the aid of a suitable non-ionic agent and stabilised by a protective colloid, e.g. potato starch gelatinised with water. The system is crystallised by cooling, and the crystals are separated from the emulsion and washed with water. The operation may be repeated if required. A typical example is shown in *Figure 8.6*, where five or six emulsion-crystallisation cycles yield an almost pure naphthalene at an over-all yield of 70 per cent compared with a less pure product at a 2 per cent yield by seven conventional fractional crystallisation steps. The high efficiency of the emulsion crystallisation is apparently due to the fact that crystal agglomeration does not occur to any great extent and the impure emulsion is readily washed away with water.

EXTRACTIVE CRYSTALLISATION

As described in Chapters 4 and 7, fractional crystallisation of a binary system that forms a eutectic mixture can only produce one of the components in pure form. Many hydrocarbon isomers or close-boiling mixtures, such as those encountered in the petroleum industry, do form eutectics, and this has presented a limiting factor in the application of conventional crystallisation techniques for these separations. However, if the solid–liquid phase relationships can be altered by introducing a third component into a binary system, it is frequently possible by a series of crystallisation steps to separate the two desired components in pure form. The added third component is usually a liquid, called the solvent, and the process is known by the name 'extractive crystallisation'.

CHIVATE AND SHAH[12] discussed the use of extractive crystallisation for the separation of mixtures of *m*- and *p*-cresol, a system in which two eutectics are formed. Acetic acid was used as the extraction solvent. Details of the relevant phase equilibria encountered in the various combinations of systems were given together with an account of the laboratory investigations. While it was shown that acetic acid is not a particularly good solvent for the separation process in question, it was clearly indicated that extractive crystallisation, provided a suitable solvent is chosen, has a large number of potential applications. DIKSHIT and CHIVATE[13] have reported ternary phase equilibria for the separation of *o*- and *p*-nitrochlorobenzenes by extractive crystallisation with *p*-dichlorbenzene, and have proposed general methods for predicting the selectivity of a solvent.

A procedure that could be adopted for an industrial extractive crystallisation process has been described by FINDLAY and WEEDMAN[14]. A hypothetical separation is considered, and the process can be followed from the phase diagram in *Figure 8.7* and the flow diagram in *Figure 8.8*. Suppose, for example, that it is required to separate completely 100 lb/h of an equal mixture of *m*- and *p*-xylene. These two isomers form a binary eutectic. For this particular process n-heptane is chosen as the solvent, and a 7:1 solvent

270

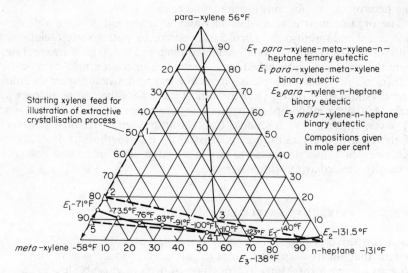

Figure 8.7. Solid-liquid phase diagram for p-*xylene–*m-*xylene–*n-*heptane system, illustrating extractive crystallisation. (After* R. A. FINDLAY *and* J. A. WEEDMAN[14])

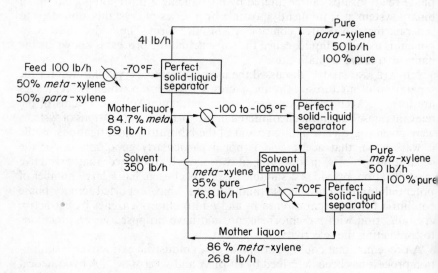

Figure 8.8. Theoretical extractive crystallisation process.

ratio based on the *m*-xylene product is used. The sequence of operations is given by points 1 to 5 in *Figure 8.7*. The 50 per cent *meta–para-* mixture (1) is first cooled to $-55\,°F$ to give pure *para-* crystals and a liquid (2) containing 80 per cent *meta-*. The crystals are removed and the required quantity of solvent (n-heptane) is added to the liquor, to give a ternary mixture (3); this is cooled to about $-105\,°F$ (4), when more *para-* is deposited. Cooling beyond about $-105\,°F$ would result in the co-deposition of both *m-* and *p*-xylene. The crystals and liquor are separated, and the n-heptane is removed from the mother liquor, e.g. by distillation, to leave a liquor (5) containing 92 per cent of *m*-xylene, which is cooled to yield pure *m*-xylene crystals and a liquor. After separation the mother liquor is returned to liquor (2), which makes the process continuous.

The flow sheet in *Figure 8.8* indicates the cycle of operations in this theoretical extractive crystallisation process. For simplicity, the solid–liquid separators are assumed to be 100 per cent efficient. The flow rates have been calculated from material balances based on the information obtained from the phase diagram in *Figure 8.7*. No industrial applications of this process have yet been reported, but, as Findlay and Weedman point out, it could be used for the recovery of valuable components of mixtures that could not be separated by simple crystallisation. The process would, of course, be more expensive than a conventional crystallisation because of the large amounts of cooling necessary and the several recovery and recycle stages required.

ADDUCTIVE CRYSTALLISATION

The separation of mixtures of substances that form eutectics, and thus cannot be separated by conventional crystallisation, can often be effected by methods based on the well-known phenomenon of compound formation (see Chapter 4). A typical sequence of operations is as follows. A certain substance X is added to a given binary mixture of components A and B. A solid complex, say $A . X$, is precipitated, which leaves component B in solution. The solid and liquid phases are separated, and the complex is split into pure A and X by the application of heat or by dissolution in some suitable solvent.

The best-known example of compound formation in solvent–solute systems is the formation of hydrates, but other solvates, e.g. with methanol, ethanol and acetic acid, are known. In these cases the ratio of the molecules of the two components in the solid solvate can usually be expressed in terms of small integers, e.g. $CuSO_4 \cdot 5H_2O$ or $(C_6H_5)_2NH \cdot (C_6H_5)_2CO$ (diphenyl-amine $\cdot$ benzophenone). There are, however, several other types of molecular complex that can be formed which are not necessarily expressed in terms of simple ratios. These complexes are best considered not as chemical compounds but as strongly bound physical mixtures. Clathrate compounds are of this type; molecules of one substance are trapped in the open structure of molecules of another. Hydroquinone forms clathrate compounds with SO_2 and methanol, for example. Urea and thiourea also have the property of forming complexes, known as adducts, with certain types of hydrocarbons.

In these cases molecules of the hydrocarbons fit into 'holes' or 'channels' in the crystals of urea or thiourea; the shape and size of the molecules determine whether they will be adducted or not.

It is open to question whether adduct formation can really be considered as a true crystallisation process, but the methods of operation employed are often indistinguishable from crystallisation methods. Several different names have been given to separation techniques based on the formation of adducts, but the name adductive crystallisation[14] is probably the best and will be used here. Several possible commercial applications of adductive crystallisation have been reported in recent years[14-20].

The system m-cresol–p-cresol forms two eutectics over the complete range of composition, and the separation of the pure components cannot be made by conventional crystallisation. SAVITT and OTHMER[15] described the separation of mixtures of these two components by the use of benzidine to form a solid addition compound with p-cresol. Actually both m-cresol and p-cresol form addition compounds with benzidine, but the meta- compound melts at a lower temperature than the para-. If the process is carried out at an elevated temperature the formation of the meta- complex can be avoided. The method consists of adding benzidine to the meta–para- mixture at 110 °C. The p-cresol · benzidine crystallises out, leaving the m-cresol in solution. The crystals are filtered off, washed with benzene to remove m-cresol, and the washed cake is distilled under reduced pressure (100 mmHg) to yield a 98 per cent pure p-cresol.

In a somewhat similar manner p-xylene can be separated from a mixture of m- and p-xylene; this binary system forms a eutectic. Carbon tetrachloride produces an equimolecular solid compound with p-xylene, but not with o- or m-xylene. EGAN and LUTHY[16] reported on a plant for the production of pure p-xylene by crystallising meta–para- xylene mixtures in the presence of carbon tetrachloride. Up to 90 per cent of the para- isomer was recovered by distillation after splitting the separated solid complex. The meta- isomer was recovered by fractionally crystallising the CCl_4-free mother liquor. Perfect separation of p-xylene is not possible, because the ternary system CCl_4/m-xylene/CCl_4·p-xylene forms a eutectic, but fortunately the concentration of the complex CCl_4·p-xylene in this eutectic is very low. Several commercial clathration processes for the separation of m-xylene from C_8 petroleum reformate fractions using a variety of complexing agents were put into operation in the early 1960s[19].

Separation processes based on the formation of adducts with urea and thiourea have been described by several authors[14-20]. Urea forms addition complexes with straight-chain, or nearly straight-chain, organic compounds such as paraffin and unsaturated hydrocarbons (>6 carbon atoms), acids, esters and ketones. Thiourea forms rather less stable complexes with branched-chain hydrocarbons and naphthenes, e.g. cyclohexane. For example, if a saturated aqueous solution of urea in methanol is added to an agitated mixture of cetane and iso-octane, a solid complex of cetane·urea is formed almost immediately, deposited from the solution in the form of fine needle crystals. The iso-octane is left in solution. After filtration and washing the complex is heated or dissolved in water, and pure cetane is recovered by distillation. If thiourea had been used instead of urea, iso-octane could have been recovered, the cetane being left in solution.

A simplified flow diagram of a possible urea-adductive crystallisation process[14] is shown in *Figure 8.9*. The hydrocarbon feedstock is fed continuously to a stirred reactor, where it is contacted with a solution of urea in a suitable solvent. The reacted slurry, containing tiny crystals of urea adducts with n-paraffins and/or straight-chain olefines, passes to a solid–liquid separator. The rest of the separation and recovery stages in this hypothetical

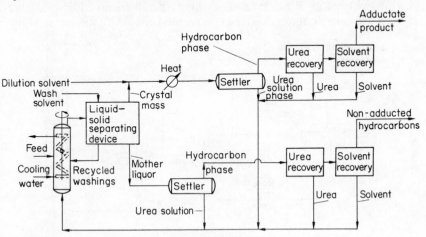

Figure 8.9. Flow diagram for urea-adduct process. (After R. A. FINDLAY *and* J. A. WEEDMAN[14])

process can be traced from the flow diagram. The major large-scale application of urea-adductive crystallisation so far reported is the dewaxing of lubricating oils[18].

DESALINATION BY FREEZING

One of the great engineering challenges of the present day is the search for the ideal process to produce fresh water from sea-water. Desalination by crystallisation offers several possible routes, and the latest position may be judged from the papers presented at a recent symposium[21]. The main freezing processes being pursued at the present time are

1. Vacuum flash freezing
2. Hydrate freezing
3. Immiscible refrigerant freezing

The first of these processes utilises the cooling effect of vaporisation. Sea-water is sprayed into a low-pressure chamber, where some of the water vapour flashes off and the brine partially freezes. The ice–brine slurry is separated, and the ice crystals are washed and remelted. Commercial units based on this principle are being operated[22, 23].

The hydrate processes utilise the fact that certain substances can form inclusion compounds with water, loosely called 'hydrates'. These crystalline substances are separated from the residual brine and decomposed to recover the hydrating agent. Propane is one of the most promising hydrating agents, although dichlorofluoromethane, Freon-12, has also been used with success[22, 23]. One of the important advantages of the hydrate processes is that they operate close to ambient temperature (10–15 °C) and energy costs are minimised. One of the main disadvantages is the difficulty in growing crystals with good filtration and washing characteristics. Light, feathery crystals are quite common, and the compacted beds have a low permeability.

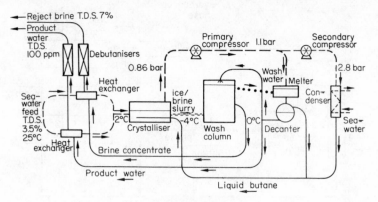

Fig. 8.10. *Immiscible refrigerant (n-butane) freeze desalination process (U.K. A.E.R.E., Harwell)*

The success of the hydrate processes, therefore, appears to hinge on developing a reliable method of controlling the crystal habit.

The indirect freezing processes, i.e. those in which brine is crystallised in some form of heat exchanger, have largely been abandoned, but the direct contact between brine and refrigerant is a very promising technique[22]. The direct contact between liquid n-butane (which does not form a 'hydrate') and sea-water has been widely investigated, and the process being developed in the U.K.[24] promises to be commercially successful (*Figure 8.10*). Pre-cooled brine (3·5 per cent dissolved salts) is fed to the crystalliser and liquid butane, which is sparged into the slurry, boils under reduced pressure (0·86 bar). Agitation is not usually necessary in the crystalliser in view of the vigorous boiling action of the butane. The slurry of ice crystals flows over a weir and is pumped to the wash column. Butane vapour leaving the crystalliser is compressed (1·1 bar) before entering the ice melter. Here it is liquefied by direct contact with the ice and after separation from the water it is pumped back into the crystalliser as a liquid. The compressor, therefore, provides the energy input to the system.

One of the keys to the success of this process is the washing stage: a countercurrent washing process with the equivalent of a large number of theoretical stages is necessary. A hydraulic piston wash column provides a simple method of both separating the bulk of the mother liquor from the ice and providing countercurrent flow for washing the ice crystals free from residual brine. Crystals enter the wash column as a slurry and cake together as they progress upwards countercurrently to a recycle of fresh wash water. Cake permeability, therefore, is all-important. Ice is removed at the top of the column by a rotating scraper blade and is remelted by direct contact with compressed butane gas. After separation and debutanisation a water containing 100 p.p.m. dissolved solids is produced.

COUNTERCURRENT FRACTIONAL CRYSTALLISATION

The process of fractional crystallisation in a column crystalliser by the countercurrent contact between crystals and their melt was first patented by P. M. Arnold in 1951. This technique, together with later refinements, has now evolved on the commercial scale as the Phillips process, in which an end-fed column utilises an oscillatory liquid flow to transport the crystals (*Figure 8.11a*). The early direct reciprocating piston units have now been replaced by indirect pulsed columns. A centre-fed crystalliser developed by

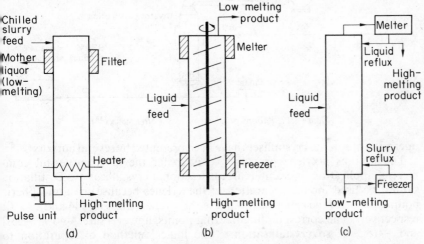

Figure 8.11. Column crystallisers: (a) end-fed pulse column (Arnold type), (b) centre-fed column with spiral conveyor (Schildknecht type). (c) centre-fed column with reflux at both ends

H. Schildknecht in 1961 utilises a spiral conveyor to transport the solids through the purification zone (*Figure 8.11b*), and attempts have been made by the Coal Tar Research Association and Newton Chambers, Ltd., to develop a commercial-scale plant.

The principles of column crystallisation are shown in *Figure 8.11*. In the end-fed pulse unit (a) the slurry feedstock enters at the top of the column and the crystals fall countercurrently to a pulsed upflow of melt. There is a heat and mass interchange between the solid and liquid phases, and pure crystals migrate to the lower zone, where they are remelted to provide a high-purity liquid for the upflow stream. The centre-fed column (b) takes a liquid feed-stock. Crystals are conveyed upwards by means of the screw conveyor and the melt flows downwards. The mode of action is similar to that in a centre-fed distillation column. Laboratory columns of this type are frequently operated batchwise, under total reflux, but reflux may be provided at both ends (c) when the unit functions as a complete fractionator. Analyses of the mode of

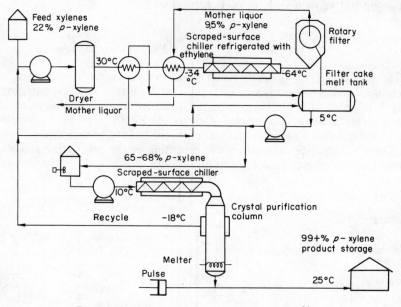

Figure 8.12. Phillips p-*xylene process.* (*After* MCKAY[34])

operation of column crystallisers have been presented by several authors[25-32].

The Phillips p-xylene process[33, 34] is so far the most successful commercial application of countercurrent fractional crystallisation. Distillation cannot be used for the separation of the xylenes because the atmospheric boiling points of the *ortho-*, *meta-* and *para-* isomers are 139, 144 and 138 °C, respectively. The corresponding freezing points, however, are -46.9, -25.2 and $+15.5$ °C, so crystallisation is the logical method of separation to employ.

A schematic diagram[34] of the Phillips process for the production of p-xylene is shown in *Figure 8.12*. In the first stage a rotary filter, ideally suited for separating crystals from a dilute slurry, concentrates feed for the purification column. The mother liquor flows in countercurrent heat exchange with the crude xylene feed and then on to storage. The filter cake is melted by heat exchange with crude xylene and becomes feed for the second stage. The concentrated xylene feed to the second stage is cooled in a scraped-surface chiller, cooled with ethylene, to about -18 °C to form a slurry containing about 40 per cent by weight of crystals, which passes into the pulsed purification column. A product containing >99 per cent of p-xylene emerges from the melter section at the base of the column. A filter at the top of the column prevents crystals being carried out with the mother liquor. Part of the mother liquor stream is returned to the filter-cake melt tank to control the composition of the second-stage feed. The remainder is recycled to the first-stage feed.

The Phillips process has also been applied to the freeze concentration of beer and other beverages[34].

A counter-flow rotating-blade column of the type used in solid–liquid extraction could be adapted[35] for use as a continuous fractional crystalliser. The column (*Figure 8.13*) consists of a series of perforated plates with sectors whose openings are set alternatively opposite one another. Each plate is constantly swept free of settled crystals by the blade. The upflowing solution becomes depleted in the less soluble component. The column can be operated with reflux at both ends. MATZ[35] reports data for the fractional crystallisation

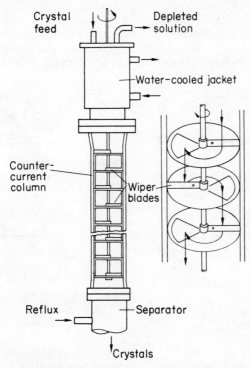

Figure 8.13. Rotating-blade plate crystalliser for countercurrent fractional crystallisation[35]

of $Ba(NO_3)_2$ from aqueous solutions of barium and lead nitrates in a 520 mm long 45 mm diam. column with 15 plates. Lead nitrate is the more soluble component (see *Figure 7.10*) and a separation factor ϕ may be defined as

$$\phi = \frac{[Pb(NO_3)_2 \text{ in solution}][Ba(NO_3)_2 \text{ in crystals}]}{[Pb(NO_3)_2 \text{ in crystals}][Ba(NO_3)_2 \text{ in solution}]} \quad (8.4)$$

A mean value of $\phi = 11\cdot5$ may be expected. In the most favourable case reported by Matz, $1\cdot9$ theoretical stages were obtained, i.e. $3\cdot67$ theoretical stages per metre in the separation section, giving a stage efficiency of $100\,(1\cdot9/15) = 12\cdot7\%$. The maximum production rate was 12 kg/h of Ba-rich crystals per square metre of column cross-section with a mean crystal residence time of 8 min. The crystal size was 20–80 µm, and the crystal purity varied from 96% for a 65% feedstock to $93\cdot5\%$ for a 50% feed. Reflux ratios (ratio of $Ba(NO_3)_2$ in the returned solution phase to $Ba(NO_3)_2$ in the product) of about 10 were used.

OTHER MELT CRYSTALLISATION PROCESSES

Newton Chambers Process

A direct-freezing crystallisation process for the purification of benzene has been developed by the Benzole Producers, Ltd., the Coal Tar Research Association and Newton Chambers, Ltd. An impure benzene feedstock is

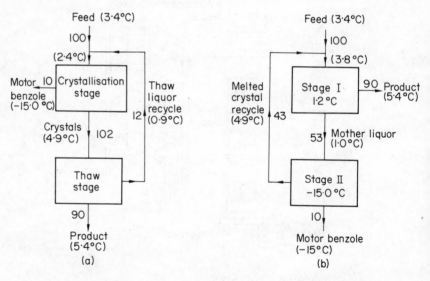

Figure 8.14. The Newton Chambers process: (a) thaw–melt, (b) two-stage. The temperatures in parentheses indicate the freezing points of the various streams. The numbers indicate typical mass flow rates for a mass feed rate of 100

mixed with a refrigerated brine and the resultant slurry is centrifuged to yield benzene crystals and a mixture of brine and mother liquor. After settling, the brine is returned to the refrigeration plant and the mother liquor is either reprocessed or rejected. The success of the method depends on the efficiency of removal of the impure mother liquor adhering to the benzene crystals. Several methods of operation have been proposed[36].

In the thaw method the benzene crystals are washed in the centrifuge with brine at a temperature slightly above the freezing point of the crystals. This melts or 'thaws' some of the benzene crystals, and the melt and warm brine wash away the adhering impure mother liquor. The thaw liquor can be recycled. A typical example of the method is shown in *Figure 8.14a*, where a comparatively rich feedstock (freezing point 3·4 °C) is separated into a pure benzene fraction (f.p. 5·4 °C) and a 'motor benzole' fraction (f.p. 15·0 °C). The latter is used in the U.K. in the formulation of high-octane aromatic motor fuels. Multi-stage operation is also possible, and *Figure 8.14b* shows a two-stage scheme which increases the yield. The first crop of crystals is taken as the product and a further crop is taken from the mother liquor at whatever temperature will give the desired reject fraction quality and product yield. The purity of the crystals from the second stage, melted for the recycle, should not be less than that of the original feedstock.

The Proabd Refiner

The Proabd refiner[37, 38] is essentially a controlled batch cooling process in which a static liquid feedstock is progressively crystallised on to cooling surfaces. If practicable, cooling is continued until the entire charge is solid. As solidification proceeds, the remaining liquid contains progressively less crystallisable component until, in most cases, the eutectic composition is reached. The crystallised solid is then slowly melted by reheating. The impure fraction melts first and drains out of the crystalline mass. As melting proceeds, the melt run-off becomes progressively richer in the desired component. Any number of fractions can be taken off during the melting stage.

A typical flow diagram is shown in *Figure 8.15*. The equipment consists of

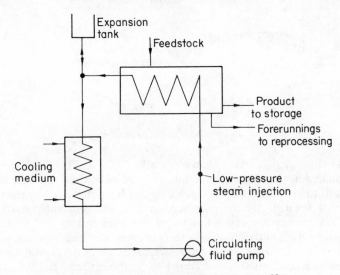

Figure 8.15. Flow diagram of the Proabd refiner[37]

a rectangular tank, containing heat-exchange surfaces such as fin-tubes, filled with feedstock to a depth of about 4 ft. It is essential that the temperature be uniform throughout the tank and that every portion of the charge be within a short distance of a heat exchange surface. During the cooling period the temperature of the circulating heat exchange fluid is raised as it passes through the crystalliser, and during the heating period it is lowered. It is necessary to keep the inlet–outlet differential temperature less than 0·5 °C by ensuring a rapid circulation of heat exchange fluid. For the processing of naphthalene, between 90 and 25 °C, the ideal heat exchange fluid is water with injection of low pressure steam for the heating stage.

Brodie Purifier

A countercurrent melt purification process developed by BRODIE[38] in association with Union Carbide Australia Limited is shown in diagrammatic form in *Figure 8.16*. Feedstock enters the recovery section 2 through inlet 6

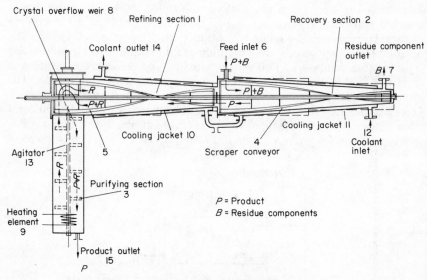

Figure 8.16. The Brodie purifier

and is slowly cooled as it flows towards outlet 7 (the 'cold end'). The product is precipitated as the feed stream cools, the rate of cooling allowing time for the crystals to grow to a size at which they can be transported by the scraper conveyor 4 towards the refining section 1, countercurrently to the feed stream. The conveyor speed is low to prevent back-mixing and to maintain gentle temperature and quality gradients between the feed inlet and the 'cold end'. The cross-sectional area of the vessel is reduced as the crystal and liquor flows diminish towards the cold end. This maintains the axial flow velocity of these streams and prevents back-mixing.

The crystal steam enters the refining section 1 and passes through it countercurrently to a reflux stream, passing over the weir 8 into the purifying section 3, down which the crystals settle as a slurry maintained in a state between sedimentation and fluidisation by the agitator 13. The agitator speed is too low to cause back-mixing of impurities towards the product outlet 15 (the 'hot end'). On reaching the heating element 9, the crystals are melted to form the product P and a reflux R, the latter passing upward through the crystal bed to enter and pass through the refining section over the weir 8. The reflux stream is slowly cooled as it passes through this section, the cross-section of which is also reduced as the reflux stream diminishes within it by recrystallisation. The coolant enters inlet 12 and passes through jackets 11 and 10 to the outlet 14.

In practice the ideal conical forms of sections 1 and 2 are approximated by cylinders of reduced diameter as indicated by the dotted lines in *Figure 8.16* These cylinders can be mounted in cascade, the liquor and crystal streams flowing from one to the other over weirs in a manner similar to their flow between sections 1 and 3.

Utility requirements of a Brodie purifier operating commercially for the production of *p*-dichlorbenzene of 99·99% purity, recovering 88% of the *p*- isomer from a mixture of chlorobenzenes containing 75% *para*- are claimed to be [38]: steam 0·11 kg/kg of product; electricity for power and heating 0·044 kWh/kg; refrigeration 340 kJ/kg. Similar performances are reported for separations of other eutectic systems where the feed composition allows for approximately similar recovery values. The process has been successfully applied in pilot plants for naphthalene, benzene, acetic acid and *p*-xylene.

Rotary Drum Crystallisers

Rotary drum crystallisers are used on an industrial scale for the recovery of salts such as sodium sulphate from waste liquors. Although no large-scale industrial units have yet been reported for the fractional crystallisation of melts, the technique appears to be quite promising for this purpose, and the method has been widely studied on the laboratory scale [39-41].

A typical laboratory scheme is shown in *Figure 8.17*. The crystalliser

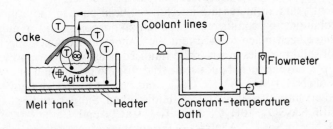

Figure 8.17. Scheme for a laboratory rotary drum crystalliser [39]

consists of a horizontally mounted hollow cylinder, partially immersed in the melt, with coolant entering and leaving the inside through hollow trunnions. As the drum rotates, crystals formed on the chilled surface emerge from the melt, are air-cooled to some extent, and then removed by a doctor knife. Good agitation of the melt near the drum surface is necessary for effective heat and mass transfer rates.

Typical laboratory units used so far for fractional crystallisation measure 10–25 cm diam. and 15–30 cm long, rotated at speeds in the range 0·1 to 1 rev/min. In an industrial plant several drums would probably be operated in series, since most systems would contain impurities at levels which a single drum could not effectively remove. A multi-drum plant proposed[40] for naphthalene–benzoic acid separation, based on laboratory data, indicated that a feed rate of 430 kg/h of a 95% naphthalene feedstock would be separrated into 330 kg/h of 99·5% naphthalene and 102 kg/h of 22·8% naphthalene in a unit consisting of four 2 m diam., 4 m long drums. Separation rates in the range 25–50 kg/h per square metre of drum surface were expected.

SPRAY CRYSTALLISATION

The term 'spray' crystallisation is really a misnomer; strictly speaking, crystals are not grown by this method – solid is simply deposited from a very concentrated solution by a technique similar to that used in spray drying. Size and shape of the solid particles depend to a large extent on those of the spray droplets. The spray method is often employed when difficulties are encountered in the conventional crystallisation techniques, or if a product with better storage and handling properties can be produced.

HOLLAND[42] described a spray crystalliser for the production of anhydrous sodium sulphate. Below 32·4 °C sodium sulphate crystallises from solution in the form of the decahydrate; above this temperature the anhydrous salt is formed. However, anhydrous sodium sulphate has an inverted temperature–solubility characteristic (see p. 30), and trouble is encountered with scale formation on the heat transfer surfaces of conventional evaporating crystallisers when operating temperatures in excess of 32·4 °C are used. In the plant described by Holland a concentrated solution of sodium sulphate was sprayed, or splashed, in the form of tiny droplets into a chamber through which hot gases flowed. The gases entered at about 900 °C. The continuously operated unit produced a powdered anhydrous product.

In recent years an increasing interest has been paid to the utility of the spray technique for commercial production purposes, and a number of studies have been made to ascertain the important variables affecting the growth of solid particles in gas–solid fluidised beds. METHENY and VANCE[43], for example, have reported pilot-scale trials for the production of crystalline granules of $(NH_4)_2SO_4$, NaCN, $CaCl_2$ and NaOH by the spraying solution onto seed crystals introduced into a fluidised bed. Similar studies have been made with $Na_2SO_4 \cdot 10H_2O$, $Al(NO_3)_3$, $CaCl_2$ and $Ca(NO_3)_2 \cdot 4H_2O$[44], NH_4NO_3[45] and NaCl[46]. Analyses of the complex kinetics of particle growth in fluidised beds have been made by several authors[45-47].

Prilling

A type of spray crystallisation is becoming widely employed in the manufacture of fertiliser chemicals such as ammonium nitrate and urea. The name 'prilling' is used to describe this process, and the solid spherical granules are called prills. Ammonium nitrate, which presents a potential explosion hazard, is produced commercially in particulate form by three main methods, viz. conventional crystallisation, graining and prilling[48]. The first process carried out, for example, in vacuum classifying crystallisers requires rather expensive equipment; but as the operating temperatures never exceed 65–70 °C, the process is quite safe. Slightly rounded crystals of about 16 to 35-mesh (1·18–0·5 mm) are generally produced. In the graining process, solutions of the salt are evaporated in open pans to a moisture content of about 2 per cent, and this material, which is almost a molten salt, is passed to graining kettles, where it is solidified into rounded granules, 12 to 35-mesh (1·7–0·5 mm), under the action of heavy ploughs. The high operating temperatures in this process, ~ 150 °C, render the process liable to be hazardous. Both equipment and operating costs are high.

The prilling process is cheapest in capital and operating costs, but intermediate between conventional crystallisation and graining in potential hazards. The method of operation is as follows[48]. A solution of ammonium nitrate is concentrated to about 5 per cent by weight of water and sprayed at about 140 °C into the top of large towers, 100 ft high and 20 ft in diameter. Cool air, at about 25 °C, enters the base of the tower and flows upwards, countercurrently to the falling droplets. The droplets, suddenly chilled when they meet the air stream, solidify. The prills are removed from the bottom of the tower, at about 80 °C, but as they still contain about 4–5 per cent moisture they are dried at a temperature not exceeding 80 °C to ensure that no phase transition occurs; they may then be dusted with diatomaceous earth or some other coating agent. The prills made by this process are reported to have twice the crushing strength of those formed by the spray drying of fused, water-free ammonium nitrate.

Ammonium nitrate exhibits four enantiotropic phase changes between -18 and 125 °C (see p. 17) and the transition between orthorhombic forms III and IV, which occurs at around 32 °C, involves a volume change of $\pm 3·6$ per cent. After a few cycles through this transition temperature ammonium nitrate prills fracture and may even disintegrate completely. The use of additives that raise the transition temperature, such as magnesium nitrate, can minimise the problem, and recent work by RUSSO[49] indicates that substances such as aluminium silicate, which act as nucleating agents during prilling, produce highly stable fine-grained prills.

An interesting method of prilling has been developed[50] for the processing of calcium nitrate in a form suitable for use as a fertiliser. Calcium nitrate, which is hygroscopic, is produced as a by-product in the manufacture of nitrophosphate fertiliser. The process consists of crystallising droplets of calcium nitrate in the form of prills in a mineral oil to which seed crystals have been added. A spray of droplets of the concentrated solution, formed by allowing jets of the liquid at 140 °C to fall on to a rotating cup, fall into an oil-bath kept at 50–80 °C. The prills are removed, centrifuged to remove surplus oil, and packed into bags. Because of the thin film of oil which remains on the particles, the material is less hygroscopic than it would normally be, there is no dust problem on handling, and the tendency to cake on storage is minimised.

SUBLIMATION

So far, in the discussion of industrial crystallisation processes, the deposition of a solid phase from a supersaturated or supercooled liquid phase only has been considered. The crystallisation of a solid substance can, however, be induced from a supersaturated vapour in the process generally known as 'sublimation'. Strictly speaking, the term sublimation refers to the phase change

$$solid \rightarrow vapour$$

without the intervention of the liquid phase. In its industrial application, however, the term usually refers to the condensation process as well, i.e.

$$solid \rightarrow vapour \rightarrow solid$$

In practice, for heat transfer reasons, it is often desirable to vaporise the substance from the liquid state, so the complete series of phase changes in an industrial sublimation process can be

$$solid \rightarrow liquid \rightarrow vapour \rightarrow solid$$

It is on the condensation side of the process that the appearance of the liquid phase is prohibited. The supersaturated vapour must condense directly to the crystalline solid state. Two recent reviews have been made[51, 52] of the industrial applications of sublimation techniques.

A list of substances amenable to purification by sublimation is given in *Table 8.1*. In addition, the sublimation of ice in the process of freeze drying has become an important operation in the biological and food industries[53, 54].

Table 8.1. MATERIALS AMENABLE TO PURIFICATION BY SUBLIMATION[52]

Aluminium chloride	Naphthalene
Anthracene	β-Naphthol
Anthranilic acid	Phthalic anhydride
Anthraquinone	o-Phthalimide
Benzanthrone	Pyrogallol
Benzoic acid	Salicylic acid
Calcium	Sulphur
Camphor	Terephthalic acid
Chromium chloride	Titanium tetrachloride
Ferric chloride	Thymol
Iodine	Uranium hexafluoride
Magnesium	Zirconium tetrachloride

The mechanism of a sublimation process can be described with reference to the pressure–temperature phase diagram in *Figure 8.18*. The significance of the *P–T* diagram applied to one-component systems has already been discussed in Chapters 3 and 4. The phase diagram is divided into three regions, solid, liquid and vapour, by the sublimation, vaporisation and fusion curves. These three curves intersect at the triple point *T*. The position of the triple point in the diagram is of the utmost importance; if it occurs at a pressure above atmospheric, the solid cannot melt under normal atmospheric conditions, and true sublimation, i.e. solid → vapour, is easy to achieve. The triple point for carbon dioxide, for example, is $-57\,^{\circ}C$ and 5 atm, so liquid CO_2 cannot result from the heating of solid CO_2 at atmospheric pressure: the solid simply vaporises. However, if the triple point occurs at a pressure less than atmospheric, certain precautions are necessary if the phase changes solid → liquid → vapour → *liquid* → solid are not to

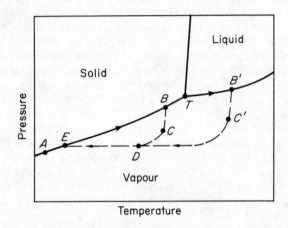

Figure 8.18. True and pseudo-sublimation cycles

take place. If the solid → liquid stage is permitted before vaporisation, the operation is often called 'pseudo-sublimation'.

In *Figure 8.18* both true and pseudo-sublimation cycles are depicted. For the case of a substance with a triple point at a pressure greater than atmospheric, true sublimation occurs. The complete cycle is given by path *ABCDE*. The original solid *A* is heated to some temperature represented by point *B*. The increase in the vapour pressure of the substance is traced along the sublimation curve from *A* to *B*. The condensation side of the process is represented by the broken line *BCDE*. As the vapour passes out of the vaporiser into the condenser, it may cool slightly, and it may become diluted as it mixes with some inert gas such as air. Point *C*, therefore, representing a temperature and partial pressure slightly lower than point *B*, can be taken as the condition at the inlet to the condenser. After entering the condenser the vapour mixes with more inert gas, and the partial pressure of the substance and its temperature will drop to some point *D*. Thereafter the vapour cools essentially at constant pressure to the conditions represented by point E, the temperature of the condenser.

When the triple point of the substance occurs at a pressure less than atmospheric, the heating of the solid may easily result in its temperature and vapour pressure exceeding the triple point conditions. The solid will then melt in the vaporiser; path A to B' in *Figure 8.18* represents such a process. However, great care must be taken in the condensation stage; the partial pressure of the substance in the vapour stream entering the condenser must be reduced below the triple point pressure to prevent initial condensation to a liquid. The required partial pressure reduction can be brought about by diluting the vapours with an inert gas, but the frictional pressure drop in the vapour lines is generally sufficient in itself. Point C' in *Figure 8.18* represents the conditions at the point of entry into the condenser, and the condensation path is represented by $C'DE$.

Sublimation techniques can be classified conveniently into three types: simple, vacuum and entrainer. In simple sublimation the solid material is heated and vaporised, and the vapours diffuse towards the condenser, the driving force for diffusion being the partial pressure difference between the vaporising and the condensing surfaces. The vapour path between vaporiser and condenser should be as short as possible to reduce the resistance to flow. Simple sublimation has been practised for centuries; ammonium chloride, iodine and 'flowers' of sulphur have all been sublimed in this manner, often in the crudest of equipment.

Vacuum sublimation is a natural follow-on from simple sublimation. The transfer of vapour from the vaporiser to the condenser is enhanced by reducing the pressure in the condenser, which thus increases the partial pressure driving force. Iodine, pyrogallol and many metals have been purified by this type of process. The exit gases from the condenser usually pass through a cyclone or scrubber to protect the vacuum-raising equipment and to minimise the loss of product.

An inert gas may be blown into the vaporisation chamber of a sublimer to increase the rate of flow of vapours to the condensing equipment and thus increase the yield. Such a process is known as 'entrainer' or 'carrier' sublimation. Air is the most commonly used entrainer, but superheated steam can be employed for substances that are relatively insoluble in water, e.g. anthracene. When steam is used as the entrainer, the vapours may be cooled and condensed by direct contact with a spray of cold water. In this manner an efficient recovery of the sublimate is made, but the product is obtained in the wet state.

The use of an entrainer in a sublimation process has many desirable features. It enhances the flow of vapours from the sublimer to the condenser, as already mentioned; it also provides the heat needed for sublimation, and thus an efficient means of temperature control is provided. The technique of entrainer sublimation, whether by gas flow over static solid particles or through a fluidised bed, is ideally suited to continuous operation.

The purification of salicylic acid provides a good example of the use industrially of entrainer sublimation. Air may be used as the carrier gas; but as salicylic acid can be decarboxylated in hot air, a mixture of air and CO_2 is often preferred. The process shown in *Figure 8.19* is carried out batchwise. A 5–10 per cent mixture of CO_2 in air is recycled through the plant, passing over heater coils before passing over the containers, e.g. bins or trays, holding the impure salicylic acid in the vaporiser. The vapours then pass to a series of air-cooled chambers, where the sublimed salicylic acid is deposited. A trap removes any entrained sublimate before the gas stream is returned to the heater. Make-up CO_2 and air are introduced to the system as required, and the process continues until the containers are emptied of all volatile matter.

A typical example of a continuous sublimation plant is shown in *Figure 8.20*. The impure material is pulverised in a mill, and hot air or any other suitable

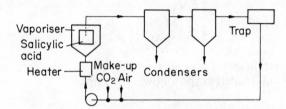

Figure 8.19. *An entrainer sublimation process used for salicylic acid purification*

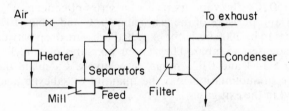

Figure 8.20. *A typical continuous sublimation unit*

gas mixture blows the fine particles, which readily volatilise, into a series of separators, e.g. cyclones, where non-volatile solid impurities are removed; a filter may also be fitted in the vapour lines to remove final traces of impurity. The vapours then pass to a series of condensers. The exhaust gases can be recycled or passed to atmosphere through a cyclone or wet scrubber.

Sublimation Yield

The yield from an entrainer sublimation process can be estimated as follows. If the inert gas mass flow rate is denoted by G, the mass rate of sublimation, S, will be given by

$$\frac{G}{S} = \frac{\rho_G p_G}{\rho_s p_s}$$

(8.5)

where p_G and p_S are the partial pressures of the inert gas and vaporised substance, respectively, in the vapour stream and ρ_G and ρ_S are their vapour densities at the temperature of vapour stream. The total pressure, P, of the system will be the sum of the partial pressures of the components

$$P = p_G + p_S$$

so equation (8.5) can be written

$$S = G\left(\frac{\rho_S}{\rho_G}\right)\left(\frac{p_S}{P - p_S}\right) \tag{8.6}$$

or, in terms of the molecular weights of the inert gas, M_G, and the material being sublimed, M_S,

$$S = G\left(\frac{M_S}{M_G}\right)\left(\frac{p_S}{P - p_S}\right) \tag{8.7}$$

The use of equation (8.7) for predicting the yield from an entrainer sublimation process is illustrated in the following example.

Example

It is proposed to purify salicylic acid (m.p. 159 °C) by entrainer sublimation with air at 150 °C. The vapours pass to a series of condensers, the internal temperature and pressure of the last being 40 °C and 1 bar (10^5 N/m^2). The air flow rate is to be 2000 kg/h, the expected pressure drop between vaporiser and last condenser, 15 mbar. The vapour pressures of salicylic acid at 150 and 40 °C are 14·4 and 0·023 mbar, respectively. Calculate the maximum possible rate of sublimation and the quantity of salicylic acid remaining uncondensed in the exit gases.

Solution

The maximum rate of sublimation is calculated by assuming that the air stream leaving the vaporiser is saturated with salicylic acid vapour.
    Molecular weights: salicylic acid = 138, air = 29.

Vaporisation stage (150 °C)

The total pressure, P, in the vaporiser is equal to 1 bar plus the pressure drop (15 mbar). Therefore $P = 1·015$ bar; $p_S = 0·0144$ bar, and $G = 2000$ kg/h. Substituting these values in equation (8.7),

$$S = 2000\left(\frac{138}{29}\right)\left(\frac{0·0144}{1·015 - 0·0144}\right)$$

$$= 137 \text{ kg/h}$$

Condensation stage (40 °C)

At this temperature $p_S = 2.3 \times 10^{-5}$ bar, total pressure $P = 1$ bar. Again, using equation (8.7),

$$S = 2000 \left(\frac{138}{29}\right)\left(\frac{2.3 \times 10^{-5}}{1 - 2.3 \times 10^{-5}}\right)$$

$$= 0.23 \text{ kg/h}$$

Therefore the rate of sublimation is 137 kg/h, while the loss from the condenser exit gases is only 0.23 kg/h.

It must be noted, however, that this maximum yield will only be obtained if the air is saturated with salicylic acid vapour at 150 °C, and saturation will only be approached if the air and salicylic acid are contacted for a sufficient period of time at the required temperature. A fluidised-bed vaporiser may approach these optimum conditions; but if air is simply blown over bins or trays containing the solid, saturation will not be reached and the actual rate of sublimation will be less than calculated. In some cases the degree of saturation achieved may be as low as 10 per cent of the possible value. Little comment is necessary, therefore, on the importance of designing an efficient vaporiser.

Only the minimum loss of product in the exit gases can be calculated. Any other losses due to entrainment, which might be considerable, will depend on the design of the condenser and cannot be calculated theoretically. An efficient scrubber can, of course, minimise or eliminate these losses.

Sublimer condensers usually take the form of large air-cooled chambers which usually give very low heat transfer rates, probably not greater than 1 or 2 Btu/h ft^2 °F (say 5–10 W/m^2 K). This is only to be expected because sublimate deposits on the condenser walls and acts as a lagging. In addition, vapour velocities within the chambers are generally very low. Quenching of the vapours with cold air in the chamber may increase the rate of heat removal, but excessive nucleation is liable to occur. If this happens, the material is deposited as a fine snow. Studies with phthalic anhydride[55] have shown that the efficiency of condensation by quenching under optimum conditions may be as high as 96 per cent. The condenser walls may be kept clear of solid by use of internal scrapers or brushes, or even swinging weights. All vapour lines in sublimation units should be of a wide bore, adequately insulated and, if necessary, provided with a heating source to minimise blockage due to the build-up of sublimate.

One of the main hazards of air-entrainment sublimation is the risk of fire; many substances that are considered to be quite safe in their normal state can produce explosive mixtures with air. All the electrical equipment, lights, etc., should be flame-proof, and every precaution should be taken to avoid the accidental production of a spark. Personnel should wear rubber-soled shoes, and all handling and maintenance tools should be made of non-sparking materials. In some cases the vapour lines and condenser chambers can become charged with static electricity, so it is essential that these items of equipment be earthed efficiently.

In sublimation plants it is usually necessary to have a battery of condensers linked in series. The internal temperatures of the condensers will decrease down the line, and it is generally found that the crystal form of the product alters from condenser to condenser. In the first chamber, where the temperature is fairly high, the rate of crystal growth may exceed the rate of condensation, and large crystals, frequently acicular, will grow. As the vapours proceed through the condensers the crystal form will reduce in size, and a fine fluffy product may be deposited in the last. In the absence of a market demand for different crystal formations, a blending operation will be necessary. If the first condenser in the battery is too cold, the entering vapours become heavily supersaturated and spontaneous nucleation will occur, with the result that a light flocculent mass (snow) is deposited. The operating conditions necessary for the control of crystal growth in sublimers can only be found from experience.

<h2 style="text-align:center">WASHING AND LIXIVIATION</h2>

The actual crystals produced in a crystalliser are themselves essentially pure; yet after they have been separated from their mother liquor and dried, the resultant mass may be relatively impure. The reason for this is that the removal of mother liquor is frequently inadequate. Crystals retain a small quantity of mother liquor on their surfaces by adsorption, and a larger amount within the voids of the mass due to capillary attraction. If the crystals are irregular, the amount of mother liquor retention within the crevices may be considerable; crystal clusters and agglomerates are notorious in this respect.

All crystallisation processes that utilise a liquid working medium must thus be followed by an efficient liquid–solid separation. Centrifugal filtration can reduce the mother liquor content to about 1 per cent by weight in certain cases, but regular crystalline masses are usually discharged from a centrifuge containing 5–10 per cent. For irregular fine crystals, however, even centrifuging may yield a product containing up to 50 per cent by weight of mother liquor. It is extremely important, therefore, not only to use the most efficient type of filter, but also to produce regular crystals in the crystalliser.

After filtration the product is usually given a wash to reduce still further the amount of impure mother liquor retained. On a centrifuge, for example, the filter cake is best washed with the basket rotating at a speed slower than that used during the filtration stage. Hoses and watering cans are often employed for this purpose, but these methods are most inefficient. Uniform washing is best achieved by the use of a series of nozzles that direct an even spray over the depth of the cake. A suitable wash liquor tank can be installed above the machine so that, on the operation of a simple on–off valve, each basket load receives a fixed quantity of wash.

Cakes of soluble crystalline substances should not be too thick if a washing operation is required; otherwise the wash liquor becomes saturated long before it passes through the mass, and the mother liquor impurities are not removed effectively; prolonged washing would, of course, reduce the final yield. The washing operation is followed by maximum-speed spinning to remove the wash liquor. A two-way outlet can be fitted to the centrifuge casing so that the strong mother liquor and weak wash liquor are directed to separate collection points.

If the crystals are very soluble in the mother solvent, another liquid in which the substance is relatively insoluble may be used for washing purposes. The wash liquid should be miscible with the mother solvent. For example, water could be used for washing substances crystallised from methanol, and methanol could be used for washing substances crystallised from benzene. Unfortunately, this 'two-solvent' method usually means that a solvent recovery unit is required. Alternatively, a wash liquor consisting of a cold saturated solution of the pure substance in the pure mother solvent could be used if the crystals were appreciably soluble. The contaminated wash liquor could then be recycled or re-used in some other way.

In extreme cases when very pure crystals are required, or when simple washing is inadequate, two mother liquor removal stages may be necessary. The wet crystals could be removed from the filter, redispersed in pure cold solvent, and then filtered off again. There may, of course, be an appreciable loss of yield after such a washing or lixiviation process, but this would be less than that after a complete recrystallisation.

When crystallisation has been carried out by the reaction method, the crystalline material may be relatively insoluble in the working solvent. The mother liquor, however, may contain large quantities of soluble material, and simple filtration and washing may be quite inadequate for its complete removal, especially if the crystalline particles are very fine. For example, barium sulphate can be precipitated as a crystalline product by mixing hot solutions of barium chloride and sodium sulphate, but the $BaSO_4$ crystals are usually very small indeed, and difficulties are encountered in both the filtration and washing operations. In this case the sodium chloride in solution could be removed by a washing and decantation technique carried out either batchwise or continuously.

The wash liquor requirements for decantation washing can be deduced rapidly by a graphical method proposed by KIRBY[56], who has also described a submerged filter for increasing the efficiency of the process[57]. The theory of decantation washing is similar to the theory of leaching proposed by DONALD[58]. The assumptions are that all the soluble impurity is in solution and that no concentration gradients exist in the agitated vessel.

Let I_0, I_1, and I_n be the impurity contents of the material (kg of impurity per kg of product) initially, after stage 1 and after stage n, respectively. Let F be the fraction of liquid removed at each decantation. Then mass balances over the various stages give

stage 1:
$$I_1 = I_0 - FI_0$$
$$= I_0(1-F)$$

stage 2:
$$I_2 = I_0(1-F)^2$$

stage n:
$$I_n = I_0(1-F)^n \tag{8.8}$$

Equation (8.8) may be rewritten in the form

$$\ln(I_n/I_0) = n \cdot \ln(1-F) \tag{8.9}$$

Equation (8.9) is plotted in *Figure 8.21a*.

For washing on a continuous basis, where fresh wash liquid enters the vessel continuously and liquor is continually being withdrawn through a filter screen, a mass balance over the unit gives

$$V dI = -I dW$$

or

$$\frac{W}{V} = -\ln \frac{I_n}{I_0} \tag{8.10}$$

where I_0 and I_n are the initial and final impurity concentrations, and V and W the volumes of the liquor in the vessel and wash water, respectively. Equation (8.10) is plotted in *Figure 8.21b* for various values of V.

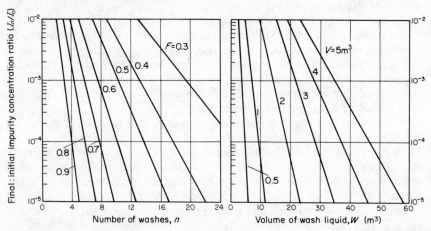

Figure 8.21. Graphical methods for estimating (a) the number of washes in a batch process, and (b) the volume of wash liquid required in a continuous process. (F = fraction of liquid removed at each stage, V = volume of liquor in the vessel, m³)

From equations (8.9) and (8.10) we get

$$n \cdot \ln(1-F) = -\frac{W}{V} \tag{8.11}$$

or, rearranging and dividing by F,

$$\frac{W}{nFV} = \frac{-\ln(1-F)}{F} \tag{8.12}$$

In batch washing the quantity nFV represents the amount of wash liquor used. The following example illustrates the use of the above equations and of *Figure 8.21*.

Example

100 kg of calcium carbonate is produced by the reaction

$$CaCl_2 + Na_2CO_3 \rightarrow CaCO_3 + 2NaCl$$
$$111 \quad + \quad 106 \qquad\qquad 100 \quad + \quad 117$$

(a) Determine the number of washes required to reduce the NaCl concentration to 0·01 per cent in the final product. The fraction of liquid removed at each decantation is determined by observing the settling of the slurry in a graduated cylinder. In this case, assume $F = 0·7$.

(b) Calculate the volume of wash water required if the reaction mixture, containing 2 m^3 of liquor, is to be washed continuously to produce the same product as in (a).

(c) Compare the wash water requirements of the batch and continuous processes.

Solution

(a) $I_0 = 117/100 = 1·17$, $I_n = 0·0001$, $I_n/I_0 = 8·55 \times 10^{-5}$

From *Figure 8.21a*, or by use of equation (8.9), for this value of I_n/I_0 and $F = 0·7, n = 7·8$. Therefore eight washes would be required.

(b) For $V = 2·0$ m^3 and a value of $I_n/I_0 = 8·55 \times 10^{-5}$, by use of *Figure 8.20b* or equation (8.10), the volume of wash water required, $W = 18·7$ m^3.

(c) The ratio of water consumption for the two methods can be determined from equation (8.12) written in the form

$$\frac{W_{cont}}{W_{batch}} = \frac{-\ln(1-F)}{F} = \frac{\ln 0·3}{0·7} = 1·72$$

In this case, therefore, the continuous method would require 72 per cent more water than the batch method.

CAKING OF CRYSTALS

One of the most troublesome properties of crystalline materials is their tendency to bind together, or cake, on storage. Most crystalline products are required in a free-flowing form; they should, for example, flow readily out of containers, e.g. sugar and table salt, or be capable of being distributed evenly over surfaces, e.g. fertilisers. Handling, packaging, tabletting and many other operations are all made easier if the crystalline mass remains in a particulate state. Caking not only destroys the free-flowing nature of the product but also necessitates some crushing operation, either manual or mechanical, before it can be used.

The causes of caking may vary for different materials. Crystal size, shape, moisture content, the pressure under which the product is stored, temperature and humidity variations during storage and storage time can all contribute to the compacting of a granular crystalline product into a solid lump. In general caking is caused by the crystal surfaces becoming damp; the solution which is formed evaporates later and unites adjacent crystals with a cement of recrystallised solid. Crystal surfaces can become damp in a number of ways— the product may contain traces of solvent left behind due to inefficient drying, or the moisture may come from external sources.

Take, for example, the case of a water-soluble substance. If the partial pressure of water vapour in the atmosphere is greater than the vapour pressure that would be exerted by a saturated aqueous solution of the pure substance at that temperature, water will be absorbed by the crystals. If, later, the atmospheric moisture content is reduced to give a partial pressure below the vapour pressure of the saturated solution, the crystals will dry out and bind together. Small fluctuations in atmospheric temperature and humidity, sufficient to bring about these changes, can occur several times in one day.

A simple test for determining the atmospheric humidities at which a particular mass of crystals will absorb moisture consists of placing samples of the crystals in desiccators containing atmospheres of different known moisture content. Solutions of various strengths of sulphuric acid may be placed in the base of the desiccator; the equivalent relative humidities of the resulting atmospheres can be calculated from vapour pressure data. Atmospheres of constant relative humidity can also be obtained by using saturated solutions of various salts. *Table 8.2* gives a list of the percentage relative humidities of the atmospheres above various saturated salt solutions at 15 °C.

The percentages listed in *Table 8.2* can also be considered to be the critical humidities of the salts at 15 °C; if the atmospheric humidity is greater than

Table 8.2. PERCENTAGE RELATIVE HUMIDITIES OF THE ATMOSPHERES ABOVE SATURATED SOLUTIONS OF VARIOUS PURE SALTS AT 15 °C

Substance	Formula	Stable phase at 15°C	Percentage relative humidity
Lead nitrate	$Pb(NO_3)_2$	anhyd.	98
Disodium phosphate	Na_2HPO_4	$12H_2O$	95
Sodium sulphate	Na_2SO_4	$10H_2O$	93
Sodium bromate	$NaBrO_3$	anhyd.	92
Dipotassium phosphate	K_2HPO_4	anhyd.	92
Potassium nitrate	KNO_3	anhyd.	92
Sodium carbonate	Na_2CO_3	$10H_2O$	90
Zinc sulphate	$ZnSO_4$	$7H_2O$	90
Potassium chromate	K_2CrO_4	anhyd.	88
Barium chloride	$BaCl_2$	$2H_2O$	88
Potassium bisulphate	$KHSO_4$	anhyd.	86
Potassium bromide	KBr	anhyd.	84
Ammonium sulphate	$(NH_4)_2SO_4$	anhyd.	81
Ammonium chloride	NH_4Cl	anhyd.	79
Sodium chloride	$NaCl$	anhyd.	78
Sodium nitrate	$NaNO_3$	anhyd.	77
Sodium acetate	$C_2H_3O_2Na$	$3H_2O$	76
Sodium chlorate	$NaClO_3$	anhyd.	75
Ammonium nitrate	NH_4NO_3	anhyd.	67
Sodium nitrite	$NaNO_2$	anhyd.	66
Magnesium acetate	$Mg(C_2H_3O_2)_2$	$4H_2O$	65
Sodium bromide	$NaBr$	$2H_2O$	58
Sodium dichromate	$Na_2Cr_2O_7$	$2H_2O$	52
Potassium nitrite	KNO_2	anhyd.	45
Potassium carbonate	K_2CO_3	$2H_2O$	43
Calcium chloride	$CaCl_2$	$6H_2O$	32
Lithium chloride	$LiCl$	H_2O	15

the critical, the salt becomes hygroscopic. The term 'deliquescence' is usually reserved for the case where the substance absorbs atmospheric moisture and continues to do so until it becomes completely dissolved in the absorbed water. This will only occur if the vapour pressure of the salt solution always remains lower than the partial pressure of the water vapour in the atmosphere. Calcium chloride, with a critical humidity of 32 per cent at 15 °C, is a well-known example of a deliquescent salt. The term 'efflorescence' refers to the loss of water of crystallisation from a salt hydrate; this occurs when the vapour pressure exerted by the hydrate exceeds the partial pressure of water vapour in the atmosphere. Sodium sulphate decahydrate, with a critical humidity of 93 per cent, is an example of an efflorescent salt.

Commercial crystalline salts frequently exhibit hygroscopy at atmospheric humidities lower than those given for the pure salts in *Table 8.2*. Usually impurities present in the product cause the trouble. For example, traces of calcium chloride in sodium chloride will render the crystals damp at very low atmospheric humidities. Removal of the hygroscopic impurity would be the answer, but this is not always economical. Coating of the crystals with a fine inert dust will often prevent the mass becoming damp; table salt, for instance, can be coated with magnesium carbonate or calcium aluminium silicate.

A number of precautionary measures can be taken to minimise the possibility of caking. One obvious method would be to pack the perfectly dry crystals in a dry atmosphere, store them in an air-tight container and prevent any pressure being applied during storage. These desirable conditions, however, cannot always be obtained. Caking can also be minimised by reducing the number of contacts between the crystals, and this can be done by endeavouring to produce granular crystals of a uniform size. The crystals should be as large as possible; the smaller they are the larger will be the exposed surface area per unit mass. The actual size, however, is only of secondary importance in this respect. Shape and uniformity of the crystals, on the other hand, are extremely important factors affecting caking, as indicated in *Figure 8.22*.

The minimum caking tendency is shown by uniform granular crystals (*Figure 8.22a*). Even if caking did occur, the mass could easily be broken up because of the open structure and the relatively few points of contact per unit volume. Non-uniform granular crystals (*Figure 8.22b*) are more prone to caking; the voids in between the large granules are filled with the smaller particles, which results in a larger number of points of contact per unit volume, and the caked mass will not be broken up as easily as in the former case. Although the crystals depicted in *Figure 8.22c* are uniform, they are also elongated and may tend to cake badly. Here there are areas, as well as points, of contact; the needles can pack together and set hard on caking. The condition shown in *Figure 8.22d* is even worse. The crystals are both elongated and non-uniform; they can pack together very tightly into a mass with negligible voidage. When this sort of mass cakes, it is often quite impossible to transform it back to its original particulate state. Plate-like crystals also have bad caking characteristics.

Shape and uniformity of the particles also affect the behaviour of the product under storage. If the crystals are packed in bags and stacked on top of one another, the pressure in the bags near the bottom of the pile tends to force the crystals into closer contact; with non-uniform crystals this compaction may be quite severe and, in extreme cases, many of the crystals may be crushed. If the solubility of the salt in water increases with an increase in pressure, traces of solution may be formed under the high local pressure at the points of contact. The solution will then tend to flow into the voids, where the pressure is lower, and crystallise. Storage of crystalline materials under pressure should always be avoided if possible.

Controlled crystallisation, coupled with some form of classifying action in the crystalliser, helps to produce crystals of uniform size. The production of granular crystals, however, may demand the careful control of other conditions of crystallisation to modify the crystal habit; the rate of cooling, the degree of supersaturation and the pH of the crystallising solution can exert

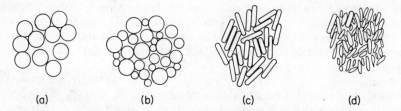

| (a) | (b) | (c) | (d) |

Figure 8.22. Effect of particle shape on the caking of crystals: (a) large uniform granular crystals (good); (b) non-uniform granular crystals (poor); (c) large uniform elongated crystals (poor); (d) non-uniform elongated crystals (very bad)

considerable influence. The deliberate addition of traces of impurity, in the form of active ions or surface-active agents, may also help to produce the right type of crystal. Habit modification control has been fully discussed in Chapter 6.

The use of coating agents has already been mentioned above in connection with table salt, in the production of which magnesium carbonate and calcium aluminium silicate are frequently used. Icing sugar may be coated with 0·5 per cent of tricalcium phosphate or cornflour to prevent caking. Other anti-caking and flow conditioning agents used for industrial crystalline materials include

aluminium powder	magnesium carbonate
Acid Magenta dye	magnesium oxide
calcium silicate	paraffin wax
calcium aluminium silicate	phosphate rock
cellulose gums	silica dust
chalk	silica, hydrated
diatomaceous earth	sodium silico-aluminate
fuller's earth	surfactants
kaolin	synthetic resins
kieselguhr	tricalcium phosphate
magnesium aluminium silicate	

Finely divided substances used as dusting agents must have a good 'covering power' so that very small quantities will produce the desired effect. Several comprehensive accounts have been given of the methods employed for conditioning crystals[48, 59-63] and of testing procedures[60, 62-65] to determine the flow properties and caking tendency of crystalline materials.

The use of habit modification as a means of minimising crystal caking is quite common. This was clearly demonstrated by WHETSTONE[59], who proposed a novel method for preventing caking in water-soluble salts by treating the crystals with a very dilute solution of a habit-modifying dyestuff, e.g. Acid Magenta. Should the crystals cake at some later stage, the cement of crystalline material, now consisting of a modified habit, will be very weak and the crystal mass will easily break down on handling. This method proved effective with salts such as ammonium nitrate, ammonium sulphate and potassium nitrate. The use of a large number of trace additives at the crystallisation stage, and their effects on habit and caking of a number of inorganic salts including NaCl, KCl, K_2SO_4, NH_4Cl, NH_4NO_3 and $(NH_4)_2SO_4$ have been reported by PHOENIX[66].

REFERENCES

1. ENYEDY, G., 'Continuous vacuum crystallisation process: material balance for maximum yield', Chem. Engng Prog. Symp. Ser., 58 (37) (1962), 10
2. GRIFFITHS, H., 'Mechanical crystallisation', J. Soc. chem. Ind. Lond., 44 (1925), 7T
3. THOMPSON, A. R. and MOLSTAD, M. C., 'Solubility and density isotherms for potassium and ammonium nitrates in isopropanol solutions', Industr. Engng Chem., 37 (1945), 1244
4. GEE, E. A., CUNNINGHAM, W. K. and HEINDL, R. A., 'Production of iron-free alum', Industr. Engng Chem., 39 (1947) 1178
5. THOMPSON, A. R. and BLECHARCZYK, S. S., 'Fractional crystallisation of natural brines by ethanol precipitation', Engng Res. Dev. Rep. No. 2 (1959), Univ. Rhode Island
6. HOPPE, H., 'Salt recovery by the Somet process', Chem. Process Engng, 49 (12) (1968), 61
7. PURDON, F. F. and SLATER, V. W., Aqueous Solution and the Phase Diagram, 1946. London; Arnold
8. GARRETT, D. E., 'Industrial crystallisation at Trona', Chem. Engng Prog., 54 (12) (1958), 65
9. BAMFORTH, A. W., Industrial Crystallisation, 1965. London; Leonard Hill
10. STEVENSON, D. G., 'Theory of the design and operation of continuous precipitators', Trans. Instn chem. Engrs, 42 (1964), 316
11. HOLECI, I., 'Emulsion crystallisation of organic substances', CSSR SNTL Tech. Digest, 71 (1965), 515
12. CHIVATE, M. R. and SHAH, S. M., 'Separation of m-cresol by extractive crystallisation', Chem. Engng Sci., 5 (1956), 232
13. DIKSHIT, R. C. and CHIVATE, M. R., 'Separation of nitrochlorobenzenes by extractive crystallisation', Chem. Engng Sci., 25 (1970), 311; 26 (1971), 719
14. FINDLAY, R. A. and WEEDMAN, J. A., 'Separation and purification by crystallisation', in Advances in Petroleum Chemistry and Refining, Vol. I (ed. K. A. KOBE and J. J. MCKETTA), 1958. New York; Interscience
15. SAVITT, S. A. and OTHMER, D. F., 'Separation of m- and p-cresols from their mixtures', Ind. Engng Chem., 44 (1952), 2428
16. EGAN, C. J. and LUTHY, R. V., 'Separation of xylenes', Ind. Engng Chem., 47 (1955), 250
17. FINDLAY, R. A., 'Adductive crystallisation', in New Chemical Engineering Separation Techniques (ed. H. M. SCHOEN), 1962. New York; Interscience
18. HOPPE, A., 'Dewaxing with urea', in Advances in Petroleum Chemistry and Refining (ed. J. J. MCKETTA), Vol. 8, 1964. New York; Interscience

19. SHERWOOD, P., 'Separating aromatic hydrocarbons by clathration', Br. chem. Engng, 10 (1965), 382
20. SANTHANAM, C. J., 'Adductive crystallisation processes', Chem. Engng, 73 (12) (1966), 165
21. Third Int. Symp. on Fresh Water from the Sea, European Federation of Chemical Engineering, Dubrovnik, 1970
22. BARDUHN, A. J., 'Desalination by crystallisation processes', Chem. Engng Prog., 63 (1) (1967), 98
23. ORCUTT, J. C., 'Desalination by freezing', Chapter 17 in reference 25
24. DENTON, W. H. et al., 'Experimental studies on the immiscible refrigerant (butane) freezing process', p. 51 in reference 21
25. M. ZIEF and W. R. WILCOX (Eds.), Fractional Solidification, 1967. New York; Dekker
26. ALBERTINS, R., GATES, W. C. and POWERS, J. E., 'Column crystallisation', Chapter 11 in reference 25
27. GATES, W. C. and POWERS, J. E., 'Determination of the mechanisms causing and limiting separations by column crystallisation', A.I.Ch.E.Jl., 16 (1970), 648
28. BETTS, W. D., FREEMAN, J. W. and MCNEIL, D., 'Continuous fractional crystallisation', J. appl. Chem. Lond., 17 (1968), 180
29. ANIKIN, A. G., Dokl. Akad. Nauk SSSR, 151 (1963), 1139
30. PLAYER, M. R., 'Mathematical analysis of column crystallisation', Ind. Eng. Chem., Process Des. Dev., 8 (1969), 210
31. HENRY, J. D. and POWERS, J. E., 'Experimental and theoretical investigation of continuous flow column crystallisation', A.I.Ch.E.Jl., 16 (1970), 1055
32. BOLSAITIS, P., 'Continuous column crystallisation', Chem. Engng Sci., 24 (1969), 1813
33. MCKAY, D. L. and GOARD, H. W., 'Continuous fractional crystallisation', Chem. Engng Prog., 61 (11) (1965), 99; 'Crystal purification column with cyclic solids movement', Ind. Eng. Chem., Process Des. Dev., 6 (1967), 16
34. MCKAY, D. L., 'Phillips fractional solidification process', Chapter 16 in reference 25
35. MATZ, G., Kristallization, 2nd Ed., 1969. Berlin; Springer-Verlag
36. MOLINARI, J. G. D., 'Newton Chambers Process', Chapter 14 in reference 25
37. MOLINARI, J. G. D., 'The Proabd refiner', Chapter 13 in reference 25. See also Société Proabd. Brit. Pat. 837 295 (1959), 899 799 (1961)
38. BRODIE, J. A., 'A continuous multi-stage melt purification process', Mechanical and Chemical Transactions, Institution of Engineers, Australia (May 1971), 37
39. CHATY, J. C. and O'HERN, H. A., 'An engineering study of the rotary drum crystalliser', A.I. Ch.E.Jl., 10 (1964), 74
40. CHATY, J. C., 'Rotary-drum techniques', Chapter 15 in reference 25
41. SVALOV, G. N., 'Calculation of drum crystallisers', Int. chem. Engng, 9 (1969), 606; also Br. chem. Engng, 15 (1970), 1153
42. HOLLAND, A. A., 'A new type of evaporator', Chem. Engng, 58 (1) (1951), 106
43. METHENY, D. E. and VANCE, S. W., 'Particle growth in fluidised bed dryers', Chem. Engng Prog., 58 (6) (1962), 45
44. MARKVART, M., VANECEK, V. and DRBOHLAV, R., 'The drying of low melting point substances and evaporation of solutions in fluidised beds', Br. chem. Engng, 7 (1962), 503
45. GRIMMETT, E. S., 'Kinetics of particle growth in the fluidised bed calcination process', A.I. Ch.E.Jl., 10 (1964), 717
46. LEE, B. S., CHU, J. C., JONKE, A. A. and LAWROSKI, S., 'Kinetics of particle growth in a fluidised bed calciner', A.I.Ch.E.Jl., 8 (1962), 53
47. KHURANA, K. L. and GUPTA, P. SEN, 'Particle growth in gas-solid fluidised beds', Indian chem. Engr, 12 (1970), 27
48. SHEARON, W. H. and DUNWOODY, W. B., 'Ammonium nitrate', Ind. Engng Chem., 45 (1953), 496
49. RUSSO, V. J., 'Stabilisation of ammonium nitrate prills', Ind. Eng. Chem., Prod. Res. Dev., 7 (1968), 69
50. VAN DEN BERG, P. J. and HALLIE, G., 'New developments in granulation techniques', Proc. Fertil. Soc., 1960
51. MATZ, G., 'Die Sublimation im Rahmen der thermischen Trennverfahren', Chemie-Ingr-Tech., 38 (1966), 299
52. HOLDEN, C. A. and BRYANT, H. S., 'Purification by sublimation', Separation Sci., 4 (1) (1969), 1

53. DAVIES, J. D., 'Freeze drying biological materials', *Process Biochem.,* **3** (6) (1968), 11 ; **3** (7) (1968), 48

54. HOLDSWORTH, S. D., 'Heat and mass transfer in freeze drying', *Process Biochem.,* **3** (11) (1968), 59

55. CIBOROWSKI, J. and SURGIEWICZ, J., 'Studies on sublimation condensing by mixing', *Br. chem. Engng,* **7** (1962), 763

56. KIRBY, T., 'Wash liquor requirements', *Chem. Engng,* **66** (8) (1959), 169

57. KIRBY, T. and FEORINO, J., 'Lamp-shade filter to increase the efficiency of batch decantation washing', *Chem. Engng Prog.,* **55** (11) (1959), 174

58. DONALD, M. B., 'Percolation leaching in theory and practice', *Trans. Instn chem. Engrs, Lond.,* **15** (1937), 77

59. WHETSTONE, J., 'Anti-caking treatment for ammonium nitrate', *Ind. Chemist,* **25** (1949) 401. See also *Discuss. Faraday Soc.,* **5** (1949), 254, 261

60. IRANI, R. R., CALLIS, C. F. and LIU, T., 'Flow conditioning and anti-caking agents', *Ind. Engng Chem.,* **51** (1959), 1285

61. IRANI, R. R., VANDERSALL, H. L. and MORGENTHALER, W. W., 'Water vapour sorption in flow conditioning and cake inhibition', *Ind. Engng Chem.,* **53** (1961), 141

62. BURAK, N., 'Chemicals for improving the flow of powders', *Chemy Ind.,* **1966,** 844

63. CARR, R. L., 'Particle behaviour, storage and flow', *Br. Chem. Engng,* **15** (1970), 1541

64. ADAMS, J. R. and ROSS, W. H., 'Relative caking tendencies of fertilisers', *Ind. Engng Chem.,* **33** (1941), 121

65. WHYNES, A. L. and DEE, T. P., 'The caking of granular fertilisers: a laboratory investigation', *J. Sci. Fd. Agric.,* **8** (1957), 577

66. PHOENIX, L., 'How trace additives inhibit the caking of inorganic salts', *Br. Chem. Engng,* **11** (1966), 34

9

Crystallisation Equipment

THERE are many ways in which the various types of industrial crystallisers can be classified. Self-explanatory headings such as 'batch' and 'continuous' or 'agitated' and 'non-agitated' equipment may be used, but these descriptions are too generalised for most purposes. The units may be classified according to the degree of control that can be exercised over the final product size; names such as 'controlled' and 'uncontrolled' or 'classifying' and 'non-classifying' crystallisers are commonly encountered.

Another method of classification is based on the way in which the necessary supersaturation of the solution or melt is achieved, e.g. by cooling, evaporation or reaction between two phases. If cooling is effected by flash evaporation at low pressure, the term 'vacuum' crystalliser may be applied. The manner in which the growing crystals are brought into contact with the supersaturated liquor has also been suggested for classification purposes[1]; according to this system the three headings 'non-agitated', 'circulating liquor' and 'circulating magma' crystallisers are used.

In the following section the various items of equipment will be divided into three main groups, cooling, evaporating and vacuum crystallisers, but reference will be made to the other classification systems. Comprehensive accounts of industrial crystallisation techniques and equipment have been given by several authors[2-5].

COOLING CRYSTALLISERS

OPEN TANK CRYSTALLISERS

In its simplest form an open tank crystalliser consists of a smooth-walled vessel with no mechanical moving parts, which permits a large surface area of solution to come into contact with the atmosphere. A hot concentrated solution of the solute is poured into the tank, where it cools by natural convection and evaporation over a given period of time, often as long as a few days. A gentle circulation of the liquor usually occurs owing to temperature and concentration gradients within the vessel. Cooling may be aided by the passage of cold air over the surface or by bubbling cold air through the liquid.

Certain melts and aqueous solutions can be processed in this manner, and he batch may be given an occasional stir with a hand paddle to prevent the deposition of a hard crystalline block on the bottom of the crystalliser. Sometimes thin rods or metal strips are hung in the solution; crystals grow on the rods and are thus prevented from falling to the bottom of the crystalliser. No seeding is required because nucleation usually occurs quite readily owing to the presence of atmospheric dust and crystalline fragments left from previous batches.

The crystal magma can be transferred by hand, or mechanical grab, to a filtration unit. Alternatively, a large proportion of the mother liquor may be drained off at the bottom of the tank, the quantity of liquid to be handled by the filter thus being minimised. In batteries of large effluent tanks used for by-product recovery, sufficient settling time may be allowed to permit continuous entry of feedstock and continuous outflow of crystal-free mother liquor; when a given tank is full of crystalline product, it is bypassed and emptied.

No control over the crystal size is possible in open tank crystallisers, but experience will indicate the required size of the vessel and the cooling time necessary to produce the desired type of crystalline mass. Because of the low rate of cooling, a high percentage of large interlocking crystals is usually obtained, but variation in size will be considerable, the crystals ranging from fine dust to large lumps. The irregularity of the crystals results in a high retention of mother liquor within the mass after filtration. Consequently the dried crystals will almost invariably be impure.

For small production of commercial grade crystals, and for by-product recovery purposes, the open tank method of crystallisation may be the most economical one. The capital outlay is usually small, and maintenance and operating costs are generally negligible. On the other hand, handling labour costs can be rather high, in terms of cost per unit mass of product, compared with other crystallisation methods. Reagent-grade chemicals are usually manufactured in open pan crystallisers.

AGITATED TANKS AND VESSELS

Undesirable temperature gradients in an open tank crystalliser can be minimised if a mechanical stirrer is employed; smaller and more uniform crystals will be deposited and the cooling time cycle may be reduced. A somewhat purer product will be obtained as a result of the smaller retention of mother liquor by the crystalline mass, and more efficient washing of the crystals is possible during the filtration stage of the process. Vertical baffles may be used to prevent swirling of the magma and induce desirable flow patterns, but they should terminate below the liquor level to prevent scaling and incrustations.

The agitated vessel may be equipped with a water jacket, or coils, to aid the cooling process. Jackets are preferred to coils because coils tend to become coated with a hard crystalline deposit which reduces very considerably the rate of heat transfer. For very high cooling duties, however, jackets are usually inadequate, and coils have to be used to provide the necessary heat transfer area. One useful rule to remember is that temperature differences greater than 10 °C between the cooling surface and the liquor are undesirable, because high local degrees of supersaturation lead to excessive nucleation and crystal deposition. The estimation of heat transfer rates from coils in agitated vessels is the subject of a recent review[6].

If a cooling jacket is employed, the inner surfaces of the crystalliser should be as smooth and as flat as possible. This minimises crystal build-up on the cold surfaces or, when deposition does occur, facilitates its removal. Crystals should never be chipped away from the wall of a crystalliser, because tiny scratches on the surface become undesirable 'seed centres'. Melting or dissolution is the only safe method of crystal scale removal. Polished stainless steel and glass-lined mild steel are good construction materials for the inner surfaces of crystallisers.

CHANDLER[7] reports an investigation to determine the rate of deposition of crystals on cooling surfaces. The rate is very sensitive to the supersaturation at the heat transfer surface, but relatively insensitive to the nature of the surface. Increased solution turbulence tends to lessen the probability of deposition. Ultrasonic vibrations can prevent crystal incrustation, as shown by DUNCAN and WEST[8], who report that cold heat exchanger surfaces subjected to vibrations in the range 10–100 kHz remain free of crystalline deposits. Tests were carried out with p-xylene crystallising from its isomers and ice from brine. Two mechanisms were suggested, depending on the power level of the ultrasonic vibration. If the power level is high enough to cause cavitation in the liquid phase, the action appears to be the same as that in conventional ultrasonic cleaning, which depends on shock waves generated by the collapse of bubbles. At lower power levels the effect is attributed to the phenomenon of acoustic micro-streaming at the solid–liquid interface.

Air cooling–agitation may often be used with advantage in cases where severe incrustation of cooling surfaces can arise. Cooling is effected predominantly by evaporation.

Although operating and maintenance costs for an agitated cooler are higher than for a simple tank crystalliser, they are small in comparison with the financial advantages gained by the quicker throughput and better product. Labour costs for handling the crystals, however, may still be rather high. The main disadvantage of both tank and agitated crystallisers is that the equipment is usually bulky and occupies much valuable floor space.

No general design for tank crystallisers, stirred or otherwise, can be proposed; the vessels will, according to the needs of the particular process, vary from small shallow pans to very large cylindrical tanks. When non-aqueous solvents are involved, the equipment will take the form of an enclosed agitated vessel fitted with coils or a cooling jacket and a vent leading to a water-cooled condenser.

A typical operating sequence for a batch agitated crystalliser is as follows. The hot solution is cooled as rapidly as possible until supersaturation is achieved. The cooling rate is then reduced to prevent the solution becoming labile, and a predetermined quantity of suitable seed crystals is added.

Seeding may not be necessary in all cases, but it is generally safer to seed than to rely on spontaneous nucleation, which may be uncontrolled and excessive. Once crystallisation commences, the temperature of the charge will tend to rise owing to the liberation of the heat of crystallisation. The cooling rate is therefore adjusted to allow crystallisation to occur at a fixed, or slowly decreasing, temperature. Slow cooling is continued until a large proportion of the solute has been deposited on the seeds; thereafter the rate of cooling can be increased until the required discharge temperature has been achieved. Trial and error on the actual plant will decide the finer details of the procedure. For the crystallisation of substances that oxidise in contact with air, an inert or reducing atmosphere may be introduced into the closed vessel. A sulphur dioxide atmosphere, for example, is used for

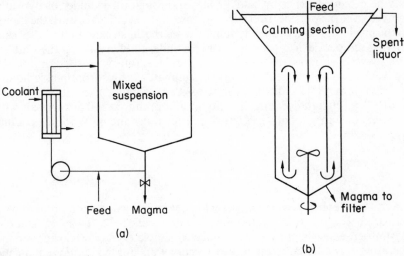

Figure 9.1. Agitated tank crystallisers: (a) external circulation through a heat exchanger, (b) internal circulation with a draft tube

hydroquinone crystallisation in agitated vessels to prevent the darkening of the product.

Good mixing within the crystalliser and high rates of heat transfer between the liquor and coolant can be achieved by the use of external circulation. Because of the high liquor velocity in the tubes, low temperature differences suffice for cooling, and scaling may be considerably reduced. The unit shown in Figure 9.1a can be used for batch or continuous operation.

Internal circulation and mixing can be improved in an agitated vessel crystalliser if the impeller is surrounded by a shroud (the so-called draft-tube agitator). A crystalliser incorporating this principle is shown in Figure 9.1b. Units of this type, called Pachuca growth-type crystallisers, are used for the large-scale production of borax from natural brines at Trona, California[9]. The large tall cylindrical vessels have an upper classifying section, which permits an out-flow of liquor without any carryover of crystals, other than excess nuclei, and maintains a predetermined slurry density in the growth zone. Internal cooling may be provided if necessary. The agitator may be under- (as shown) or over-driven and the flow of magma through the draft tube may be upward or downward, as desired. Some of the considerations to be made in arriving at the dimensions, speed and location of the agitator are described in Chapter 11.

Rotary Crystallisers

Rotating cylinders, similar in some respects to those used as rotary driers or kilns, have been used for the crystallisation of solutions. The cylinder slopes slightly from the feed liquor inlet down to the crystal magma outlet. Cooling may be provided either by cold air blown through the cylinder or by water sprayed over the outside. In the former case internal baffles disturb the liquor on the inside wall and cause it to rain through the air stream. In the latter case internal scraper devices prevent excessive build-up of crystal on the walls.

BAMFORTH[2] quotes the use of an air-cooled rotary crystalliser for the recovery of sulphuric acid and ferrous sulphate crystals from pickle liquor. A 9·5 m long, 1·4 m diam. cylinder with a cooling capacity of 81 kW (7×10^4 kcal/h) handled 3 m^3/h of liquor and produced 500 μm crystals with a power consumption of 4 kW to drive the cylinder and 4 kW for the cooling fan.

Twin Crystalliser

NÝVLT[5] has described a crystalliser which, it is claimed, gives a product with an extremely close size distribution. The unit, called the 'Twin' or 'Double' crystalliser, consists of two interconnected simple crystallisers, each section operating at a different temperature (*Figure 9.2*). Hot feed liquor enters and mixes with the circulating contents of the crystalliser, which pass downwards through the water-cooled draft tube, A, under the influence of an agitator. Part of the cooled magma passes under the adjustable gate, B, into the second compartment of the crystalliser, where it mixes with the circulating magma in the second draft tube, C, operated at a lower temperature. A back-flow of magma occurs above the gate. Large crystals migrate to the bottom of the crystalliser and are discharged. The mother liquor exit is located behind the baffle. If both sections of the crystalliser operate with supersaturated solution, the system functions essentially as crystallisers in series. If one section operates slightly above the saturation temperature, the excess fines are dissolved and coarse crystals with rounded edges are produced.

TROUGH CRYSTALLISERS

Crystallisers consisting of a long shallow trough with an internal agitator and a cooling system are frequently encountered in industries where concentrated and viscous solutions are to be crystallised. The use of the trough crystalliser is, of course, not restricted to viscous solutions only. A well-known piece of equipment belonging to this class is the Swenson–Walker crystalliser[6] illustrated in *Figure 9.3*. One unit of this crystalliser consists of a semi-cylindrical trough about 2 ft wide and 10–15 ft long, generally fitted with a water-cooled jacket. The trough may be open or closed, depending upon

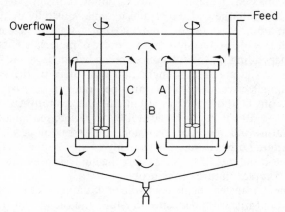

Figure 9.2. The 'twin' or 'double' crystalliser[5]

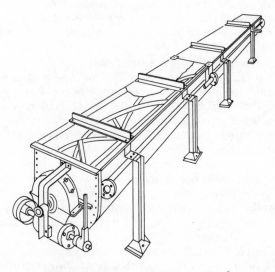

Figure 9.3. Swenson–Walker crystalliser[6]

whether additional atmospheric cooling is required or not. A helical stirrer rotates at a slow speed (5–10 rev/min) inside the trough to aid the growth of the crystals by lifting them and then allowing them to fall back through the solution. The system is kept in gentle agitation and the crystals are conveyed along the trough. This crystalliser, therefore, may be considered as belonging to the 'circulating magma' type.

A number of units can be joined together to provide as long a crystallisation path as desired. Alternatively, to save space, the units can be mounted above one another, but excessive nucleation may occur at the point where the discharge from one unit falls into the lower one, resulting in the production of many fine crystals. Cold water, or refrigerated brine in the later stages, can be passed through the jackets countercurrently to the flow of crystals. At the discharge end the crystal magma is usually delivered direct to a filtration unit. Moderately sized and fairly uniform crystals can be obtained from this type of crystalliser. A variety of materials of construction can be used depending on the substances being handled.

A 40 ft run is the maximum length that can safely be driven from one stirrer shaft: an effective heat transfer area per foot run of crystalliser of the above dimensions is about 3 ft^2 (~ 0.9 m^2/m) and overall heat transfer co-efficients of 10–25 Btu/h . ft^2 °F (~ 50–150 W/m^2 K), based on a logarithmic mean temperature difference between solution and cooling water, may be expected[11]. Among other things, the viscosity of the batch exerts a considerable influence on the rate of heat transfer; the analysis presented by KOOL[12] indicates the complex nature of the problem.

Very low heat transfer coefficients will be obtained if the inner cooling surface is not kept free from deposited crystals, and the helical stirrer can be arranged in close contact with the vessel wall to act as a scraper. Scraping action produces crystal attrition, however, and a high percentage of fines will result in the finished product. For many purposes this is most undesirable, and the helix may have to be located at a distance of $\frac{1}{2}$–1 in away from the wall. Production rates of 20 ton/day of salts such as sodium phosphate and sodium sulphate are possible from a single unit.

Crystallisers similar to the Swenson–Walker type with semi-cylindrical (U-type) or nearly cylindrical (O-type) cross-section have been used for many years in the sugar industry for the crystallisation of concentrated molasses. Units fitted with water-cooled jacket are still employed, but these show poor heat transfer characteristics; scraper-stirrers are not permitted on account of the damage they do to the crystals. Accordingly, many other types of cooling arrangement have been tried[10]. For example, separate banks of water-cooled tubes have been inserted in between slowly rotating paddles in the trough. Alternatively, the water-cooled tubes can function as stirrers (*Figure 9.4*), but they must be of sturdy construction to prevent water leakage. Both stationary and rotating cooling discs have been employed quite successfully in trough crystallisers.

Another continuous trough crystalliser of the circulating magma type is the Wulff-Bock unit, sometimes called a crystallising cradle or rocking crystalliser (*Figure 9.5*). This consists of a long shallow trough, which can be rocked on supporting rollers. The solution to be crystallised is fed in at one end and the crystals are discharged at the other end, continuously. Transverse baffles may be fitted inside the trough to prevent longitudinal surging of the liquor, so the charge flows in zigzag fashion along the unit. The slope of the trough, towards the discharge end, is varied according to the required residence time of the liquor in the crystalliser.

One of the great advantages of the Wulff–Bock crystalliser is the absence of moving parts within the crystallisation zone. Corrosion problems, there-fore, are greatly minimised; the trough can easily be rubber-lined or simi-larly protected if corrosive liquors have to be handled. The method is not suitable for organic solvent systems or for the crystallisation of oxidisable substances, because the crystalliser is open to the atmosphere. A number of

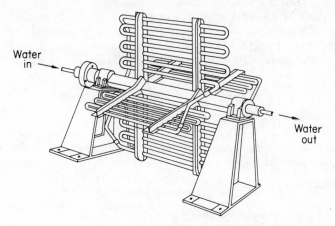

Water in →

Water out

Figure 9.4. Rotating-tube cooling system for a trough crystalliser[13]

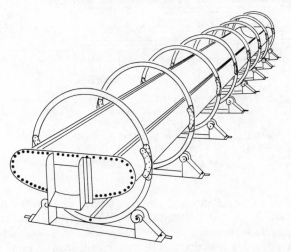

Figure 9.5. Wulff–Bock crystalliser. (After H. GRIFFITHS[14])

units may be joined together; assemblies up to 30 m in total length have been installed. The power requirement for driving a typical 50 m by 1 m unit is about 1·1 kW.

No external cooling is employed; heat is lost by natural convection to the atmosphere. High degrees of supersaturation, therefore, are not encountered at any point within the unit and crystallisation occurs slowly. The gentle agitation prevents crystal attrition as well as formation of a skin of crystals at the surface of the liquor. Large, uniform crystals can be grown in Wulff–Bock crystallisers, and production rates of 3 ton/day of 12 mm crystals have been reported[11]. Potassium chloride, potassium permanganate, sodium acetate, sodium thiosulphate, sodium sulphate and sodium sulphite have been produced commercially in this manner.

Only a few Wulff–Bock crystallisers now remain in service. They have fallen out of favour for a number of reasons: their large size and low through-put, variable performance depending on ambient conditions, and so on. Nevertheless, this crystalliser is still one of the best for the industrial pro-duction of very large regular crystals.

SCRAPED-SURFACE CRYSTALLISER

A typical double-pipe crystalliser, or scraped-surface chiller, is the Votator apparatus shown in *Figure 9.6*. The unit is essentially a double-pipe heat

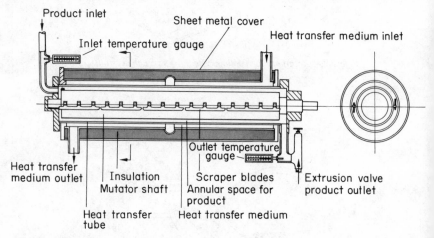

Figure 9.6. Double-pipe crystalliser (Votator apparatus). By courtesy of A. Johnson & Co. (London) Ltd.

exchanger fitted with an internal scraping device to keep the heat transfer surfaces clean. It was first developed by Vogt in 1928 as an ice-cream freezer. The annulus between the two concentric tubes contains the cooling fluid, which moves countercurrently to the crystallizing solution flowing in the central pipe. Located in the central pipe is a shaft upon which scraper blades, generally spring-loaded, are fixed. The solution is pumped through the unit and, due to the high degree of turbulence, fairly high heat transfer coefficients can be recorded.

Scraped-surface crystallisers are employed for the crystallisation of paraffin wax and the processing of viscous materials such as lard and margarine. They have also been applied to the freeze concentration of foodstuffs such as fruit juices, vinegar, tea and coffee[15], but very small crystals are generally produced, since a considerable degree of attrition and nucleation is unavoidable. Indeed in some processes scraped-surface units are often installed for the sole purpose of providing nuclei for the growth zone of another crystalliser. In hydrocarbon separation processes, e.g. the recovery of p-xylene from solvent xylene mixtures (p. 276), ethylene or propane-cooled units may be used in series with open tank crystallisers: the chiller effects heat removal and nucleation; the tank permits the slow growth of the crystals. One advantage of the double-pipe crystalliser is that the amount of liquor hold-up is very low.

The units, which can be fabricated in many materials, vary in diameter from about 75 to 600 mm and in length from about 0·3 to 3 m. For use as heat exchangers rotor speeds up to 2000 rev/min may be used, but for crystallisers much lower speeds (10–50 rev/min) are employed[16, 17]. Heat transfer studies on scraped surface heat exchangers indicate that rotor speeds, blade clearances and liquor properties exert a considerable effect[18]. For the high-speed heat exchangers, heat transfer coefficients up to 4 kW/m² K have been reported, but much lower coefficients (50–700 W/m² K) are expected for crystalliser operation.

KRYSTAL COOLING CRYSTALLISER

Towards the end of the First World War, investigations were carried out in Norway by Isaachsen and Jeremiassen into the problems associated with the continuous production of large uniform crystals, in particular of sodium chloride. These investigations subsequently led to the development, by Jeremiassen, of a method for obtaining a stable suspension of crystals within the growth zone of a crystalliser. The practical application of this method has been incorporated in a continuous classifying crystalliser known by the names Oslo, Jeremiassen or Krystal apparatus[19]. The latter name is now almost exclusively used, and the manufacturing rights of Krystal crystallisers in the United Kingdom, British Commonwealth and many other countries are held by the Power Gas Corporation, Ltd. The Struthers Wells organisation manufacture these crystallisers in the U.S. BAMFORTH[2] has given a concise account of the design and uses of these versatile and compact units.

There are several basic forms of the Krystal apparatus, e.g. the cooling (*Figure 9.7*), the evaporating and the vacuum crystalliser (*Figure 9.20*), but all units based on the original Jeremiassen process have one feature in common – a concentrated solution, which is continuously cycled through the crystalliser, is supersaturated in one part of the apparatus, and the supersaturated solution is conveyed to another part, where it is gently released into a mass of growing crystals. These units, therefore, belong to the 'circulating liquor' type of crystalliser.

The operation of the Krystal cooling crystalliser (*Figure 9.7*) may be described as follows. A small quantity of warm concentrated feed solution (0·5 to 2 per cent of the liquor circulation rate) enters the crystalliser vessel at point A, located directly above the inlet to the circulation pipe B. Saturated solution from the upper regions of the vessel, together with the small amount of feed liquor, is circulated by pump C through the tubes of heat exchanger D, which is cooled rapidly by a forced circulation of water or brine. On cooling, the solution becomes supersaturated, but not sufficiently for spontaneous nucleation to occur, i.e. metastable, and great care is taken to prevent it entering the labile condition. The temperature difference between the process liquor and coolant should not normally exceed 2 °C. The supersaturated solution flows down pipe E and emerges from the outlet F, located near the bottom of the crystalliser vessel, directly into a mass of crystals growing in the

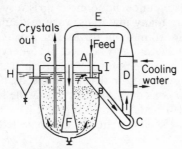

Figure 9.7. Krystal cooling crystalliser

vessel. The rate of liquor circulation is such that the crystals are maintained in a fluidised state in the vessel, and classification occurs by a hindered settling process. Crystals that have grown to the required size fall to the bottom of the vessel and are discharged from outlet G, continuously or at regular intervals. Any excess fine crystals floating near the surface of the solution in the crystalliser vessel are removed in a small cyclone separator H, and the clear liquor is introduced back into the system through the circulation pipe. A mother liquor overflow pipe is located at point I.

Like all other cooling crystallisers, this unit can only be used to advantage when the solute shows an appreciable reduction in solubility with decrease in temperature. Examples of some of the salts that can be crystallised in this manner are sodium acetate, sodium thiosulphate, saltpetre, silver nitrate, copper sulphate, magnesium sulphate and nickel sulphate. BAMFORTH[2] reports the production of 7 ton/day of 10×5 mm sodium thiosulphate crystals in a 2 m diam. 6 m high vessel with 200 m² heat exchange surface and an over-all heat transfer coefficient of 930 W (800 kcal).

PULSE COLUMN

A pulsed-column cooling crystalliser, based on an invention by A. E. Zdansky[20], has recently been developed by Giovanola Frères, S.A. Hot saturated solution enters the top of a tall column that is divided into sections by means of a series of conical (apex downward) perforated plates. Also within the column are a number of vertical water-cooled tubes of the 'cold finger' type, which give a flow of water countercurrent to the downward-flowing solution. The feedstock is pulsed down the column at the rate of 1 pulse/s and the bottom discharge outlet is opened briefly at 6–10 s intervals for ejecting the cooled magma. The pulse action tends to redisperse the crystal nuclei so that they continue to grow, and the unit acts as a classifier. Production rates of 12–15 ton/day of 0·5 mm adipic acid crystals have been reported. Other substances produced on a commercial scale include dicyandiamide, aminocaprylic acid, fumaric acid, monochloracetic acid, barium chloride and borax.

DIRECT CONTACT COOLING

The simplest form of direct contact cooling is effected by blowing air into a hot crystallising solution. Cooling takes place predominantly by evaporation, and the air also serves as a means of agitation.

In recent years a considerable amount of attention has been given to the possibility of using direct contact refrigeration for crystallisation processes. Success has already been reported in the field of desalination (p. 274) and processes are being developed for hydrocarbon separations, e.g. p-xylene. The refrigerant may be immiscible with the process fluid, e.g. liquid butane and brine in the desalination process, or immiscible, e.g. CO_2 in the freeze-separation of liquid hydrocarbons. Miscibility between refrigerant and process fluid presents no difficulty if their relative volatility is very high; separation can then be effected by simple techniques. Other miscible or partially miscible direct contact refrigerants for hydrocarbon systems include liquid ethylene, ethane and the fluorocarbons (Freons).

A continuous cooling crystalliser that uses the principle of direct heat exchange between the crystallising solution and an immiscible coolant is shown in *Figure 9.8*[2, 21]. One example of the use of this Cerny crystalliser is the production of calcium nitrate tetrahydrate. Aqueous feedstock enters at the top of the crystalliser at 25 °C and cools as it flows countercurrently to an upflow of immiscible coolant (e.g. petroleum at − 15 °C) introduced as droplets into a draft tube. The low-density coolant collects in the upper layers, but the high-density aqueous solution circulates up the draft tube and down the annulus, keeping the small crystals in suspension. Crystals >0·4 mm settle to the lower regions and are discharged in the magma at about − 5 °C. A slow-speed agitator prevents consolidation of the crystals in the magma outlet. The coolant is passed to a cyclone, to remove traces of aqueous solution, and recycled.

EVAPORATING CRYSTALLISERS

When the solubility of a solute in a solvent is not appreciably decreased by a reduction in temperature, supersaturation of the solution can be achieved by removal of some of the solvent. Evaporation techniques for the crystallisation of salts have been used for centuries, and the simplest method, the utilisation of solar heat, is still a commercial proposition in many parts of the world. An example of the modern application of solar evaporation in the production of salts from Dead Sea waters[22, 23]. Some 40 square miles of the shallow water around the periphery of the 360 square mile lake is divided by dykes into compartments where the complex mixtures of salts crystallise. The Dead Sea is unusual because it contains a highly concentrated brine (~ 27 per cent) with a high percentage of magnesium salts (~ 14 per cent) (see *Table 9.1*). One of the principal salts crystallised is carnallite ($KCl \cdot MgCl_2 \cdot 6H_2O$). By comparison, ordinary sea-water is quite dilute (3·3 per cent dissolved salts) and its main soluble constituent is sodium chloride (2·6 per cent). Magnesium chloride only accounts for 0·3 per cent in sea-water.

Fishery salt, required in the form of hard saucer-shaped crystals, 6–12 mm diameter, is produced in large quantities by the evaporation of brine in long, shallow open pans heated by direct fire, hot gases or steam coils. Evaporation may be aided by a gentle draught of air over the surface of the

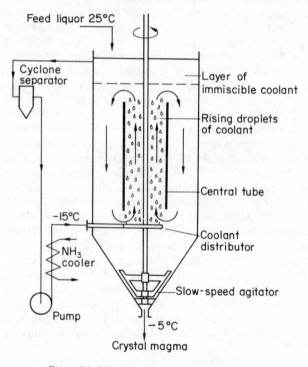

Figure 9.8. The Cerny direct-coolant crystalliser[2]

pans. Crystals that appear on the surface of the brine continue to grow until they become too heavy to be supported on the surface, whereupon they fall to the bottom of the pan and are removed by rakes. Labour costs per pound of product are high and the efficiency of heat utilisation is low, yet no other method has so far been discovered that will produce this peculiar type of hard crystal necessary for use in the fisheries industry.

HESTER and DIAMOND[24] described the production of grainer salt, similar in shape to fishery salt but smaller in size, in open troughs 36 m long, 5 m wide and 0·6 m deep. The brine is kept at about 99 °C by recirculation at

200–400 m³/h through external steam-heated heat exchangers. Each pan produces about 30 ton/day of the hopper-shaped crystals. This coarse salt, which has a high surface area per unit mass, must be handled with care to prevent breakage. After screening, several grades are produced, the coarsest being 6- to 28-mesh (0·5–3 mm). Grainer salt, which has a high rate of solubility in water, is used in the manufacture of cheese and butter.

For most other purposes common salt is produced from brine in enclosed calandria evaporators — appropriately called salting evaporators. No control over the crystal size is possible and fine crystals, rarely larger than 1·5 mm, are formed. The use of reduced pressure in an evaporator to aid the removal of solvent, to minimise the heat consumption or to decrease the operating temperature of the solution, is common practice. Calandria or coil evaporators, often in multi-effect series, are used in the sugar and common salt industries. Sometimes, as in sugar refining, the concentrated solution leaving the last effect of an evaporator system is charged into a separate evaporator, where the liquor is 'struck', by skilled control of the vacuum, to produce a mass of small regular crystals.

Table 9.1. COMPOSITION OF DEAD SEA WATER AT A DEPTH OF 75m

Salt	Percentage
Sodium chloride	7·2
Potassium chloride	1·3
Magnesium chloride	13·7
Calcium chloride	3·8
Magnesium bromide	0·6
Calcium sulphate	0·1
Total dissolved solids	26·7

Density = 1·24 g/cm³

Salt and sugar crystallisation may be regarded as rather special cases, not because their governing principles are in any way different from the general principles of evaporative crystallisation, but because of the considerable amount of technical 'know-how' retained in the various large commercial organisations. Several publications[24-26] deal in some detail with the practices adopted in these industries. A brief account will be given here of some of the equipment commonly used as evaporating crystallisers, and their methods of operation.

Most evaporation units are steam-heated, although resistance heating and heating by passing hot gases through liquors can be used. Steam coils, once widely used in sugar pans, are gradually becoming less favoured. A typical evaporator body used in evaporative crystallisation is the short-tube vertical type, heated by steam, which condenses on the outside of the tubes (*Figure 9.9*). A steam chest, or calandria, with a large central downcomer allows the magma to circulate through the tubes; during operation the tops of the tubes are just covered with liquor. To increase the rate of heat transfer, especially in dealing with viscous liquors, a forced circulation of liquor may be effected by installing an impeller in the downcomer.

MULTIPLE-EFFECT EVAPORATION

Low pressure steam, i.e. <4 bar, is normally used in evaporators, and frequently by-product steam ($\sim 1 \cdot 5$–2 bar) from some other process is

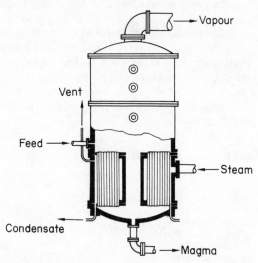

Figure 9.9. A typical crystallising evaporator containing a calandria with a large central downcomer

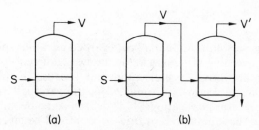

Figure 9.10. Principle of multiple-effect evaporation: (a) single-effect; (b) double-effect

employed. Nevertheless, 1 kg of steam cannot evaporate more than 1 kg of water from a liquor, and for very high evaporation duties the use of process steam as the sole heat source can be very costly. However, if the vapour from one evaporator is passed into the steam chest of a second evaporator, a great saving can be achieved. This is the principle of the method of operation known as multiple-effect evaporation (*Figure 9.10*). As many as six effects may be used in practice.

It is beyond the scope of this book to deal in any detail with the subject of multiple-effect evaporation; all standard chemical engineering textbooks contain adequate accounts of this method of operation, but two important points may be made here. First, multiple-effect evaporation increases the *efficiency* of steam utilisation (kg of water evaporated per kg of steam used) but reduces the *capacity* of the system (kg of water evaporated). The well-known equation for heat transfer may be written

$$Q = UA\Delta T$$

where Q is the rate of heat transfer, U the over-all heat transfer coefficient, A the area of the heat exchanger, and ΔT the driving force, the temperature difference across the heat transfer septum. The area A is usually fixed, so the variables to consider are Q, U and ΔT. Q will be reduced by heat losses from the equipment, U by sluggish liquor movement and scaling in the tubes of the calandria, ΔT by the increase in the boiling point of the liquor, as it gets more and more concentrated. These and many other factors prevent the achievement of the ideal condition

1 effect: 1 kg steam $\rightarrow$ 1 kg vapour
2 effects: 1 kg steam $\rightarrow$ 2 kg vapour
3 effects: 1 kg steam $\rightarrow$ 3 kg vapour, etc.

Nevertheless a close approach to this ideality can often be produced. Some of the evaporators in a multiple-effect system are frequently operated under reduced pressure to reduce the boiling point of the liquor and thereby increase the available ΔT.

Before leaving this brief account of multiple-effect evaporation, mention may be made of the various methods of feeding that can be employed. *Figure 9.11* shows, in diagrammatic form, the possible feed arrangements. In all cases fresh process steam enters effect number one. S denotes live steam, F the feed solution, V vapour passing to the condenser system, and L the thick liquor, or crystalline magma, passing to a cooling system or direct to a centrifuge. Pumps are indicated on these diagrams to indicate the number required for each of the systems. In all cases the vapours flow in the direction of effects $1 \rightarrow 2 \rightarrow 3 \rightarrow$, etc.

In the forward-feed arrangement (*Figure 9.11a*) liquor as well as vapour pass from effect $1 \rightarrow 2 \rightarrow 3 \rightarrow$, etc. As the last effect is usually operated under reduced pressure, a pump is required to remove the thick liquor; a feed pump is also required. The transfer of liquor between the intermediate effects is automatic as the pressure decreases in each successive effect. Suitable control valves are installed in the liquor lines. The disadvantages of a forward-feed arrangement are (a) the feed may enter cold and consequently require a considerable amount of live steam to heat it to its boiling point; (b) the thick liquor, which flows sluggishly, is produced in the last effect, where the available ΔT is lowest; (c) the liquor pipelines can easily become steam-blocked as the liquor flashes into the evaporator body.

In the backward-feed arrangement (*Figure 9.11b*) the thick liquor is produced in effect 1, where the ΔT is highest; the liquor is more mobile on account of the higher operating temperature. Any feed preheating is done

316

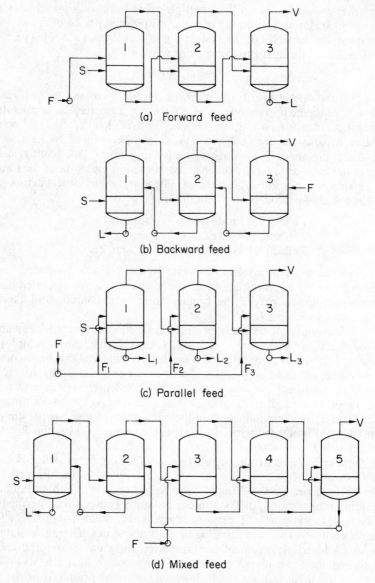

(a) Forward feed

(b) Backward feed

(c) Parallel feed

(d) Mixed feed

Figure 9.11. Various feeding arrangements in multiple-effect evaporators

in the last effect, where low-quality steam (vapour) is being utilised. More pumps will be required in backward feeding than in forward feeding; the liquor passes into each effect in the direction of increasing pressure. No feed pump is necessary, as the last effect is under reduced pressure. Liquor does not flash as it enters an evaporator body, so small-bore liquor lines can be used. Backward feeding is best for cold feed liquor.

In the parallel-feed arrangement (*Figure 9.11c*) one feed pump is required, and predetermined flows F_1, F_2 and F_3 are passed into the corresponding effects. A thick liquor product is taken from each effect; for this purpose pumps are generally necessary. Parallel feeding is is often encountered in crystallisation practice, e.g. in the salt industry, and it is useful if a concentrated feedstock is being processed.

In *Figure 9.11d* a 5-effect system is chosen to illustrate the mixed-feed arrangement. Many different sequences are possible; the one demonstrated is as follows. Feed enters an intermediate effect (number 3 in this case) wherefrom the liquor flows in forward-feed arrangement to the last effect, is then pumped to effect number 2 and from there flows in backward-feed arrangement to the first effect. Some of the advantages of this method of feeding are (a) fewer pumps are required compared with backward feeding; (b) the final evaporation is effected at the highest operating temperature; and (c) frothing and scaling problems are claimed to be minimised. Caustic soda evaporation is often carried out on a mixed-feed basis.

FORCED CIRCULATION EVAPORATORS

Two widely used forced circulation evaporator-crystallisers are shown in *Figure 9.12*. They are both essentially circulating magma units, with external heat exchangers, operated under reduced pressure. In the Swenson crystalliser (*Figure 9.12a*) the crystal magma is circulated from the conical base of the evaporator body through the vertical tubular heat exchanger and reintroduced tangentially into the evaporator below the liquor level to create a swirling action and prevent flashing. Feedstock enters on the pump inlet side of the circulation system. In the APV–Kestner unit (*Figure 9.12b*), usually called a long-tube salting evaporator, the feedstock also enters the circuit on the pump inlet side. The liquor level in the separator (the evaporator body) is kept above the top of the heat exchanger to prevent boiling in the tubes. Liquor enters tangentially, giving a swirl, but a baffle in the separator creates a relatively quiescent growth zone. The larger crystals settle to the conical base and are discharged through the salt box, or by some other suitable means. The fine crystals are recirculated through the heat exchanger. Both these crystallisers are widely used for a variety of substances such as NaCl, $(NH_4)_2SO_4$, Na_2SO_4, $FeSO_4$, $NiSO_4$, citric acid, etc. In general, small crystals (< 0.5 mm) are produced.

KRYSTAL EVAPORATING CRYSTALLISER

The principles of the Krystal process, already referred to in connection with cooling crystallisers, can also be applied to evaporative crystallisation.

318

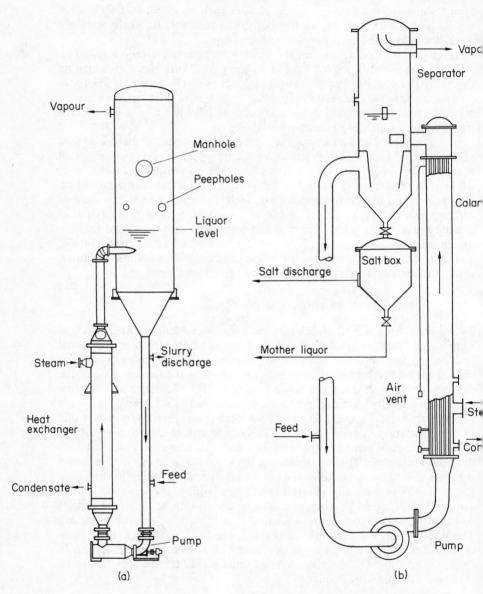

Vapour

Manhole

Peepholes

Liquor
level

Slurry
discharge

Steam

Heat
exchanger

Condensate

Feed

Pump

(a)

Vapo

Separator

Calar

Salt discharge

Salt box

Mother liquor

Air
vent

Feed

St

Cor

Pump

(b)

Figure 9.12. Forced circulation evaporator–crystallisers: (a) Swenson, (b) APV–Kestner

Three forms of the Krystal evaporating crystalliser are shown in *Figure 9.13*. The construction of these crystallisers, commonly used in multiple-effect systems, is of the 'closed' form, i.e. the vaporiser is directly connected with the crystalliser body to form a sealed unit.

There are two basic types, the waisted (a) and the monolithic (b and c). One advantage of the latter is that the conical downcomer, which joins the vaporiser and crystalliser sections and contains highly supersaturated (metastable) solution, is insulated from external conditions by mother liquor within the crystalliser. The circulating liquor inlet and outlet, A and B, are connected through a heat exchanger and pump. Liquor velocities of 1·5– 2 m/s are commonly used through the tubes to minimise crystal depositions.

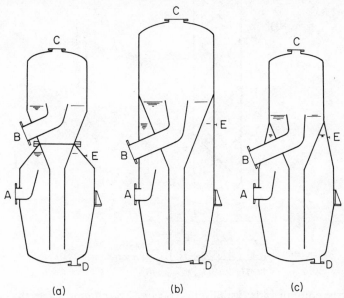

Figure 9.13. Krystal evaporator types[2]: (a) Waisted, (b) and (c) monolithic. A and B, circulating liquor outlet and inlet; C, vapour outlet; D, crystal magma outlet; E, liquor overflow

BAMFORTH[2] has given operating details of several Krystal plants producing ammonium sulphate. Krystal evaporating crystallisers have also been used for the manufacture of sodium chloride, sodium dichromate, ammonium nitrate and oxalic acid.

AIR-CONTACT EVAPORATION

Zahn Hose Crystalliser

An air-cooled crystalliser, suitable for reasonably small production requirements such as recovery of salts from waste liquors, is the Zahn Hose depicted in *Figure 9.14*. It is usually constructed of soft rubber, although any other suitable synthetic material can be used. A unit 4 m high and of 0·6 m diam.

will handle 500 l/h of feed liquor. Hot feedstock flows down the hose over internal baffles located to give a good distribution of liquor over the whole cross-sectional area. Air is drawn upwards, countercurrently to the liquor, by means of the extractor fan mounted at the top. Evaporative cooling takes

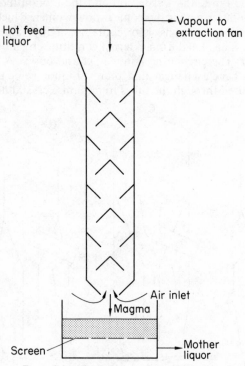

Figure 9.14. The Zahn hose crystalliser

place in the column and crystallisation occurs. The magma leaves the bottom at near the ambient temperature and drops into a screen box, from which crystals may be removed manually. The liquor hold-up in the system is very low, and start-up and shut-down times are very short. The space requirement is small, and units are often installed in the open air, suspended from an outside wall of a building.

Wetted-wall Evaporative Crystalliser

A somewhat unusual application of the wetted-wall column, frequently used in gas–liquid mass transfer operations, has been reported by CHANDLER[27]. A hot concentrated solution is fed into a horizontal pipe, and cold air is blown in concurrently at a velocity of about 100 ft/s. The liquid stream spreads over the internal surface of the pipe and cools, mainly by evaporation (see *Figure 9.15*). The crystal slurry and air leave from the same end of the pipe. Only small crystals can be produced by this method, and because of the evaporative loss of solvent only aqueous solutions can be handled. Nevertheless the equipment required is extremely simple and cheap, apart from the blower.

Though the pilot plant work described by Chandler was carried out on a relatively small unit and confined to the crystallisation of sodium chloride, the wetted-wall crystalliser could be scaled up to larger sizes and used for other systems. The potential throughput of this small unit can be quite high,

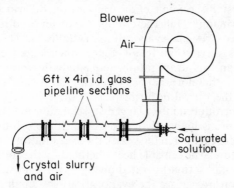

Figure 9.15. Arrangement of a wetted-wall evaporative crystalliser[27]

Table 9.2. APPROXIMATE THEORETICAL CAPACITY OF A 4 in DIAM. WETTED-WALL EVAPORATIVE CRYSTALLISER, BASED ON A 3000 lb/h SOLUTION FLOW RATE AND A SOLUTION TEMPERATURE DROP FROM 100 TO 67 °C (AFTER J. L. CHANDLER[27])

Solute	Crystal yield, lb/h of anhydrous salt			Deposited crystalline phase
	Due to cooling	Due to evaporation	Total	
NaCl	66	60	126	anhyd.
$CuSO_4$	912	115	1027	$5H_2O$
$CaCl_2$	570	240	810	$2H_2O$
$CuCl_2$	480	161	641	$2H_2O$
$Al_2(SO_4)_3$	750	137	887	$18H_2O$
$BaCl_2$	313	90	403	$2H_2O$
$Ba(NO_3)_2$	352	52	404	anhyd.
K_2SO_4	150	37	187	anhyd.
KNO_3	3480	375	3855	anhyd.
$CH_3 \cdot COOK$	1560	625	2185	$\frac{1}{2}H_2O$
$K_2SO_4 \cdot Al_2(SO_4)_3$	3600	227	3827	$24H_2O$
$MgSO_4$	165	65	230	$6H_2O$
$MgCl_2$	300	111	411	$6H_2O$
$MnCl_2$	159	175	334	$2H_2O$
$(NH_4)_2SO_4$	398	157	555	anhyd.
NH_4Cl	580	117	697	anhyd.
Na_3PO_4	1350	180	1530	$12H_2O$
Na_2HPO_4	486	155	641	$2H_2O$
NaH_2PO_4	1830	372	2202	anhyd.
$CH_3 \cdot COONa$	811	258	1069	anhyd.

depending on the temperature–solubility characteristics of the solute–solvent system, as indicated in *Table 9.2*.

VACUUM CRYSTALLISERS

The term 'vacuum' crystallisation is capable of being interpreted in many ways; any crystalliser that is operated under reduced pressure could be called a vacuum crystalliser. Some of the evaporators described above could be classified in this manner, but these units are better, and more correctly, described as reduced-pressure evaporating crystallisers. The true vacuum crystalliser operates on a slightly different principle: supersaturation is achieved by simultaneous evaporation and adiabatic cooling of the feed solution. These units, therefore, act as both evaporators and coolers.

To demonstrate the operating principles of these units, consider a hot saturated solution introduced into a lagged vessel maintained under vacuum. If the feed temperature is higher than that at which the solution would boil under the low pressure existing in the vessel, the feed solution cools adiabatically to this temperature. The sensible heat liberated by the solution, together with any heat of crystallisation liberated owing to the deposition of crystals at the lower temperature, causes the evaporation of a small amount of the solvent, which in turn results in the deposition of more crystals owing to the increased concentration.

In a continuously operated vacuum crystalliser the feed solution should reach the surface of the liquor in the vessel quickly, otherwise evaporation and cooling will not take place, because, owing to the hydrostatic head of solution, the boiling point elevation becomes appreciable at the low pressures (7–20 mbar, 5–15 mm Hg) used in these vessels, and the feed solution will tend to migrate down towards the bottom outlet. Care must be taken, therefore either to introduce the feed near the surface of the liquor in the vessel or to provide some form of agitation.

BATCH VACUUM CRYSTALLISER

Figure 9.16 shows a typical batch vacuum crystalliser[10], which consists of a lagged vertical cylindrical vessel with a top vapour outlet leading to a condenser and vacuum unit, and a conical bottom section fitted with a discharge outlet. The vessel is charged to a predetermined level with a hot concentrated solution, and the propeller agitators and vacuum and condensing equipment are put into operation. As the pressure inside the vessel is reduced, the solution begins to boil and cool until the limit of the condensing equipment is reached. In order to increase the capacity of the condenser, and thereby increase the crystal yield, the vapour leaving the vessel can be compressed before condensation by the use of a steam-jet booster (see *Figure 9.17*). The vacuum equipment usually consists of a two-stage steam ejector.

The swirling motion produced by the agitators helps to maintain the batch at a fairly uniform temperature and to keep the crystals suspended in the liquor. Crystalline deposits around the upper portions of the inner walls of the vessel cause little inconvenience; as the unit is operated batch-wise, the next charge will redissolve the deposit. When the batch reaches the

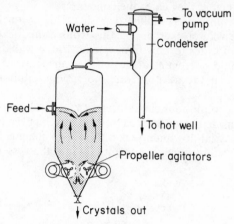

Figure 9.16. Swenson batch crystalliser[10]

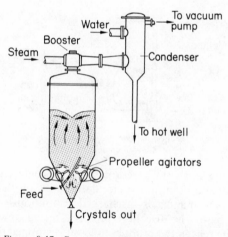

Figure 9.17. Swenson continuous vacuum crystalliser[10]

required temperature, i.e. the desired degree of crystallisation, it is discharged to a filtration unit. Small crystals, rarely much larger than about 60-mesh (250 μm), are obtained from this type of crystalliser.

SWENSON CONTINUOUS VACUUM CRYSTALLISER

The batch unit described above can be adapted for continuous operation[10], as shown in Figure 9.17. Hot concentrated feed solution is introduced continuously through an insulated nozzle at a velocity such that it is delivered to the surface of the liquor in the vessel. The agitators perform the same duty as in the batch unit. The product is discharged continuously through the bottom outlet. A steam-jet booster compresses the vapours leaving the vessel before they enter the condenser.

As in the batch unit, a crystalline deposit builds up on the upper walls of the vessel; one way of overcoming this troublesome feature is to allow a small quantity of water to flow film-wise down the walls. A water rate of 200–400 l/h is found adequate for vessels 2 m in diameter; and as this is less than the normal vaporisation rate, no serious dilution of the charge occurs.

Another type of continuous crystalliser, of the circulating magma type, in which agitation inside the vessel is caused by recycling the charge, is shown in *Figure 9.18*. An axial-flow pump takes the crystal magma from the conical base portion and reintroduces it tangentially into the vessel, just below the

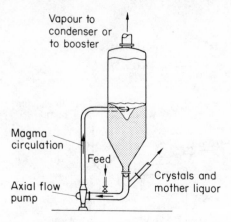

Figure 9.18. Swenson continuous vacuum crystal-liser with agitation by pump circulation[28]

liquor level, creating a swirling action. Hot concentrated feed solution enters continuously at a point in the circulating pipe on the suction side of the pump, and crystal magma is discharged continuously from an outlet below the conical base. The discharge pipe is directed upwards to prevent blockage if for some reason the product take-off is stopped for a short while. Vapours leave the vessel through a top outlet, as described for the previous units. Slightly larger-sized crystals, 20–30 mesh (of the order of $\sim 700\ \mu m$), are reported for this modified unit[28].

SWENSON DRAFT-TUBE BAFFLE (DTB) CRYSTALLISER

The three vacuum crystallisers described above cannot be used for the production of large uniform crystals. Even though certain salts have been grown to about 1 mm size in a forced-circulation unit such as that shown in *Figure 9.18*, the achievement of this size is rarely an easy matter. A more typical product size is about 60-mesh (250 μm), with an appreciable proportion of material finer than 100-mesh (150 μm). The main difficulties are as follows. First, propeller agitation or pump circulation can be rather violent, and mechanical shock is a well-known method of inducing nucleation, so excessive quantities of unwanted crystal nuclei can be produced within the circulating magma. Second, the sudden flashing of vapour from the circulating liquor as soon as it enters the crystalliser causes the liquor to become supersaturated; if it becomes labile, spontaneous nucleation will occur. On the other hand, even if the labile zone is not penetrated, the degree of supersaturation may be too high for the required growth rate on the crystal seeds located near the surface of the liquor in the vessel. Vigorous agitation, of

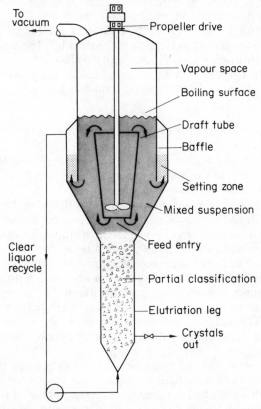

Figure 9.19. The Swenson draft-tube baffled (DTB) crystalliser

course, would bring more small crystals into the highly supersaturated region, but the disadvantages of violent circulation still remain.

A vacuum unit capable of producing large crystals is the Swenson draft-tube baffled (DTB) crystalliser (Figure 9.19). A comprehensive account of this and the older types of Swenson crystallisers has been given by NEWMAN and BENNETT[1], who discuss their application to the crystallisation of potassium chloride. A change-over from a unit such as that shown in *Figure 9.17* to a DTB type increased the quantity of crystals larger than 0·6 mm from 0·1 to 94 per cent.

The operating principles of the DTB crystalliser can be followed from *Figure 9.19*. An agitator is located centrally in the lower section of a draft tube, which is centred by support vanes to prevent swirling of the liquor. Feedstock is injected into the draft tube. The gentle movement of the slurry up the draft tube to the free surface reduces flashing and produces only about $\frac{1}{2}$ °C supercooling. The boiling action is also spread uniformly over the exposed surface. Because of the non-violent flashing, no salt build-up occurs on the upper walls of the vessel. An internal baffle in the crystalliser forms an annular space in which agitation effects are virtually absent, which provides a settling zone that permits regulation and control over the removal of excess nuclei. High magma densities can be permitted and longer residence times are possible, so the growth of large crystals is feasible. The crystalliser may be fitted with an elutriation leg (as illustrated) if a classified discharge of crystals is required. The DTB crystalliser is best suited for the production of crystals with settling velocities > 3 mm/s in their mother liquor.

KRYSTAL VACUUM CRYSTALLISER

Krystal vacuum crystallisers can be of the 'open' (*Figure 9.20*) or 'closed' types (*Figure 9.13*). In the former the crystallisation zone is at atmospheric pressure. In the latter all parts of the equipment are under reduced pressure. The essential feature of a vacuum crystalliser is that no external heat is supplied to the circulating liquor. BAMFORTH[2] describes a number of applications of these vacuum units. Two different methods for operating Krystal crystallisers are shown in *Figure 9.20* with (a) a classified suspension (circulating liquor) and (b) a mixed suspension (circulating magma). In these particular units, used for the production of ammonium nitrate[29–31], holes in the downcomer (barometric leg) divert about 10 per cent of the supersaturated liquor to the upper regions of the crystal bed to promote growth in that zone. The manifold in the vaporiser (see *Figure 9.20b*) was designed to prevent salt deposition by directing the circulating magma against the wall.

Classified operation, while capable of producing large regular crystals, limits productivity because both the liquor velocity and the mass of crystals in suspension have to be restricted to keep the fines level below the pump inlet. Modification to magma circulation can improve the productivity considerably, because higher circulation rates and magma densities can be employed. Furthermore, the suspension volume is increased because magma circulates through the vaporiser and downcomer. In this type of operation, however, the bulk classifying action is lost, and it is necessary to provide a secondary elutriation zone in the suspension to permit segregation and removal of excess nuclei. Fines can be redissolved with live steam and the resulting solution fed to the vaporiser.

One of the major factors in the successful operation of any controlled suspension crystalliser is the incorporation of a suitable fines trap. The earlier the excess nuclei are collected and destroyed the more efficient will be the process. In practice, fines are most economically removed when they reach $\frac{1}{10}$ th to $\frac{1}{20}$ th of the average product size. The work of SAEMAN[29-32] has been of particular value in demonstrating the importance of this aspect of crystalliser design. He suggests that the key to effective size control is the segregation time of nuclei in the fines trap, and that size classification by

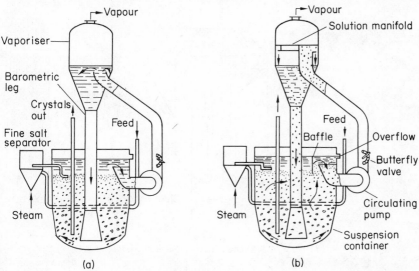

Figure 9.20. A Krystal vacuum crystalliser showing two different methods of operation: (a) classified suspension, (b) mixed suspension[29]

hydraulic elutriation cannot be effective unless the segregation time requirements are also satisfied. A number of possible arrangements of auxilliary baffles to facilitate fines segregation in several types of crystalliser have been described[32].

TURBULENCE CRYSTALLISER

A vacuum crystalliser recently developed by Standard Messo[33] is shown in Figure 9.21. Two concentric pipes, an outer 'ejector tube' with a circumferential slot and an inner 'guide tube', create two liquor flow circuits. Circulation is created by a variable speed agitator in the guide tube. The principle of the Oslo crystalliser is utilised in the growth zone, A; partial classification occurs in the lower regions and fine crystals segregate in the upper. There is a fast upward flow of liquor in the guide tube and a downflow in the annulus; this causes liquor to be drawn in through the slot, which sets up a secondary flow circuit in the lower region of the vessel.

Feedstock is introduced into the guide tube and passes into the vaporiser section, where flash evaporation takes place. Nucleation, therefore, occurs

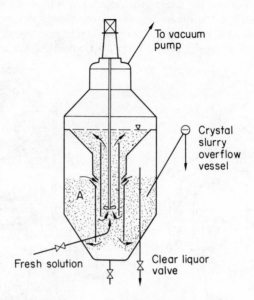

Figure 9.21. Standard Messo Turbulence Crystalliser

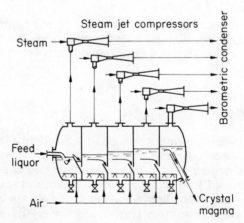

Figure 9.22. Standard Messo Multistage Vacuum Crystalliser

in this region, and the nuclei are swept into the primary circuit. Mother liquor can be drawn off by means of a control valve, a control over the salt slurry density thus being effected. Turbulence crystallisers, in 3- or 4-stage array, have been widely used in the production of such salts as potassium chloride and ammonium sulphate.

MULTI-STAGE VACUUM CRYSTALLISER

A vacuum crystalliser designed by Standard Messo to provide a number of cooling stages in one vessel is shown in *Figure 9.22*. The horizontal cylinder is divided into several compartments by vertical baffles that permit underflow of magma from one section to another but isolate the vapour spaces. Each vapour space is kept at its operating pressure by a thermocompressor, which discharges to a barometric condenser.

Hot feedstock is sucked into the first compartment, which is operated at the highest pressure and temperature (say 100 mbar and 45 °C). Flash evaporation and cooling occur, and the resulting crystal slurry passes into the successive compartments, where the pressure is successively reduced and evaporation and cooling continue. In the last compartment the temperature and pressure may be 10 °C and 10 mbar, for example. Agitation is provided by the boiling action in the compartments supplemented by air spargers. Mother liquor or magma is withdrawn from the last stage through a barometric leg or by means of a pump. Industrial units in a variety of sizes have been installed for the recovery of sodium sulphate from spin-bath liquors, the regeneration of pickling liquors, etc.

NATURAL CIRCULATION CRYSTALLISER

A new type of circulating magma crystalliser with an internal draft tube (*Figure 9.23*) has recently been reported by MATUSEVICH[3, 34]. The unit has no circulating pump; movement of the magma is promoted by convection aided by a central draft tube. Circulation is started initially by means of air injected up the draft tube and is continued by passing in hot concentrated feedstock, which mixes with the circulating magma. The difference in density between the warm liquor inside the draft tube and the colder liquor outside provides the driving force for circulation. The boiling action in this reduced-pressure crystalliser is very gentle and salt incrustation is minimised. The results of a number of pilot plant trials on different versions of the basic crystalliser on a number of salts including $NaNO_3$, $K_2Cr_2O_7$, $(NH_4)_2SO_4$ and $Na_2Cr_2O_7 \cdot 2H_2O$ have been reported[3, 34].

CONDENSING SYSTEMS

The vapours leaving a vacuum crystalliser or evaporator have to be condensed, either in a conventional tubular heat exchanger or by direct contact with a spray of water in a jet or contact condenser. The latter method is probably the most widely used. The low-temperature vapours produced in the flash chamber are usually compressed by means of a steam booster to facilitate their condensation in the cooling system. The temperature in the flash chamber depends on that of the available cooling water; the partial pressure of water vapour in the flash chamber is equal to the vapour pressure of water at the temperature of the mixture of cooling water and condensate

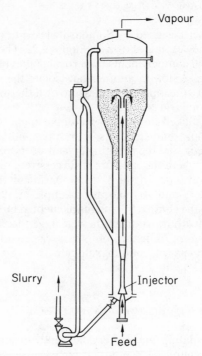

Figure 9.23. Natural circulation crystal-liser[3, 34]

leaving the jet condenser. BAMFORTH[2] has given a comprehensive account of the various types of ancillary equipment used in crystallisation practice.

CHOICE OF CRYSTALLISER

Solutions that require crystallising generally fall into one of two categories: those that will deposit appreciable quantities of crystals on cooling, and those that will not. The temperature–solubility relationship between the required product and the solvent is therefore of prime importance in the choice of a crystalliser. For solutions in the second category, an evaporating crystalliser would normally be used, although the salting-out method could be employed in specific cases. For solutions that deposit appreciable quantities of crystals on cooling, the choice of equipment will lie between a simple cooling crystalliser and a vacuum crystalliser.

Unfortunately, few performance and cost data are available for the various units commonly used in practice. SEAVOY and CALDWELL[10] discussed in some detail the relative merits of a cooling (Swenson–Walker type) and a vacuum crystalliser (Swenson type), and MCCABE[11] presented American production and installation costs for several units. A more recent cost survey, again of American origin, has been made by GARRETT and ROSENBAUM[35] who compare several standard types of Krystal, growth-type Pachuca, forced-circulation and mechanical crystallisers. Broadly speaking, for rough estimation purposes the six-tenths rule.

$$\text{Capital cost} \propto (\text{capacity})^{0.6}$$

can be applied for capacities in the range 10 to 1000 ton/day.

BAMFORTH[2] also deals with the economics of crystallisation, using for comparison purposes a typical batch tank crystalliser operated on an evaporation–cooling cycle and a small continuous Krystal evaporating crystalliser operated under reduced pressure. Even with a production as small as 1 ton of crystals per day, the continuous unit was found to be more economical in operating costs; but its greatest advantage was the fact that, whereas about 40 per cent of the liquor from the tank crystalliser required reworking, the quantity for the continuous unit was only about 7 per cent. Labour costs per pound of product for both units were similar but decrease considerably for the continuous unit with increasing production capacity. For example, a continuous crystalliser producing, say, 1 ton/day would require substantially the same labour force as a unit producing, say, 100 ton/day.

There are, of course, many other advantages associated with crystallisers that also act as classifiers: product size is more regular; efficient filtration and washing are facilitated, thus tending to give a purer product; drying costs are reduced because of the lower amount of retained solvent; and screening of the product is frequently unnecessary.

Although some of the simpler cooling crystallisers, especially the open tanks, are relatively inexpensive, the initial cost of a mechanical unit can be fairly high. The maintenance costs of a mechanical crystalliser can also be quite appreciable. On the other hand, no costly vacuum-producing or condensing equipment is required. Very dense crystal slurries can be handled in cooling units not requiring liquor circulation. Unfortunately, the inner cooling surfaces often become coated with a hard crust of crystals, and the outer water-sides of these surfaces readily become fouled, with the result that cooling efficiency is reduced considerably. Similar problems can also be encountered in the operation of evaporating crystallisers; coils and calandrias often become coated with scale. As the true vacuum crystallisers have no heat-exchange surfaces, they do not suffer from these scaling problems, but they cannot be used when the liquor has a high boiling point elevation. Vacuum and evaporating crystallisers require a considerable amount of head room, but the floor space needed is usually very much less than that required by evaporating pans and cooling tanks to produce the same amount of finished product.

Once a particular class of crystalliser is decided upon, the choice of a specific unit will depend upon such factors as the initial and operating costs, the space availability, the type and size of crystals required, the physical characteristics of the feed liquor and crystal slurry, the need for corrosion resistance, and so on. The production rate and supply of feed liquor to the crystalliser will usually be the deciding factors in the choice between a batch and continuous unit. Production rates > 1 ton/day, or feed liquor rates $> 20 \, m^3/h$ are best handled on a continuous basis.

INFORMATION REQUIRED

It is useful to know what sort of information a crystalliser manufacturer needs before he can begin to specify or design a piece of equipment. Although most large fabricators have facilities to determine the necessary physical data, the customer can usually provide this information much more precisely, quickly, and cheaply.

A detailed description of the product should be given, including its full chemical name and formula, specifying if a hydrate is required or not. A realistic purity specification should be laid down. The production rate should be given, e.g. as ton/yr or kg/h or preferably as both. Of course, other convenient units can be used. Very generous allowances should be made for maintenance and other shutdown periods.

The shape and size of the crystalline product should be specified. An actual sample of the desired crystals, if available, is very helpful. It is important to remember that the more rigid the size specification the more difficult the crystalliser design becomes. Vague statements such as 'about 30-mesh' should not be made, as they have no useful meaning, but sieve mesh sizes (e.g. ASTM or BS standard sieves) may be quoted as allowable upper and lower size limits. A size specification such as '90 per cent between 20- and 40-mesh sieves' would be quite acceptable; but in view of the present-day trend to relegate sieve mesh sizes to a subsidiary role, it is much better to quote the actual aperture sizes of the sieves, e.g. 90 per cent between 850 and 425 micron sieves. A full account of this topic and of the methods used for the specification and size grading of crystals has been provided in some detail in Chapter 11.

A solubility curve of the product in the liquor should be drawn. Care should be taken in doing this, because solubility data reported in the literature usually refer to pure solutes and solvents. These may be quite inapplicable. If impurities are known to be present in the working liquors, the relevant solubility data should be measured.

The working temperature range for batch-operated crystallisers, or the operating temperature for continuous units, has to be specified. The optimum temperature of operation is not an easy quantity to determine; it is dependent on, for example, the required crystal yield and the energy expenditure incurred in the operation of heaters, coolers, vacuum pumps, etc. However, since the crystal growth process is temperature dependent, the quality of the crystalline

product can depend significantly on the temperature of operation.
The following liquor and solids data are also useful.

Liquor: Feedstock analysis and pH; density of the feedstock at the feed
temperature; boiling-point elevation of the saturated solution; density and
viscosity of the saturated solution over the working range of temperature;
maximum allowable undercooling (metastable zone width) in the presence of
crystals in agitated solution.

Crystals: True density of the substance; heat of crystallisation; settling
velocity of crystals of the desired size and shape in saturated solutions over a
range of temperatures; crystal growth and nucleation rates as functions of
supersaturation and temperature.

The design of crystallisers is discussed in Chapter 10.

REFERENCES

1. NEWMAN, H. H. and BENNETT, R. C., 'Circulating magma crystallisers', *Chem. Engng Prog.*, **55** (3) (1959), 65
2. BAMFORTH, A. W., *Industrial Crystallisation*, 1965. London; Leonard Hill
3. MATUSEVICH, L. N., *Crystallisation from Solutions* (in Russian), 1968. Moscow
4. MATZ, G., *Kristallisation*, 2nd Edn., 1969. Berlin; Springer-Verlag
5. NÝVLT, J., *Crystallisation from Solution*, 1970. London; Butterworths
6. SETH, K. K. and STAHEL, E. P., 'Heat transfer from helical coils immersed in agitated vessels', *Ind. Engng Chem.*, **61** (6) (1969), 39
7. CHANDLER, J. L., 'Effect of supersaturation and flow conditions on the initiation of scale formation', *Trans. Instn. chem. Engrs*, **42** (1964), 24
8. DUNCAN, A. G. and WEST, C. D., 'Prevention of incrustation on crystalliser heat exchanger surfaces by ultrasonic vibrations', *U.K. AERE Harwell Research Report* R 6482 (1970)
9. GARRETT, D. E., 'Industrial crystallisation at Trona', *Chem. Engng Prog.*, **54** (12) (1958), 65
10. SEAVOY, G. E. and CALDWELL, H. B., 'Vacuum and mechanical crystallisers', *Ind. Engng Chem.*, **32** (1940), 627
11. MCCABE, W. L., 'Crystallisation', in *Chemical Engineers' Handbook* (ed. J. H. PERRY), 1950. New York; McGraw-Hill
12. KOOL, J., 'Heat transfer in scraped vessels and pipes handling viscous materials', *Trans. Instn chem. Engrs*, **36** (1958), 253
13. DE BRUYN, G. C., 'Crystallisation of massecuites by cooling', in *Principles of Sugar Technology* (ed. P. HONIG), Vol. 2, 1959. Amsterdam; Elsevier
14. GRIFFITHS, H., 'Mechanical crystallisation', *J. Soc. chem. Ind.*, **44** (1925), 7T
15. DOOLEY, J. R. and LINEBERRY, D. D., 'Freeze concentration of various foods', *Chem. Engng Prog. Symp. Ser.*, **62**, No. 69 (1966), 111
16. ARMSTRONG, A. J., 'Scraped surface crystallisers', *Chem. Process Engng*, **51** (11) (1970), 59
17. ARMSTRONG, R. M., 'The scraped shell crystalliser', *Br. chem. Engng*, **14** (1969), 647
18. SKELLAND, A. H. P., 'Correlation of scraped-film heat transfer in the Votator', *Chem. Engng Sci.*, **7** (1958), 166
19. JEREMIASSEN, F. and SVANOE, H., 'Supersaturation control attains close crystal sizing', *Chem. metall. Engng*, **39** (1932), 594
20. 'Continuous flow crystalliser', *Engineer*, **206** (1958) 984
21. CERNY, J., Brit. Patent, 932 215 (1963)
22. NOVOMEYSKY, M. A., 'The Dead Sea—a storehouse of chemicals', *Trans. Instn chem. Engrs*, **14** (1936), 60
23. NADEL, S., 'Harvesting more Israel potash', *Engng Mining J.*, **166** (10) (1965), 84
24. HESTER, A. S. and DIAMOND, H. W., 'Salt manufacture', *Ind. Engng Chem.*, **47** (1955), 672
24a. KAUFMANN, D. W. (Ed.), *Sodium Chloride*, 1960. New York; Reinhold
25. HONIG, P. (Ed.), *Principles of Sugar Technology*, Vol. 2, 1959. Amsterdam; Elsevier
26. LYLE, O., *Technology for Sugar Refinery Workers*, 3rd Ed., 1957. London; Chapman and Hall

27. CHANDLER, J. L., 'The wetted-wall column as an evaporative crystalliser', *Br. chem. Engng,* **4** (1959), 83

28. MCCABE, W. L. and SMITH, J. C., *Unit Operations of Chemical Engineering,* 1956. New York; McGraw-Hill

29. SAEMAN, W. C., 'Crystal size distribution in mixed suspensions', *A.I.Ch.E.Jl,* **2** (1956), 107

30. MILLER, P. and SAEMAN, W. C., 'Continuous vacuum crystallisation of ammonium nitrate', *Chem. Engng Prog.,* **43** (1947), 667

31. SAEMAN, W. C., MCCAMY, I. W. and HOUSTON, E. C., 'Production of ammonium nitrate by continuous vacuum crystallisation', *Ind. Engng Chem.,* **44** (1952), 1912

32. SAEMAN, W. C., 'Scale-up and performance relations for continuous crystallisers', in *Separation Processes in Practice* (ed. R. F. CHAPMAN), 1961. New York; Reinhold

33. MESSING, T., 'Development in industrial crystallisation', *Br. chem. Engng,* **14** (1969), 641

34. MATUSEVICH, L. N. and ODINOSOV, V. A., 'A new type of circulation vacuum crystalliser', *Int. Chem. Engng.* **6** (1966), 687

35. GARRETT, D. E. and ROSENBAUM, G. P., 'Crystallisation', *Chem. Engng* (Aug. 11, 1958), 127

10

Crystalliser Operation and Design

MANY of the difficulties facing designers of industrial crystallisers arise from the shortage of basic data. However, not only are published data scarce, they are so frequently unreliable. It is not uncommon to find different investigators reporting crystal growth rates for the same substance differing by an order of magnitude or more. In such cases it is often impossible to select the appropriate value for a given situation, usually because some important parameter has not been specified, or perhaps not even measured. Reliable nucleation data applicable to industrial systems are seldom available in the literature.

The unit operation of crystallisation is governed by some very complex interacting variables. It is a simultaneous heat and mass transfer process with a strong dependence on fluid and particle mechanics. It takes place in a multi-phase, multi-component system. It is concerned with particulate solids whose size and size distribution, both incapable of unique definition, vary with time. The solids are suspended in a solution which can fluctuate between a so-called metastable equilibrium and a labile state, and the solution composition can also vary with time. The nucleation and growth kinetics, the governing processes in this operation, can often be profoundly influenced by mere traces of impurity in the system; a few parts per million may alter the product beyond all recognition.

It is, perhaps, no wonder that crystallisation has been called an art rather than a science. Nevertheless, crystallisers have to be designed, constructed and operated successfully, despite the insecure foundations on which they are built. The object of this chapter is to describe some of the recent developments in the area of crystalliser design and operation and to demonstrate how simple laboratory scale tests, some of which have been described earlier, can provide basic design information.

CRYSTAL SIZE DISTRIBUTION (CSD)

One of the earliest investigations aimed at studying the size distribution of crystals in a continuous crystalliser was made by MONTILLON and BADGER[1], who studied the growth of $Na_2SO_4 \cdot 10H_2O$ and $MgSO_4 \cdot 7H_2O$ in a continuous crystalliser. Shortly afterwards MCCABE[2] analysed the problem of crystal size distribution (CSD) and developed what is now known as the ΔL law. In the development of this law McCabe made the following assumptions: (a) all crystals have the same shape; (b) they grow invariantly, i.e. the growth rate is independent of crystal size; (c) supersaturation is constant throughout the crystalliser; (d) no nucleation occurs; (e) no size classification occurs in the crystalliser; (f) the relative velocity between crystals and liquor remains constant.

For a full account of the derivation of the ΔL law reference should be made to the original paper[2], but the main steps are briefly as follows. The mass of one crystal of size L is given by $\alpha \rho L^3$, where α is a volume shape factor and ρ is the solid density. The number of crystals, dN, of size L in a mass dW is thus $dN = dW/\alpha \rho L^3$. Assuming no nucleation, the number of seeds dN_s of size L_s is equal to the number of product crystals dN_p of size L_p, i.e.

$$\frac{dW_s}{\alpha \rho L_s^3} = \frac{dW_p}{\alpha \rho L_p^3} = \frac{dW_p}{\alpha \rho (L_s + \Delta L)^3}$$

where ΔL is the growth increment. Therefore

$$dW_p = \left(\frac{L_s + \Delta L}{L_s}\right)^3 dW_s$$

$$W_p = \int_0^{W_s} \left(1 + \frac{\Delta L}{L_s}\right)^3 dW_s \tag{10.1}$$

where W_p is the product crystal yield obtained from an initial mass of seed crystals, W_s. The most commonly used unit for L, the characteristic linear dimension of a crystal, is the equivalent sieve aperture size (see Chapter 11).

McCabe recognised that the ideal conditions assumed in this derivation are unlikely to be attained in a real crystalliser. Temperature and supersaturation gradients are unavoidable; invariant growth is comparatively rare; different crystal faces usually grow at different rates; crystal growth rates may be dependent on solution velocity, in which case large crystals may grow faster than small; and so on. However, an interesting feature of the derivation is that the dependence of growth on supersaturation need not be known. Despite its limitations, the ΔL law still provides an interesting, if oversimplified, approach to the development of a CSD.

One use of the law is demonstrated by the following example. If a quantity of seed crystals of known CSD is added to a crystalliser, it is possible to estimate the CSD of the final product by the following procedure.

1. Find the value of ΔL compatible with the product yield, W_p, calculated from solubility data and the mass of added seeds, W_s. This involves a trial and error evaluation of equation (10.1), which may be integrated graphically (plot W_s versus $(1 + \Delta L/L_s)^3$; the area enclosed under $W_s = 0$ and $W_s =$ mass of seeds added is equal to W_p).

2. Plot the integral curve of equation (10.1) (W_p versus W_s) using the correct value of ΔL.

3. Plot the screen analysis curve of the product crystals (W_p versus L_p) remembering that $L_p = L_s + \Delta L$. For comparison purposes, this plot may be made on the same diagram as the screen analysis curve of the seed crystals (W_s versus L_s).

The tedious trial and error solution of equation (10.1) to find the appropriate value of ΔL, as outlined in step 1, may be simplified by the use of a nomograph (*Figure 10.1*). For example[3], *Table 10.1* shows a seed screen analysis based on 100 g and gives the values of $(1 + \Delta L/L_s)^3$ for a first trial using $\Delta L = 100\ \mu m$. A plot of these values against W_s, as explained in step 1, gives an area $W_p = 165$. If, for example, $W_p = 1.57\ W_s$, then the correct value of ΔL is found from the nomograph as follows. Draw a line from $\Delta L = 100\ \mu m$ through $(1 + \Delta L/L_s)^3 = 1.65$ and extend it to the L_s scale (at 520 μm). Now draw a line from this point on the L_s scale through $(1 + \Delta L/L_s)^3 = 1.57$ and extend it to the ΔL scale. The desired value of ΔL is found to be 90 μm.

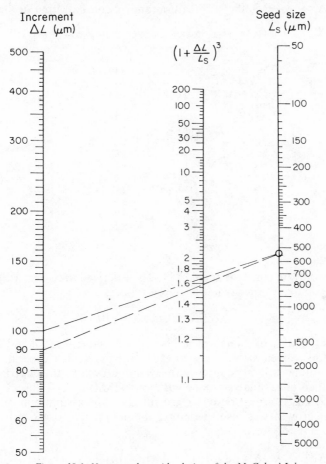

Figure 10.1. Nomograph to aid solution of the McCabe ΔL law

Column 4 in *Table 10.1* can now be completed and a plot of W_s versus $(1 + \Delta L/L_s)^3$ yields the correct value of $W_p = 157$. It is interesting to note that the focal point on the L_s scale is near the maximum of the size distribution curve of the seeds plotted on a differential basis (between 500 and 600 μm).

The postulation of the ΔL law, despite its limitations, was a most significant step forward; it paved the way for a better understanding of the complex behaviour of crystallisers. The next important contribution was made by BRANSOM, DUNNING and MILLARD[4], who, like McCabe, dealt with a mixed-suspension mixed-product removal (MSMPR) crystalliser, but considered the more realistic case where spontaneous nucleation and subsequent growth occurs. They also assumed a 'first-order' growth law

$$\frac{dL}{dt} = k\Delta c^l \tag{10.2}$$

where $\Delta c = c - c^*$ is the supersaturation, i.e. concentration driving force, k

Table 10.1. SOLUTION OF THE ΔL LAW (EQUATION 10·1)

L_s, μm	W_s, g	$(1+\Delta L/L_s)^3$ $\Delta L = 100\,\mu m$	$\Delta L = 90\,\mu m$
1400	0	1·23	1·21
1180	0·1	1·28	1·25
1000	3·0	1·33	1·30
850	15·7	1·41	1·35
710	28·7	1·49	1·43
600	54·5	1·60	1·52
500	74·1	1·74	1·64
425	87·4	1·91	1·78
355	93·7	2·12	1·97
300	97·3	2·40	2·20
250	98·3	2·78	2·52
212	99·5	3·25	2·89
180	100	3·88	3·38

is a growth rate constant and $l = 1$†. Their derived mass–size CSD relation may be written in the form of

$$W(L) = AL^3 \exp(-BL) \tag{10.3}$$

$W(L)$ is the mass of crystals of size between L and $L+\Delta L$. A and B are constants for a given system. Equation (10.3) was verified experimentally for the laboratory-scale crystallisation of cyclonite (cyclomethylene trinitramine).

The theoretical analysis of a continuous MSMPR crystalliser by SAEMAN[5] was also based on a first-order growth law. The assumption of size independent growth was also made and the cumulative size distribution was expressed in the form

$$\int_0^L W(L)\,dL = K\left\{6 - \left[6 + 6\left(\frac{L}{L_T}\right) + 3\left(\frac{L}{L_T}\right)^2 + \left(\frac{L}{L_T}\right)^3\right]\exp\left(-\frac{L}{L_T}\right)\right\} \tag{10.4}$$

K is a constant and L_T is the size of a crystal grown from a nucleus (zero size) in the 'draw-down' time, T, i.e. the over-all residence time. T may be defined as the ratio of the mass of crystals in suspension to the crystal production rate, or volume of suspension to the volumetric feed rate. Saeman also

† In this section (pp. 338–351) the symbol l is used for the 'order' of the growth process (see Chapters 5 and 6) instead of the more usual symbol, n, which is used here to denote numbers of crystals.

compared the case of classified product removal (CPR) with MPR operation. He showed, for example, that the age of a product crystal was four times the draw-down time for CPR and three times for MPR.

BRANSOM[6] later considered the more general case where crystal growth may be size dependent and 'non-first-order' with respect to supersaturation:

$$\frac{dL}{dt} = kL^b \Delta c^l \tag{10.5}$$

k is a constant, $b \geqslant 0$ and $l \geqslant 1$. The resulting CSD equation for a continuous MSMPR crystalliser, as shown by RANDOLPH[7], becomes

$$W(L) = AL^{3-b} \exp(-BL^{1-b}) \tag{10.6}$$

where A and B are constants. For the case of size independent growth ($b = 0$), equation (10.6) reduces to equation (10.3).

BEHNKEN, HOROWITZ and KATZ[8] presented a mathematical formulation for determining the particle size distribution in mixing vessels where the particles grow or shrink according to a prescribed differential law. Steady state size distributions were developed and procedures were given for following transients. The analysis was made in abstract terms and illustrated with reference to a catalytic fluidised reactor. CIOLAN[9] later extended this approach specifically for the case of a crystalliser.

POPULATION BALANCE

For a complete description of the CSD in a continuously operated crystalliser it is necessary to quantify the nucleation and growth kinetics, and to apply the conservation laws of mass, energy and numbers of crystals. The recognition of this latter requirement, the concept of the *population balance*[10,11], has been a major contribution to the subject of crystalliser analysis and design. The CSD depends on the rates of nucleation and growth, and also on the generation or destruction of particles by 'birth' and 'death' processes, e.g. breakage and agglomeration. All particles must ultimately be accounted for.

To facilitate formulation of the population balance[12-15] it is necessary to define a continuous variable to represent the discrete distribution. The density function used is a representation of the number of crystals in a given size range. The population density, n, is defined by

$$\lim_{\Delta L \to 0} \frac{\Delta N}{\Delta L} = \frac{dN}{dL} = n \tag{10.7}$$

where ΔN is the number of particles in the size range $\Delta L(L_1$ to $L_2)$ per unit volume of system. The value of n depends on the value of L at which the interval dL is taken, i.e. n is a function of L (*Figure 10.2a*) and has the dimensions of number per unit size (length) per unit volume of system i.e. it is expressed in units such as number/(μm) (litre). The number of crystals in the size range L_1 to L_2 is thus given by

$$\Delta N = \int_{L_1}^{L_2} n \, dL \tag{10.8}$$

Application of the population balance is best demonstrated with reference to a continuous MSMPR crystalliser[14]. The assumptions made are (a) steady state operation, (b) no particles in the feed, (c) no attrition of the particles. Then a population balance (input = output) in a system of volume, V, for an arbitrary size range, L_1 to L_2, and time interval, Δt, is

$$n_1 V G_1 \Delta t = n_2 V G_2 \Delta t + Q \bar{n} \Delta L \Delta t \tag{10.9}$$

where Q = volumetric feed and discharge rate, G = linear growth rate ($\mathrm{d}L/\mathrm{d}t$), n = population density (numbers per unit length per unit slurry volume), $\bar{n}$ = average population density. The left-hand input term represents

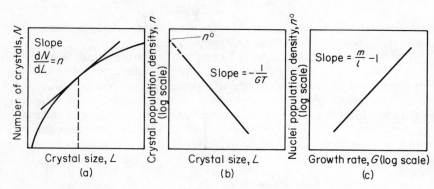

Figure 10.2. Population plots characterising the crystal size distribution, and the nucleation and growth kinetics for a continuous MSMPR crystalliser

the number of crystals growing into the arbitrary size range over the time interval. The right-hand output terms represent the number of crystals (a) growing out of the size range, and (b) in the size range when removed from the crystalliser. Equation (10.9) rearranges to

$$\frac{V(G_2 n_2 - G_1 n_1)}{\Delta L} + Q \bar{n} = 0 \tag{10.10}$$

and in the limit, $\Delta L \to 0$,

$$V \frac{\mathrm{d}(nG)}{\mathrm{d}L} + Qn = 0 \tag{10.11}$$

Letting $V/Q = T$, the draw-down or mean residence time, and assuming that growth is independent of size (ΔL law) i.e. $\mathrm{d}G/\mathrm{d}L = 0$, then

$$\frac{\mathrm{d}n}{\mathrm{d}L} + \frac{n}{GT} = 0 \tag{10.12}$$

Integration of equation (10.12) letting n^0 denote the population density of zero-size particles (the nuclei population density), gives

$$\int_{n^0}^{n} \frac{\mathrm{d}n}{n} = \int_0^L -\frac{\mathrm{d}L}{GT} \tag{10.13}$$

i.e.

$$n = n^0 \exp\left(-L/GT\right) \tag{10.14}$$

Equation (10.14) is the fundamental relationship between L and n characterising the CSD. A plot of $\log n$ versus L should give a straight line with an intercept at $L = 0$ equal to n^0 and a slope of $-1/GT$ (*Figure 10.2b*). Therefore, if the residence time T is known, the crystal growth rate, G, can be calculated. Furthermore the data can be used to give information on the nucleation and growth kinetics. For example, the nucleation rate, dN^0/dt, can be expressed as a function of the supersaturation, Δc:

$$\frac{dN^0}{dt} = k_1 \Delta c^m \tag{10.15}$$

The crystal growth rate G can be expressed in a similar manner:

$$G = \frac{dL}{dt} = k_2 \Delta c^l \tag{10.16}$$

Now

$$\frac{dN^0}{dt} = \left.\frac{dN}{dt}\right|_{L=0} = \left.\frac{dN}{dL}\right|_{L=0} \cdot \frac{dL}{dt}$$

so the nucleation rate may be expressed in terms of the growth rate by

$$\frac{dN^0}{dt} = n^0 G \tag{10.17}$$

Therefore

$$\frac{dN^0}{dt} = k_3 G^{m/l} \tag{10.18}$$

or

$$n^0 = k_4 G^{\frac{m}{l}-1} \tag{10.19}$$

So a plot of $\log n^0$ versus $\log G$ should give a straight line of slope $\frac{m}{l}-1$ or a plot of $\log(dN^0/dt)$ versus $\log G$ should give a line of slope m/l. Thus the kinetic order of nucleation, m, may be evaluated if the kinetic order of growth, l, is known.

The simple population balance (input = output) must be modified for application to an industrial crystalliser, since crystal breakage and spurious nucleation will undoubtedly occur. The general case population balance, therefore, may be written:

$$\text{input} = \text{output} + \text{accumulation}$$

$$\frac{Q_i n_i}{V} + B = \frac{Q_o n_o}{V} + D + \left[\frac{\partial n}{\partial t} + \frac{\partial(nG)}{\partial L} + n\frac{\partial(\ln V)}{\partial t}\right] \tag{10.20}$$

where B and D are birth and death functions, respectively, e.g. representing crystal breakage (i = inlet, o = outlet) RANDOLPH[16] has proposed several birth and dealth models and demonstrated their effect on the CSD.

Equation (10.14) is a number–size distribution relationship. Other distributions can be obtained in the following manner. From equations (10.8) and (10.14) the number of crystals, N, up to size L is given by

$$N = \int_0^L n \, dL$$

$$= \int_0^L n^0 \exp(-L/GT) \, dL$$

$$= n^0 GT[1 - \exp(-L/GT) \tag{10.21}$$

Equation (10.21) is the *zeroth moment* of the distribution, and for large values of $L(L \to \infty)$ it reduces to the total number of crystals in the system:

$$N_T = n^0 GT \tag{10.22}$$

The *first moment* of the distribution is the cumulative length (all the crystals laid side by side):

$$\mathscr{L} = \int_0^L nL \, dL = \int_0^L n^0 L \exp(-L/GT) \, dL$$

$$\mathscr{L} = n^0 GT \quad \{GT[1 - \exp(-L/GT)] - L \exp(-L/GT)\}$$

and for $L \to \infty$

$$\mathscr{L}_T = n^0 (GT)^2 \tag{10.23}$$

The *second moment* gives the surface area:

$$A = \beta \int_0^L nL^2 \, dL$$

where β is a surface shape factor. For $L \to \infty$

$$A_T = 2\beta n^0 (GT)^3 \tag{10.24}$$

The *third moment* gives the mass:

$$W = \alpha\rho \int_0^L nL^3 \, dL$$

where α is a volume shape factor and ρ is the crystal density. For $L \to \infty$

$$W_T = 6\alpha\rho n^0 (GT)^4 \tag{10.25}$$

These equations can be simplified in dimensionless form using a relative size by putting $X = L/GT$, i.e. the ratio of crystal size to the size of a crystal that has grown for a period equal to the draw-down time. Thus

$$\frac{N}{N_T} = N(X) = 1 - \exp(-X) \tag{10.26}$$

$$\frac{\mathscr{L}}{\mathscr{L}_T} = \mathscr{L}(X) = 1 - (1+X)\exp(-X) \tag{10.27}$$

$$\frac{A}{A_T} = A(X) = 1 - \left(1 + X + \frac{X^2}{2}\right)\exp(-X) \tag{10.28}$$

$$\frac{W}{W_T} = W(X) = 1 - \left(1 + X + \frac{X^2}{2} + \frac{X^3}{6}\right)\exp(-X) \tag{10.29}$$

By use of the relative size, X, the CSD can be analysed without knowing the dependence of growth rate on supersaturation. Equation (10.29), which is identical in form with equation (10.4), represents the cumulative mass distribution. The differential mass distribution is represented by

$$\frac{dW(X)}{dX} = \tfrac{1}{6}X^3 \exp(-X) \tag{10.30}$$

These relationships are shown in *Figure 10.3*, where it is seen that the mixed product shows a dominant size fraction for $X = 3$ (compared with $X = 4$

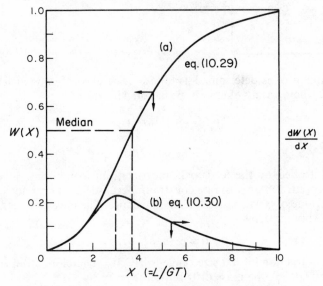

Figure 10.3. (a) Cumulative and (b) differential mass distributions for MSMPR operation

for classified product removal). The median value of X is 3·67. In other words, half the product is larger than 3·67 GT.

APPLICATION OF MASS AND POPULATION BALANCES

A solute mass balance on a MSMPR crystalliser operating continuously at steady state with a production rate W_T gives

$$Q_i c_i = Q_o(c_o + W_T) \tag{10.31}$$

where Q = volumetric flow rate, c = solution concentration, and suffixes i and o denote inlet and outlet conditions (*Figure 10.4*). Assuming $Q_i = Q_o = Q$, a particle population balance (equation 10.11) gives

$$nQ = -GV\frac{dn}{dL} \tag{10.32}$$

giving a particle distribution (equation 10.14)

$$n = n^0 \exp(-LQ/GV) \tag{10.33}$$

Under steady state conditions the particle distribution $n(L)$ must satisfy the mass and population balances simultaneously.

The relationship between the total solids content, W_T, and the particle

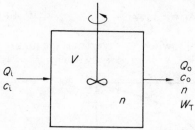

Q_i

c_i

V

n

Q_o

c_o

n

W_T

Figure 10.4. *A continuous mixed-suspension, mixed-product removal (MSMPR) crystalliser*

distribution, $n(L)$, is determined as follows. The mass of one particle of size L is $\alpha\rho L^3$, where α is a volume shape factor. Therefore

$$W_T = \alpha\rho \int_0^\infty nL^3 \, dL \tag{10.34}$$

$$= 6\alpha\rho n^0 (GT)^4 \tag{10.25}$$

W_T is controlled by the feedstock and operating conditions, so equation (10.25) is, in effect, a growth rate constraint because for a given nuclei density only one value of G will satisfy the mass balance. The mass of crystals dW in a given size range dL is

$$dW = n\alpha\rho L^3 \, dL \tag{10.35}$$

so the mass fraction in that size range is dW/W_T. Therefore, from equations (10.25) and (10.35), the mass distribution is given by

$$\frac{W(L)}{dL} = \frac{nL^3}{6n^0(GT)^4} \tag{10.36}$$

which, from the population density relationship (equation 10.14), becomes

$$\frac{W(L)}{dL} = \frac{\exp(-L/GT)L^3}{6(GT)^4} \tag{10.37}$$

The maximum of this mass distribution (the dominant size, L_D, of the CSD) is found by maximising equation (10.37):

$$6(GT)^4 \frac{d}{dL} \cdot \frac{W(L)}{dL} = 3L^2 \exp(-L/GT) - \frac{L^3}{GT}\exp(-L/GT) = 0 \tag{10.38}$$

which gives

$$L_D = 3\,GT \tag{10.39}$$

Equation (10.39) must not be taken to imply that the dominant size is independent of the nucleation rate, because for a given value of W_T the growth rate, G, depends on the nuclei density (equation 10.25).

EFFECT OF SUSPENSION DENSITY

LARSON, TIMM and WOLFF[17] considered the case where nucleation in the system is dependent on the suspension (magma) density, as expressed by a simple power law model:

$$n^0 = k\,W_T^h\,G_i^{m-1} \tag{10.40}$$

Equation (10.40) may be compared with equation (10.19) for the simpler case of nucleation being independent of the suspension density. From equations (10.25), (10.39) and (10.40) the dominant crystal size may be expressed as

$$L_D = k'W_T^{(1-h)l/m+3l}\qquad T^{m-l/m+3l} \tag{10.41}$$

It is thus possible to determine the system changes (draw-down time or suspension density) necessary to effect a change in the dominant size of the CSD, since

$$\frac{L_{D_1}}{L_{D_2}} = \left(\frac{W_{T_1}}{W_{T_2}}\right)^{(1-h)l/m+3l}\left(\frac{T_1}{T_2}\right)^{m-l/m+3l} \tag{10.42}$$

For example, for the crystallisation of ammonium dihydrogenphosphate from aqueous solution $m = 6$ and $l = 2$, and, assuming $h = 1$,

$$\frac{L_{D_1}}{L_{D_2}} = \left(\frac{T_1}{T_2}\right)^{\frac{1}{3}}$$

Thus, to double the dominant size, the retention time would have to be increased eightfold. This, of course, may not be feasible, since it would mean reducing the feed rate (throughput) to $\frac{1}{8}$th or increasing the crystalliser volume by a factor of 8. In practice the CSD could be changed more effectively by varying the relative holding times of the fine and coarse particle fractions, e.g. by fines destruction and/or size classification.

EXPERIMENTAL AND THEORETICAL STUDIES

A number of experimental studies have been made to support the utility of the population balance approach to the prediction of CSD in continuous MSMPR crystallisers. MURRAY and LARSON[18], for example, studied the crystallisation of ammonium alum from aqueous solution by precipitation with ethanol, and the results were in good agreement with the steady state model. They also operated the crystalliser under unsteady state conditions and devised a theoretical model to accommodate production rate upsets. AMIN and LARSON[19] reported data for the growth of $CaSO_4 \cdot \frac{1}{2}H_2O$ from aqueous solution in the presence of phosphoric acid in a continuous MSMPR crystalliser. GENK and LARSON[20] made a similar study with KNO_3 and KCl, and determined the effect of temperature on nucleation and growth rates.

RANDOLPH and RAJAGOPAL[21], working with K_2SO_4, developed a novel technique for generating a CSD in the 0–50 μm size range by allowing the discharge from a mixed suspension crystalliser to pass through a fine screen to retain all crystals larger than 50 μm. Data in this small-size range are very useful for determining nucleation and growth kinetics, but such measurements have been difficult to obtain with conventional sizing techniques, such as sieving. However, measurements with a Coulter particle counter (see Chapter 11, p. 396) have overcome these difficulties. The technique was further refined by CISE and RANDOLPH[22], who showed that the nucleation was secondary rather than primary in this system, and that at a given mass concentration large crystals were more effective than small for stimulating nucleation. The crystal growth rate was found to be size dependent.

In theoretical studies ABEGG, STEVENS and LARSON[23] introduced a size dependent growth rate model which allows a better fit with experimental data, while BECKER and LARSON[24] considered the effect of various modes of mixing and process geometries on the expected CSD. It is, of course, too much to expect that a single growth rate expression will be valid over the whole size range $O < L < \infty$ and for this reason ESTRIN, SAUTER and KARSHINA[25] proposed the use of different growth rate models for small and large crystals.

POPULATION BALANCE EXAMPLES

Uses of the population balances are described in a number of papers[10-24]. The two following examples are similar in style to those quoted by LARSON and RANDOLPH[14].

Example 1

The sieve analysis of the product leaving a continuous MSMPR crystalliser operating with a draw-down time $T = 0.25$ h is given in the table below.

BS sieve, μm	Size μm	ΔL, μm	$\bar{L}$, μm	$\bar{L}^3$, μm^3	dW,† g	$\alpha \rho \bar{L}^3 \Delta L$, g μm	Population density† n, μm^{-1}
710	>710	—	—	—	—	—	—
500	710–500	210	605	2.21×10^8	1.20	3.88×10^{-2}	3.09×10^1
355	500–355	145	428	7.84×10^7	2.72	9.45×10^{-3}	2.88×10^2
250	355–250	105	303	2.78×10^7	4.27	2.43×10^{-3}	1.75×10^3
180	250–180	70	215	9.94×10^6	3.15	5.80×10^{-4}	5.43×10^3
125	180–125	55	153	3.58×10^6	1.69	1.64×10^{-4}	1.03×10^4
90	125–90	35	108	1.26×10^6	0.75	3.68×10^{-5}	2.04×10^4
63	90–63	27	77	4.57×10^5	0.28	1.03×10^{-5}	2.73×10^4
45	63–45	18	54	1.58×10^5	0.12	2.36×10^{-6}	5.08×10^4
—	<45	—	—	—	—	—	—

† per litre of slurry.

dW is the mass (g) of crystals of the given size fractions $\bar{L}(\mu m)$ per litre of the exit slurry. $\bar{L}$ is the mean size of the fraction retained between two sieves and ΔL is the difference in size between the two sieve apertures. The population density, n (number μm^{-1}), in per litre of slurry is calculated from equation (10.32). The crystals, of density $\rho = 1 \cdot 77 \, g/cm^3$, are almost perfect octahedra, so the volumetric shape factor $\alpha = 0 \cdot 471$ (see p. 397).

These data are plotted as $\log n$ versus $\bar{L}$ in *Figure 10.5*. A straight line is drawn through the points by a trial and error procedure to satisfy the mass

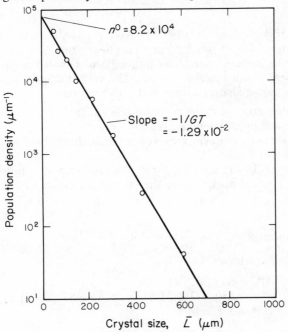

Figure 10.5. Population density plot (Example 1)

balance (equation 10.22). Extrapolation of this line to $\bar{L} = 0$ gives the nuclei population density $n^0 = 8 \cdot 2 \times 10^4 \, \mu m^{-1}$.

The slope of the line is $1/GT$ (equation 10.14). The value in the present case is $-1 \cdot 29 \times 10^{-2} \, \mu m^{-1}$, which gives a linear growth rate, G (for $T = 0 \cdot 25 \, h$) of $310 \, \mu m \, h^{-1}$ ($= 8 \cdot 6 \times 10^{-8} \, m \, s^{-1}$).

The nucleation rate, dN^0/dt (equation 10.17) $= n^0 G = 2 \cdot 54 \times 10^7 \, h^{-1}$.

Other runs, made with draw-down times of $T = 0 \cdot 5$ and $0 \cdot 75 \, h$, respectively, gave the following data:

$T,$ h	$n^0,$ μm^{-1}	$G,$ μm	$dN^0/dt,$ h^{-1}
0·25	$1 \cdot 21 \times 10^5$	845	$9 \cdot 73 \times 10^7$
0·5	$8 \cdot 2 \times 10^4$	310	$2 \cdot 54 \times 10^7$
0·75	$6 \cdot 7 \times 10^4$	152	$1 \cdot 02 \times 10^7$

A plot of $\log n^0$ versus $\log G$ (as in *Figure 10.2c*) in accordance with equation 10.19 gives a line of slope $\dfrac{m}{l} - 1$ or, alternatively, a plot of log (dN^0/dt) versus log G gives a line of slope m/l, where m and l are the 'orders' of the nucleation and growth processes, respectively (equations 10.15 and 10.16).

In the present case m/l is evaluated as 1.2.

If a value of $l = 1.5$ is taken as the 'order' of the growth process then the 'order' of the nucleation process $m = 1.5 \times 1.2 = 1.8$.

Example 2

An MSMPR crystalliser operates with a steady state nucleation of $n^0 = 10^7$ $(\mu m)^{-1}$ $(m)^{-3}$ (i.e. 10^{13} m^{-4}) a growth rate $G = 10^{-8}$ m/s, and a mixed-product removal rate (clear liquor basis) $Q = 0.6$ m^3/h. The volume of vessel (clear liquor basis) $V = 4$ m^3. The crystal density $\rho = 2660$ kg/m^3 and the volumetric shape factor $\alpha = 0.7$. Determine:

(a) the solids content in the crystalliser (kg/m^3)
(b) the crystal production rate (kg/h)
(c) the percentage of nuclei removed in the discharge by the time they have grown to 100 µm.
(d) the liquor flow rate (m^3/h) through a fines trap to remove 90 per cent of the original nuclei by the time they have grown to 100 µm.

Solution

Draw-down time $T = \dfrac{V}{Q} = \dfrac{4}{0.6} = 6.67$ h

(a) From a mass balance:

$$W_T = 6\alpha\rho n^0(GT)^4$$
$$= 6\,(0.7)\,(2660)\,(10^{13})\,(10^{-8} \times 6.67 \times 3600)^4$$
$$= 376 \text{ kg/m}^3$$

(b) Production rate, $P = W_T Q$
$$= 376 \times 0.6 = 226 \text{ kg/h}$$

(c) The crystal population decays exponentially with size (equation 10.14):

$$\frac{n}{n^0} = \exp\left(-\frac{L}{GT}\right)$$

$$= \exp\left(\frac{10^{-4}}{10^{-8} \times 6.67 \times 3600}\right)$$

$$= \exp(-0.414) = 0.66$$

Therefore 34 per cent of the original nuclei have been removed in the discharge by the time they have reached 100 µm.

(d) If 90 per cent of the nuclei are to be removed, i.e. 10 per cent allowed to remain and grow to a size > 100 µm, then, from equation (10.14)

$$\frac{n}{n^0} = \exp\left(-\frac{L}{GT}\right) = 0.1$$

i.e.

$$\frac{L}{GT} = \frac{10^{-4}}{10^{-8}\,T} = \ln\left(\frac{1}{0\cdot1}\right) = 2\cdot303$$

Therefore

$$T = 4350\text{s} = 1\cdot2\text{ h}$$

$$= \frac{V}{Q + Q_F}$$

and the volumetric flow rate through the fines trap, Q_F, is given by

$$Q_F = \frac{V}{T} - Q = \frac{4}{1\cdot2} - 0\cdot6 = 2\cdot73\text{ m}^3/\text{h}$$

EXPRESSIONS OF THE CSD

A method widely used for characterising the size distribution of a particulate mass is by means of the M.S./C.V., described in detail in Chapter 11. Briefly, the M.S. is the median size of the distribution: 50% of the material is smaller and 50% is larger than the M.S. The M.S. was originally called the M.A.

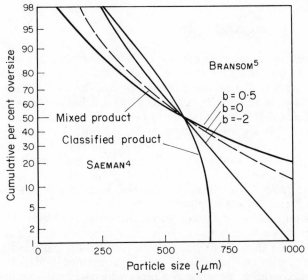

Figure 10.6. Theoretical crystal size distributions (probability plot)

(mean aperture), because crystal sizing is generally effected by sieving, but since other sizing techniques can be used, and the 'mean' size is more correctly termed a 'median', the designation M.S. is preferred.

The C.V. (coefficient of variation) is a statistical quantity, related to the standard deviation of a Gaussian distribution, which quantifies the size spread, although, as described on pp. 389–391, it is not necessary for the

CSD to be Gaussian in order to apply the concept. The higher the C.V., normally expressed as a percentage, the broader the spread. For a mono-sized distribution C.V. = 0. The M.S. and C.V. are readily obtained by plotting the results of a size analysis, on a cumulative basis, on probability graph paper (see *Figure 11.11*).

As pointed out by BENNETT[26] the CSD relationships resulting from the MSMPR models of SAEMAN[5] and BRANSOM[6] (equations 10.4 and 10.6) and the Saeman model for classified product removal can be interpreted in terms of the C.V. (see *Figure 10.6*). C.V. = 52% for a continuous MSMPR crystalliser and 20% for a CPR. For the Bransom model, C.V. changes from 78 to 32% as exponent b in equation (10.6) changes from 0·5 to -2. The case for $b = 0$ corresponds to Saeman's model of an MSMPR crystalliser with a C.V. of 52%.

Values of the C.V. reported by BENNETT[26] for a number of large-scale industrial crystallisers of the forced-circulation and DTB types were in the 30–50% range, but some materials such as sodium chloride were consistently in the 20–30% range. The agreement with theoretical predictions did not appear to be good, but, as Bennett pointed out, it was not difficult to see how conditions in an industrial crystalliser differed from those assumed in the ideal models. Industrial C.V. data have also been reported by PALMER[27] for a large number of substances crystallised in Oslo-Krystal units.

CRYSTALLISERS IN SERIES

One of the first attempts to develop CSD relationships in a series of linked crystallisers was made by ROBINSON and ROBERTS[28]. They used a probabilistic approach to the problem, assuming that the level of supersaturation was constant throughout the cascade of agitated crystallisers, and that all crystals entered the first stage (at zero size) and grew at the same rate in each vessel. No spurious nucleation was considered, and the dominant effect of the nucleation rate (in the first crystalliser of the cascade) on the resultant CSD was clearly demonstrated. This analysis also showed the advantages of operating in a linked series; a more uniform size distribution may be expected, for example. Later work by RANDOLPH, DEEPAK and ISKANDER[29] reinforced this point.

RANDOLPH and LARSON[12] considered the same problem but allowed for nucleation in each stage of the series, as did NÝVLT and co-workers[30, 31], who supported their theoretical model with results from laboratory-scale crystallisers. ABEGG and BALAKRISHNAN[32] computed CSDs for a series of continuous, perfectly mixed crystallisers with nucleation in the first stage only and with equal nucleation rates in all stages, and compared them with those reported[26] for a number of industrial crystallisers. The good agreement led the authors to suggest that large, imperfectly mixed, industrial crystallisers can be modelled mathematically by a suitable cascade of perfectly mixed units.

A recently mathematical study by HILL[33] considered the theory of continuous crystallisation (particularly of sucrose) in crystallisers connected in series, in which a suspension of growing crystals is fed forward through the system without back-mixing. The calculation of the minimum C.V. that can be obtained under stated conditions was described. Two cases specifically considered were stirred reactors of equal mean residence times connected in series, and tubular reactors in which the residence time distribution is given by the Gaussian error function. The C.V. can be reduced either by connecting reactors in series or by extending a tubular reactor in the axial direction. Series operation is always superior to extension, because it prevents mixing at the points of connection.

UNSTEADY STATE OPERATION

The steady state population balance is given by equation (10.12). A balance for the unsteady state, where crystal breakage can occur, may be written:

$$\text{input} + \text{generation} = \text{output} + \text{accumulation}$$

and the transient form of the population balance becomes

$$\frac{\partial n}{\partial L} + \frac{n}{GT} = -\frac{1}{G}\frac{\partial n}{\partial t} \tag{10.43}$$

The utility of this relationship in assessing the stability of the CSD has been discussed by LARSON and RANDOLPH[14], who showed that, if the nucleation rate is a sensitive enough function of supersaturation, it is possible for the system to exhibit a sustained cycling of the CSD.

A continuously operated crystalliser will often exhibit cyclic changes in the product CSD even though the input conditions remain constant, and several analyses of the dynamic behaviour of a continuous crystalliser have been made in recent years[34-38]. It has often been suggested that a continuous crystalliser can be self-stabilising. For example, an increase of the supersaturation leads to a higher nucleation rate; but as the total crystal surface area increases owing to the growth of the nuclei, the supersaturation decreases again, and the nucleation rate is thus reduced. The total crystal surface area decreases when product crystals are withdrawn, and this in turn causes an increase of the supersaturation and the establishment of a new corresponding nucleation rate, and so on. On this basis a mathematical treatment of continuous crystalliser operation under steady state conditions may be made[31].

However, the dynamic behaviour of a continuous well-stirred crystalliser may not be so simple under certain conditions. The above stabilising effect is subject to a considerable time lag, because newly formed nuclei have no appreciable surface for a long time. Before the stabilising action can occur, therefore, large numbers of nuclei might be formed which will later reduce the supersaturation below its steady state value. The resulting slow nucleation will lead to a decrease of the total crystal surface area below its steady state value, and this in turn will cause an increase of the supersaturation above its steady-state value, and so on. The result of this sequence of events, as shown by NÝVLT and MULLIN[37], will be the occurrence of limit cycles.

Mathematical models of continuous crystallisers with perfect and imperfect mixing, with and without product classification have shown[37] how periodic changes of supersaturation, solids content, crystal size and production rate can readily occur. Periodic behaviour is most pronounced at the beginning of the crystallisation process, and in most cases the fluctuations are subsequently damped. This means that under favourable conditions a steady state is usually reached. However, under certain conditions the damping is not effective and the cyclic behaviour may continue for a considerable time. Indeed, in some cases the steady state may not be achieved at all. The cycle period, which is comparatively long in most cases, depends on the rate at which supersaturation is created and on the relative amount of product crystals withdrawn.

The stability of the crystallising system increases with increasing crystal growth rate and magma density, and with decreasing nucleation 'order', minimum product size, and the relative amount of crystals withdrawn. Continuous seeding is shown to have the same effect as decreasing the effective nucleation order, and should lead to a stabilisation of the system.

PROGRAMMED COOLING OF BATCH CRYSTALLISERS

Batch-operated, stirred cooling crystallisers are widely used in the chemical industry, but they usually yield a poor-quality non-uniform product. This is mainly due to the use of a high cooling rate in the initial stages of the process, which results in the formation of large numbers of crystal nuclei that cannot grow to the desired size. In addition, the large temperature drop across the cooling surfaces in the early stages often causes intensive scaling and significantly decreases the cooling capacity of the crystalliser. These disadvantages, normally associated with batch cooling crystallisers, can be overcome by the application of an appropriate temperature control[39].

A supersaturation balance on a batch cooling crystalliser gives

$$\frac{-\,\mathrm{d}\Delta c}{\mathrm{d}t} = \frac{\mathrm{d}c^*}{\mathrm{d}t} + k_g A \Delta c^n + k_n \Delta c^m \qquad (10.44)$$

where c = concentration, c^* = equilibrium saturation, $\Delta c = c - c^*$, t = time, A = total crystal area, k_g and k_n are growth and nucleation contants, and n and m are the 'orders' of growth and nucleation, respectively (see Chapters 5 and 6).

Equation (10.44) describes the rate of change of supersaturation in a seeded and nucleating solution. The first term on the right-hand side corresponds to the creation of supersaturation by cooling, taking advantage of the positive temperature dependence of solubility. The second and third terms describe the desupersaturation rate caused by crystal growth and nucleation, respectively.

To minimise the production of unwanted nuclei, it is necessary to control the level of supersaturation well within the metastable zone throughout the whole cooling operation (see *Figure 8.1*) so that solute deposition occurs predominantly on the added seed crystals present. If this situation can be achieved, the mean crystal size should be increased and the product quality improved. However, in the controlled cooling depicted in *Figure 8.1d* the supersaturation, Δc, changes considerably over the operating range of temperature, rising to a maximum and then decreasing again. This means that the crystal growth does not occur at a constant level of supersaturation. In a simple ideal case, Δc could be kept constant throughout the operation, i.e. $d\Delta c/d\theta = 0$. Or if this were not possible, e.g. if Δc_{max} were temperature dependent, Δc could be allowed to change at a constant rate, i.e.

$$\frac{d\Delta c}{d\theta} = \text{constant} \tag{10.45}$$

Starting with equation (10.44) it is possible to predict the optimum cooling curve for a controlled crystalliser operating under specified conditions. When the operation takes place within the metastable zone, and growth occurs predominantly on the added seeds, the nucleation term in equation (10.44) may be ignored, i.e.

$$\frac{-d\Delta c}{dt} = b\frac{dc^*}{d\theta} + k_g A\Delta c^n \tag{10.46}$$

where b is the cooling rate $(d\theta/dt)$. The optimum cooling rate may be obtained, after making certain simplifying assumptions[39], by solving equation (10.46) for $d\theta/dt$:

$$b = \frac{k_g A\Delta c^n}{\dfrac{dc^*}{d\theta} + \dfrac{d\Delta c}{d\theta}} \tag{10.47}$$

where $dc^*/d\theta$ is the temperature dependence of solubility and $d\Delta c/d\theta$ is the temperature dependence of supersaturation. Equation (10.47) describes the optimum cooling rate at a controlled supersaturation when no nucleation occurs. The conditions for which nucleation may not be neglected have also been considered by MULLIN and NÝVLT[39].

At the commencement of a batch-cooled seeded crystallisation, the total available crystal surface area is small and the solution can only be desupersaturated at a slow rate. But as the seeds grow and their surface area increases, the rate of desupersaturation can be increased. Equation (10.47) indicates that, to avoid spurious nucleation, the solution should be cooled slowly in the early stages and faster as the crystals grow, i.e. the reverse of natural cooling (*Figure 10.7*).

Introducing the over-all linear crystal growth rate G $(= dL/dt)$ into equation (10.47), we get[39]

$$b = \frac{3W_{so}G(t)}{\left[\dfrac{d\Delta c}{d\theta} + \dfrac{dc^*}{d\theta}\right]L_{so}^3} \cdot \left[L_{so} + \sum_{i=0}^{t-1}G(i)\right]^2 \tag{10.48}$$

which describes the desired cooling curve. W_{so} = initial mass of seeds, and L_{so} = initial size of seeds.

It is important to note that the optimisation described here is concerned with (a) a given mass of seeds, W_{so}, (b) a given seed size, L_{so}, (c) a given temperature range, θ_0 to θ_f and (d) a controlled level of supersaturation within the metastable zone. The total operating time may be determined in terms of these parameters by integrating equation (10.48).

In an experimental draft-tube agitated crystalliser, provided with a programmed cooling circuit, it has been shown that controlled cooling produces larger and more uniform crystals than does uncontrolled cooling. In the case

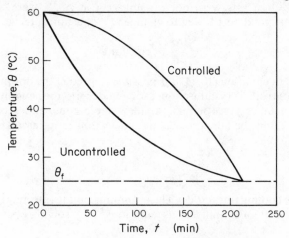

Figure 10.7. Controlled and uncontrolled cooling curves for the crystallisation of a seeded solution of potassium sulphate[39]

illustrated in Figure 10.7 (20 l of K_2SO_4 solution seeded with 50 g of 0·5 mm seeds cooled over 210 min, the supersaturation level being kept at 50 per cent of the maximum allowable) the mean size of the product crystals was 1·4 mm compared with 1·0 mm for uncontrolled cooling[39].

DESIGN OF CLASSIFYING CRYSTALLISERS

BASIC DESIGN PARAMETERS

Continuous classifying crystallisers are widely used for the industrial production of large uniform crystals of a wide variety of substances, and these units are usually designed on an *ad hoc* basis. Design procedures have rarely been reported in the literature and over the years many 'rules of thumb', some more successful than others, have been devised to facilitate crystalliser specification.

Four basic quantities are generally needed for the design of a suspended bed crystalliser:

    Liquor circulation rate, Q
    Vessel cross-sectional area, A
    Height of the crystal suspension, H
    Over-all residence (draw-down) time, T

Although the calculations of these interrelated quantities are reasonably simple, they involve assumptions that can only be made on the basis of either relevant experience or reliable experimental data. The basic relationships are:

$$Q = \frac{\text{crystal production rate, } P}{\text{effective supersaturation, } S} \qquad (10.49)$$

$$A = \frac{\text{liquor circulation rate, } Q}{\text{liquor upflow velocity, } u} \qquad (10.50)$$

$$H = \frac{\text{volume of suspension, } V}{\text{cross-sectional area, } A} \qquad (10.51)$$

$$V = \frac{\text{mass of crystals in suspension, } W}{\text{suspension (magma) density, } d} \qquad (10.52)$$

$$T = \frac{\text{mass of crystals in suspension, } W}{\text{crystal production rate, } P} \qquad (10.53)$$

The quantity, T, which for a steady state system refers to both liquid and solid phases, must not be confused with the crystal growth time, i.e. the age of a product crystal, τ. However, the quantities that present considerable difficulty in measurement or estimation are S, u, and V or W. The last two are related through the crystal density ρ_c and the bed voidage ε:

$$\varepsilon = 1 - \frac{W}{\rho_c V} \qquad (10.54)$$

These quantities will now be considered in turn.

Effective Supersaturation, S

The choice of the working level of supersaturation should be based on reliable measurements of the metastable limits of the system which may be made in the laboratory under carefully controlled conditions. Metastable zone widths depend on many factors including the temperature, cooling rate, agitation, presence of impurities, etc., but the most important requirement is that they must be determined *in the presence of the crystalline phase* (Chapter 6) and with the actual liquor to be processed.

The effective supersaturation, S, is that which is released during the passage of the liquor through the crystalliser, i.e., the difference between the inlet and outlet supersaturations.

Solution Upflow Velocity, u

The solution upflow velocity, u, is not an easy quantity to specify. It is obvious that the solution upflow should balance the settling velocity of the crystals, but this is difficult to measure and virtually impossible to predict with accuracy. The trouble stems from the fact that the crystal suspension contains crystals of all sizes between product size and freshly generated nuclei. To choose the upflow velocity as the free settling velocity of a product crystal would result in the smaller crystals being swept out of the crystalliser. An upflow velocity taken as the free settling velocity of the smallest crystals held up in the bed would at best demand that the product crystals be withdrawn through an elutriating leg, and at worst fail to fluidise the bed. The safest method, but admittedly the most tedious, is to measure the upper and lower fluidisation velocities for the particular suspension and to use an intermediate value between these two limits. But the success of this method depends on the ability to specify accurately the suspension density or bed voidage to be used in the crystalliser (see pp. 408–415).

Suspension Volume, V

There is no generally successful method for estimating the suspension volume, V. Several methods are available and more than one should always be used in an attempt to arrive at a reasonable design value. A few of the more popular methods – and some are simply rough rules – include the following.

1. The suspension height in a classifying crystalliser is frequently between one and two times the diameter of the vessel, which may be calculated from equation (10.50). This generalisation, of course, can only give a very rough guide. It is well known, for example, that the production rate of a classifying crystalliser under given conditions is determined by its cross-sectional area and not by its height.

2. The separation intensity, S.I., defined[40] as the mass of equivalent 1 mm crystals produced in 1 m^3 of crystalliser volume in 1 h generally lies in the range 50 to 300 kg/m^3 h. At temperatures near 30 °C the lower limit is more frequent; an increase in temperature leads to higher values. For crystals larger (but not smaller) than 1 mm the relationship

$$\text{S.I.} = LP/V \qquad (10.55)$$

may be used where L is the product size (mm), P the production rate (kg/h), and V the suspension volume (m^3).

In other words, L is inversely proportional to the crystal production rate, P.

3. The over-all residence (draw-down) time, T, in many classifying crystallisers is about 2 h. Thus the mass of crystals in suspension, W, may be determined from the crystal production rate, P, by equation (10.53). After the appropriate value for the suspension density or bed voidage, ε, has been chosen the suspension volume may be calculated from equations (10.52) or (10.54).

Product Crystal Growth Time, τ

The true crystal time, τ, i.e. the time it takes a crystal seed to grow to product size residence can be estimated in several different ways.

1. Theoretical considerations[5] suggest that the true crystal residence time, τ, i.e. the age of a product crystal, in a classifying crystalliser is four times the over-all residence time (draw-down time), T. Thus from point (3) above τ is about 8 h.

2. The crystallisation rate in a classifying crystalliser is frequently equiva-

lent to a mean deposition of 10–15% of the suspended crystal per hour. This is the same thing as saying that the product crystal growth time is 7–10 h.

3. τ can be estimated from the linear crystal growth rate, $G(= \mathrm{d}L/\mathrm{d}t)$, which can be measured directly or calculated from over-all growth data (Chapter 6), by

$$\tau = \frac{L_p - L_0}{G} \tag{10.56}$$

where L_p is the product crystal size and L_0 is the size of the smallest crystals held up in the crystal bed. However, to make this calculation it is necessary to know or estimate the mean level of supersaturation in the crystallisation zone in order to obtain the appropriate value of G.

4. τ can be estimated from

$$\tau = \frac{L_p}{2\,G_p}\left(\frac{L_p^2}{L_0^2} - 1\right) \tag{10.57}$$

which has been derived[31] by assuming total desupersaturation of the solution passing through the crystal bed. In this equation G_p is the maximum linear crystal growth rate, i.e. at the bottom of the crystallisation zone (product exit, liquor inlet). This equation will be discussed later.

Suspension (Magma) Density, d, or Bed Voidage, ε

Most of the above methods require the choice of a working value of the suspension density or bed voidage, ε. Little general guidance can be given here. PULLEY[41], for example, suggests that values of ε lie between 0·5 at the bottom and 0·975 at the top of the suspension. However, in most cases over-all mean values in the range 0·8 to 0·9 are usually found in practice. This is confirmed if we make a linear interpolation for a mean voidage, ε_{mean}, and mean size, L_{mean}, in the suspension between Pulley's values of $\varepsilon_p = 0·5$ for the product size L_p and $\varepsilon_0 = 0·975$ for the smallest retained crystals, L_0:

$$\varepsilon_{mean} = \varepsilon_p + \frac{\varepsilon_0 - \varepsilon_p}{L_p - L_0}\cdot L_{mean} \tag{10.58}$$

The mean crystal size in a perfectly classified suspension is given by

$$L_{mean} = 0·63\,L_p \tag{10.59}$$

and, putting this value into equation (10.58), we find that the mean bed voidage, ε_{mean}, varies from 0·8 for $L_0 \to 0$ to 0·9 for $L_0 = 0·25\,L_p$.

It should be clear that most of the design methods described above depend considerably upon experience. They cannot be applied indiscriminately.

AN IDEAL CLASSIFIED BED CRYSTALLISER

In this analysis[42] several simplifying assumptions are made.

1. No uncontrolled nucleation occurs in the crystalliser; a constant input of $\mathscr{N}$ seed crystals per unit time is assumed, the size of the seeds being L_0 (the

smallest crystals to be found in the bed). This is a commonly used simplification[5, 12, 41] and is equivalent to the practical case where all excess nuclei are withdrawn and dissolved.

2. The crystal bed is perfectly classified and can be subdivided into layers comprising equal crystal sizes (i.e. equal retention times). This simplification facilitates calculation of the crystal surface area in the various sectors of the fluidised bed. Even if there were some intermixing between the 'layers', this simplification imposes no insuperable restriction, because the presence of undersize crystals in a given layer is to some extent compensated by the presence of oversize crystals.

3. The crystal growth kinetics are sufficiently described by the equation

$$\frac{dW}{dt} = k_g A \Delta c^n \tag{10.60}$$

This type of empirical relationship has been discussed fully in Chapter 6.

4. The crystal shape can be represented by a geometrical body characterised by a single size, L, and surface and volume shape factors, α and β, respectively (see equations 10.64 and 10.65).

5. Steady state operation is assumed with all local values of the parameters remaining constant with time.

A layer-by-layer materials balance leads to the general relationship

$$Q(\Delta c_p - \Delta c) = \alpha \rho \, \mathcal{N}(L_p^3 - L^3)$$

or:

$$\Delta c = \Delta c_p - \frac{\alpha \rho \, \mathcal{N}}{Q}(L_p^3 - L^3) \tag{10.61}$$

where Δc_p is the supersaturation in the lowest layer containing product crystals of size L_p.

The crystal production rate is

$$P = \alpha \rho \, \mathcal{N} L_p^3 \tag{10.62}$$

so equation (10.61) may be rewritten

$$\Delta c = \Delta c_p - \frac{P}{Q}\left(1 - \frac{L^3}{L_p^3}\right) \tag{10.63}$$

The total crystal surface area and mass in the crystalliser containing N crystals of size L are

$$A = \beta N L^2 \tag{10.64}$$

$$W = \alpha \rho N L^3 \tag{10.65}$$

and from equation (10.65) we have

$$dW = 3\alpha \rho N L^2 dL \tag{10.66}$$

Substituting these values into equation (10.60) and rearranging, we obtain

$$G = \frac{dL}{dt} = \frac{\beta}{3\alpha \rho} \cdot k_g \Delta c^n \tag{10.67}$$

Further substitution of Δc from equation (10.63) into equation (10.67) leads to

$$G = \frac{\beta}{3\alpha\rho} \cdot k_g \Delta c_p^n \left[1 - \frac{P}{Q\Delta c_p} \left(1 - \frac{L^3}{L_p^3} \right) \right]^n \qquad (10.68)$$

Generally the supersaturation of the solution leaving a classified bed of crystals is not zero but has a positive value given by

$$\Delta c_0 = \Delta c_p - \frac{P}{Q} \left(1 - \frac{L_0^3}{L_p^3} \right) \qquad (10.69)$$

In most cases $L_0^3 \ll L_p^3$ and, hence,

$$\Delta c_0 = \Delta c_p (1 - \gamma) \qquad (10.70)$$

where

$$\gamma = 1 - \frac{\Delta c_0}{\Delta c_p} = \frac{P}{Q\Delta c_p} \qquad (10.71)$$

From equation (10.68) it follows that

$$G = G_p \left[1 - \gamma \left(1 - \frac{L^3}{L_p^3} \right) \right]^n \qquad (10.72)$$

where

$$G_p = \frac{\beta}{3\alpha\rho} \cdot k_g \Delta c_p^n \qquad (10.73)$$

If it is assumed that the crystal growth rate does not depend on the crystal size (or if a mean value is taken in cases where it does), then, by use of the substitution

$$x = L_p \cdot \sqrt[3]{\left(\frac{1-\gamma}{\gamma} \right)} \qquad (10.74)$$

it can be shown[42] that:
For $\gamma = 1$ (total desupersaturation): equation (10.72) becomes

$$G = G_p \left(\frac{L}{L_p} \right)^{3n} \qquad (10.75)$$

and therefore

$$\tau = \frac{L_p}{(3n-1)G_p} \cdot \left[\left(\frac{L_p}{L_0} \right)^{3n-1} - 1 \right] \qquad (10.76)$$

Equation (10.76) reduces to equation (10.57) for the case of $n = 1$.
If $L_0 = 0 \cdot 1\, L_p$, then
for $n = 1$, $\tau = 49\, L_p/G_p$ (10.77)
for $n = 2$, $\tau = 2 \times 10^4\, L_p/G_p$ (10.78)
If $L_0 = 0 \cdot 3\, L_p$, then
for $n = 1$, $\tau = 4\, L_p/G_p$ (10.79)
for $n = 2$, $\tau = 82\, L_p/G_p$ (10.80)

For $\gamma = 0.9$ (90% desupersaturation):
From equation (10.74) we get: $x = 0.48\,L_p$, and if $L_p \gg L_0$ then

$$\text{for } n = 1, \qquad \tau = 7.1\,L_p/G_p \tag{10.81}$$
$$\text{for } n = 2, \qquad \tau = 50.7\,L_p/G_p \tag{10.82}$$

For $\gamma = 0.5$ (50% desupersaturation):
From equation (10.74) $x = L_p$, and if $L_p \gg L_0$ then

$$\text{for } n = 1, \qquad \tau = 1.67\,L_p/G_p \tag{10.83}$$
$$\text{for } n = 2, \qquad \tau = 2.89\,L_p/G_p \tag{10.84}$$

For $\gamma \to 0$ (very low desupersaturation):

$$G = G_p \tag{10.85}$$

and therefore

$$\tau = \frac{L_p - L_0}{G_p} \tag{10.86}$$

There still remains the question of the appropriate choice of the working supersaturation, Δc_p. This must be lower than the critical value set by the metastable limit, corresponding to the relevant hydrodynamic and other conditions. Metastable limits are normally measured in terms of the maximum allowable undercooling, $\Delta\theta_{max}$, of the system and this may be re-calculated into the appropriate concentration units from the temperature–solubility coefficient:

$$\Delta c_{max} = \left(\frac{dc^*}{d\theta}\right)\Delta\theta_{max} \tag{10.87}$$

Alternatively, if the maximum allowable supersaturation is the same as the value corresponding to a perfectly stirred crystalliser, Δc_{max} may be calculated from a crystal number balance:

$$\mathcal{N} = \frac{P}{\alpha\rho L_p^3} = \frac{k_n \Delta c_{max}^m}{\alpha\rho L_n^3} \tag{10.88}$$

where L_n is the size of a crystal nucleus. It is assumed that the nucleation kinetics are expressed by an empirical relationship of the form $dW/dt = k_n\Delta c_{max}^m$ (see Chapter 6). Equation (10.88) gives

$$\Delta c_{max} = \left[\frac{P}{k_n(L_p/L_n)^3}\right]^{1/m} \tag{10.89}$$

However, equation (10.89) represents the balance over the lowest layer of product crystals and the total number of crystals produced at this super-saturation level would probably be too high. Therefore, it is reasonable to take into account values of supersaturation which are lower than Δc_{max}. If the supersaturation level is too high, more crystal nuclei are produced and the crystalliser should be provided with a fines trap. On the other hand, too low a supersaturation leads to a low suspension density and to an irregular or periodic behaviour of the crystalliser[37].

From a materials balance over the fluidised bed it follows that

$$W = \sum_{L_0}^{L_p}\alpha\rho N_i L_i^3 \simeq \frac{P\tau L^{-3}}{L_p^3} \tag{10.90}$$

where

$$\bar{L}^3 = \frac{1}{L_p - L_0} \int_{L_0}^{L_p} L^3 \, dL = \tfrac{1}{4}[(L_p^4 - L_0^4)/(L_p - L_0)] \qquad (10.91)$$

or, for $L_p \gg L_0$,

$$\bar{L} \simeq 0.63 \, L_p \qquad (10.92)$$

which is identical with equation (10.59) and, hence,

$$W = \frac{P\tau}{4 \, \bar{L}_p^3} \cdot \left(\frac{L_p^4 - L_0^4}{L_p - L_0}\right) \simeq \frac{P\tau}{4} \cdot \left(\frac{L_p}{L_p - L_0}\right) \qquad (10.93)$$

or, for $L_p \gg L_0$,

$$W = \tfrac{1}{4}P\tau \qquad (10.94)$$

which is identical with the statement[5] that the true crystal residence time, τ, in a classifying crystalliser is four times the over-all residence time, T: combining equations (10.53) and (10.94) we obtain

$$\tau = 4 \, T \qquad (10.95)$$

APPLICATION OF THE DESIGN EQUATIONS

By way of example, the design methods described above will now be applied to the crystallisation of potassium sulphate at 20 °C. The calculations are carried out for a production rate of 1000 kg/h of 1 mm crystals. The data used are:

$$k_g = 0.75 \text{ kg m}^{-2} \text{ s}^{-1} (\Delta c)^{-n}, \text{ with } n = 2.0$$
$$k_n = 2 \times 10^8 \text{ kg s}^{-1} \Delta c^{1-m}, \text{ with } m = 8.3$$

The size of the smallest crystals remaining in the fluidised bed, L_0, is chosen arbitrarily as 0.3 mm. This represents a free settling velocity of about 4 cm/s. The size of a crystal nucleus, L_n, for use in equation (10.89) is not the size of a 'critical' nucleus. Sub-microscopic stable nuclei probably grow at very fast rates (*Figure 5.18*), so L_n is taken here as the size of the smallest crystals detected in the system (0.1 mm, a value obtained by two independent experimental techniques[42]).

The other physical constants used are:

crystal density, $\rho = 2660 \text{ kg/m}^3$
solution density, $\rho_s = 1082 \text{ kg/m}^3$
solution viscosity, $\eta = 1.2 \times 10^{-3} \text{ N s/m}^2$
solubility, $c^* = 0.1117 \text{ kg/kg water}$

The calculation steps are as follows.

1. The working supersaturation, Δc_p, is estimated from values of the maximum allowable supersaturation Δc_{max} calculated from equation (10.89). Thus

$$\Delta c_{max} = \left(\frac{1000}{3600 \times 2 \times 10^8 \times 10^3}\right)^{1/8.3} = 0.037 \text{ kg K}_2\text{SO}_4/\text{kg water}$$

In practice the working supersaturation level in a classifying crystalliser has to be kept much lower than this limit, and for calculation purposes the inlet supersaturation value will be arbitrarily reduced to about 30% of Δc_{max}. In this example, therefore, a value of $\Delta c_p = 0.01$ will be used, i.e. 0.0089 kg/kg of solution.

2. The solution circulation rate, Q, is calculated from the materials balance, equation (10.71).

3. The maximum linear growth rate, G_p, is calculated from equation (10.73), assuming a value of $\beta/\alpha = 6$.

4. The crystal growth time, τ, for different relative desupersaturations, γ, is calculated from equations (10.76), (10.82), (10.84) and (10.86).

5. The mass of suspended crystals, W, is given by equation (10.93).

6. The suspension volume, V, is calculated from equation (10.54), a value of $\varepsilon = 0.85$ being used.

7. The solution up-flow velocity for very small crystals (e.g. <0.1 mm) may sometimes be calculated without serious error from modified Stokes' law relationships. However, strictly speaking, these should not be used for particle Reynolds numbers in excess of about 0.3 and for particle shapes other than spheres. Other expressions are available for larger crystals but considerable errors may be incurred for non-granular shapes.[43] It is usually much safer to measure these velocities in the laboratory. For the present calculations, a measured value of 4 cm/s is used.

8. The crystalliser cross-sectional area, A, is given by equation (10.50), and the crystalliser diameter, D, is

$$D = \sqrt{\left(\frac{4A}{\pi}\right)} \tag{10.96}$$

9. The crystal bed height, H, is given by equation (10.51).

10. The separation intensity, S.I., is calculated from equation (10.55).

The results are summarised below.

Desupersaturation, γ	1.0	0.9	0.5	0.1
Maximum growth rate, G_p (m/s) $\times 10^8$	5.6	5.6	5.6	5.6
Up-flow velocity, u (m/h)	144	144	144	144
Circulation rate, Q (m³/h)	104	115	208	103
Crystal residence time, τ (h)	408	252	14.4	3.5
Mass of crystals, W (kg)	145 000	90 000	5100	1250
Suspension volume, V (m³)	364	225	12.8	3.15
Cross-sectional area, A (m²)	0.72	0.80	1.45	7.2
Crystalliser height, H (m)	505	281	8.8	0.44
Crystalliser diameter, D (m)	0.96	1.01	1.36	3.02
H/D	525	280	6.5	0.15
Separation intensity S.I.	3	4.5	78	320
Economically possible	no	no	yes	no

From these results it can be seen that the cross-sectional area, A, of the crystalliser depends linearly on the relative desupersaturation, γ. It may also be shown that the production rate depends linearly on the crystalliser area out is independent of the crystalliser height. This means that to change the height: diameter ratio, keeping the desupersaturation constant, it is necessary to change the production rate of the crystalliser. On the other hand, if the production rate is to remain constant, the crystalliser height may be adjusted by changing the seed or product crystal sizes L_0 or L_p.

A similar design procedure has been proposed by MULLIN and NÝVLT[42] for a mixed-suspension, classified product removal crystalliser.

Discussion

1. For a classified bed crystalliser the calculations show that for potassium sulphate, i.e. for a substance with a moderate growth rate that depends on the square of the supersaturation, a crystalliser allowing a relatively low desupersaturation would be most suitable. For a 50 per cent desupersaturation, the volume of such a crystalliser to produce 1000 kg/h should be about 12·8 m³ but the height: diameter ratio of the crystalliser is too high (6·5). In fact, to achieve a height: diameter ratio of about 2 would require a suspension volume of about 6 m³ and desupersaturation of about 30 per cent.

2. Two factors affect the height of a classified bed crystalliser: the relative desupersaturation (high values of γ require deep crystal beds), and the difference between the size of the smallest crystals held back in the bed, L_0, and the size of the desired product crystals, L_p (the greater the difference the greater the depth of the bed).

3. The height of a classified bed crystalliser depends neither of the working supersaturation nor on the production rate.

4. The cross-sectional area of a classified bed crystalliser depends linearly on the production rate, the working supersaturation and the relative desupersaturation.

5. It follows from paragraphs 2 to 4 that in principle it should be possible, for a given system, to relate the relative desupersaturation, production rate and the height: diameter ratio of the crystalliser. A specification of any of these quantities (e.g. taking a 2 : 1 height : diameter ratio as being reasonable) allows the specification of the other two.

The following comments may be made on the different methods sometimes used for a rough estimation of the suspension volume.

(a) The fact that the suspension height in a classifying crystalliser is frequently between one and two times the diameter of the vessel arises from the general aim to minimise the construction material requirement per unit volume of crystalliser. Therefore, after an acceptable height : diameter ratio has been chosen, the vessel diameter will be specified by the production rate and the height by the degree of desupersaturation.

(b) The separation intensity, S.I., enables a rough estimation to be made of the production rate or crystalliser volume. Its value depends on the ability of substances to produce large crystals, as pointed out by GRIFFITHS[40]. The normal range of S.I. is 50–300 kg/m³ h for 1 mm equivalent crystals, the lower values relating to the lower operating temperatures.

(c) The product crystal growth time in many classifying crystallisers is in the range 5 to 15 h; thus a value of $\tau = 8$ h is a reasonable value to use for preliminary calculations.

(d) As a rough guide to the relative desupersaturation, γ, substances with a first-order growth rate dependence on supersaturation ($n = 1$) may allow a desupersaturation of up to about 90 per cent, whereas for those with a second-order dependence the allowable desupersaturation may be lower than 50 per cent. It must be emphasised, however, that the relative desupersaturation is also a function of the crystalliser height, and the actual value of γ chosen for the operating conditions will determine the dimensions of the crystalliser.

SCALE-UP AND OPERATING PROBLEMS

The design of a large-scale crystalliser from data obtained on a small-scale unit is never a simple matter. After installation, trial and error procedures are generally necessary before the correct operating conditions are determined for the production of a given product from the crystalliser. One must not give the impression that crystalliser design is largely a matter of good luck and guess work, but it must be appreciated that the scaling-up of crystallisation equipment is more difficult than that for any of the other unit operations of chemical engineering, SAEMAN[44] put it very neatly when he said: 'It may appear that if all the dominant variables and the major interrelations of the variables are known, then scale-up may be reduced to a sound basis, *but scale-up itself is a dominant variable...*'

If a small crystalliser produces the required type of product, then the proposed large crystalliser must simulate a large number of different conditions obtaining in the small unit. The five most important conditions are

1. Identical flow characteristics of liquid and solid.
2. Identical supersaturation in all equivalent regions of the crystallisers
3. Identical initial seed sizes
4. Identical magma densities
5. Identical contact times between growing crystals and supersaturated liquor

The scaling-up of agitation equipment has long been recognised as a difficult problem, and the two dimensionless numbers most frequently encountered in the analysis of stirrers and agitators are the Reynolds number, Re, and the Froude number, Fr. The former gives the ratio of the inertia and viscous forces, the latter the ratio of centrifugal acceleration and acceleration due to gravity g. For a stirrer blade of diameter d, rotating at n revolutions per unit time in a liquid of density ρ and absolute viscosity η, these two numbers may be written

$$Re = \frac{\rho n d^2}{\eta} \text{ and } Fr = \frac{n^2 d}{g} \qquad (10.97)$$

For scale-up purposes, values of both (Re) and (Fr) should be kept constant but this is exceptionally difficult, if not impossible in most cases. For example if the stirrer diameter is increased by a factor of 4, the stirrer speed must be decreased by a factor of 2 if the Froude number is to be kept constant

$n \propto 1/\sqrt{d}$), but by a factor of 16 if the Reynolds number is to remain the same ($n \propto 1/d^2$).

If the conditions in two agitated systems are to be similar in all respects, conditions of geometrical, kinematic and dynamic similarity must be ensured. By the application of the technique of dimensional analysis the dimensionless equation can be derived[45]

$$\frac{P}{\rho n^3 d^5} = \phi(Re, Fr) \tag{10.98}$$

P is the power input to the stirrer. The group $P/\rho n^3 d^5$ is known as the Power number, P_N. If it is assumed that $P \propto \rho n^3 d^5$ then it can be seen, for example, that if the speed of a given stirrer is doubled, the power input would have to be increased eightfold. Again, if the stirrer diameter is doubled, its speed remaining constant, the power input would have to be increased by a factor of 32.

The significance of these dimensionless groups can now be discussed in relation to the problems of crystalliser scale-up. NEWMAN and BENNETT[46] showed that it is much easier to scale-up successfully a crystalliser of the draft-tube DTB type (e.g. as in *Figure 9.19*) than one of the forced-circulation type (e.g. as in *Figure 9.18*). *Figure 10.8* shows the variables that have to be

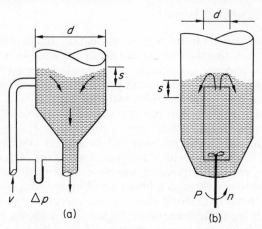

(a)

(b)

Figure 10.8. Scale-up parameters for (a) forced-circulation (b) draft-tube crystallisers[46]

considered in each case. For the DTB type, the Reynolds, Froude and Power numbers are given by equations (10.97) and (10.98). For the forced-circulation type, the corresponding dimensionless numbers are given by

$$Re = \frac{\rho v d}{\eta}, \qquad Fr = \frac{v^2}{dg}, \qquad P_N = \frac{g\Delta p}{\rho v^2} \tag{10.99}$$

where d = diameter of the crystalliser body, v = inlet velocity of the liquor, Δp = pressure drop. Another dimensionless ratio d/s was also defined, where s denotes the submergence required to suppress vapour formation in the superheated liquid entering the vessel (forced circulation) or rising to the boiling surface (DTB).

When both viscous and gravitational forces apply, e.g. in the forced-circulation crystalliser where vortexing and toroidal circulation occur, both the Reynolds and Froude numbers are important factors. Because $Fr \propto v^2$ $Re \propto v$, and $P_N \propto 1/v^2$, it is very difficult to reproduce the same hydraulic regime in different vessels. The submergence ratio is also an important scale-up factor. In the DTB type of crystalliser, however, vortexing does not occur and the Froude number and submergence ratio are not important variables. Equivalent hydraulic regimes can be produced in large equipment with the same Power number at very different Reynolds numbers. The agitator, of course, must operate at a speed well below that at which attrition and nucleation are induced. It is claimed that a DTB crystalliser is more easily scaled up than a forced-circulation type[46].

SAEMAN[44] has pointed out, however, that scale-up of crystallisers on the basis of similarity alone is unduly restrictive. It is generally more important to build into the crystalliser a large number of degrees of freedom, so that with this flexibility it is possible, by trial and error procedures, to arrive at the optimum operating conditions.

FINES REMOVAL

Most crystallisers suffer from excessive nucleation, and to produce reasonably large crystals it is necessary to remove the excess fines. In terms of mass, the small crystals appear insignificant; the cumulative mass of crystals in a mixed suspension varies as the fourth power of their size. Thus, for example crystals up to $\frac{1}{2}$ the product size constitute only $(\frac{1}{2})^4$, one-sixteenth, of the total mass in suspension. But in terms of *numbers* the small crystals exert a dominant effect (see *Figure 8.2*).

For crystallisers operating with full or partial classification, the fine crystals tend to circulate with the liquor, leaving the larger crystals in the crystallisation zone. It is possible, therefore, to remove the excess fines by circulating the liquor through a fines trap. With mixed suspensions, internal traps are preferred to external ones and the energy to operate them can be drawn directly from the suspension. Such a trap is shown in *Figure 10.9*.

The trap efficiency increases as the size of fines collected decreases, and the liquor velocity must be kept low to allow the small crystals to settle out However, fines should be segregated as quickly as possible because they continue to grow until they are removed, and the time required for segregation decreases as the flow rate through the trap increases. So a balance must be struck.

Segregation time requirements in a mixed suspension may be derived as follows. If V represents the suspension volume, F the concentration of fines and R the flow rate of suspension through the trap, the rate at which fines are removed may be written

$$-\frac{dF}{dt} = \frac{RF}{V}$$

(10.100)

and integrating between the limits of time zero and t

$$F = F_o \exp(-Rt/V) \qquad (10.101)$$

where F_o is the original concentration of fines, t is the time for a nucleus to grow to the maximum size tolerated by the fines trap, and V/R is the time required to circulate the contents of the crystalliser once through the trap. The ratio F_o/F, therefore, represents the 'excess of fines'. For excesses of 100 and 1000, the values of Rt/V are 4·6 and 6·9, respectively, which indicates the number of times the crystalliser contents must be circulated through the

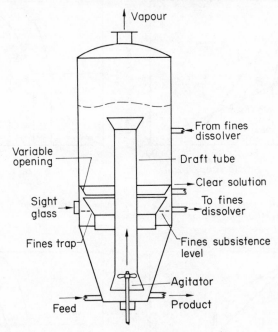

Figure 10.9. Vacuum crystalliser with internal self-regulating fines trap. (After W. C. SAEMAN[44])

fines trap in time t, during which a nucleus grows to the threshold size. With this information the size of the fines trap may be established[47].

For efficient operation the separation of fines should be at the smallest feasible size, but at a large enough rate to materially decay the population density in the size range of removal[48]. Figure 10.10 shows the effect of an internal fines removal system, such as that shown in Figure 10.9. The small-sized crystals undergo a very rapid decay up to the maximum size of particles in the fines trap. Thereafter the decay proceeds at the rate appropriate to mixed product removal operation. It is possible to extrapolate the population density decay line (Figure 10.10a) back to zero to obtain an effective nuclei density, and this is much lower than the actual value produced by the crystalliser. The mean product size (Figure 10.10b) is increased by fines destruction. Several detailed analyses of the influence of fines trapping on the performance and dynamic response of continuous crystallisers have recently been presented[49-52].

PRODUCT SIZE

Product classification is frequently adopted in crystalliser practice, e.g. with an elutriating device, hydrocyclone or wet screen located in a recycle leg of

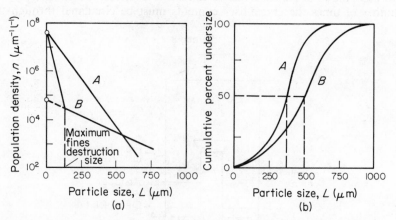

Figure 10.10. Effect of fines removal on (a) the population density and (b) the CSD[48]. A, no fines destruction; B, fines destruction

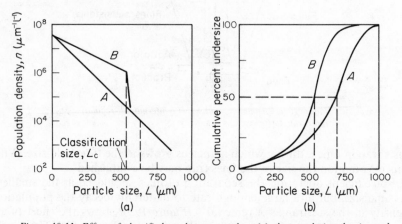

Figure 10.11. Effect of classified product removal on (a) the population density and (b) the CSD[48]. A, Mixed product removal; B, classified product removal

the crystalliser. However, as RANDOLPH[48] has pointed out, it is not generally appreciated that although this practice may narrow the size range, it leads to the production of a smaller product mean size. In other words, it decreases both C.V. and M.S. (p. 349). This behaviour is clearly seen in *Figure 10.11*.

The cases of mixed and classified product removal are compared on the population density plot (*Figure 10.11a*). With classified product operation the accelerated removal that occurs for sizes in excess of the classification size, L_c, can be seen. The net result, however, is the production of a smaller mean product size (*Figure 10.11b*).

Maximum Product Size

There is quite a lot of misunderstanding about the maximum size of crystals which can be obtained from a crystalliser. Theoretically there is no limit to the product crystal size, but there is generally a practical limit. It is common experience that some crystals will not grow beyond a certain size under the conditions normally encountered in an industrial crystalliser, and there is as yet no single clear-cut answer to this problem.

Some crystals have such low growth rates that excessive residence times would be necessary to produce large crystals. For example, at a growth rate of 10^{-7} m/s a nucleus would grow to 2 mm in 3 h, but at 10^{-9} m/s it would require 12 days. Both of these growth rates are within the range commonly encountered with inorganic salts growing in aqueous solution (see *Table 6.2*). Of course, the growth rate could be increased by raising the operating level of supersaturation, but nucleation is extremely sensitive to supersaturation and excessive nucleation prevents the production of large crystals (*Figure 8.2*).

The presence of impurities in the system can also have a significant effect. It is well known that crystallisation of copper and cadmium sulphates from plating bath liquors, to which gelatin has been added, produces crystals no larger than 1 μm, yet both these salts can readily be crystallised from normal aqueous solution as large crystals.

Some crystals appear to become prone to attrition beyond a certain critical size. Sometimes a mosaic type of growth occurs once the crystals exceed this size and this tends to make them friable. However, it is worth considering that mosaic or polycrystalline growth may not only render the crystals mechanically weak but may even make the crystals thermodynamically unstable (the Gibbs–Thomson effect – see p. 222). In this case they would tend to dissolve at the edges and grain boundaries, i.e. at regions of very small radius and ultimately achieve a rounded shape. It is possible, therefore, that opaque egg-shaped crystals, produced in many industrial crystallisers, are as much the result of sequences of crystallisation–dissolution as of attrition.

REFERENCES

1. MONTILLON, G. H. and BADGER, W. L., 'Rate of growth of crystals in aqueous solution', *Ind. Engng Chem.,* **19** (1927), 809
2. MCCABE, W. L., 'Crystal growth in aqueous solutions', *Ind. Engng Chem.,* **21** (1929), 30 and 112
3. HOOKS, I. J. and KERZE, F., 'Nomograph predicts crystal sizes', *Chem. metall. Engng,* **53** (7) (1946), 140
4. BRANSOM, S. H., DUNNING, W. J. and MILLARD, B., 'Kinetics of crystallisation in solution', *Discuss. Faraday Soc.,* **5** (1949), 83 and 96
5. SAEMAN, W. C., 'Crystal size distributions in mixed suspensions', *A.I.Ch.E.Jl.,* **2** (1956), 107

6. BRANSOM, S. H., 'Factors in the design of continuous crystallisers', *Br. chem. Engng*, **5** (1960), 838

7. RANDOLPH, A. D., 'Mixed suspension, mixed product removal crystalliser as a concept in crystalliser design', *A.I.Ch.E.Jl.*, **11** (1965), 424

8. BEHNKEN, D. W., HOROWITZ, J. and KATZ, S., 'Particle growth processes', *Ind. Engng Chem. Fundam.*, **2** (1965), 212

9. CIOLAN, I., 'Distribution des dimensions des cristeaux dans les appareils de cristallisation', *Génie chim., Paris*, **95** (1966), 1381

10. RANDOLPH, A. D., 'A population balance for countable entities', *Can. J. chem. Engng*, **42** (1964), 280

11. HULBURT, H. M. and KATZ, S., 'Some problems in particle technology: a statistical mechanical formulation', *Chem. Engng Sci.*, **19** (1964), 555

12. RANDOLPH, A. D. and LARSON, M. A., 'Transient and steady state size distribution in continuous mixed suspension crystallisers', *A.I.Ch.E.Jl.*, **8** (1962), 639

12a. RANDOLPH, A. D. and LARSON, M.A., *Theory of Particulate Processes*, 1971. New York; Academic Press

13. CANNING, T. F. and RANDOLPH, A. D., 'Some aspects of crystallisation theory: systems that violate McCabe's ΔL law', *A.I.Ch.E.Jl.*, **13** (1967), 5

14. LARSON, M. A. and RANDOLPH, A. D., 'Size distribution analysis in continuous crystallisation', *Chem. Engng Prog. Symp. Ser.*, **65**, No. 95 (1969), 1

15. CANNING, T. F., 'Interpreting population density data from crystallisers', *Chem. Engng Prog.*, **66** (7) (1970), 80

16. RANDOLPH, A. D., 'Effect of crystal breakage on crystal size distribution in a mixed suspension crystalliser', *Ind. Engng Chem. Fundam.*, **8** (1) (1969), 58

17. LARSON, M. A., TIMM, D. C. and WOLFF, P. R., 'Effect of suspension density on crystal size distribution', *A.I.Ch.E.Jl.*, **14** (1968), 448

18. MURRAY, D. C. and LARSON, M. A., 'Size distribution dynamics in a salting out crystalliser', *A.I.Ch.E.Jl.*, **11** (1965), 728

19. AMIN, A. B. and LARSON, M. A., 'Crystallisation of calcium sulphate from phosphoric acid', *Ind. Eng. Chem., Process Des. Dev.*, **7** (1) (1968), 133

20. GENK, W. J. and LARSON, M. A., 'Temperature effects on growth and nucleation rates in mixed suspension crystallisation', *A.I.Ch.E. Symposium, Denver, 1970*

21. RANDOLPH, A. D. and RAJAGOPAL, K., 'Direct measurement of crystal nucleation and growth rate kinetics in a backmix crystal slurry', *Ind. Engng Chem. Fundam.*, **9** (1) (1970), 165

22. CISE, M. D. and RANDOLPH, A. D., 'Some aspects of crystal growth and nucleation of K_2SO_4 in a continuous-flow seeded crystalliser', *A.I.Ch.E. Symposium, Denver, 1970*

23. ABEGG, C. F., STEVENS, J. D. and LARSON, M. A., 'Crystal size distributions in continuous crystallisers when growth rate is size-dependent', *A.I.Ch.E.Jl.*, **14** (1968), 118

24. BECKER, G. W. and LARSON, M. A., 'Mixing effects in continuous crystallisation', *Chem. Engng Prog. Symp. Ser.*, **65**, No. 95 (1969), 14

25. ESTRIN, J., SAUTER, W. A., and KARSHINA, G. W., 'On size dependent growth rate expressions', *A.I.Ch.E.Jl.*, **15** (1969), 289

26. BENNETT, R. C., 'Product size distribution in commercial crystallisers', *Chem. Engng Prog.*, **58** (9) (1962), 76

27. PALMER, K., 'Product size performance from Oslo-Krystal crystallisers', p. 190 in *Industrial Crystallisation*, Instn. chem. Engrs Symposium, London, 1970

28. ROBINSON, J. N. and ROBERTS, J. E., 'A mathematical study of crystal growth in a cascade of agitators', *Can. J. chem. Engng*, **35** (1957), 105

29. RANDOLPH, A. D., DEEPAK, C. and ISKANDER, M., 'On the narrowing of particle size distribution in staged vessels with classified product removal', *A.I.Ch.E.Jl.*, **14** (1968), 827

30. NÝVLT, J., SKRIVANEK, J. and MOUDRY, F., *Colln. Czech. chem. Commun.*, **30** (1966), 1759

31. NÝVLT, J., *Industrial Crystallisation from Solutions*, 1971. London; Butterworths

32. ABEGG, C. F. and BALAKRISHNAN, N. S., 'The tanks-in-series concept as a model for imperfectly mixed crystallisers', *Chem. Engng Prog. Symp. Ser.*, **67**, No. 110 (1971), 88

33. HILL, S., 'Residence time distribution in continuous crystallisers', *J. appl. Chem., Lond.*, **20** (1970), 300

34. RANDOLPH, A. D. and LARSON, M. A., 'Analogue simulation of dynamic behaviour in a mixed crystal suspension', *Chem. Engng Prog. Symp. Ser.*, **61**, No. 55 (1965), 147

35. SHERWIN, M. B., SHINNAR, R. and KATZ, S., 'Dynamic behaviour of the well-mixed isothermal crystalliser', *A.I.Ch.E.Jl.*, **13** (1967), 1141

36. TIMM, D. C. and LARSON, M. A., 'Effects of nucleation kinetics on the dynamic behaviour of a continuous crystalliser', *A.I.Ch.E.Jl.,* **14** (1968), 452

37. NÝVLT, J. and MULLIN, J. W., 'Periodic behaviour of continuous crystallisers', *Chem. Engng Sci.,* **25** (1970), 131

38. HAN, C. D., 'A control study on isothermal mixed crystallisers', *Ind. Eng. Chem. Process Des. Dev.,* **8** (2) (1969), 150

39. MULLIN, J. W. and NÝVLT, J., 'Programmed cooling of batch crystallisers', *Chem. Engng Sci.,* **26** (1971), 369

40. GRIFFITHS, H., 'Crystallisation', *Trans. Instn Chem. Engrs,* **25** (1947), XIV

41. PULLEY, C. A., 'The Krystal crystalliser', *Ind. Chemist* **38** (1962), 63, 127, 175

42. MULLIN, J. W. and NÝVLT, J., 'Design of classifying crystallisers', *Trans. Instn chem. Engrs,* **48** (1970), 8

43. MULLIN, J. W. and GARSIDE, J., 'Voidage-velocity relationships in the design of suspended-bed crystallisers', *Br. chem. Engng,* **15** (1970), 773

44. SAEMAN, W. C., 'Crystallisation equipment design', *Ind. Engng Chem.,* **8** (1961), 612

45. RUSHTON, J. H., COSTICH, E. W. and EVERITT, H. J., 'Power characteristics of mixing impellers', *Chem. Engng Prog.,* **46** (1950), 395, 467

46. NEWMAN, H. H. and BENNETT, R. C., 'Circulating magma crystallisers', *Chem. Engng Prog.,* **55** (3) (1959), 65

47. JORDAN, D. G., 'Crystallisation' Chapter 9 in *Chemical Process Development,* Part II, 1968. New York; Wiley (Interscience)

48. RANDOLPH, A. D., 'How to approach problems of crystallisation', *Chem. Engng,* May 4 (1970), 80

49. NAUMAN, E. B. and SZABO, T. T., 'Non-selective fines destruction in recycle crystallisers', *Chem. Engng Prog. Symp. Ser.,* **67,** No. 110 (1971), 108

50. NAUMAN, E. B., 'Selective fines destruction in recycle crystallisers', *Chem. Engng Prog. Symp. Ser.,* **67,** No. 110 (1971), 116

51. GUPTA, G. and TIMM, D. C., 'Predictive-corrective control for continuous crystallisation', *Chem. Engng Prog. Symp. Ser.,* **67,** No. 110 (1971), 121

52. LEI, S. J., SHINNAR, R. and KATZ, S., 'The regulation of a continuous crystalliser with fines trap', *Chem. Engng Prog. Symp. Ser.,* **67,** No. 110 (1971), 129

11

Size Classification of Crystals

THE MOST widely employed physical test applied to a crystalline product is the one by means of which an estimate may be made of the particle size distribution. Product specifications invariably incorporate a clause that defines, often quite stringently, the degree of fineness or coarseness of the material. For most industrial purposes the demand is for a small range of particle size; regularity results in the crystalline product having good storage and transportation properties, a free-flowing nature and, above all, a pleasant appearance. Manufacturers are well aware of the enhanced selling potential of a nice-looking product.

Size analyses are indispensable in the routine control of crystallisation plant. Economical performances of such operations as filtration, washing and drying largely depend on particle size. Here, again, uniformity of size plays an important role. The production of excessive quantities of crystals smaller or greater than the desired size entails the installation of a screening operation; and unless there is an outlet for these 'fines' or 'roughs', a wasteful recrystallisation process may have to be considered. If there is a demand for many different size grades of the same crystalline product, complex screening arrangements have to be made.

SAMPLING

The physical and chemical characteristics of a bulk quantity of crystalline material are determined by means of tests on small samples. These test samples must be truly representative of the bulk quantity; otherwise any results obtained will be grossly misleading or completely useless. Inefficient sampling followed by detailed analyses in the laboratory constitutes a waste of everyone's time and effort.

Sampling, which is a highly specialised skill, should be carried out by conscientious, well-trained personnel who are fully aware of the tests that are to be made on the sample, without having any direct interest in the outcome of the analyses. Far too often the task is delegated to the most junior laboratory assistant, who regards it as a chore, or to a busy process operator whose main aim is to send a 'good' sample – from his point of view – to the analyst.

Although a sample should represent, as closely as possible, all the charac-
eristics of the bulk quantity, it is impossible to achieve the ideal condition
of a sample being identical in all respects with the parent lot. In general,
however, it may be stated that the larger the sample taken, the greater will
be the probability that the error in the value of the measured property is
small. The complex theories of sampling, based on the laws of probability,
have been discussed by HASSIALIS and BEHRE[1], who also present a comprehensive account of a large number of possible sampling methods. A new B.S.
specification also deals with this topic[2].

The actual technique employed for sampling will depend on many factors,
such as the nature of the bulk quantity of material, its location, the properties
to be tested, the accuracy required in the test, and so on. Difficulties may be
encountered in the sampling of solids in containers after transportation,
owing to the partial segregation of fine and coarse particles; the fines tend
to migrate towards the bottom of the container, and thorough remixing may
be the only answer. Similar problems caused by segregation may be met in
the sampling of solids flowing down chutes or through outlets.

Automatic sampling by mechanical means is to be preferred to sampling
by hand, and is also the method best suited to continuous processes. Hand
sampling, widely employed for batchwise produced materials, is a time-
consuming operation and prone to error, but its use cannot always be
avoided. One common method of hand sampling involves the use of a
sample gun or 'thief'; this simply consists of a piece of pipe with a sharp
bottom edge, which is plunged vertically downwards into the full depth of
the material. It is then withdrawn and the sample removed. This operation
can be performed at fixed or random intervals in the bulk quantity. Sampling at intervals by means of scoops or shovels, known as grab sampling, is
also widely used, but serious errors can be encountered when dealing with
non-homogeneous materials.

The above methods and many others are designed to produce a bulk
sample representative of the bulk quantity. Bulk samples may range from
about 1 kg for small batches to 50 kg or more for large tonnage lots. The
next step is the preparation of a series of test samples that are each representative of the bulk sample. This operation may be carried out by hand or with
the aid of a sample divider.

The best-known hand method is that of coning and quartering. The
sequence of operations, carried out on a clean, smooth surface, or on black
glossy paper for small quantities in the laboratory, is shown in *Figure 11.1*.
The bulk sample is thoroughly mixed and piled into a conical heap. The
pile is then flattened and the truncated cone divided into four equal quarters
(*Figure 11.1c*). This may be done, for example, with a sharp-edged wooden or
sheet metal cross pressed into the heap. One pair of opposite quarters are
rejected, the other pair are thoroughly mixed together and piled into a
conical heap, the procedure being repeated until the required laboratory
sample is obtained.

If several tests are to be carried out on the reduced laboratory sample, a
number of test samples may be prepared by scooping the crystalline material
into a series of sample tins in the following manner[3]. If, for example, four

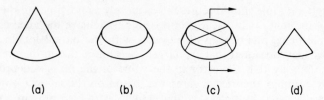

(a)   (b)   (c)   (d)

Figure 11.1. Method of coning and quartering: (a) bulk sample in a conical heap; (b) flattened heap; (c) flattened heap quartered; (d) two opposite quarters mixed together and piled into a conical heap

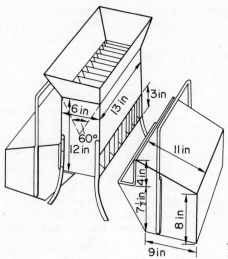

Figure 11.2. A riffle sample divider. (From B.S. 1796³)

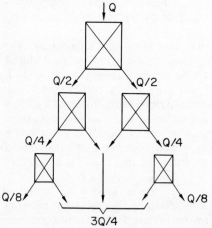

Figure 11.3. Battery of riffles used for reduction of a large quantity of material

test samples are required, four tins are placed in a row. A quantity of the laboratory sample is scooped up and, as near as can be judged, one-quarter of the scoopful is discharged into each tin. Fresh scoopfuls are taken, and the tins are filled in the sequence

$$1\text{st scoopful:} \quad 1, 2, 3, 4$$
$$2\text{nd scoopful:} \quad 2, 3, 4, 1$$
$$3\text{rd scoopful:} \quad 3, 4, 1, 2, \text{etc.}$$

so that no tin is filled exclusively with material from a particular section of the scoopful.

A simple and useful sample divider is the riffle. This apparatus usually takes the form of a box divided into a number of compartments with bottoms sloping about 60° to the horizontal, the slopes of alternate compartments being directed towards opposite sides of the box. Thus, when the bulk sample is poured through the riffle, it is divided into two equal portions. One type of riffle[3] is shown in *Figure 11.2*. The dimensions of a riffle depend on the size of the particles and the feed rate of material — an even flow of solids must be spread over the whole inlet area. When very large bulk samples have to be reduced to small test quantities, a battery of riffles decreasing in size may be employed, as shown diagrammatically in *Figure 11.3*. This arrangement is also suitable for the continuous or intermittent sampling of materials flowing out of hoppers or other items of process plant.

LABORATORY SIEVING

STANDARD SIEVES

Standard sieves for particle size analysis usually consist of a circular metal frame, of 8 or 12 in diam. (200 or 350 mm in metric sizes) with a square-aperture woven wire cloth rigidly mounted in the bottom. For aperture sizes larger than 1 mm, perforated plate sieves (round or square holes) are also available. The bottom rim of each sieve fits snugly inside the upper rim of the next sieve, thus forming an effective dust seal. A top lid and bottom receiving pan are usually provided for the nest of sieves.

Sieves were formerly designated by the mesh number, N, i.e. the number of wires per inch, which is the same thing as the number of apertures per inch (see *Figure 11.4*). However, this practice has been abolished by international agreement and standard sieves should now be designated by their aperture size in millimetres or microns (μm).

The basic dimensions of a wire screen are the wire diameter, w, the sieve aperture, a, and the open area, A. The relationships between these and the mesh number, N, are

$$N = \frac{1}{a+w} \tag{11.1}$$

$$A = \left(\frac{a}{a+w}\right)^2 \tag{11.2}$$

If N represents the meshes per inch, then a and w must also be expressed in inches. From equation (11.1) it can be seen that the mesh number is a function of both the aperture size and the wire diameter, and this has caused much confusion in the past. Many different standard sieve scales are in existence and most of these use different wire diameters for a given grade of mesh. Thus, although all 10-mesh sieves, for example, have 10 apertures and 10 wires to the inch, the aperture widths may be quite different, because each standard sieve series employs a slightly different wire diameter (see *Table 11.1*). It is clear, therefore, that the safest method and most sensible of classifying a sieve is simply by the size of its apertures.

The percentage sieving area, A, of a perforated plate sieve can be calculated from the relationships

$$A = 100\, a^2/p^2 \quad \text{for square holes} \tag{11.3}$$

and

$$A = 90\cdot7\, a^2/p^2 \quad \text{for round holes} \tag{11.4}$$

where a is the aperture size and p is the pitch of the apertures.

For size testing purposes it is desirable to have a set of sieves in which the screen apertures bear some sort of relationship to each other. As early as 1867 Rittinger, in his work on ore dressing, had suggested that a useful scale would be one in which the ratio of the aperture widths of adjacent sieves was the square root of two ($\sqrt{2} = 1\cdot414$).

There are many advantages of $\sqrt{2}$ series of sieves. For instance, the aperture areas double at each sieve proceeding from the fine to the coarse end of the scale, and by omitting certain sieves in the range it is possible to get a sequence in which the aperture widths increase in the ratios $2:1$, $3:1$ or $4:1$. It was later recognised by Richards in the United States that a better sieve scale would be one based on a fourth root of two ratio ($\sqrt[4]{2} = 1\cdot189$); in other words, a second $\sqrt{2} =$ series inserted between the sieves of the first $\sqrt{2}$ series. In this way a much closer sizing of particles is possible.

On the continent of Europe it has long been the practice to use a metric scale of aperture sizes based on a 'tenth root of ten' ($\sqrt[10]{10} = 1\cdot259$) or 'preferred number' series[4]. In this type of sieve series the aperture widths double at every fourth sieve.

The two most widely used standard sieve scales are the American[5] and British[6] (*Tables 11.1–11.4*). These are basically $\sqrt[4]{2}$ progressions, are compatible with the proposed international (ISO) scale[7] based on a series of preferred numbers. The American Tyler scale (*Table 11.1*), which follows the $\sqrt[4]{2}$ principle, is based on the standard 200-mesh screen with 0·0021 in diam. wires giving an aperture of 0·0029 in (74 μm), but there are no specified tolerances attached to this commercial sieve series.

Most standard sieve specifications lay down permissible aperture tolerances. The apertures should be measured in both the warp and weft directions of the wire cloth, and the average aperture width must comply with the limits imposed. Typical tolerances range from ±3 per cent for apertures larger than 1000 μm aperture to ±10 per cent for the smallest sieve aperture in the range (38 μm). In view of these permitted tolerances, therefore, it is quite

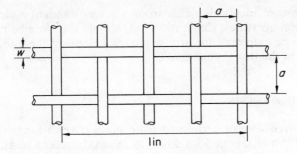

Figure 11.4. Mesh number, N (here = 4), wire diameter, w, and mesh aperture, a

Table 11.1. COMPARISON BETWEEN THE BRITISH AND U.S. STANDARD AND TYLER COMMERCIAL WIRE MESH SIEVE SCALES SHOWING THE DANGER OF USING THE 'MESH NUMBER' AS A SIEVE DESIGNATION

Mesh number* (obsolete)	B.S. 410	A.S.T.M. E11 (microns)†	Tyler	Mesh number* (obsolete)	B.S. 410	A.S.T.M. E11 (microns)†	Tyler
3	5600			45		355	
3½	4750	5600	5613	48			295
4	4000	4750	4699	50		300	
5	3350	4000	3962	52	300		
6	2800	3350	3327	60	250	250	246
7	2360	2800	2794	65			208
8	2000	2360	2362	70		212	
9			1981	72	212		
10	1700	2000	1651	80		180	175
12	1400	1700	1395	85	180		
14	1180	1400	1168	100	150	150	147
16	1000	1180	991	115			124
18	850	1000		120	125	125	
20		850	833	140		106	
22	710			150	106		104
24			701	170	90	90	88
25	600	710		200	75	75	74
28			589	230		63	
30	500	600		240	63		
32			495	250			61
35		500	417	270		53	53
36	425			300	53		
40		425		325		45	43
42			351	350	45		
44	355			400	38	38	38

* The definition of 'mesh number' is the number of apertures per inch in the sieve mesh. This obsolete designation leads to confusion because different standards specify different wire diameters.

† 1 micron (μm) = 10^{-6} m.

possible for two 'identical' sieves from a given standard series to have different mean apertures. Calibration tests should therefore be made if the sieves are to be employed for precise analyses. Some of these tests are discussed fully in the British Standard[6].

MICRO SIEVES

Standard woven wire or perforated plate sieves cover a range of aperture sizes from 125 mm (about 5 in) down to 38 µm. In recent years, this wide

Table 11.2. BRITISH STANDARD SIEVES. WIRE MESH SERIES.
B.S. 410: 1969

Nominal aperture size, mm	Nominal wire diameter, mm	Nearest mesh number	Nominal aperture size, µm	Nominal wire diameter, µm	Nearest mesh number
*16·0	3·15		850	500	18
13·2	2·80		*710	450	22
*11·2	2·50		600	400	25
9·50	2·24		*500	315	30
*8·00	2·00		425	280	36
6·70	1·80		*355	224	44
*5·60	1·60	3	300	200	52
4·75	1·60	3½	*250	160	60
*4·00	1·40	4	212	140	72
3·35	1·25	5	*180	125	85
*2·80	1·12	6	150	100	100
2·36	1·00	7	*125	90	120
*2·00	0·90	8	106	71	150
1·70	0·80	10	*90	63	170
*1·40	0·71	12	75	50	200
1·18	0·63	14	*63	45	240
*1·00	0·56	16	53	36	300
			*45	32	350
			38	30	400

Sieves marked with an asterisk (*) correspond to those proposed for an International Standard (ISO) Scale. It is recommended that these sieves be used for test data that are intended for international publication.

range for particle sizing has been further extended down to about 5 µm by the introduction of micro-sieves, although they usually demand special techniques[8, 8a].

SIEVE TESTS

Briefly, a sieve test is commenced by placing a weighed quantity of the solid material on to the top sieve in a nest which comprises a number of

Table 11.3. BRITISH STANDARD SIEVES. PERFORATED PLATE SERIES.
B.S. 410: 1969

Nominal aperture size, mm	Preferred plate thickness, mm	Minimum width of bridge, mm	Preferred pitch, mm
Round or square holes			
*125 ⎫		17·5	160
106 ⎪		13·0	132
*90·0 ⎪		11·0	112
⎬ 3·0			
75·0 ⎪		10·0	95·0
*63·0 ⎪		8·5	80·0
53·0 ⎭		7·0	67·0
*45·0 ⎫		5·5	56·0
37·5 ⎪		5·0	47·5
*31·5 ⎪		4·25	40·0
⎬ 2·0			
26·5 ⎪		3·5	33·5
*22·4 ⎪		2·8	28·0
19·0 ⎪		2·3	23·6
*16·0 ⎭		2·0	20·0
13·2 ⎫		1·9	17·0
*11·2 ⎪		1·4	14·0
⎬ 1·5			
9·50 ⎪		1·35	12·2
*8·00 ⎭		1·3	10·6
6·70 ⎫		1·3	9·3
*5·60 ⎬ 1·0		1·2	8·0
4·75 ⎪		1·1	6·9
*4·00 ⎭		1·0	6·0
Round holes only			
3·35 ⎫		0·90	5·15
*2·80 ⎪		0·85	4·50
2·36 ⎬ 1·0		0·75	3·87
*2·00 ⎭		0·70	3·35
1·70 ⎫		0·60	2·90
*1·40 ⎪		0·55	2·50
1·18 ⎬ 0·50		0·50	2·18
*1·00 ⎭		0·45	1·90

Sieves marked with an asterisk (*) correspond to those proposed for an International Standard (ISO) Scale. It is recommended that these sieves be used for test data that are intended for international publication.

Table 11.4. U.S. STANDARD WOVEN WIRE SIEVES.
ASTM SPECIFICATION E11–1970

Sieve designation Standard, mm	Alternate, in	Nominal sieve opening, mm	in	Preferred wire diameter, mm
*125·0	5·0	125	5·00	8·0
106·0	4·24	106	4·24	6·40
100·0	4·0	100	4·00	6·30
*90·0	$3\frac{1}{2}$	90	3·50	6·08
75·0	3	75	3·00	5·80
*63·0	$2\frac{1}{2}$	63	2·50	5·50
53·0	2·12	53	2·12	5·15
50·0	2	50	2·00	5·05
*45·0	$1\frac{3}{4}$	45	1·75	4·85
37·5	$1\frac{1}{2}$	37·5	1·50	4·59
*31·5	$1\frac{1}{4}$	31·5	1·25	4·23
26·5	1·06	26·5	1·06	3·90
25·0	1	25·0	1·00	3·80
*22·4	$\frac{7}{8}$	22·4	0·875	3·50
19·0	$\frac{3}{4}$	19·0	0·750	3·30
*16·0	$\frac{5}{8}$	16·0	0·625	3·00
13·2	0·530	13·2	0·530	2·75
12·5	$\frac{1}{2}$	12·5	0·500	2·67
*11·2	$\frac{7}{16}$	11·2	0·438	2·45
9·50	$\frac{3}{8}$	9·50	0·375	2·27
*8·00	$\frac{5}{16}$	8·00	0·312	2·07
6·70	0·265	6·70	0·265	1·87
6·30	$\frac{1}{4}$	6·30	0·250	1·82
	Sieve No.			
*5·60	$3\frac{1}{2}$	5·60	0·223	1·68
4·75	4	4·75	0·187	1·54
*4·00	5	4·00	0·157	1·37
3.35	6	3.35	0.132	1.23
*2·80	7	2·80	0·111	1·10
2·36	8	2·36	0·0937	1·00
*2·00	10	2·00	0·0787	0·900
1·70	12	1·70	0·0661	0·810
*1·40	14	1·40	0·0555	0·725
1·18	16	1·18	0·0469	0·650
*1·00	18	1·00	0·0394	0·580
microns				
850	20	0·850	0·0331	0·510
*710	25	0·710	0·0278	0·450
600	30	0·600	0·0234	0·390
*500	35	0·500	0·0197	0·340
425	40	0·425	0·0165	0·290
*355	45	0·355	0·0139	0·247
300	50	0·300	0·0117	0·215
*250	60	0·250	0·0098	0·180
212	70	0·212	0·0083	0·152
*180	80	0.180	0.0070	0.131

Table 11.4. CONTINUED

Sieve designation Standard, microns	Sieve number	Nominal sieve opening, mm	Nominal sieve opening, in	Preferred wire diameter, mm
150	100	0·150	0·0059	0·110
*125	120	0·125	0·0049	0·091
106	140	0·106	0·0041	0·076
*90	170	0·090	0·0035	0·064
75	200	0·075	0·0029	0·053
*63	230	0·063	0·0025	0·044
53	270	0·053	0·0021	0·037
*45	325	0·045	0·0017	0·030
38	400	0·038	0·0015	0·025

Sieves marked with an asterisk (*) correspond to those proposed for an International Standard (ISO) Scale. It is recommended that these sieves be used for test data that are intended for international publication.

sieves, arranged in order of decreasing aperture, mounted on a bottom collecting pan. A top lid may be fitted to minimise loss of dust. The nest is shaken for a given period and the sieves are then removed, one by one, and shaken individually either for a specified time or until the rate at which particles pass through the sieve falls below a certain specified rate. The procedure is repeated for each of the sieves in turn. All the collected fractions are then weighed. The loss of material for accurate work should not exceed 0·5 per cent; for routine analyses in control laboratories up to 2 per cent loss may be permitted.

The preparation of the test sample is an important preliminary operation. The weight of the material must not change during the sieving operation. Damp materials must be dried in an oven, then be allowed to cool in the atmosphere before sieving. If the material is hygroscopic, the sample must be oven-dried, cooled in a desiccator and sieved with minimum exposure to the atmosphere. For 8 in (200 mm) diam. B.S. sieves the recommended test sample weights[2] are

$$25 \text{ g for materials of density } < 1 \cdot 2 \text{ g/cm}^3$$
$$50 \text{ g for materials of density } \quad 1 \cdot 2 - 3 \cdot 0$$
$$100 \text{ g for materials of density } > 3 \cdot 0$$

These sample weights may be modified in special cases. For coarse solids, say $> 300 \, \mu m$, the above weights may be increased, provided that the weight retained on any one sieve does not exceed 50 per cent of the test sample weight. Again, the sample weight may be increased to better the accuracy of determining small percentages of relatively coarse particles. Other recommendations for the weight of a test sample are:

Mean particle size, mm	Test sample weight, g
4·0–2·0	1000
2·0–1·0	500
1·0–0·5	200
0·5–0·25	100
<0·25	50

For materials that contain a high percentage of fine particles, a preliminary removal of dust may be made before carrying out the main sieve test. This is done with a nest consisting of a medium screen, e.g. 500 µm, the finest sieve to be used in the main test, and the bottom collecting pan; the reason for the medium mesh is to protect the fine mesh from the abrasive action of the coarse particles. The sample is sieved, as described below, for a suitable period, and the fractions are removed and weighed. The material retained on the medium and fine meshes can then be taken as the test sample for the main test using the complete nest of fine sieves.

Sieve testing can be carried out by hand or on a machine designed to hold the nest of sieves and perform the shaking, rotating and tapping operations. Many different types of such machines are available from suppliers of laboratory equipment. Hand sieving may be carried out as follows[3]. Hold the nest of sieves, together with the lid and collector pan, in the left hand, including the sieve surfaces downwards towards the left at an angle of 30° to the horizontal. Tap the higher side of the sieve frame six or eight times with the hand or a piece of wood. While maintaining the inclination of the sieve, shake the nest to and fro several times, also rotating it in the plane of the gauze through an angle of about 60°. These alternate tapping and shaking operations are repeated for 5 min. The top sieve is removed, inverted over a clean piece of paper, and tapped gently to remove the material retained on it. The bottom surface of the gauze, i.e. the surface that is now uppermost while the sieve is held in its inverted position, may be brushed gently to aid the cleaning operation. All the discharged material is transferred back to the cleaned sieve mounted on a collector pan and submitted to a sieving operation, as described above, for 2 min. If the amount of material passing through the sieve in this 2 min period is less than 0·2 per cent of the original test sample weight, sieving on this sieve may be considered complete; if the amount passing exceeds 0·2 per cent, the procedure is repeated until the 'end point' is achieved. All the weighed material from the collector pan is then transferred on to the material retained on the top sieve of the remaining nest, which is submitted to another 5 min sieving operation. The 'end point' rate test is again applied to the top sieve, and the whole procedure is repeated until all the sieves in the nest have been dealt with.

For machine sieving the following procedure may be adopted. The complete nest is shaken on the machine for 5 minutes. After this period the top sieve is removed and submitted to the cleaning process and 'end point' test by hand, as described above. The reduced nest, with the top sieve containing the material that passed through the former top sieve during the end point test, is again shaken on the machine for 5 min, and so on.

Wet sieving may be used for very fine powders, smaller than, say, 100 µm. or in cases where the particles are fragile and will not withstand the vigorous shaking motions of dry sieving. Small quantities of the solid particles are thoroughly mixed, with an added wetting agent if necessary, in the chosen liquid, e.g. water, alcohol or light petroleum fraction. The slurry is then washed on to a fine screen held near the surface of the liquid in a bowl. The load on the sieve is swirled and jigged until all the fine particles have passed through the gauze. The oversize fraction can be dried and submitted to a dry sieve test, as described above. The fine solids which pass through the gauze into the liquid in the bowl can either be removed, dried and weighed, or submitted to a further sizing test, e.g. by sedimentation or elutriation (p. 393).

The results of a sieve test can be reported, in tabular form, as shown in *Table 11.5*, for fractional percentages and for cumulative percentages, distinguishing between oversize and undersize. The quantity marked with an asterisk (*) should be annotated 'including a loss, during test, of x per cent'. The method of sieving should also be stated.

ANALYSIS OF SIEVE TEST DATA

Once a sieve test has been performed, and the results have been tabulated in an approved manner, there remains the task of assessing the size characteristics of the tested material and of extracting the maximum amount of information from the data. While a table may record all the measured quantities, this form of expression is not always the best one; the magnitudes of the various quantities may be readily visualised, but certain trends may be completely obscured in a mass of figures. The real significance of a sizing test can most readily be judged when the data are expressed graphically. From such a pictorial representation trends in the data are easily detected, and the prediction of the expected behaviour of the material on sieves other than those used in the test can often be made with a reasonable degree of accuracy.

Many different forms of graphical expression may be employed, and the use and applicability of some of these methods will be demonstrated with reference to the results of the sieve test given in *Table 11.6*. These data are chosen for their fairly wide spread over the size range of approximately 100

Table 11.5. TABULAR REPRESENTATION OF SIEVE TESTS

(a) Fractional percentages		Per cent by weight	(b) Cumulative percentages		
Sieve aperture, mm			Size, mm	On (oversize)	Through (undersize)
Through	On				
—	355	1·0	355	1·0	99·0
355	250	20·0	250	21·0	79·0
250	180	40·5	180	61·5	38·5
180	125	30·0	125	91·5	8·5*
125	—	8·5*			
		100·0			

Table 11.6. SIEVE TEST DATA USED FOR THE CONSTRUCTION OF
FIGURES 11.5–11.11

B.S. mesh number	Sieve aperture, μm	Fractional weight per cent retained	Cumulative weight per cent oversize	Cumulative weight per cent undersize
7	2360	1·2	1·2	98·8
10	1700	2·9	4·1	95·9
14	1180	18·8	22·9	77·1
18	850	28·8	51·7	48·3
25	600	22·0	73·7	26·3
36	425	11·1	84·8	15·2
52	300	6·0	90·8	9·2
72	212	3·9	94·7	5·3
100	150	1·8	96·5	3·5
150	106	1·3	97·8	2·2
>150	—	2·2	—	—

to 2500 μm to illustrate the scope of the methods of plotting. The data are listed in the three recommended ways, viz. the percentage by weight of the fractions retained on each sieve, and the cumulative percentages by weight of oversize and undersize material. Alternate sieves in the B.S. series have been used, i.e. meshes with aperture widths decreasing approximately in the ratio $\sqrt{2}:1$. The equivalent B.S. mesh numbers are translated into aperture widths in microns.

Four types of graph paper are commonly used, depending on the sort of information that is required: (a) the ordinary squared or arithmetic, (b) the log-linear or semi-log, where one of the axes is marked off on a log scale and the other on an arithmetic scale, (c) the log-log, where both axes are marked off on a logarithmic scale, and (d) the arithmetic-probability, where one axis is marked off on a probability scale, the intervals being based on the probability integral. Log-probability and double-logarithmic (RRS) grids also find use in special cases.

In *Figure 11.5a* the weight percentages of the fractions retained on each successive sieve used in the test are plotted against the widths of the sieve apertures (in microns). The lines joining the points have no significance; they merely complete the graph or frequency polygon. The sharp peak in the distribution curve occurs at 18-mesh (850 μm). This point, however, represents the fraction that passes 14-mesh (1180 μm) and is retained on 18-mesh, so it could be plotted at the 'mean' size $(1180+850)/2 = 1015$ μm. Alternatively, the results may be represented in the form of a frequency histogram, as shown in *Figure 11.5b*, depicting the size range of each collected fraction. For example, 18·8 per cent of the test sample passed through the 10-mesh and was retained on the 14-mesh (see *Table 11.6*). Thus the 18·8 per cent horizontal is drawn between the 1180 and 1700 μm positions on the histogram.

From both diagrams in *Figure 11.5* the general picture of the over-all spread of particle size can be seen quite clearly. However, the simple arithmetic method of plotting suffers from the disadvantage of producing a congested picture in the regions of the fine mesh sieves. If all the available sieves in the B.S. range had been used, i.e. those belonging to the $\sqrt[4]{2}$ series, the conditions in the small-size region of this type of plot would have become quite obscure. The fine mesh data, of course, could be replotted on an enlarged scale, but the over-all picture would then be lost.

The cumulative weight percentages are plotted against aperture size in *Figure 11.6*. This type of plot often permits the drawing of smooth curves through the plotted points, although the accuracy with which the curves can

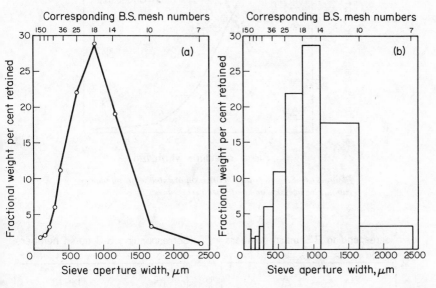

Figure 11.5. Sieve test data plotted on arithmetic graph paper—percentage by weight of the fractions retained between two given sieves in the B.S. series: (a) the frequency polygon; (b) the frequency histogram

be drawn can be very poor if not all the sieves in the range have been used. The two curves in *Figure 11.6* are mirror images of each other. The advantage of this method of plotting is that estimates can be made of the percentages of the material that would be retained on or pass through any sieve, standard or non-standard, with apertures in the range tested. It could be predicted, for example, that about 87 per cent of the original material would pass through a B.S. 12-mesh (1400 μm) or that about 40 per cent would be retained on a B.S. 16-mesh (1000 μm). However, as in *Figure 11.5*, the data in the region of the small aperture sieves are congested, and interpolation is difficult.

When the test data are plotted on semi-log paper (*Figure 11.7a*), with the aperture widths recorded on the logarithmic scale, the points in the coarse sieve region are brought closer together, and those in the fine sieve region located further apart than in the corresponding simple arithmetic plot. In

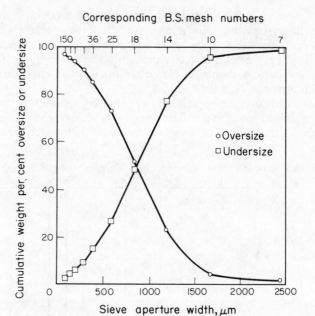

Figure 11.6. Sieve test data plotted on arithmetic graph paper—cumulative oversize and undersize percentages

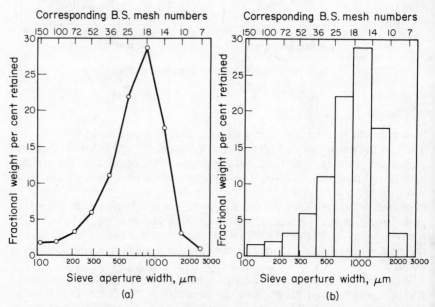

Figure 11.7. Sieve test data plotted on semi-log graph paper—percentage by weight of fractions retained between two sieves in the B.S. series: (a) the frequency polygon; (b) the frequency histogram

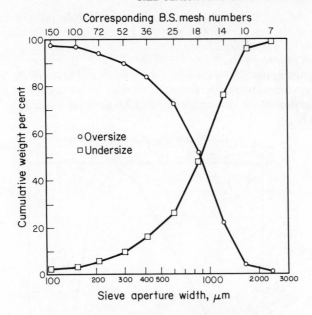

*Figure 11.8. Sieve test data plotted on semi-log graph paper—
cumulative oversize and undersize percentages*

fact, the successive points are more or less equally spaced along the horizontal
scale. The frequency histogram (*Figure 11.7b*) is composed of columns of
approximately equal widths. The picture of the size spread in the fine sieve
region is thus clarified, and data obtained on the complete $\sqrt[4]{2}$ series of
sieves can be plotted on graphs of this type without congestion.

The semi-log graphs for the cumulative oversize and undersize are plotted
in *Figure 11.8*; here again the curves are mirror images. Interpolation in the
fine sieve region is facilitated owing to the even spread of the plotted points.
It can be estimated, for instance, that about 87 per cent of the original
material would be retained on a 44-mesh (355 µm) or that about 7 per cent
would pass through a 60-mesh (250 µm).

In *Figure 11.9* the weight percentages of the fractions retained on the given
sieves are plotted against the aperture size on a log-log basis. This type of
plot usually tends to bring the points in the fine sieve region into a straight
line. The data in *Figure 11.9* are fitted by a straight line between about 600
and 150 µm (25- to 100-mesh). A linear correlation of the data indicates the
application of an equation of the type

$$p = c \cdot a^n \tag{11.5}$$

where p is the weight percentage of material retained on a sieve of aperture a
after passing through the next larger sieve used in the test. The constant c,
mathematically speaking, is the value of p when $a = 1$, but this cannot be
interpreted literally, because the linear relationship would not be valid in the
region of 1 µm particle size. Therefore c is just a constant depending on the

material; the higher the value of c, the finer is the material. The exponent n, the slope of the line, gives a measure of the particle size spread; for close-sized particles n tends to infinity.

The cumulative percentages of oversize and undersize particles are plotted against aperture width on a log-log basis in *Figure 11.10*. In this type of plot the cumulative undersize data tend to lie on a straight line over a wide range of particle size, about 100 to 1200 µm in this case. The undersize

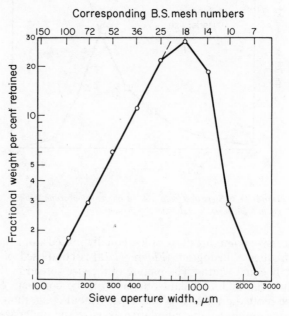

Figure 11.9. Sieve test data plotted on log-log graph paper — fractional weight percentages retained between two sieves in the B.S. series

and oversize curves are clearly not mirror images, and oversize data are rarely correlated on this basis. For the undersize line the relationship

$$P = b \cdot a^m \qquad (11.6)$$

holds over the linear region. P is the total percentage of the original material that would pass through a screen of aperture a. The exponent m, the slope of the line, gives a measure of the spread; the lower the value of m, the greater the size distribution. The constant b indicates the fineness or coarseness of the particles: the higher the value of b, the finer the material. The log-log method of plotting of undersize data is extremely useful because rough checks may be made on the size distribution by the use of only two, or possibly three, test sieves. Material of the type considered in *Table 11.4* could be size-checked with 1000, 500 and 250 µm sieves, for example. Sieves of any standard series can be used, provided that their apertures are accurately known, and the data can readily be translated into the sizes of any other required sieve specification.

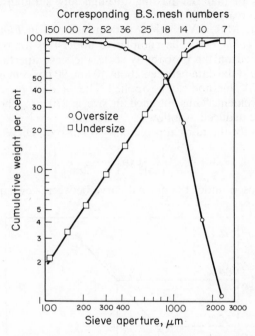

Figure 11.10. Sieve test data plotted on log-log graph paper—cumulative oversize and undersize percentages

Normal and Log-normal Distributions

Probability plots have often been suggested for particle size analysis, particularly in connection with the assessment of comminution processes. The following method, proposed by POWERS[10] for use in the sugar industry, has much to be said in its favour for size specification of crystalline materials. This method employs arithmetic-probability graph paper (one scale divided into equal intervals, the other marked off according to the probability integral) and provides a simple means of recording the crystal size distribution in terms of two members only—the *mean aperture,* M.A., and a statistical quantity the *coefficient of variation,* C.V., expressed as a percentage. The significance of these terms is as follows.

The M.A. is the sieve aperture size through which 50 per cent of the particles pass. This really designates the *median size* of the distribution (50 per cent is larger than the median, and 50 per cent is smaller); and as this method of expression is not restricted to sizing by sieves, it is proposed to use the symbol M.S. (median size). A size distribution expressed as

M.S./C.V. = 600/30 indicates that the median size is 600 μm and the coefficient of variation is 30 per cent. From a knowledge of these two numbers the percentages of material passing through any standard sieve can be estimated with a reasonable degree of accuracy.

The use of the M.S./C.V. method is demonstrated in *Figure 11.11* with the data from *Table 11.6*. The cumulative undersizes (or oversizes if preferred) are plotted on the probability scale, the sieve aperture sizes on the arithmetic scale. If the data between about 10 and 90 per cent lie on a straight line, the M.S./C.V. method can be applied. The data in *Figure 11.11* comply with this requirement. Thus the median size is 870 μm. The coefficient of variation can be deduced as follows.

The equation for the normal probability (Gaussian) curve is

$$f(a) = \frac{1}{\sigma\sqrt{(2\pi)}} \exp\left[-\frac{(a-\bar{a})^2}{2\sigma^2} \right] \tag{11.7}$$

If the area enclosed under the normal curve between sieve apertures $a = 0$

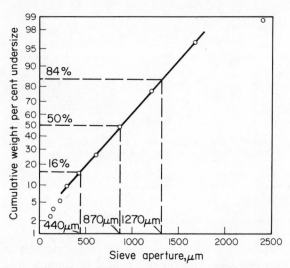

Figure 11.11. Sieve test data plotted on arithmetic/probability graph paper (cumulative undersize percentages), illustrating the use of the M.S./C.V. method of analysis

to ∞ is taken as unity, the area enclosed between $a = 0$ and $a = \bar{a}+\sigma$, where $\bar{a}$ is the median (50 per cent) size M.S., is 0·8413. This value is obtained from tables of the normal probability function. Therefore the area enclosed between $a = \bar{a}+\sigma$ and $a = \infty$ is $1-0\cdot8413 = 0\cdot1587$. The value of σ, the standard deviation, can be obtained from the arithmetic probability diagram by reading the value of a at 84·13 per cent (84 per cent is accurate enough for this purpose) and subtracting the value of $\bar{a}$. Alternatively, the value of a at 15·87 (or 16) per cent can be subtracted from $\bar{a}$, i.e.

$$\sigma = a_{84\%} - \bar{a} = \bar{a} - a_{16\%}$$

These two values of σ may not coincide, so a mean value can be taken as

$$\sigma = \frac{a_{84\%} - a_{16\%}}{2}$$

The coefficient of variation, as a percentage, is given by

$$\text{C.V.} = \frac{100\sigma}{\bar{a}} = \frac{100\sigma}{a_{50\%}} \tag{11.8}$$

or

$$\text{C.V.} = \frac{100(a_{84\%} - a_{16\%})}{2a_{50\%}} \tag{11.9}$$

From *Figure 11.11*

$$\text{C.V.} = \frac{100(1270 - 440)}{2(870)} = 48 \text{ per cent}$$

and the size distribution can be specified, in terms of M.S./C.V., as 870/48.

If the crystalline material being produced is known to obey this straight-line rule, routine analysis of the product is greatly simplified. Only two sieves, carefully standardised, need be used when checks are made on the product specification. It is important to understand that the size distribution does not have to be Gaussian for the M.S./C.V. method to apply. Many skew distributions also give the necessary linear relationship between about 10 and 90 per cent, but in these cases the M.S. will not coincide with the modal diameter at the peak of the distribution curve.

For skew distributions that are not approximately Gaussian over the 10–90 per cent region, plotting on log-probability paper (one scale logarithmic, the other marked off according to the logarithmic probability function) may give a better correlation. The equation of the logarithmic normal curve is

$$f(\log a) = \frac{1}{\log \sigma' \sqrt{2\pi}} \exp\left[-\frac{(\log a - \log \bar{a}')^2}{2\log^2 \sigma'} \right] \tag{11.10}$$

where $\bar{a}'$ is the geometric mean size and σ' is the geometric standard deviation.

The log-normal distribution gives a curved skewed towards the larger sizes, and it frequently gives a good representation of particle size distributions from precipitation and comminution processes. Furthermore, the log-normal distribution is often used because it overcomes the objection to the normal (Gaussian) distribution function which implies the existence of particles of negative size.

Another distribution function gaining popularity for characterising crystal size distributions is the gamma function, expressed as

$$f(a) = a^x \exp(ax/y)\Gamma(x+1)(y/x)^{x+1} \tag{11.11}$$

The parameters x and y, which give measures of the 'skewness' and 'size' of the distribution respectively, can be related to the crystallisation process. The median size, M.S. ($= \bar{a}$) and standard deviation, σ, may be calculated from

$$\bar{a} = y(x+1)/x \quad \text{and} \quad \sigma = y(x+1)^{\frac{1}{2}}/x$$

Therefore from equation (11.8) the coefficient of variation is given by

$$C.V. = 100(x+1)^{-\frac{1}{2}}$$

Rosin–Rammler–Sperling (RRS) Plot

Some skew distributions, particularly those of comminuted materials, can be fitted by the RRS function. This relationship, based on one originally derived from probability considerations, may be written

$$P = 100 \exp\left[-(a/a')^n\right] \qquad (11.12)$$

where P = cumulative percentage oversize, a = particle size, and a' is a

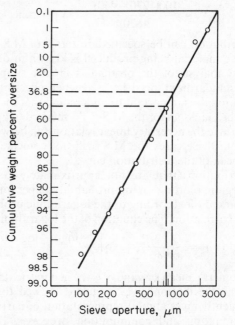

Figure 11.12. Sieve test data plotted on Rosin–Rammler–Sperling (RRS) graph paper

statistical mean size corresponding to $P = 36\cdot8\%$ (100/e, where e $= 2\cdot718$, the base of natural logarithms). Equation (11.12) indicates a linear relationship between $\log . \log (100/P)$ and $\log a$. The slope of the line, n, has been called a 'uniformity factor', i.e. $n = \tan \theta$, where θ is the angle between the RRS line and the horizontal. As the size distribution narrows towards a mono-sized dispersion, $n \to \infty$; as it broadens, $n \to 0$.

Special double logarithmic graph paper suitable for this type of plotting is available and the data from Table 11.6 are shown on such a plot in Figure 11.12. From this plot it is possible to determine the 16, 50 and 84% cumulative percentages needed to calculate the M.S./C.V., as described above.

Alternatively, the distribution can be characterised by the uniformity factor, n. In the example shown the median size $a_{50\%} = 870 \, \mu m$ (the same as determined in *Figure 11.11*), the statistical mean size $a' = 1000 \, \mu m$, and the uniformity factor $n = 1.8$.

PARTICLE SIZE, VOLUME AND SURFACE AREA

It is not possible to measure or define absolutely the size of an irregular particle, and perfectly regular crystalline solids are rarely, if ever, encountered. The terms length, breadth, thickness or diameter applied to irregular particles are meaningless unless accompanied by further definition, because so many different values of these quantities can be measured. The only precise properties that can be defined for a single solid particle are the volume and surface area, but even the measurement of these quantities may present insuperable experimental difficulties. All particle size measurements are made by indirect methods: some property of the solid body which can be related to size is measured.

Despite these difficulties of definition and measurement it is most convenient, for classification purposes, if a single-length parameter can be ascribed to an irregular solid particle. The most frequent expression used in connection with particle size is the 'equivalent diameter', i.e. the diameter of a sphere that behaves exactly like the given particle when submitted to the same experimental procedure. Several of these equivalent diameters may be defined. For example, a particle that just passes through a sieve aperture is classified according to the diameter of a sphere that would also just pass through. The term equivalent sieve aperture diameter, $d_{s.a.}$, is usually applied in this case.

The present chapter is concerned primarily with sieving operations, which are almost exclusively used for grading commercial crystals, but a few brief references can be made to some of the other methods that can be employed for size classification in the sub-sieve range[11, 12]. Sedimentation and elutriation methods, for instance, are based on Stokes' law, which relates the free falling velocity u of a particle in a fluid medium of density ρ_f and viscosity η to the diameter d_{St} of a solid sphere of the same density ρ_s as the particle, by means of the equation

$$d_{St} = \sqrt{\frac{18\eta u}{(\rho_s - \rho_f)g}} \qquad (11.13)$$

This equivalent diameter, d_{St}, is usually referred to as the Stokes' diameter. Water or any other suitable liquid may be used as the fluid medium in sedimentation processes; liquids or gases can be used as elutriating media. Both operations can be carried out fractionally to get a picture of the range of size distribution in the original sample.

A simple sedimentation technique, which readily lends itself to the determination of crystal size distribution in the range 1–50 μm, is the Andreasen pipette method (*Figure 11.13*). Although it is generally better to prepare a fresh suspension of the crystals under test in a suitable inert liquid, it is possible to classify crystals suspended in their own mother liquor. If the difference in density between the particles and suspending liquid is <0·5 g/cm³, special care must be taken to avoid convection currents. The method, briefly, is as follows.[11]

A homogeneous suspension of the crystalline material in a suitable liquid is prepared in the graduated sedimentation cylinder of capacity ~600 cm³ (*Figure 11.13*). Small samples (e.g. 10 cm³) of the suspension are withdrawn through the fixed pipette, at a known depth, h, below the liquid level, at chosen time intervals. The samples, including the one taken at zero time, are analysed for total *suspended solids* content by a suitable method. Ideally the suspension should be dilute (<3 per cent) and a dispersion agent may be

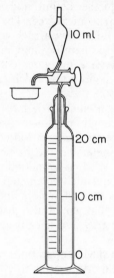

Figure 11.13. Andreasen fixed pipette method[11]

needed to prevent agglomeration: for particles in insoluble water a 0·1 per cent solution of sodium pyrophosphate is generally suitable.

A sample taken at time t will contain no particles larger than size d, calculated from Stokes' law which may be written

$$d = \left[\frac{18h\eta}{(\rho_s - \rho_f)gt} \right]^{\frac{1}{2}} \tag{11.14}$$

Thus by taking samples at suitable intervals, e.g. 0, 5, 10, 20, 40, 80 . . . min, the size distribution of the original suspension may be evaluated. For routine analysis only one or two samples may be needed to characterise the particles. If t is measured in minutes, h in cm, ρ in g/cm³ and η in centipoise, then the particle size, d, in μm, is given by

$$d = 17·5 \left[\frac{h}{(\rho_s - \rho_f)t} \right]^{\frac{1}{2}} \tag{11.15}$$

For a given sample, n, the cumulative weight percentage, P_n, of particles smaller than the limiting Stokes' diameter, d_n, for the time interval, t_n, may be calculated from the weight, W_n, of the suspended solids in the fraction by

$$P_n = 100\left[\frac{W_n}{W}\cdot\frac{V}{V_n}\right] \tag{11.16}$$

where W = original test sample weight (g), i.e. weight of solids originally suspended in the apparatus, V = volume of the suspension (cm^3) and V_n = volume of sample taken via pipette (cm^3).

A typical analysis is given in *Table 11.7*, where the size distribution of precipitated calcium carbonate ($\rho_s = 2\cdot7\,g/cm^3$) is measured by sedimentation at 20 °C in water containing 0·1 per cent sodium pyrophosphate ($\rho_f = 1\cdot0\,g/cm^3$ and $\eta = 1\cdot0\,cP$, i.e. $10^{-3}\,N\,s/m^2$). In this test the $CaCO_3$ was determined volumetrically by adding N/5 HCl to each sample, boiling to remove CO_2 and back-titrating with N/10 NaOH. A 'blank' was run on the suspending liquid. Alternatively, in this case a gravimetric method could have been used, i.e. by evaporation to dryness.

Microscopic techniques can also be used for the size estimation of very fine particles. The actual size recorded is the diameter of a circle of the same area as the projected image of the particle viewed in a direction perpendicular to its plane of maximum stability. The particle image is compared with

Table 11.7. MEASUREMENT OF THE PARTICLE SIZE DISTRIBUTION OF A SAMPLE OF PRECIPITATED CALCIUM CARBONATE BY USE OF THE ANDREASEN PIPETTE METHOD

Time, t (min)	Pipette depth, h (cm)	Stokes' diameter, d (μm)	$CaCO_3$ in fraction*, W_n (g)	Cumulative percentage undersize
0	20·0	—	0·231†	—
5	19·6	28	0·147	64
10	19·2	19	0·108	47
20	18·7	13	0·0763	33
40	18·3	9·1	0·0508	22
80	17·9	6·4	0·0299	13
160	17·4	4·4	0·0184	8
320	17·0	3·1	0·00924	4

* Sample volume $V_n = 10\,cm^3$.
† Test sample weight $W = 14\cdot3$ g in suspension volume $V = 620\,cm^3$.

graduated circles on an eyepiece or graticule. The symbol $d_{p.a.}$ may be ascribed to this 'projected area' diameter. The approximate useful size ranges for the above methods are

Sieving (including micro-sieves)	$d_{s.a.}$:	16 mm to 10 μm
Elutriation	d_{St}:	50 to 5 μm
Gravity sedimentation	d_{St}:	50 to 1 μm
Visible light microscopy	$d_{p.a.}$	100 to 0·5 μm

As a rough guide in the comparison of size analyses by these various methods the following relationship may be used:

$$d_{s.a.}:d_{St}:d_{p.a.} \sim 1:0\cdot94:1\cdot4$$

In many cases the degree of fineness of a particulate mass is better expressed in terms of the available surface area of the particles than as an equivalent diameter. Particle area is a very important factor to be considered when chemical reactions or other mass transfer operations are to be performed with the solid substance. For very fine particles, surface area measurements may be made by turbidimetric, adsorption, permeability and many other techniques[12-14].

The permeability method has proved valuable for the determination of the degree of fineness of powdered substances in the size range 1–100 µm. Briefly, a known quantity of air is forced through a small bed of the fine solids under a constant pressure drop, and the flow time is recorded. The theory is based on the laminar flow of fluids through porous beds, and the specific surface S (cm^2/g) of the material is calculated from the Kozeny equation

$$S^2 = \frac{\Delta P}{ku\eta L\rho^2} \cdot \frac{\varepsilon^3}{(1-\varepsilon)^2} \tag{11.17}$$

where ΔP = pressure drop across the bed; ε = voidage of the bed; L = depth of the bed; η = viscosity of the air; u = empty-tube velocity; ρ = density of the solid material; k is a constant (Kozeny's constant), which has a value equal to about 5·0 for granular solids.

Several different types of permeability cell are available[12, 14, 15].

A relatively new particle sizing technique is provided by the Coulter particle counter[16]. This is an instrument which utilises an electrical sensing device capable of detecting particles suspended in an electrolyte. A sample of the suspension is drawn through a small orifice, with an electrode on either side, immersed in the suspension. The amplitudes of the voltage pulses generated are proportional to the particle volumes. The pulses are amplified, measured and counted, which enables the particle size distribution to be evaluated.

A good account of the use of the Coulter counter is given by ALLEN[14]. The application of the technique to the sizing of crystals, and a comparison with sieving techniques, has been described by ROSEN and HULBURT[17].

SHAPE FACTORS

For particles in the sieve size range, an estimate of the surface area may be made from the results of a sieve analysis, but this estimate may be liable to gross error. A precise calculation of the volume or surface area of a solid body of regular geometric shape can only be made when its length, breadth and thickness are known. For crystals, or indeed particulate solids in general, these three dimensions can never be precisely measured. Therefore, before a brief account is given of some of the methods of calculation available, a word of warning is necessary. It must be fully appreciated that the precision of calculation is always far greater than that of measurement of the various quantities used in the mathematical expressions. An equation, especially a complex one, always has a look of absolute dependability, but in this particular connection it most certainly leads to a sense of false security. All calculated volume or surface area data must be used with caution.

Most calculation methods are based on one dimension of the particle, usually the equivalent diameter. If this dimension is obtained from a sieve analysis, it will be the sieve aperture diameter, $d_{s.a.}$; but as crystals are never true spheres, this diameter will be the second largest dimension of the particle. *Figure 11.14* demonstrates some particle shapes that would, in a sieve analysis, all yield the same value for $d_{s.a.}$. One source of error is thus clearly seen.

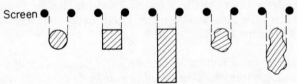

Figure 11.14. Various particle shapes that would all be classified under the same sieve-aperture diameter

For a single particle, the size of which is defined by some length parameter or diameter, d, the following relationships can be applied:

$$\text{volume} \qquad v = f_v d^3 \tag{11.18}$$

$$\text{mass} \qquad m = f_v \rho d^3 \tag{11.19}$$

$$\text{surface area} \quad s = f_s d^2 \tag{11.20}$$

The constants f_v and f_s may be called volume and surface shape factors, respectively. In previous sections (e.g. pp. 193 and 358) these have been given the symbols α and β.

For spherical (diameter $= d$) and cubical (length of side $= d$) particles

$$\alpha = f_v = \tfrac{\pi}{6} \text{ (sphere) and 1 (cube)}$$
$$\beta = f_s = \pi \text{ (sphere) and 6 (cube)}$$

The shape factors are readily calculated for other regular geometrical solids. For an octahadron, for example, with d representing the length of an edge, $v = \sqrt{(2)}d^3/3$ and $s = 2\sqrt{(3)}d^2$, therefore

$$\alpha = f_v = v/d^3 = \sqrt{(2)}/3 = 0\cdot471$$
$$\beta = f_s = s/d^2 = 2\sqrt{3} = 3\cdot46$$

From equations (11.18) and (11.20) two basic ratios may be defined:

$$\text{surface: volume} \quad \frac{s}{v} = \frac{f_s d^2}{f_v d^3} = \frac{F}{d} \tag{11.21}$$

$$\text{surface: mass} \quad \frac{s}{m} = \frac{f_s d^2}{f_v \rho d^3} = \frac{F}{\rho d} \tag{11.22}$$

Equation (11.22) defines the important quantity known as the specific surface, i.e. the surface area per unit mass of solid. Some authors, unfortunately, have also called the surface:volume ratio a specific surface, but this definition is not widely adopted.

The constant $F\,(= f_s/f_v = \beta/\alpha)$ may be called the over-all, surface–volume or specific surface shape factor. For spheres and cubes, $F = 6$. For other shapes $F > 6$. For an octahedron $f_v = \sqrt{(2)}/3, f_s = 2\sqrt{3}$ and $F = 7\cdot35$. Values of $F \sim 10$ are frequently encountered in comminuted solids, and much higher values may be found for flakes and plate-like crystals. If the particles are elongated or needle-shaped, their volume and surface area may be calculated on the assumption that they are cylindrical; length and diameter may be measured microscopically, or the diameter can be taken as the equivalent sieve aperture diameter.

For example, a crystal with a length : breadth : height ratio of $5:2:1$ would be characterised in a sieving operation by its second largest dimension, i.e. $d = 2$. Therefore $f_v = v/d^3 = 10/8 = 1\cdot25$, $f_s = a/d^2 = 34/4 = 8\cdot6$ and $F = f_s/f_v = 6\cdot88$.

The determination of volume shape factors for particles smaller than about 500 μm becomes extremely difficult, since it may be necessary to count and weigh several thousand particles. However, the following method[18] may be used to simplify the procedure. Prepare a sample of the particles by sieving between two close sieves. Clean the finer of the two sieves (the retaining sieve) and attach a strip of adhesive tape, of known weight and dimensions, to its underside. Place a quantity of the particles on the sieve and shake the sieve for several minutes. Peel off the adhesive tape, which will now have hundreds or thousands of particles in a regular matrix (more or less one per sieve aperture). The number of particles per unit area can be determined from the designation of the sieve mesh (see equations 11.1 and 11.2). For example, a 100-mesh B.S. sieve (150 μm apertures with 100 μm wires) contains 1550 apertures per cm². The adhesive strip can then be weighed and the average mass of one particle can be determined.

MEAN PARTICLE SIZE

In a total mass, M, of uniform particles, each of mass m and equivalent diameter d, the number of particles, n, is given by

$$n = \frac{M}{m} = \frac{M}{f_v \rho d^3} \tag{11.23}$$

and the total surface area, Σs, by

$$\Sigma s = ns = \frac{f_s M d^2}{f_v \rho d^3} = \frac{FM}{\rho d} \tag{11.24}$$

However, before equations (11.23) and (11.24) can be applied to masses of non-uniform particles, some average value of the equivalent diameter must be defined. A few of the many suggested methods are described below.

The simplest of all average diameters is the arithmetic mean. For example, if sieving has been carried out between two sieves of aperture a_1 and a_2, the average particle diameter is given by

$$\bar{d}_a = (a_1 + a_2)/2 \tag{11.25}$$

This description is quite adequate for two consecutive sieves in the $\sqrt[4]{2}$ series, but it can be absolutely meaningless for two sieves at extreme ends of the mesh range. Another simple average diameter is the geometric mean, defined by

$$d_g = \sqrt{(a_1 a_2)} \tag{11.26}$$

Values of $\bar{d}$ calculated from equation (11.26) are smaller than those given by equation (11.23), but for two close sieves the difference is not great.

The volume mean diameter (or weight mean if the particle density is constant) is widely used:

$$\bar{d}_v = \frac{\Sigma nd^4}{\Sigma nd^3} = \frac{\Sigma(Md)}{\Sigma M} \tag{11.27}$$

When the surface area of the particles is an important property the surface mean diameter can be employed, defined by

$$\bar{d}_s = \frac{\Sigma nd^3}{\Sigma nd^2} = \frac{\Sigma M}{\Sigma(M/d)} \tag{11.28}$$

where n and M are the number and mass, respectively, of all particles of equivalent diameter d.

The root mean square diameter is also frequently used when surface properties are important. This statistical quantity is defined by

$$\bar{d}_{rms} = \sqrt{\left(\frac{\Sigma nd^2}{\Sigma n}\right)} = \sqrt{\left[\frac{\Sigma(M/d)}{\Sigma(M/d^3)}\right]} \tag{11.29}$$

Values of the over-all mean diameters calculated from equations (11.28) and (11.29) can differ considerably (see *Table 11.8*), yet for a mass of particles with a wide size distribution there is no general agreement as to the preferred method.

Table 11.8. CALCULATION OF OVER-ALL 'MEAN' DIAMETERS

Size range (µm)	Mean size of fraction, d (µm)	Mass of fraction, M (g)	Md	M/d	M/d^3
850–600	725	11·8	8550	0·0163	$0·031 \times 10^{-6}$
600–425	512	18·6	9520	0·0363	$0·139 \times 10^{-6}$
425–300	362	38·5	13 900	0·1064	$0·812 \times 10^{-6}$
300–212	256	22·7	5810	0·0887	$1·353 \times 10^{-6}$
212–150	181	8·4	1520	0·0464	$1·417 \times 10^{-6}$
		100·0	39300	0·2941	$3·752 \times 10^{-6}$

$\bar{d}_v = 39300/100 = 393$ µm.
$\bar{d}_s = 100/0·2941 = 340$ µm.
$\bar{d}_{rms} = (0·2941/3·75 \times 10^{-6})^{\frac{1}{2}} = 280$ µm.

Two other statistical diameters are often encountered, viz. the modal and median diameters; both are determined from frequency plots (size interval versus number of particles in each interval). The modal diameter is the diameter at the peak of the frequency curve, whereas the median diameter defines a mid-point in the distribution — half the total number of particles are smaller than the median, half are larger. If the distribution curve obeys the Gaussian or Normal Error law, the median and modal diameters coincide.

In connection with particle size measurement more than 20 different 'average' diameters have been proposed; and while several have certain points in their favour in special cases, none has yet been found to be generally satisfactory. Therefore all calculations based on an average diameter are prone to appreciable error, and it is recommended that such calculated quantities be clearly annotated with the method of calculation so that the results of different workers can be compared.

INDUSTRIAL SCREENING

Crystalline products are generally marketed in graded sizes, and consequently dried crystals are frequently submitted to a screening process before final packaging. One of the functions of the classifying crystallisers, described in Chapter 9, is to produce directly a product of the required size, and thus to eliminate the necessity for screening, which proves to be a wasteful, costly bottleneck process. But even the products from these crystallisers may require a gentle screening to remove pieces of scale, or fine dust resulting from attrition during filtration, drying, handling and other post-crystallisation operations.

The objects of most crystal-screening processes are to make a 'cut' in the original material, to remove the roughs and fines and to leave as the main product a mass of fairly regular granular crystals. There are cases where regularity of crystal size is not desired. Drugs and other fine chemicals required for tabletting purposes, for instance, must have a reasonable spread of particle size to reduce the quantity of air entrapped in the mass when it is compressed in the die of the tabletting machine; a certain proportion of fines is required to fill the voids of the larger particles. In some industrial processes complicated screening followed by blending operations may be necessary to produce the final product.

Screening on an industrial scale is quite different from the laboratory procedures described above for sieve testing purposes. In the latter operation screening is continued to an end point, i.e. until no more, or very little, of the material passes through the given screen. In industrial practice there is neither the time nor indeed the necessity to approach this degree of perfection. The operation is usually continuous, feed material flowing at a steady rate on to the shaking or vibrating screen and remaining on the screening surface for a relatively short time. The passage of particles through the sieve apertures is impeded by the motion of the screen and by the presence of other particles. Particle interference coupled with the short residence time of material on the screen lead to imperfect separation. The size of the sieve apertures is also an important factor; sieves finer than about 100-mesh are rarely used industrially, on account of their low throughput and liability to clogging or 'binding'.

TYPES OF SCREENS

Industrial screens may be fabricated in mild steel, stainless steel, brass, phosphor bronze, Monel metal and many other alloys to suit special requirements. Stainless steel is probably the most widely favoured, as it is corrosion-resistant and does not suffer severely from abrasion. Fine screens in a sieving machine are usually supported on a more open screen with stronger wires to prevent distortion of the fine mesh by the weight of material flowing over it. The presence of a supporting mesh, however, restricts the effective screening area.

The nominal aperture of a screen determines the diameter of a spherical particle that will just pass through or be retained on the screen. This condition, however, only applies if the screen is laid horizontally and the particle is presented to the aperture in a vertically downward direction. For industrial screening purposes many factors have to be considered before a mesh size can be specified for a given grading duty. For instance, the screens are usually laid in a sloping position, and the effective aperture of the screen can be much less than the nominal aperture. Again, crystals are not spherical, nor are they presented to the apertures in a vertically downward direction; passage through the screen is effected by a combination of jostling and pushing. Elongated or irregular particles can easily block the sieve apertures, rendering the screen ineffective. A similar state of affairs is encountered when the feed material contains a high proportion of particles of near-mesh size.

Most wire screens have square apertures, but specially woven screens with elongated apertures are occasionally employed. These latter types are useful for the sieving of needle crystals as they are less prone to blinding; they are not suitable, however, for the grading of tabular or platy crystals. The throughput or capacity of an elongated aperture screen is greater than that of an equivalent square aperture screen, but the sharpness of size grading which can be effected is generally inferior.

SCREENING EQUIPMENT

No attempt will be made here to describe in any detail the vast number of screening units that are employed in industrial size grading. For an account of the construction and operation of these machines reference should be made to handbooks dealing with materials-handling equipment. The treatment of this subject given by TAGGART[19] is most comprehensive. Broadly speaking, industrial screens may be classified according to the motion, if any, of the screening surface, and it is on this basis that the following notes are made.

Stationary screens, punched and slotted plates, and parallel bars (grizzlies) may be used for the coarse grading of particles larger than about 10 mm. They may be laid in a horizontal or sloping position, and the material is passed on to or down the screening surface. The passage of the material through the screen may be assisted by raking, and rough lumps may be broken up on the more robust assemblies.

Revolving screens, or trommels, consist of a horizontal cylindrical screen, of wire or perforated plate, which rotates within a casing. Feed material enters at one end of the cylinder, fines pass through the screen into the collector casing, and the coarse material leaves at the other end of the cylinder. The longitudinal axis of rotation usually slopes towards the discharge end to facilitate the movement of material through the trommel. The meshes or holes may increase in size at set intervals along the cylindrical screen, and several product grades can be collected. Alternatively, two or more concentric screens of different aperture may be used. Conical trommels, with horizontal axes of rotation, are also available. Rotating screens of nominal aperture smaller than about 3 mm are rarely used.

Shaking, gyrating and vibrating screens are most frequently employed for the size grading of crystalline products, and literally dozens of different machines are in common use. Although their operating mechanisms differ considerably, they are all aimed at producing a jerky but continuous movement of material over the whole available screening surface. The screens are usually laid in a sloping position, about 15–30° to the horizontal, but some, particularly those on the circular gyrators, may be laid almost horizontally. Several decks of screens may be mounted above one another in the same unit, which permits the separation of a number of different size fractions. The vibrating screens, in which the vibrations (about 10 to 50 per second) may be produced electrically or mechanically, e.g. by an off-balance flywheel, are less prone to blinding than are the shakers and gyrators, and generally give larger throughputs per unit area of screen surface.

SCREENING CAPACITY AND EFFICIENCY

The material fed to a screen can be considered to be composed of two fractions, an oversize fraction consisting of particles that are too large to pass through the screen apertures, and an undersize fraction consisting of particles that are too small to be retained on the screen. The screening efficiency, or effectiveness of separation, therefore, should indicate the degree of success obtained in the segregation of these two fractions. In an industrial screening operation a 'clean' separation is never achieved; undersize particles are invariably left in the oversize fraction, mainly because the material does not remain on the sieve for a sufficiently long period, and oversize particles may be found in the undersize fraction if the screen mesh is non-uniform, punctured or inadequately sealed around its edges.

The capacity of a screen is the feed rate at which it performs the specified duty. In general, other factors remaining constant, the capacity, i.e. throughput, decreases as the required degree of separation, i.e. efficiency, increases. Some compromise, therefore, has to be made between capacity and efficiency. Various expressions of capacity are used. For a given screen, the specification kg/min or ton/h may be quite adequate. Capacity may also be quoted as mass per unit time per unit area of screening surface, but care must be taken here, because the length and width of the screen may be independent factors.

The term 'efficiency' applied to screening processes is not easy to define. In fact, there is no generally accepted definition of the term, and various industries adopt the one that most simply and adequately meets their needs. The following analysis indicates a few of the expressions commonly employed. The sieve-test data shown diagrammatically in *Figure 11.15* point out the differences between perfect and actual screening operations. *Figure 11.15a* gives the sieve analysis (cumulative oversize fractions plotted against sieve aperture) of the feed material. Therefore, for an effective screen aperture a^*, fraction O represents the oversize particles and fraction U represents the undersize. For perfect separation, the sieve analyses of these two fractions would be as shown in *Figure 11.15b*; no particles smaller than a^* appear in the oversize fraction, none larger than a^* in the undersize fraction. In

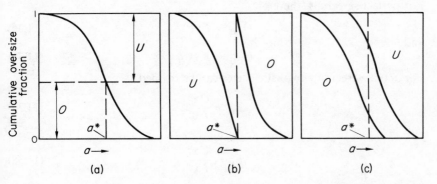

Figure 11.15. Cumulative oversize diagrams: (a) feedstock; (b) perfect screening; (c) actual screening

practice, however, unwanted particle sizes do appear in the undersize and oversize flow streams (*Figure 11.15c*).

A screening operation can be considered as the separation of a feedstock F into an oversize top product O and an undersize bottom product U, i.e.

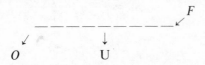

Therefore, if F, O and U represent the masses of these flow streams, an over-all mass balance gives

$$F = O + U \qquad (11.30)$$

If x_F^o, x_O^o and x_U^o represent the mass fractions of oversize material, i.e. particles larger than a^*, in the feed, overflow and underflow, respectively, and x_F^u, x_O^u and x_U^u represent the corresponding undersize mass fractions, then a balance on the oversize material gives

$$Fx_F^o = Ox_O^o + Ux_U^o \tag{11.31}$$

and a balance on the undersize material gives

$$Fx_F^u = Ox_O^u + Ux_U^u \tag{11.32}$$

Thus, from equations (11.30)–(11.32),

$$\frac{O}{F} = \frac{x_F^o - x_U^o}{x_O^o - x_U^o} = \frac{x_U^u - x_F^u}{x_U^u - x_O^u} \tag{11.33}$$

and

$$\frac{U}{F} = \frac{x_O^o - x_F^o}{x_O^o - x_U^o} = \frac{x_F^u - x_O^u}{x_U^u - x_O^u} \tag{11.34}$$

For perfect screening, therefore,

$$Ox_O^o = Fx_F^o \tag{11.35}$$

and

$$Ux_U^u = Fx_F^u \tag{11.36}$$

For actual screening, two efficiencies can be defined:

$$E_O = \frac{Ox_O^o}{Fx_F^o} \tag{11.37}$$

and

$$E_U = \frac{Ux_U^u}{Fx_F^u} \tag{11.38}$$

Equation (11.37) gives a measure of the success of recovering oversize particles in the overflow stream, equation (11.38) of undersize material in the underflow stream. For perfect screening both E_O and E_U will be unity. Equations (11.37) and (11.38) require values of the mass flow rates F, O and U, but substitution from equations (11.31) and (11.32) can eliminate these quantities:

$$E_O = \frac{x_O^o(x_F^o - x_U^o)}{x_F^o(x_O^o - x_U^o)} \tag{11.39}$$

$$= \frac{(1 - x_O^u)(x_U^u - x_F^u)}{(1 - x_F^u)(x_U^u - x_O^u)} \tag{11.40}$$

and

$$E_U = \frac{x_U^u(x_F^u - x_O^u)}{x_F^u(x_U^u - x_O^u)} \tag{11.41}$$

$$= \frac{(1 - x_U^o)(x_O^o - x_F^o)}{(1 - x_F^o)(x_O^o - x_U^o)} \tag{11.42}$$

An over-all screen effectiveness, E, can be defined as the product of E_O and E_U. Table 11.9 illustrates the calculation of these efficiencies.

These data are plotted in Figure 11.16, where it can be seen that the values of x_F^o, x_O^o and x_U^o at $a^* = 460$ µm are 0·58, 0·86 and 0·14, respectively.

Table 11.9. SIEVE ANALYSES (EFFECTIVE SCREEN APERTURE $a^* = 460\mu m$)

B.S. mesh	Aperture, μm	Cumulative weight fraction oversize		
		Feedstock	Overflow	Underflow
18	850	0·02	0·06	—
22	710	0·12	0·29	—
25	600	0·26	0·52	0·02
30	500	0·45	0·78	0·08
36	425	0·68	0·90	0·22
44	355	0·81	0·96	0·46
52	300	0·90	1·00	0·68
72	212	0·98	—	0·84
<72	—	1·00	—	1·00

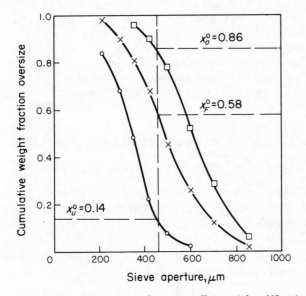

Figure 11.16. Determination of screening efficiency ($a^* = 460\,\mu m$)

Therefore, from equations (11.37) and (11.40),

$$E_O = \frac{0·86(0·58 - 0·14)}{0·58(0·86 - 0·14)} = 0·91$$

and

$$E_U = \frac{(1 - 0·14)(0·86 - 0·58)}{(1 - 0·58)(0·86 - 0·14)} = 0·80$$

and the over-all effectiveness $E = E_O E_U = 0·91 \times 0·80$

$$= 0·728 \text{ or } 73 \text{ per cent}$$

A simpler form of equation (11.39) is often used to express the recovery of true undersize material in the overflow fraction, on the assumption that all the underflow stream consists of undersize material, i.e. $x_U^u = 1$:

$$E_O' = \frac{x_F^u - x_O^u}{x_F^u(1 - x_O^u)}$$ (11.43)

or

$$E_O' = \frac{x_O^o - x_F^o}{x_O^o(1 - x_F^o)}$$ (11.44)

The above assumption, of course, is not always valid.

Formulae of the above types have been criticised because they do not take into consideration the difference between an easy separation duty and a difficult one. For example, consider the sieving of two feedstocks A and B through, say, an 18-mesh screen (850 µm). In both cases all the particles are smaller than 850 µm, but A contains a high proportion of particles smaller than, say, 250 µm, B a high proportion in the range 700–850 µm. Clearly, A can be sieved with ease, whereas the sieving of B is a difficult operation. TAGGART[19] gives an account of several efficiency calculations based on the near-mesh particle contents of various flow streams.

Some of the factors that can affect the capacity of a screen and the efficiency of separation are

(a) *Feedstock properties:* particle shape and size; bulk density; moisture content; abrasion resistance.

(b) *Screen characteristics:* percentage open area; aperture size and shape; length and width of screen; angle of inclination; material of construction; uniformity of mesh.

(c) *Operating conditions:* feed rate; method of feeding; depth of layer on screen; frequency of vibration; amplitude of vibration; direction of vibration.

Several authors [19-21] have discussed the influence of these variables on screening operations, but there is still ample scope for further investigation in this relatively unexplored field.

FLUID-PARTICLE SUSPENSIONS

AGITATED VESSELS

There is, as yet, no completely satisfactory method for predicting the solids suspension characteristics of agitated vessels, although a considerable amount of attention has been paid to this intractable problem in recent years[22-25].

It has long been known, for example, that the impeller geometry and speed, coupled with the vessel geometry and physical properties of the particulate system, are the important parameters. Several authors have presented relationships for predicting the minimum impeller speed to keep all the particles in suspension, i.e. to prevent particles resting on the bottom of the vessel. However, the spatial distribution of particles throughout the

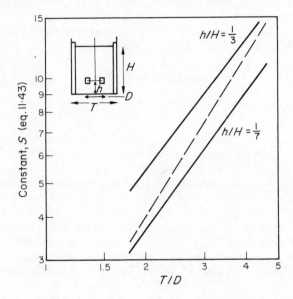

Figure 11.17. Suspension of crystals in a baffled turbine agitated vessel—value of the constant S to be used in equation (11.45). Broken line indicates data of ZWEITERING[22]. (After A. W. NIENOW[24])

vessel is not easily predicted or assessed, but, as shown by NIENOW[24], it is undoubtedly a function of the type of impeller and vessel geometry.

The equation due to ZWEITERING[22] has been shown to be of wide applicability for determining the minimum impeller speed, N, to suspend all the particles smaller than a certain size, d:

$$N = S[v^{0.1}d^{0.2}x^{0.13}D^{-0.85}(g\Delta\rho/\rho)^{0.45}] \qquad (11.45)$$

where v = kinematic viscosity of the liquid in m²/s; d = nominal particle size in m; x = weight fraction of solids in the system; D = impeller diameter in m; g = gravitational acceleration in m/s²; ρ = density of the liquid in kg/m³; and N = impeller speed in rev/s. S is a dimensionless constant and values are reported[22] on graphs of S versus T/D with h/H as a parameter for a large number of different agitators (propeller, paddle, vaned disc turbine and fan disc turbine). T = vessel diameter, h = agitator clearance from bottom of vessel, H = liquid depth.

NIENOW[24] has studied the suspension of several inorganic salts in their aqueous solutions. For the suspension of crystals in a flat-bottomed vessel with baffles, using a flat-bladed turbine, he suggested the relationship

$$N \propto d^{0.21} x^{0.12} D^{-2.2} \Delta \rho^{0.43} \qquad (11.46)$$

which is compatible with the more complex equation (11.45) since ZWEITERING's data indicate that the factor S is proportional to $(T/D)^{1.5}$, i.e. $N \propto D^{-2.35}$.

Figure 11.17 indicates a range of values of S suggested for use with equation (11.45) for the case of a flat-bladed turbine impeller.

DRAFT TUBES

The use of the draft tube (or shrouded stirrer, as it used to be called) in mechanically agitated vessels is becoming widespread in the chemical industry for suspending solids in liquids. Despite the increasing popularity of these devices, however, virtually no published information is yet available concerning their design or performance. *Figures 9.1a and 9.19* indicate diagrammatically the general mode of action of this type of stirrer: a propeller acts as a pump within the draft tube located centrally in the vessel. The liquid or suspension is sucked into the draft tube and then ejected into the annular region between the wall and tube. The flow pattern may be 'up the centre' or 'down the centre'. In the former case any settled solids are sucked from the bottom of the vessel and returned to the suspension at the top. In this way the whole suspension is circulated (e.g. as in a circulating magma crystalliser). In the latter case the solids are kept suspended in the annular region, the larger particles remaining in the lower regions. Only liquor, together with any very small particles present, is circulated through the draft tube. This arrangement is suitable if the vessel is also required to function as a classifier (e.g. a circulating liquor crystalliser).

The important design parameters for a draft-tube agitated vessel are the diameters of the vessel and draft tube, the height of the draft tube and its position in the vessel, and the type and speed of the agitator. Dead spaces within the vessel should be avoided if efficient particle suspension is to be achieved. Another problem in draft-tube agitated vessels is that the liquid emerging from the draft tube under the action of the impeller has a radial component of velocity. This creates an undesirable swirling motion in the annulus and a vortex is formed at the free liquid surface. Vertical baffles located on the vessel wall can overcome this difficulty and produce a stable fluidised bed. There is an optimum vessel: tube diameter ratio for a given vessel and duty, and the clearance of the draft tube from the base is important; a high clearance gives a low exit-liquid velocity and poor fluidisation, a small clearance gives highly turbulent conditions and the high pressure drop at the tube exit results in a greatly decreased efficiency of the pumping characteristics of the draft tube.

FLUIDISED BEDS

A large number of industrial crystallisers operate on the suspended-bed principle, in which a mass of crystals is fluidised in an upward flowing stream of liquor. One of the important parameters needed in the design of such crystallisers is the upflowing liquor velocity necessary to keep the crystals in suspension, and this is not easy to predict accurately. The crystals, being present in large quantities, are subjected to hindered settling. Further complications can arise if the crystals have irregular shape.

Despite the fact that it has often been recommended in the literature, Stokes' law should not be used to predict the crystal suspension velocity as it can lead to gross errors. If certain simplifying assumptions are made it is possible, with the aid of conventional fluidisation theory[26], to predict the behaviour of a crystal suspension, but with so many inherent errors it is often just as convenient and certainly more reliable to measure these characteristics as shown by MULLIN and GARSIDE[27].

Free-fall Velocities

The drag force, F, acting on a particle totally immersed in an infinite fluid is conventionally defined by the equation

$$F = c_D A_{p\frac{1}{2}} \rho u^2 \tag{11.47}$$

where c_D is the drag coefficient, which is a function of the particle Reynolds number ($Re_p \rho u d/\eta$) and the particle shape.

The free-fall or terminal settling velocity of a particle in a fluid, u_0, can be calculated by equating the drag force (equation 11.47) with the gravitational force. Thus for a spherical particle

$$c_D A_{p\frac{1}{2}} \rho u_0^2 = \frac{\pi}{6} d^3 (\rho_s - \rho) g$$

i.e.

$$u_0 = \left[\frac{4d(\rho_s - \rho)g}{3\rho c_D} \right]^{\frac{1}{2}} \tag{11.48}$$

or

$$c_D = \frac{4d(\rho_s - \rho)g}{3\rho u_0^2} \tag{11.49}$$

For particle Reynolds numbers less than about 0.3, laminar flow conditions exist and Stokes (1851) showed analytically that, in this region, the drag force for spheres is given by

$$F = 3\pi u \eta d \tag{11.50}$$

Hence, for laminar conditions

$$c_D = \frac{3\pi u \eta d}{\frac{\pi}{4} d^2 \cdot \frac{1}{2}\rho u^2} = \frac{24}{Re_p} \tag{11.51}$$

and

$$u_0 = \frac{gd^2(\rho_s - \rho)}{18\eta}$$ (11.52)

For particle Reynolds numbers exceeding about 1000, the value of c_D becomes constant at approximately 0·44. The free-fall velocity in this region is thus given by

$$u_0 = 1·74 \left[\frac{gd(\rho_s - \rho)}{\rho} \right]^{\frac{1}{2}}$$ (11.53)

Unfortunately the usual range of interest for crystalliser design lies between these two regions. For instance, the particle Reynolds numbers for individual

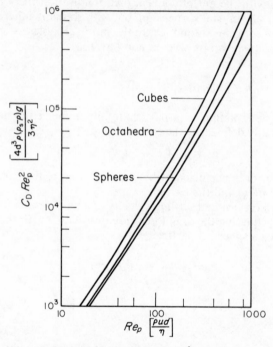

Figure 11.18. Variation of $c_D Re_p^2$ with R_p

crystals of potassium sulphate freely suspended in a saturated solution at 50 °C are about 3 and 550 for sizes of 0·1 and 2·0 mm, respectively. Within this intermediate region c_D is usually obtained from empirically derived values, but it cannot be used directly to determine the free-fall velocity, u_0, because this term occurs in both c_D and Re_p. A trial and error procedure can be used but this is tedious and it is more convenient to combine c_D and Re_p, as $c_D Re_p^2$, to eliminate u_0. Values of $c_D Re_p^2$ are usually presented in graphical form (see *Figure 11.18*).

The use of the above relations is straightforward for the case of spheres. Crystals, however, are rarely spherical and irregular particles have drag characteristics that can differ appreciably from those of spheres. Drag coefficients for non-spherical particles have been measured by a number of authors, but to some extent these values are difficult to use, since different shape factors have been defined[27].

Crystalline products are usually sized by sieving, so ideally the length term to be used should be that derived directly from the sieving operation, i.e. the second largest dimension of the particle (*Figure 11.14*).

Measured terminal settling velocities of aluminium potassium sulphate (potash alum) and potassium sulphate crystals[27] give an indication of the behaviour of different crystal shapes under free-fall conditions. The alum crystals were near-perfect octahedra, whereas those of potassium sulphate were elongated with a length to breadth ratio of about 3. Single crystals from a closely sized sieve fraction were allowed to fall through a stagnant saturated

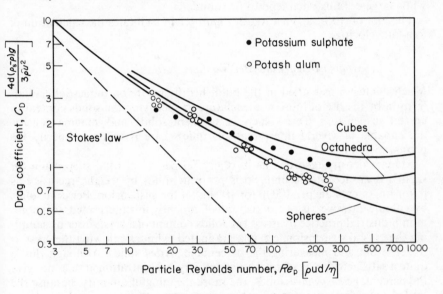

Figure 11.19. Variation of c_D with Re_p

solution of the appropriate salt contained in a 70 mm internal diameter column. The time of fall through a fixed distance was measured for ten or more crystals of a given size and the average velocity calculated.

The measured free-fall velocities, u_0, were corrected for the wall effect to give the free-fall velocities in an infinite medium, $u_{0(\infty)}$, with the equation[28]

$$u_{0(\infty)} = u_0(1 + 2\cdot1\, d/D) \tag{11.54}$$

where D is the column diameter. Other correction factors have been quoted by various authors, but all these give very similar results for values of d/D below about 0·1.

The drag coefficients, calculated from $u_{0(\infty)}$ by use of equation (11.49) are plotted against the particle Reynolds number in *Figure 11.19*. The length term used in both these parameters is the arithmetic mean of the two adjacent sieve sizes between which the crystals were graded. Also shown in *Figure 11.19* are the curves for spheres and two different isometric particles.

At low Reynolds numbers the drag coefficients for different shapes differ little from those for spheres. The effect of shape is more pronounced at high Reynolds numbers, where the drag coefficients are appreciably higher than the corresponding values for spheres.

It is frequently observed that the larger crystals produced in commercial crystallisers are noticeably more rounded than smaller crystals. This effect may result from attrition and the influence of the higher relative liquor velocities experienced by larger crystals, or as a result of crystallisation–dissolution sequences (p. 369). In practical cases, therefore, the value of c_D may seldom differ appreciably from that appropriate for spheres; small crystals correspond to low Reynolds numbers, where shape has little effect, while large crystals often tend to be rounded.

The plot of $c_D Re_p^2$ against Re_p in *Figure 11.18* facilitates the calculation of free-fall velocities.

Magma Density and Bed Voidage

Much confusion is caused in the published literature on industrial crystallisation by the use of loose nomenclature in specifying the solids content of crystal suspensions. Terms such as 'per cent solids' and 'magma density' are capable of several interpretations unless care is taken to specify the precise meaning.

Magma density recorded as lb of crystal per ft^3 of total suspension (or kg/m^3) is an acceptable unit. So is per cent solids by weight, recorded as lb (or kg) of crystals per 100 lb (or 100 kg) of the suspension. Per cent solids by volume, however, is one of the units frequently misinterpreted. It is common industrial practice to assess the solids content of a crystalliser by sampling the crystal suspension, e.g. in a graduated cylinder, allowing the crystals to settle out and measuring the percentage *settled* solids. Although this is quite a satisfactory way of assessing the slurry concentration, it does not give the percentage *suspended* solids, the more meaningful quantity, because the settled volume contains a significant proportion of liquid. Settled spheres of uniform size, for example, contain a void space of about 40 per cent, but considerable deviations from this value can occur for irregular particles.

Probably the most satisfactory method of defining slurry concentration is to specify the over-all voidage of the mixed suspension, ε, i.e. the fraction of the total system that is liquid. The solids fraction is thus $1 - \varepsilon$. The relationship between the system voidage and the settled solids expressed as a fraction, s, is given by

$$\varepsilon = 1 - s(1 - \varepsilon_s) \tag{11.55}$$

where ε_s = voidage of the settled solids (~ 0.4 for unsized spheres)

Superficial Liquor Velocities

RICHARDSON and ZAKI[28] have shown that in a liquid fluidised bed the superficial liquor velocity, u_s, and the bed voidage, ε, are related through the equation

$$u_s = u_i \varepsilon^z \qquad (11.56)$$

The value of u_i is very nearly equal to the free-fall velocity of a single particle in an infinite fluid, $u_{0(\infty)}$, except when the ratio of particle size to column diameter is large, and, in general,

$$\log u_{0(\infty)} = \log u_i + \frac{d}{D} \qquad (11.57)$$

The exponent z varies with the particle Reynolds number $(u_{0(\infty)}\rho d/\eta)$ and the ratio d/D, and for the range of Reynolds numbers of interest in crystalliser design is given by

$$1 < Re_p < 200: \quad z = (4\cdot4 + 18d/D)\,.\,Re_p^{-0\cdot1} \qquad (11.58)$$

$$200 < Re_p < 500: \quad z = 4\cdot4 Re_p^{-0\cdot1} \qquad (11.59)$$

$$Re_p > 500: \quad z = 2\cdot4 \qquad (11.60)$$

Although these equations have been shown to be accurate for spheres, there is little published information on the behaviour of non-spherical particles. Richardson and Zaki give a factor by which z must be multiplied when non-spherical particles are being handled, but this was only verified

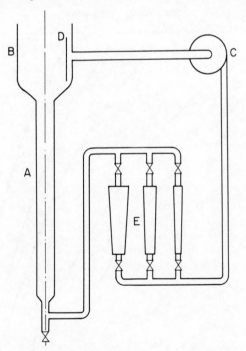

Figure 11.20. Apparatus for measuring velocity–voidage relationships: A, fluidisation section; B, calming section; C, pump; D, baffle plate; E, flow meters

414

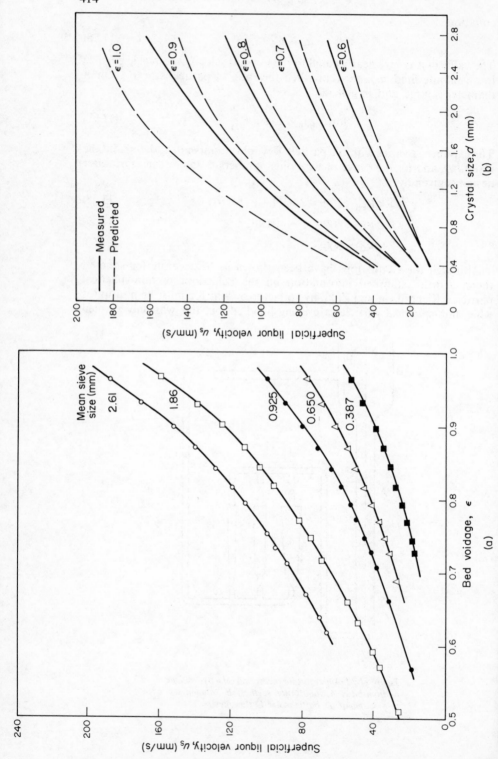

for Reynolds numbers above 500. A further difficulty is that when d/D is not zero a discontinuity occurs in the calculated value of z at $Re_p = 200$, but in spite of these deficiencies the equations of Richardson and Zaki can be used to predict the performance of crystal suspensions.

Velocity–voidage relationships for crystals may be measured[27] in the simple laboratory apparatus shown in *Figure 11.20* comprising a glass column fluidisation section, 25 mm diam. and 380 mm long surmounted by a 75 mm diam. calming section to prevent crystals being carried over into the pump. A scale attached to the side of the fluidisation section is used to measure crystal bed height. A given weight of crystals, graded into closely sized sieve fractions, is introduced into the fluidisation zone. The solution circulation rate is varied and the crystal bed height measured at different solution velocities. From this height the voidage is calculated.

Some typical results for the variation of voidage with superficial velocity for several sizes of potassium sulphate crystals in saturated solution at 20 °C are shown in *Figure 11.21a*. Equations (11.56)–(11.58) are used to calculate the variation of superficial liquor velocity with crystal size and voidage. The measured values of $u_{0(\infty)}$ are used together with the values of z for spheres. *Figure 11.21b* compares calculated and experimental values and the agreement is reasonable, although errors of up to 20% occur for the larger sizes.

REFERENCES

1. HASSIALIS, M. D. and BEHRE, H. A., 'Sampling and testing' (Section 19 in reference 19)
2. *Sampling of Chemical Products*, B.S. 0000: 1972 (Part 1. Introduction and general principles; Part 2. Sampling of gases; Part 3. Sampling of liquids; Part 4. Sampling of solids). London; British Standards Institution
3. *Methods for the Use of B.S. Fine-mesh Test Sieves*, B.S. 1796: 1952. London; British Standards Institution
4. *Preferred Numbers*, B.S. 2045: 1965. London; British Standards Institution. *See also* MULLIN, J. W., 'Preferred numbers', *Engineer, Lond.*, **220** (1965), 630; and 'Preferred numbers in engineering', *Chem. Engr, Lond.*, (1970), 53
5. *Wirecloth Sieves for Testing Purposes*, A.S.T.M. E11–70. Philadelphia; American Society for Testing Materials
6. *Test Sieves*, B.S. 410: 1969. London; British Standards Institution
7. *Woven Wirecloth and Perforated Plates in Test Sieves: Nominal Sizes of Apertures*. ISO Recommendation R 565: 1967. Zurich; International Organisation for Standardisation.
8. COLON, F. J., 'Size analyses with micro-mesh sieves', *Chemy Ind.*, **1965**, 263
8a. MULLIN, J. W., 'Particle size analysis with micro-sieves using ultrasonic vibrations', *Chemy Ind.*, **1971**, 1435
9. MULLIN, J. W., 'The analysis of sieve test data', *Ind. Chemist*, **36** (1960), 272
10. POWERS, H. E. C., 'Determination of the grist of sugars', *Int. Sug. J.*, **50** (1948), 149
11. *Determination of Particle Size of Powders*, B.S. 3406: 1963 (Part 1, Sub-division of samples; Part 2, Liquid sedimentation; Part 3, Air elutriation; Part 4, Optical microscope). London; British Standards Institution
12. *Determination of Specific Surface of Powders*, B.S. 4359: 1971 (Part 1, Nitrogen adsorption (B.E.T.); Part 2, Air permeability; Part 3, Calculation from particle size distribution). London; British Standards Institution
13. HERDAN, G., *Small Particle Statistics*, 2nd Ed, 1960. London; Butterworths
14. ALLEN, T., *Particle Size Measurement*, 1968. London; Chapman and Hall
15. HILL, S. and MULLER, E. G., 'Routine control of the fineness of icing sugar', *Int. Sug. J.*, **60** (1958), 194
16. Coulter Electronics, Inc., Chicago, Ill., U.S.A. and Dunstable, Beds., U.K.

17. ROSEN, H. N. and HULBURT, H. M., 'Size analysis of irregular shaped particles in sieving: comparison with the Coulter counter', *Ind. Eng. Chem. Fundam.*, 9 (1970), 65

18. HARRIS, C. C., 'A method for the routine measurement of particle shape factor in the sieve range', *Nature, Lond.*, **187** (1960), 402

19. TAGGART, A. F., *Handbook of Mineral Dressing*, 1945. New York; Wiley

20. FOWLER, R. T. and LIM, S. C., 'Influence of various factors on the effectiveness of separation on a vibrating screen', *Chem. Engng Sci.*, **10** (1959), 163

21. RENDELL, M. and MULLIN, J. W., 'The flow of particles through a screen aperture', in *Behaviour of Granular Materials*, I. Chem. E. London. Symposium Series, **29** (1968), 55

22. ZWEITERING, Th. N., 'Suspension of solid particles in liquids by agitators', *Chem. Engng Sci.*, **8** (1958), 244

23. KOLAR, V., 'Suspending solid particles in liquids by means of mechanical agitation', *Colln Czech. chem. Comm.*, **26** (1961), 613

24. NIENOW, A. W., 'Suspension of solid particles in turbine agitated baffled vessels', *Chem. Engng Sci.*, **23** (1968), 1453

25. NARAYANAN, S., BHATIA, V. K., GUHA, D. K. and RAO, M. N., 'Suspension of solids by mechanical agitation', *Chem. Engng Sci.*, **24** (1969), 223

26. ZENZ, F. A. and OTHMER, D. F., *Fluidisation and Fluid Particle Systems*, 1960. New York; Reinhold.

26a. DAVIDSON, J. F. and HARRISON, D., *Fluidisation*, 1971. London; Academic Press

27. MULLIN, J. W. and GARSIDE, J., 'Velocity-voidage relationships in the design of suspended bed crystallisers', *Br. chem. Engng*, **15** (1970), 773

28. RICHARDSON, J. F. and ZAKI, W. N., *Trans. Instn. chem. Engrs*, **32** (1954), 35

Appendix

Table A.1. INTERNATIONAL ATOMIC WEIGHTS

Name	Symbol	At. No.	At. Wt.	Valency	Name	Symbol	At. No.	At. Wt.	Valency
Actinium	Ac	89	227	—	Neodymium	Nd	60	144·27	3
Aluminium	Al	13	26·98	3	Neon	Ne	10	20·183	0
Antimony	Sb	51	121·76	3, 5	Nickel	Ni	28	58·71	2, 3
Argon	A	18	39·944	0	Niobium	Nb	41	92·91	3, 5
Arsenic	As	33	74·91	3, 5	Nitrogen	N	7	14·008	3, 5
Astatine	At	85	[210]	1, 3, 5, 7	Osmium	Os	76	190·2	2, 3, 4, 8
Barium	Ba	56	137·36	2	Oxygen	O	8	16·000	2
Beryllium	Be	4	9·013	2	Palladium	Pd	46	106·4	2, 4
Bismuth	Bi	83	209·00	3, 5	Phosphorus	P	15	30·975	3, 5
Boron	B	5	10·82	3	Platinum	Pt	78	195·09	2, 4
Bromine	Br	35	79·916	1, 3, 5, 7	Polonium	Po	84	210	
Cadmium	Cd	48	112·41	2	Potassium	K	19	39·100	1
Calcium	Ca	20	40·08	2	Praseodymium	Pr	59	140·92	3
Carbon	C	6	12·011	2, 4	Promethium	Pm	61	[145]	3
Cerium	Ce	58	140·13	3, 4	Protactinium	Pa	91	231	
Cesium	Cs	55	132·91	1	Radium	Ra	88	226·05	2
Chlorine	Cl	17	35·457	1, 3, 5, 7	Radon	Rn	86	222	0
Chromium	Cr	24	52·01	2, 3, 6	Rhenium	Re	75	186·22	—
Cobalt	Co	27	58·94	2, 3	Rhodium	Rh	45	102·91	3
Copper	Cu	29	63·54	1, 2	Rubidium	Rb	37	85·48	1
Dysprosium	Dy	66	162·51	3	Ruthenium	Ru	44	101·1	3, 4, 6, 8
Erbium	Er	68	167·27	3	Samarium	Sm(Sa)	62	150·35	2, 3
Europium	Eu	63	152·0	2, 3	Scandium	Sc	21	44·96	3
Fluorine	F	9	19·00	1	Selenium	Se	34	78·96	2, 4, 6
Francium	Fr	87	[223]	1	Silicon	Si	14	28·09	4
Gadolinium	Gd	64	157·26	3	Silver	Ag	47	107·880	1
Gallium	Ga	31	69·72	2, 3	Sodium	Na	11	22·991	1
Germanium	Ge	32	72·60	4	Strontium	Sr	38	87·63	2
Gold	Au	79	197·0	1, 3	Sulphur	S	16	32·066 ± 0·003	2
Hafnium	Hf	72	178·50	4	Tantalum	Ta	73	180·95	5
Helium	He	2	4·003	0	Technetium	Tc	43	[99]	6, 7
Holmium	Ho	67	164·94	3	Tellurium	Te	52	127·61	2, 4, 6
Hydrogen	H	1	1·0080	1	Terbium	Tb	65	158·93	3
Indium	In	49	114·82	3	Thallium	Tl	81	204·39	1, 3
Iodine	I	53	126·91	1, 3, 5, 7	Thorium	Th	90	232·05	4
Iridium	Ir	77	192·2	3, 4	Thulium	Tm	69	168·94	3
Iron	Fe	26	55·85	2, 3	Tin	Sn	50	118·70	2, 4
Krypton	Kr	36	83·8	0	Titanium	Ti	22	47·90	3, 4
Lanthanum	La	57	138·92	3	Tungsten	W	74	183·86	6
Lead	Pb	82	207·21	2, 4	Uranium	U	92	238·07	4, 6
Lithium	Li	3	6·940	1	Vanadium	V	23	50·95	3, 5
Lutecium	Lu	71	174·99	3	Xenon	Xe	54	131·3	0
Magnesium	Mg	12	24·32	2	Ytterbium	Yb	70	173·04	2, 3
Manganese	Mn	25	54·94	2, 3, 4,	Yttrium	Y	39	88·92	3
Mercury	Hg	80	200·61	1, 2	Zinc	Zn	30	65·38	2
Molybdenum	Mo	42	95·95	3, 4, 6	Zirconium	Zr	40	91·22	4

Table A.2. SOME PHYSICAL PROPERTIES OF PURE WATER

Temp., °C	Density, g/cm³	Viscosity, cP	Surface tension, dyn/cm	Vapour pressure, mmHg	Dielectric constant	Refractive index*
0	0·99987	1·787	75·6	4·580	87·7	
5	0·99999	1·516	74·9	6·538	85·8	
10	0·99973	1·306	74·2	9·203	83·8	1·33410
15	0·99913	1·138	73·5	12·78	82·0	1·33377
20	0·99823	1·002	72·8	17·52	80·1	1·33335
25	0·99707	0·8903	72·0	23·75	78·3	1·33287
30	0·99568	0·7975	71·2	31·82	76·6	1·33228
40	0·99224	0·6531	69·6	55·34	73·2	1·33087
50	0·98807	0·5467	67·9	92·56	69·9	1·32930
60	0·9832	0·4666	66·2	149·5	66·8	1·32754
70	0·9778	0·405	64·4	233·8	63·9	1·32547
80	0·9718	0·355	62·6	355·3	61·0	1·32323
90	0·9653	0·316	60·7	525·9	58·3	1·32086
100	0·9584	0·283	58·9	760·0	55·7	1·31819

* Absolute index for sodium light.
Density: $1 \text{ g/cm}^3 = 1000 \text{ kg/m}^3$.
Viscosity: $1 \text{ cP} = 10^{-3} \text{ N s/m}^2 \,(= 10^{-3} \text{ kg/m})$
Surface tension: $1 \text{ dyn/cm} = 10^{-3} \text{ N/m} \,(= 10^{-3} \text{ J/m}^2)$.
Pressure: $1 \text{ mm Hg} = 133 \cdot 3 \text{ N/m}^2 \,(1 \text{ bar} = 10^5 \text{ N/m}^2)$.

Table A.3. SOLUBILITY PRODUCTS

Substance	Solubility product	Temperature, °C	Substance	Solubility product	Temperature, °C
Aluminium hydroxide	4×10^{-13}	15	Lead sulphide	3.4×10^{-28}	18
Aluminium hydroxide	1.1×10^{-15}	18	Lithium carbonate	1.7×10^{-3}	25
Barium carbonate	7×10^{-9}	16	Magnesium ammonium phosphate	2.5×10^{-13}	25
Barium chromate	1.6×10^{-10}	18	Magnesium carbonate	2.6×10^{-5}	12
Barium fluoride	1.7×10^{-6}	18	Magnesium fluoride	7.1×10^{-9}	18
Barium iodate ($2H_2O$)	6.5×10^{-10}	25	Magnesium hydroxide	1.2×10^{-11}	18
Barium oxalate ($2H_2O$)	1.2×10^{-7}	18	Magnesium oxalate	8.57×10^{-5}	18
Barium oxalate ($\frac{1}{2}H_2O$)	2.18×10^{-7}	18	Manganese hydroxide	4×10^{-14}	18
Barium sulphate	0.87×10^{-10}	18	Manganese sulphide	1.4×10^{-15}	18
Cadmium sulphide	3.6×10^{-29}	18	Mercuric sulphide	4×10^{-53} to 2×10^{-49}	18
Calcium carbonate (calcite)	0.99×10^{-8}	15	Mercurous bromide	1.3×10^{-21}	25
Calcium fluoride	3.4×10^{-11}	18	Mercurous chloride	2×10^{-18}	25
Calcium oxalate (H_2O)	1.78×10^{-9}	18	Mercurous iodide	1.2×10^{-28}	25
Calcium sulphate	1.95×10^{-4}	10	Nickel sulphide	1.4×10^{-24}	18
Calcium tartrate ($2H_2O$)	0.77×10^{-6}	18	Potassium acid tartrate $[K^+][HC_4H_4O_6^-]$	3.8×10^{-4}	18
Cobalt sulphide	3×10^{-26}	18	Silver bromate	3.97×10^{-5}	20
Cupric iodate	1.4×10^{-7}	25	Silver bromide	4.1×10^{-13}	18
Cupric oxalate	2.87×10^{-8}	25	Silver carbonate	6.15×10^{-12}	25
Cupric sulphide	8.5×10^{-45}	18	Silver chloride	1.56×10^{-10}	25
Cuprous bromide	4.15×10^{-8}	18–20	Silver chromate	9×10^{-12}	25
Cuprous chloride	1.02×10^{-6}	18–20	Silver cyanide $[Ag^+][Ag(CN)_2^-]$	2.2×10^{-12}	20
Cuprous iodide	5.06×10^{-12}	18–20	Silver hydroxide	1.52×10^{-8}	20
Cuprous sulphide	2×10^{-47}	16–18	Silver iodide	1.5×10^{-16}	25
Cuprous thiocyanate	1.6×10^{-11}	18	Silver sulphide	1.6×10^{-49}	18
Ferric hydroxide	1.1×10^{-36}	18	Silver thiocyanate	0.49×10^{-12}	18
Ferrous hydroxide	1.64×10^{-14}	18	Strontium carbonate	1.6×10^{-9}	25
Ferrous oxalate	2.1×10^{-7}	25	Strontium fluoride	2.8×10^{-9}	18
Ferrous sulphide	3.7×10^{-19}	18	Strontium oxalate	5.61×10^{-8}	18
Lead carbonate	3.3×10^{-14}	18	Strontium sulphate	3.81×10^{-7}	17.4
Lead chromate	1.77×10^{-14}	18	Zinc oxalate	1.35×10^{-9}	18
Lead fluoride	3.2×10^{-8}	18	Zinc sulphide	1.2×10^{-23}	18
Lead iodate	1.2×10^{-13}	18			
Lead iodide	7.47×10^{-9}	15			
Lead oxalate	2.74×10^{-11}	18			
Lead sulphate	1.06×10^{-8}	18			

Table A.4. SOLUBILITIES OF INORGANIC SALTS IN WATER (g OF ANHYDROUS COMPOUNDS PER 100 g OF WATER)

Compound	Formula	Solubility, °C								Stable hydrates. 0–25 °C
		0	10	20	30	40	60	80	100	
Aluminium chloride	$AlCl_3$	31·3	46		47					6
sulphate	$Al_2(SO_4)_3$	60	33·5	36·2	40·4	46·1	59·2	73·0	89·1	18
nitrate	$Al(NO_3)_3$		68	74	82	89	106	132	160	9
Ammonium alum	$(NH_4)_2Al_2(SO_4)_4$	2·1	5·0	7·7	11·0	14·9	26·7		109·7(95°)	24
bicarbonate	NH_4HCO_3	12	16	21	27	35	decomp.			
bromide	NH_4Br	60·6	68·0	75·5	83·5	91·0	108	126	146	
chloride	NH_4Cl	29·7	33·4	37·2	41·4	45·8	55·2	65·6	77·3	
dihydrogen phosphate	$NH_4H_2PO_4$	22·0	28·0	36·5	45·8	56·6				
iodide	NH_4I	154	163	172	181	191	209	230	250	
nitrate	NH_4NO_3	118	150	192	242	297	421	580	870	
oxalate	$(NH_4)_2C_2O_4$	2·1	3·1	4·4	6·0	8·0	14			1
perchlorate	NH_4ClO_4	11·7	16·4	21·8	27·9	34·4	50·0	69·6		
sulphate	$(NH_4)_2SO_4$	71·0	73·0	75·4	78·0	81·0	88·0	95·3	103·3	
thiocyanate	NH_4CNS	121		162						
vanadate	NH_4VO_3			4·8	8·4	13·2				
Barium acetate	$Ba(C_2H_3O_2)_2$	58	63	72	75	79	74	74	74	3
bromide	$BaBr_2$	98	100	104	107	112	124	140	160	2
chlorate	$Ba(ClO_3)_2$	20·3	27·0	33·8	41·7	49·6	66·8	84·8	105	1
chloride	$BaCl_2$	31·6	33·2	35·7	38·2	40·7	46·4	52·4	58·3	2
hydroxide	$Ba(OH)_2$	1·6	2·5	3·9	5·6	8·2	21	101		8
iodide	BaI_2	170	186	203	220	232	247	261	272	6
nitrate	$Ba(NO_3)_2$	5·0	7·0	9·2	11·6	14·2	20·3	27·0	34·2	
Beryllium chloride	$BeCl_2$	68		73	77	79				4
nitrate	$Be(NO_3)_2$	98		107	110		177			4
sulphate	$BeSO_4$	35	37	39	41	44	54	67	85	4
Boric acid	H_3BO_3	2·7	3·6	5·0	6·6	8·7	14·8	23·8	40·3	
Cadmium bromide	$CdBr_2$	56·3	75·5	96·5	128	152			160	4
chloride	$CdCl_2$	90	135	134	132	135	136	140	147	2½
iodide	CdI_2	80	83	86	90				128	
nitrate	$Cd(NO_3)_2$	120	140	140		220	400		660	4
sulphate	$CdSO_4$	76·5	76·0	76·6		78·5	83·7		60·8	3

Salt	Formula									Ref.
Caesium chloride	CsCl									
chlorate	CsClO₃	2·46	3·8	6·2	9·5	13·8	26·2	45·0	79·0	
nitrate	CsNO₃	9·3	14·9	23·0	33·9	47·2	83·8	134	197	
perchlorate	CsClO₄	0·1	1·0	1·6	2·6	4·0	7·3	14·4	30·0	
sulphate	Cs₂SO₄	167	173	179	184	190	200	210	220	2
Calcium acetate	Ca(C₂H₃O₂)₂	37·4	36·0	34·7	33·8	33·2	32·7	33·5	29·7	6
bicarbonate	Ca(HCO₃)₂	16·2	16·4	16·6	16·8	17·1	17·5	18·0	18·4	8
chloride	CaCl₂	59·5	65·0	74·5	102		137	147	159	8
iodide	CaI₂	192	196	204	220	240		359	430	4
nitrate	Ca(NO₃)₂	102	115	129	153	196			363	2
sulphate	CaSO₄	0·18	0·19	0·20	0·21	0·21	0·20	0·18	0·16	6
Cobalt ammonium sulphate	Co(NH₄)₂SO₄	6·0	9·2	12·6	17·5	21·8	32·7	49·0		6
bromide	CoBr₂	92		110	156	156	226		257	6
chloride	CoCl₂	42	46	50	56		92	97	104	1
iodide	CoI₂	138	160	185	234	300		400		6
nitrate	Co(NO₃)₂	85	89	97	110	126	167	211	83	7
sulphate	CoSO₄	25·5	30·0	36·2	41·8	48	60	70	98	2
Copper (Cupric) chloride	CuCl₂	69	71	74	76	81	179	208	250	6
nitrate	Cu(NO₃)₂	81·8	95·3	125	160	160				5
sulphate	CuSO₄	14·3	17·4	20·7	25·0	28·5	40·0	55·0	75·4	
Ferric ammonium sulphate	Fe₂(SO₄)₃(NH₄)₂SO₄	18	18	32	59					24
chloride	FeCl₃	74·4	81·9	91·8		120		526	540	6
Ferrous ammonium sulphate	FeSO₄(NH₄)₂SO₄	12·5		26·4		32·9				6
bromide	FeBr₂	102		115	122	128	144	160	177	6
chloride	FeCl₂	61	64	68	73	77	89	100	106	6, 4
potassium sulphate	FeSO₄K₂SO₄	20	25	32	39	45	59			6
sulphate	FeSO₄	15·6	20·5	26·5	32·9	40·2				7
Lead acetate	Pb(C₂H₃O₂)₂	19·7	29·2	44·1	69·5	116				3
bromide	PbBr₂	0·45	0·62	0·85	1·2	1·5		3·3	4·8	
chloride	PbCl₂	0·67	0·81	1·0	1·2	1·5		2·6	3·3	
nitrate	Pb(NO₃)₂	39	48	57	66	75		115	139	2
Lithium bromide	LiBr	143	160	177	191	205	224	245	266	
carbonate	Li₂CO₃	1·54	1·43	1·33	1·25	1·17	1·01	0·85	0·72	3
chloride	LiCl	64	70	80	90		102	112	125	1
hydroxide	LiOH	12·6	12·7	12·8	13·0	13·2	13·9	15·4	17·5	3
iodide	LiI	151	158	165	172	180			480	3
nitrate	LiNO₃	48	60	76	63	72	93	125	227	3
perchlorate	LiClO₄		51	56						3
sulphate	Li₂SO₄	35	35	34		33			29	1

continued

Table A.4. CONTINUED

Compound	Formula	Solubility, °C								Stable hydrates, 0–25 °C
		0	10	20	30	40	60	80	100	
Magnesium bromide	$MgBr_2$	92·0	95·0	96·5	99·2	101·6	107·5	113·7	120·6	6
chloride	$MgCl_2$	52·8	53·5	54·5	56·0	57·5	61·0	66·0	73·0	6
iodide	MgI_2	120		140		174				8
nitrate	$Mg(NO_3)_2$	66·5				84·7			137$^{(90°)}$	6
sulphate	$MgSO_4$		30·9	35·5	40·8	45·5	55·1	64·2	74	7
Manganous chloride	$MnCl_2$	63·4	68·1	73·9	80·7	88·6	109	113	115	4
nitrate	$Mn(NO_3)_2$	50·5	54·1	58·8	67·4					6
sulphate	$MnSO_4$	53·2	60·0	64·5	66·4	68·8	55·0	48·0	34·0	7
Mercuric bromide	$HgBr_2$	0·3	0·4	0·6	0·7	1·0	1·7	2·8	4·9	
chloride	$HgCl_2$	4·66	5·43	6·59	8·14	10·2	17·4	30·9	58·3	
Mercurous perchlorate	$Hg_2(ClO_4)_2$	282		368	420	457	500	540	600	4
Nickel ammonium sulphate	$Ni(NH_4)_2(SO_4)_2$	1·6	4·0	6·5	9·0	12·0	17·5			6
bromide	$NiBr_2$	112	122	131	138	144	152	154	155	6
chloride	$NiCl_2$	54	60	64	69	73	82	87	88	6
iodide	NiI_2	124	135	147	157	174	184	187		6
nitrate	$Ni(NO_3)_2$	80	88	96	109	122	163			6
sulphate	$NiSO_4$	26	32	37	43	47	55	63	77	6, 7
Potassium acetate	$KC_2H_3O_2$	217	234	256	284	323	350	380		1½
aluminium sulphate	$K_2Al_2(SO_4)_4$	3·0	4·0	5·9	8·4	11·7	24·8	71·0		24
bicarbonate	$KHCO_3$	22·5	27·7	33·2	39·1	45·4	60·0			
bisulphate	$KHSO_4$	36·3		51·4		67·3			121·6	
bromate	$KBrO_3$	3·1	4·0	6·8	10·0	13·1	22·5	33·9	50·0	
bromide	KBr	53·5	58·0	64·6	70·0	74·2	84·5	96·0	102·0	
carbonate	K_2CO_3	106	108	110	114	117	127	140	156	1½
chlorate	$KClO_3$	3·3	5·0	7·0	10·5	14·0	24·5	38·5	57	
chloride	KCl	27·6	31·0	34·0	37·0	40·0	45·5	51·1	56·7	
chromate	K_2CrO_4	58·2	60·0	61·7	63·4	65·2	68·6	72·1	75·6	
dichromate	$K_2Cr_2O_7$	5	7	12	20	26	43	61	80	
dihydrogen phosphate	KH_2PO_4	15·9	18·3	22·6	27·7	33·5	50·0	70·4		
ferricyanide	$K_3Fe(CN)_6$	31	36	43	50	60	66	72	81	
ferrocyanide	$K_4Fe(CN)_6$	14	21	27	34	40	54	69	86	3
hydrogen tartrate	$KHC_4H_4O_6$	0·32	0·40	0·53	0·90	1·3	2·5	4·6	6·9	

Formula	Name									Ref.
KOH	hydroxide	97	103	112	126				178	2
KIO₃	iodate	4·7	6·2	8·1	10·3	12·2	18	25	32	
KI	iodide	128	135	144	150	160	175	190	210	
KNO₃	nitrate	13·3	20·9	31·6	45·8	63·9	110	169	247	
KNO₂	nitrite	280	290	300	310	330			413	1
K₂C₂O₄	oxalate	25·9	30·2	34·7	39·2	43·8		63·4	75	
KClO₄	perchlorate	0·75	1·1	1·8	2·6	4·4	9·0		21·8	
KMnO₄	permanganate	2·8	4·4	6·3	9·0	12·6	22·2			
K₂PtCl₆	platinichloride	0·74	0·90	1·12	1·41	1·76	2·64	3·79	5·18	1
K₂SO₄	sulphate	7·4	9·2	10·9	13·0	14·8	18·2	21·4	24·2	
K₂SO₃	sulphite			107					113	
KCNS	thiocyanate	176	189	242			150			
RbBr	**Rubidium bromide**	89		110				148	190	
RbClO₃	chlorate	2·1	3·4	5·4	8·0	11·5	22·3	9·3	65	
RbCl	chloride	70·6	77·4	83·6	89·5		136	175	128	
RbNO₃	nitrate	13·3	22·6	36·5	55·5	79·0		211	305	
RbClO₄	perchlorate		0·64	0·98	1·5	2·4	4·9		18	
Rb₂SO₄	sulphate	34·2	39·7	45·0	50·3	55·2		38·2	79·5	
AgC₂H₃O₂	**Silver acetate**	0·72	0·88	1·04	1·21	1·41	1·89	2·52		
AgNO₃	nitrate	122	170	222	300	376	525	669	952	
Ag₂SO₄	sulphate	0·57	0·70	0·80	0·89	0·98	1·15	1·30	1·41	3
NaC₂H₃O₂	**Sodium acetate**	36·3	40·8	46·5	54·5	65·5	139	153	170	
NaHCO₃	bicarbonate	6·9	8·2	9·6	11·1	12·7	16·4	decomp.		4
NaBO₂	borate (meta)	17	21	25	31	39	62	78	110	10
Na₂B₄O₇	borate (tetra) (borax)	1·2	1·8	2·7	3·9	6·0	20·3	31·5	52·5	
NaBrO₃	bromate	27	30	35	42	50	63	76	91	2
NaBr	bromide	79·5	83·8	90·5	97·2	105		118	121	10
Na₂CO₃	carbonate	7·1	12·5	21·4	38·8	48·5	46·4	45·8	45·5	
NaClO₃	chlorate	80	89	101	113	126	155	189	233	10
NaCl	chloride	35·7	35·8	36·0	36·3	36·6	37·3	38·4	39·8	2
Na₂CrO₄	chromate	31·7	50·2	88·7	88·7	96·0	115	125	126	2
Na₂Cr₂O₇	dichromate	163	170	178	196	220	275	380	430	
NaH₂PO₄	dihydrogen phosphate	58	70	85	107	138	179	207	247	
Na₄Fe(CN)₆	ferrocyanide			18	37	30	65	59	63	12
Na₂HAsO₄	hydrogen arsenate	7·3	15·5	26·5		47	82·9	85		12
Na₂HPO₄	hydrogen phosphate	1·7	3·6	7·7	20·8	51·8		92·4	102	4, 3½, 1½
NaOH	hydroxide	42·0	51·5	109	119	129	174		340	
NaIO₃	iodate	2·5	5·6	9·1	13·2		23	27	34	
NaI	iodide	159	169	179	196	210	250		302	2

continued

Table A.4. CONTINUED

Compound	Formula	Solubility, °C								Stable hydrates, 0–25 °C
		0	10	20	30	40	60	80	100	
Sodium nitrate	$NaNO_3$	73	80	88	96	104	124	148	180	—
nitrite	$NaNO_2$	72	78	85	92	98		133	163	—
oxalate	$Na_2C_2O_4$			3·7					6·33	
phosphate	Na_3PO_4	1·5	4	11	20	31	55	81	108	12
pyrophosphate	$Na_4P_2O_7$	3·2	3·9	6·2	10·0	13·5	21·8	30·0	40·3	10
sulphate	Na_2SO_4	4·8	9·0	19·4	40·8	48·8	45·3	43·7	42·5	10
sulphide	Na_2S	14·4	15·4	18·8	22·5	28·5	39	49		9
sulphite	Na_2SO_3		20·0	26·5	36	28	28	28		7
thiocyanate	$NaCNS$			139						
thiosulphate	$Na_2S_2O_3$	52	61	70	84	103	207	250	266	5
triphosphate	$Na_5P_3O_{10}$	16·3	14·9	14·3	15·2	15·9	17·7			6
Stannous chloride	$SnCl_2$	84		$270^{(15°)}$						—
iodide	SnI_2			1·0	1·2	1·4	2·1	3·0	4·0	—
sulphate	$SnSO_4$			19						
Strontium acetate	$Sr(C_2H_3O_2)_2$	36·9	41·6	42	39·5			36·1	36·4	4½
bromide	$SrBr_2$	85	93	102	112	124	150	182	223	6
chloride	$SrCl_2$	43·5	47·7	52·9	58·7	65·3	81·8	90·5	101	6
hydroxide	$Sr(OH)_2$	0·9	1·2	1·7	2·6	3·8				8
iodide	SrI_2	164		179		198			370	6
nitrate	$Sr(NO_3)_2$	40	54	70	89	90	94	98	101	4
Thallium chlorate	$TlClO_3$	2		4				37	57	
chloride	$TlCl$	0·21	0·25	0·33	0·42	0·52	0·8	1·2	1·8	—
hydroxide	$TlOH$	25·4			39·9	49·5	73·8	106	148	—
nitrate	$TlNO_3$	3·91	6·22	9·55	14·3	20·9	46·2	111	414	—
sulphate	Tl_2SO_4	2·7	3·7	4·9	6·2	7·5	10·9	14·6	18·4	
Uranyl nitrate	$UO_2(NO_3)_2$	97·5	110	125	143	169	252			6
Zinc bromide	$ZnBr_2$	390	420	440			620	640	670	2
chlorate	$ZnClO_3$	145	153	200	209	223				6,4
nitrate	$Zn(NO_3)_2$	95		118		207				6
sulphate	$ZnSO_4$	42	47	54	61	70		87	81	7

Table A.5. SOLUBILITIES OF ORGANIC SOLIDS IN WATER (g OF ANHYDROUS COMPOUND PER 100 g OF WATER)

Compound	Formula	Solubility, °C								Anhydrous melting point, °C
		0	10	20	30	40	60	80	100	
Acetamide	$CH_3 \cdot CONH_2$	138	175	230	310	440	850			81
Acetanilide	$C_6H_5NH \cdot COCH_3$		0·48	0·52	0·63	0·87	2·1	4·7	160	114
Adipic acid	$(CH_2)_4(COOH)_2$	0·8	1·0	1·9	3·0	5·0	18	70		153
Alanine (D)	$CH_3CH \cdot NH_2 \cdot COOH$	12·7	14·2	15·8	17·6	19·6	24·3	30·0	37·3	~300(d.)
Alanine (DL)	$CH_3CH \cdot NH_2 \cdot COOH$	12·1	13·8	15·7	17·9	20·3	26·3	33·9	44·0	~300(d.)
o-Aminophenol	$C_6H_4 \cdot OH \cdot NH_2$	1·7	1·9	2·0	2·2	2·4	2·7	3·0	7·0	173
m-Aminophenol	$C_6H_4 \cdot OH \cdot NH_2$		2·0	2·7	3·8	5·6	21	280	950	123
p-Aminophenol	$C_6H_4 \cdot OH \cdot NH_2$	1·1	1·3	1·6	1·9	2·3	3·6	7·9	37	184(d.)
Anthranilic acid (o)	$C_6H_4 \cdot NH_2 \cdot COOH$		0·3	0·35	0·6	0·9			95	145
Benzamide	$C_6H_5 \cdot CONH_2$		0·6	1·0	1·3	1·6	5	200	800	130
Benzoic acid	$C_6H_5 \cdot COOH$	0·17	0·20	0·29	0·40	0·56	1·16	2·72	5·88	122
Cinnamic acid*	$C_6H_5 \cdot CH:CH \cdot COOH$		0·03	0·04	0·06				0·59	133
Citric acid*	$C_3H_4 \cdot OH \cdot (COOH)_3$	96	118	146	183	215	277	372	526	153
Dicyandiamide	$NH_2 \cdot C(:NH) \cdot NH \cdot CN$	1·3	1·9	3·2	5·0	7·8	19	38		208
Fructose	$CH_2OH(CHOH)_3COCH_2OH$	75	70	80		85	90			95–105
Fumaric acid (trans)	$HOOC \cdot CH:CH \cdot COOH$	0·23	0·35	0·50	0·72	1·1	2·3	5·2	9·8	287
Glucose (α-D) (dextrose)*	$C_6H_{12}O_6$	46	70	92	120	160	280	440		146
Glutamic acid (D)	$COOH(CH_2)_2CH \cdot NH_2 \cdot COOH$	0·34	0·50	0·72	1·0	1·5	3·2	6·5	14·0	198(d.)
Glycine	$CH_2 \cdot NH_2 \cdot COOH$	14·2	18·0	22·5	27	33	45	57	70	235(d.)
Hydroquinone (p)	$C_6H_4(OH)_2$	4·0	5·4	7·2	9·6	13	35	88	198	170
o-Hydroxybenzoic acid	$C_6H_4 \cdot OH \cdot COOH$	0·13	0·15	0·20	0·28	0·42	0·91	2·26	8·12	159
m-Hydroxybenzoic acid	$C_6H_4 \cdot OH \cdot COOH$	0·35	0·55	0·86	1·3	2·0	4·5	12·4	58·7	200
p-Hydroxybenzoic acid	$C_6H_4 \cdot OH \cdot COOH$	0·25	0·35	0·53	0·80	1·25	4·29	13·7	49·9	215
Lactose*	$C_{12}H_{22}O_{11}$	12·2	15·0	19·5	25·2	33·3	57·5	102	153	202
Maleic acid	$HOOC \cdot CH:CH \cdot COOH$	39·3	50	70	90	115	178	283		130

Table A.5. CONTINUED

Compound	Formula	Solubility, °C								Anhydrous melting point, °C
		0	10	20	30	40	60	80	100	
Malic acid (DL)	$CH \cdot OH \cdot CH_2(COOH)_2$	89	105	126	150	180	270	460		128
Malonic acid	$CH_2(COOH)_2$	108	128	153	180	212	292	455	810	135(d.)
Maltose*	$C_{12}H_{22}O_{11}$	57	65	78	93	110	175	300		
Mannitol (D)	$(CH_2OH)_2(CHOH)_4$	10·4	13·7	18·6	25·2	34·6	64·4	115	197	166
Melamine	$C_3N_3(NH_2)_3$	0·12	0·18	0·27	0·42	0·71	1·5	2·8	5·0	~250
Oxalic acid†	$(COOH)_2$	3·5	6·0	9·5	14·5	21·6	44·3	84·4		189
Pentaerythritol	$C(CH_2OH)_4$	4	5	6	8	13	22	40	100	262
Phenacetin (p)	$C_2H_5O \cdot C_6H_4 \cdot NHCOCH_3$			0·07					1·43	135
Phthalic acid (o)	$C_6H_4(COOH)_2$	0·23	0·36	0·56	0·8	1·2	2·8	6·3	18·0	208
Picric acid (2,4,6)	$C_6H_2 \cdot OH \cdot (NO_2)_3$	1·0	1·1	1·2	1·5	1·9	3·1	4·6	7·2	122
Pyrocatechol (o)	$C_6H_4(OH)_2$			45·1		172	412	1120	8360	104
Raffinose‡	$C_{18}H_{32}O_{16}$	3·4	6·6	13·6	27·1	49·9	86·9	153·8		118
Resorcinol (m)	$C_6H_4(OH)_2$	66·2	85	123	170	225	390	634	1060	111
Salicylic acid (o)	$C_6H_4 \cdot OH \cdot COOH$	0·13	0·15	0·20	0·28	0·42	0·91	2·26	8·12	159
Sorbitol	$CH_2OH \cdot (CHOH)_4 \cdot CH_2OH$		179	222	275	355				95–100
Succinic acid	$(CH_2 \cdot COOH)_2$	2·8	4·4	6·9	10·5	16·2	35·8	70·8	127	183
Succinimide	$(CH_2CO)_2NH$	10	16	26	48	83	140	213		125
Sucrose	$C_{12}H_{22}O_{11}$	179	190	204	219	238	287	362	487	170–186(d.)
Sulphanilic acid (p)	$C_6H_4 \cdot NH_2 \cdot SO_3H$	0·45	0·80	1·12		2·03	3·01	4·51	6·67	>280(d.)
Tartaric acid (D or L)	$(CHOH \cdot COOH)_2$	115	126	139	156	176	220	273	343	170
Tartaric acid (racemic)*	$(CHOH \cdot COOH)_2$	8·2	12·3	18·0	25·2	37·0	64·5	98·1	138	205
Taurine	$NH_2CH_2CH_2SO_3H$	3·9	6·0	8·8	12·4	16·8	27·4	38·4	45·7	~330(d.)
Thiourea	$NH_2 \cdot CS \cdot NH_2$	4·9	8·0	13·6	20·1	30·8	71	138	238	181
Triglycine sulphate	$(NH_2CH_2COOH)_3H_2SO_4$	12	20	27	36	45	75			
Urea	$NH_2 \cdot CO \cdot NH_2$	67	85	105	135	165	250	400	730	133
Uric acid	$C_5H_4O_3N_4$	0·002	0·004	0·006	0·009	0·012	0·023	0·039	0·062	(D)

* Crystallises from water with $1H_2O$ † Crystallises from water with $2H_2O$ ‡ Crystallises from water with $5H_2O$ d. = decomposes

Table A.6. DENSITIES OF CRYSTALLINE INORGANIC SOLIDS

Substance		Formula	Crystal system	Density, g/cm^3
Aluminium	bromide	$AlBr_3 \cdot 6H_2O$		2·54
	chloride	$AlCl_3 \cdot 6H_2O$		2·40
	sulphide	Al_2S_3	hexagonal	2·02
	sulphate	$Al_2(SO_4)_3 \cdot 18H_2O$	monoclinic	1·69
Ammonium	acetate	$NH_4C_2H_3O_2$		1·17
	alum	$NH_4Al(SO_4)_2 \cdot 12H_2O$	octahedral	1·64
	arsenate (o)	$(NH_4)_2HAsO_4$		1·99
	bicarbonate	NH_4HCO_3	monoclinic	1·58
	bisulphate	NH_4HSO_4		1·78
	bisulphite	NH_4HSO_3		2·03
	bromide	NH_4Br	cubic	2·43
	chlorate	NH_4ClO_3		1·80
	chloride	NH_4Cl	cubic	1·53
	chromate	$(NH_4)_2CrO_4$	monoclinic	1·91
	cyanate	NH_4CNO		1·34
	dichromate	$(NH_4)_2Cr_2O_7$	monoclinic	2·15
	dihydrogen phosphate	$NH_4H_2PO_4$	tetragonal	1·80
	hydrogen phosphate	$(NH_4)_2HPO_4$	monoclinic	1·62
	iodate	NH_4IO_3	monoclinic	3·31
	iodide	NH_4I	cubic	2·51
	nitrate	NH_4NO_3	orthorhombic	1·73
	nitrite	NH_4NO_2		1·69
	oxalate	$(NH_4)_2C_2O_4 \cdot H_2O$	orthorhombic	1·50
	perchlorate	NH_4ClO_4	orthorhombic	1·95
	permanganate	NH_4MnO_4		2·21
	sulphate	$(NH_4)_2SO_4$	orthorhombic	1·77
	thiocyanate	NH_4CNS	monoclinic	1·31
	thiosulphate	$(NH_4)_2S_2O_3$	monoclinic	1·68
	vanadate	NH_4VO_3		2·33
Barium	acetate	$Ba(C_2H_3O_2)_2 \cdot 3H_2O$	triclinic	2·19
	bromate	$Ba(BrO_3)_2 \cdot H_2O$	monoclinic	3·99
	bromide	$BaBr_2 \cdot 2H_2O$	monoclinic	3·58
	carbonate	$BaCO_3$	orthorhombic	4·43
	chlorate	$Ba(ClO_3)_2 \cdot H_2O$	monoclinic	3·18
	chloride	$BaCl_2 \cdot 2H_2O$	monoclinic	3·10
	chromate	$BaCrO_4$	orthorhombic	4·50
	hydroxide	$Ba(OH)_2 \cdot 8H_2O$	monoclinic	2·18
	iodate	$Ba(IO_3)_2$	monoclinic	5·00
	iodide	$BaI_2 \cdot 2H_2O$	hexagonal	5·15
	nitrate	$Ba(NO_3)_2$	cubic	3·24
	oxalate	BaC_2O_4		2·66
	perchlorate	$Ba(ClO_4) \cdot 3H_2O$	hexagonal	2·74
	sulphate	$BaSO_4$	orthorhombic	4·50
	titanate	$BaTiO_3$	cubic	6·02
Beryllium	chloride	$BeCl_2$	orthorhombic	1·90
	iodide	BeI_2		4·33
	nitrate	$Be(NO_3)_2 \cdot 3H_2O$		1·56
	sulphate	$BeSO_4$		2·44
		$BeSO_4 \cdot 4H_2O$	tetragonal	1·71

continued

Table A.6. CONTINUED

Substance		Formula	Crystal system	Density, g/cm³
Borax		$Na_2B_4O_7 \cdot 10H_2O$	monoclinic	1·73
Boric	acid	H_3BO_3	triclinic	1·44
Cadmium	acetate	$Cd(C_2H_3O_2)_2 \cdot 3H_2O$	monoclinic	2·01
	borate	$Cd(BO_3)_2 \cdot H_2O$	orthorhombic	3·76
	bromide	$CdBr_2$	hexagonal	5·19
	carbonate	$CdCO_3$	trigonal	4·26
	chlorate	$Cd(ClO_3)_2 \cdot 2H_2O$		2·28
	chloride	$CdCl_2$	trigonal	4·05
		$CdCl_2 \cdot 2\frac{1}{2}H_2O$	monoclinic	3·33
	iodate	$Cd(IO_3)_2$		6·43
	iodide	CdI_2	hexagonal	5·67
	nitrate	$Cd(NO_3)_2 \cdot 4H_2O$		2·46
	sulphate	$CdSO_4 \cdot H_2O$	monoclinic	3·79
		$CdSO_4 \cdot 7H_2O$	monoclinic	2·48
	sulphide	CdS	hexagonal	4·82
Caesium	chloride	$CsCl$	cubic	3·99
	chlorate	$CsClO_3$	cubic	3·57
	iodate	$CsIO_3$	monoclinic	4·85
	iodide	CsI	monoclinic	4·51
	nitrate	$CsNO_3$	cubic	3·69
	perchlorate	$CsClO_4$	cubic	3·33
	sulphate	Cs_2SO_4	hexagonal	4·24
Calcium	bromide	$CaBr_2$	orthorhombic	3·55
		$CaBr_2 \cdot 6H_2O$	hexagonal	2·30
	carbonate (calcite)	$CaCO_3$	trigonal	2·71
	chlorate	$Ca(ClO_3)_2 \cdot 2H_2O$	orthorhombic	2·71
	chloride	$CaCl_2 \cdot 6H_2O$	trigonal	1·71
	fluoride	CaF_2	cubic	3·18
	iodide	$CaI_2 \cdot 6H_2O$	hexagonal	2·55
	nitrate	$Ca(NO_3)_2 \cdot 4H_2O$	monoclinic	1·89
	oxalate	$CaC_2O_4 \cdot H_2O$	cubic	2·2
	sulphate	$Ca(SO_4) \cdot 2H_2O$	monoclinic	2·32
	thiosulphate	$CaS_2O_3 \cdot 6H_2O$	triclinic	1·87
Cobalt	bromide	$CoBr_2 \cdot 6H_2O$		2·46
	chlorate	$Co(ClO_3)_2 \cdot 6H_2O$	cubic	1·92
	chloride	$CoCl_2 \cdot 6H_2O$	monoclinic	1·92
	iodate	$Co(IO_3)_2 \cdot 6H_2O$	octahedral	3·69
	iodide	$CoI_2 \cdot 6H_2O$	hexagonal	2·90
	nitrate	$Co(NO_3)_2 \cdot 6H_2O$	monoclinic	1·87
	sulphate	$Co SO_4 \cdot 7H_2O$	monoclinic	1·95
	sulphide	CoS_2	cubic	4·27
Copper	bromate	$Cu(BrO_3)_2 \cdot 6H_2O$	cubic	2·6
	bromide	$CuBr_2$	monoclinic	4·77
	carbonate	Cu_2CO_3	hexagonal	4·00
	chloride	$CuCl_2 \cdot 2H_2O$	orthorhombic	2·54
	iodate	$Cu(IO_3)_2$	monoclinic	5·24
	iodide	CuI	cubic	5·6
	nitrate	$Cu(NO_3)_2 \cdot 6H_2O$		2·07
	sulphate	$CuSO_4 \cdot 5H_2O$	triclinic	2·28

Table A.6. CONTINUED

Substance		Formula	Crystal system	Density, g/cm^3
Ferrous	ammonium sulphate	$FeSO_4(NH_4)_2SO_4 \cdot 6H_2O$	monoclinic	1·86
	bromide	$FeBr_2$	hexagonal	4·64
	chloride	$FeCl_2 \cdot 2H_2O$	monoclinic	2·36
		$FeCl_2 \cdot 4H_2O$	monoclinic	1·93
	sulphate	$FeSO_4 \cdot 7H_2O$	monoclinic	1·90
Ferric	ammonium sulphate	$FeNH_4(SO_4)_2 \cdot 12H_2O$	octahedral	1·71
	iodate	$Fe(IO_3)_2$		4·8
	nitrate	$Fe(NO_3)_3 \cdot 9H_2O$	monoclinic	1·68
	sulphate	$Fe_2(SO_4)_3 \cdot 9H_2O$	orthorhombic	2·10
Lead	acetate	$Pb(C_2H_3O_2)_2 \cdot 3H_2O$	monoclinic	2·55
	bromate	$Pb(BrO_3)_2 \cdot H_2O$	monoclinic	5·53
	bromide	$PbBr_2$	orthorhombic	6·63
	chlorate	$Pb(ClO_3)_2 \cdot H_2O$	monoclinic	4·04
	chloride	$PbCl_2$	orthorhombic	5·85
	chromate	$PbCrO_4$	monoclinic	6·10
	fluoride	PbF_2	orthorhombic	8·24
	iodate	$Pb(IO_3)_2$	hexagonal	6·16
	iodide	PbI_2	hexagonal	6·16
	nitrate	$Pb(NO_3)_2$	cubic	4·53
	oxalate	PbC_2O_4		5·28
	sulphate	$PbSO_4$	monoclinic	6·20
	sulphide	PbS	cubic	7·50
Lithium	bromide	$LiBr$	cubic	3·46
	carbonate	Li_2CO_3	monoclinic	2·11
	chloride	$LiCl$	cubic	2·07
		$LiCl \cdot H_2O$	tetragonal	1·78
	fluoride	LiF	cubic	2·64
	hydroxide	$LiOH \cdot H_2O$	monoclinic	1·51
	iodate	$LiIO_3$	hexagonal	4·51
	iodide	$LiI \cdot 3H_2O$	hexagonal	3·48
	nitrate	$LiNO_3$	trigonal	2·38
	oxalate	$Li_2C_2O_4$	orthorhombic	2·12
	sulphate	Li_2SO_4	monoclinic	2·22
Magnesium	acetate	$Mg(C_2H_3O_2)_2$	orthorhombic	1·42
		$Mg(C_2H_3O_2)_2 \cdot 4H_2O$	monoclinic	1·45
	ammonium sulphate	$MgSO_4 \cdot (NH_4)_2SO_4 \cdot 6H_2O$	monoclinic	2·21
	bromate	$Mg(BrO_3)_2 \cdot 6H_2O$	cubic	2·29
	bromide	$MgBr_2 \cdot 6H_2O$	hexagonal	2·00
	chlorate	$Mg(ClO_3)_2 \cdot 6H_2O$	orthorhombic	1·80
	chloride	$MgCl_2 \cdot 6H_2O$	monoclinic	1·57
	iodate	$Mg(IO_3)_2 \cdot 4H_2O$	monoclinic	3·30
	nitrate	$Mg(NO_3)_2 \cdot 6H_2O$	monoclinic	1·64
	oxalate	$MgC_2O_4 \cdot 2H_2O$		2·45
	perchlorate	$Mg(ClO_4)_2 \cdot 6H_2O$	orthorhombic	1·80
	sulphate	$MgSO_4$	orthorhombic	2·66
		$MgSO_4 \cdot 7H_2O$	orthorhombic	1·68
	sulphide	MgS	cubic	2·84
Manganese	acetate	$Mn(C_2H_3O_2)_2 \cdot 4H_2O$	monoclinic	1·59
	carbonate	$MnCO_3$	orthorhombic	3·13

continued

Table A.6. CONTINUED

Substance		Formula	Crystal system	Density, g/cm^3
Manganese	chloride	$MnCl_2 \cdot 4H_2O$	monoclinic	2·01
	nitrate	$Mn(NO_3)_2 \cdot 4H_2O$	monoclinic	1·82
	oxalate	MnC_2O_4		2·43
	sulphate	$MnSO_4 \cdot 7H_2O$	monoclinic	2·09
	sulphide	MnS	cubic	3·46
Mercuric	acetate	$Hg(C_2H_3O_2)_2$		3·27
	bromide	$HgBr_2$	orthorhombic	6·11
	chlorate	$Hg(ClO_3)_2$		5·00
	chloride	$HgCl_2$	orthorhombic	5·44
	fluoride	HgF_2	cubic	8·95
	iodide (α)	HgI_2	tetragonal	6·36
	iodide (β)	HgI_2	orthorhombic	6·09
	nitrate	$Hg(NO_3)_2 \cdot H_2O$		4·30
	sulphate	$HgSO_4$	orthorhombic	6·47
	sulphide (α)	HgS	hexagonal	8·10
	sulphide (β)	HgS	cubic	7·73
Mercurous	bromide	Hg_2Br_2	tetragonal	7·31
	chlorate	$Hg_2(ClO_3)_2$	orthorhombic	6·41
	chloride	Hg_2Cl_2	tetragonal	7·15
	fluoride	Hg_2F_2	cubic	8·73
	iodide	HgI_2	tetragonal	7·70
	nitrate	$Hg_2(NO_3)_2 \cdot 2H_2O$	monoclinic	4·79
	sulphate	Hg_2SO_4	monoclinic	7·56
Nickel	acetate	$Ni(C_2H_3O_2)_2 \cdot 4H_2O$	monoclinic	1·74
	ammonium sulphate	$Ni(NH_4)_2(SO_4)_2 \cdot 6H_2O$	monoclinic	1·92
	bromate	$Ni(BrO_3)_2 \cdot 6H_2O$	monoclinic	2·57
	bromide	$NiBr_2$	hexagonal	5·10
	carbonate	$NiCO_3$	cubic	2·6
	chlorate	$Ni(ClO_3)_2 \cdot 6H_2O$		2·07
	chloride	$NiCl_2 \cdot 2H_2O$	orthorhombic	2·58
	iodate	$Ni(IO_3)_2$		5·07
	iodide	NiI_2	hexagonal	5·83
	nitrate	$Ni(NO_3)_2 \cdot 6H_2O$	monoclinic	2·05
	sulphate	$NiSO_4 \cdot 6H_2O$	tetragonal	2·07
		$NiSO_4 \cdot 7H_2O$	orthorhombic	1·95
Potassium	acetate	$KC_2H_3O_2$	monoclinic	1·57
	aluminium sulphate	$KAl(SO_4)_2 \cdot 12H_2O$	octahedral	1·76
	bicarbonate	$KHCO_3$	monoclinic	2·17
	bisulphate	$KHSO_4$	orthorhombic	2·32
	bromate	$KBrO_3$	trigonal	3·27
	bromide	KBr	cubic	2·75
	carbonate	$K_2CO_3 \cdot 1\frac{1}{2}H_2O$	monoclinic	2·04
	chlorate	$KClO_3$	monoclinic	2·32
	chloride	KCl	cubic	1·98
	chromate	K_2CrO_4	orthorhombic	2·73
	dichromate	$K_2Cr_2O_7$	monoclinic	2·68
	dihydrogen phosphate	KH_2PO_4	tetragonal	2·30
	ferricyanide	$K_3Fe(CN)_6$	monoclinic	1·85
	ferrocyanide	$K_4Fe(CN)_6 \cdot 3H_2O$	monoclinic	1·85

Table A.6. CONTINUED

Substance		Formula	Crystal system	Density, g/cm^3
	fluoride	KF	cubic	2·48
	formate	$KCHO_2$	orthorhombic	1·91
	hydrogen oxalate	KHC_2O_4	monoclinic	2·04
	hydrogen tartrate	$KHC_4H_4O_6$	orthorhombic	1·98
	hydroxide	KOH	orthorhombic	2·04
	iodate	KIO_3	monoclinic	3·93
	iodide	KI	cubic	3·13
	nitrate	KNO_3	orthorhombic	2·11
	nitrite	KNO_2	monoclinic	1·92
	oxalate	$K_2C_2O_4 \cdot H_2O$	monoclinic	2·13
	perchlorate	$KClO_4$	orthorhombic	2·52
	permanganate	$KMnO_4$	orthorhombic	2·70
	sulphate	K_2SO_4	orthorhombic	2·66
	sulphide	K_2S	cubic	1·81
	thiocyanate	KCNS	orthorhombic	1·89
	thiosulphate	$K_2S_2O_3 \cdot \frac{1}{3}H_2O$	monoclinic	2·23
Rubidium	bromate	$RbBrO_3$	cubic	3·68
	bromide	RbBr	cubic	3·35
	chlorate	$RbClO_3$	orthorhombic	3·19
	chloride	RbCl	cubic	2·80
	chromate	Rb_2CrO_4	orthorhombic	3·52
	dichromate	$Rb_2Cr_2O_4$	triclinic	3·13
	iodate	$RbIO_3$	cubic	4·33
	iodide	RbI	cubic	3·55
	nitrate	$RbNO_3$	cubic	3·11
	perchlorate	$RbClO_4$	orthorhombic	2·80
	sulphate	Rb_2SO_4	orthorhombic	3·61
	sulphide	Rb_2S		2·91
Silver	acetate	$AgC_2H_3O_2$		3·26
	bromate	$AgBrO_3$	tetragonal	5·21
	bromide	AgBr		6·47
	chlorate	$AgClO_3$	tetragonal	4·43
	chloride	AgCl	cubic	5·56
	iodate	$AgIO_3$	orthorhombic	5·53
	iodide	AgI	cubic	6·01
	nitrate	$AgNO_3$	orthorhombic	4·35
	oxalate	$Ag_2C_2O_4$		5·03
	sulphate	Ag_2SO_4	orthorhombic	5·45
	sulphide	Ag_2S	orthorhombic	7·33
Sodium	acetate	$NaC_2H_3O_2 \cdot 3H_2O$	monoclinic	1·45
	bicarbonate	$NaHCO_3$	monoclinic	2·16
	borate (borax)	$Na_2B_4O_7 \cdot 10H_2O$	monoclinic	1·73
	borate (meta)	$NaBO_2 \cdot 4H_2O$	triclinic	1·74
	bromate	Na_2BrO_3	cubic	3·34
	bromide	$NaBr \cdot 2H_2O$	monoclinic	2·18
	carbonate	$Na_2CO_3 \cdot 10H_2O$	monoclinic	1·44
	chlorate	$NaClO_3$	cubic	2·49
	chloride	NaCl	cubic	2·17
	chromate	$Na_2CrO_4 \cdot 10H_2O$	monoclinic	1·48
	citrate	$C_6H_5O_7Na_3$	orthorhombic	1·70
	cyanide	NaCN	orthorhombic	1·66

continued

Table A.6. CONTINUED

	Substance	Formula	Crystal system	Density, g/cm^3
Sodium	dichromate	$Na_2Cr_2O_7 \cdot 2H_2O$	monoclinic	2·52
	dihydrogen phosphate	$NaH_2PO_4 \cdot 2H_2O$	orthorhombic	1·91
	fluoride	NaF	cubic	2·56
	hydrogen arsenate	$Na_2HAsO_4 \cdot 12H_2O$	monoclinic	1·74
	hydrogen phosphate	$Na_2HPO_4 \cdot 12H_2O$	orthorhombic	1·52
	hydroxide	$NaOH$	orthorhombic	2·13
	iodate	$NaIO_3$	orthorhombic	4·28
	iodide	$NaI \cdot 2H_2O$	monoclinic	2·45
	nitrate	$NaNO_3$	trigonal	2·26
	nitrite	$NaNO_2$	orthorhombic	2·17
	oxalate	$Na_2C_2O_4$	monoclinic	2·34
	perborate	$NaBO_3$	triclinic	1·73
	perchlorate	$NaClO_4$	orthorhombic	2·48
	periodate	$NaIO_4$	tetragonal	4·12
	phosphate	$Na_3PO_4 \cdot 12H_2O$	trigonal	1·62
	pyrophosphate	$Na_4P_2O_7 \cdot 10H_2O$	monoclinic	1·82
	sulphate	$Na_2SO_4 \cdot 10H_2O$	monoclinic	1·46
	sulphide	$Na_2S \cdot 9H_2O$	tetragonal	1·43
	sulphite	$Na_2SO_3 \cdot 7H_2O$	monoclinic	1·54
	thiosulphate	$Na_2S_2O_3 \cdot 5H_2O$	monoclinic	1·73
	triphosphate	$Na_5P_3O_{10} \cdot 6H_2O$	triclinic	2·10
Stannous	bromide	$SnBr_2$	orthorhombic	5·12
	chloride	$SnCl_2$	orthorhombic	3·95
	iodide	SnI_2	monoclinic	5·29
	sulphate	$SnSO_4 \cdot 2H_2O$	hexagonal	—
Strontium	acetate	$Sr(C_2H_3O_2)_2$		2·10
	bromate	$Sr(BrO_3)_2 \cdot H_2O$	monoclinic	3·77
	bromide	$SrBr_2 \cdot 6H_2O$	hexagonal	2·39
	carbonate	$SrCO_3$	orthorhombic	3·70
	chlorate	$Sr(ClO_3)_2$	orthorhombic	3·15
	chloride	$SrCl_2 \cdot 6H_2O$	trigonal	1·93
	hydroxide	$Sr(OH)_2 \cdot 8H_2O$	tetragonal	1·90
	iodate	$Sr(IO_3)_2$	triclinic	5·05
	iodide	$SrI_2 \cdot 6H_2O$	hexagonal	2·67
	nitrate	$SrNO_3 \cdot 4H_2O$	monoclinic	2·2
	sulphate	$SrSO_4$	orthorhombic	3·96
	sulphide	SrS	cubic	3·70
Thallium	bromide	$TlBr$	cubic	7·56
	carbonate	Tl_2CO_3	monoclinic	7·11
	chlorate	$TlClO_3$	orthorhombic	5·05
	chloride	$TlCl$	cubic	7·00
	iodide	TlI	cubic	7·29
	nitrate	$TlNO_3$	orthorhombic	5·56
	oxalate	$Tl_2C_2O_4$	monoclinic	6·31
	sulphate	Tl_2SO_4	orthorhombic	6·77
	sulphide	Tl_2S	tetragonal	8·46
Uranyl	acetate	$UO_2(C_2H_3O_2)_2 \cdot 2H_2O$	orthorhombic	2·89
	nitrate	$UO_2(NO_3)_2 \cdot 6H_2O$	orthorhombic	2·81

Table A.6. CONTINUED

Substance		Formula	Crystal system	Density, g/cm^3
Zinc	acetate	$Zn(C_2H_3O_2)_2 \cdot 2H_2O$	monoclinic	1·74
	bromate	$Zn(BrO_3)_2 \cdot 6H_2O$	cubic	2·57
	bromide	$ZnBr_2$	orthorhombic	4·20
	chlorate	$Zn(ClO_3)_2 \cdot 4H_2O$	cubic	2·15
	chloride	$ZnCl_2$	hexagonal	2·91
	iodate	$Zn(IO_3)_2 \cdot 2H_2O$		4·22
	iodide	ZnI_2	hexagonal	4·74
	nitrate	$Zn(NO_3)_2 \cdot 6H_2O$	tetragonal	2·07
	sulphate	$ZnSO_4 \cdot 7H_2O$	orthorhombic	1·96
	sulphide	ZnS		3·98

Table A.7. DENSITIES OF CRYSTALLINE ORGANIC SOLIDS

Substance	Formula	Crystal system	Density, g/cm^3
Acetamide	$CH_3 \cdot CONH_2$	trigonal	0·999
Acetanilide	$C_6H_5NH \cdot COCH_3$	orthorhombic	1·21
Acetylsalicylic acid	$C_6H_4 \cdot OCOCH_3 \cdot COOH$	monoclinic	1·39
Adipic acid	$(CH_2)_4(COOH)_2$	monoclinic	1·36
Alanine (DL)	$CH_3 \cdot CH \cdot NH_2 \cdot COOH$	orthorhombic	1·40
o-Aminophenol	$C_6H_4 \cdot OH \cdot NO_2$	orthorhombic	1·29
m-Aminophenol	$C_6H_4 \cdot OH \cdot NO_2$	orthorhombic	1·28
p-Aminophenol	$C_6H_4 \cdot OH \cdot NO_2$	orthorhombic	1·30
Anthracene	$C_{14}H_{10}$	monoclinic	1·25
Benzamide	$C_6H_5 \cdot CONH_2$	monoclinic	1·08
Benzoic acid	$C_6H_5 \cdot COOH$	monoclinic	1·27
Cinnamic acid (trans)	$C_6H_5 \cdot CH:CH \cdot COOH$	monoclinic	1·25
Citric acid	$C_3H_4 \cdot OH \cdot (COOH)_3 \cdot H_2O$	orthorhombic	1·54
Cresol (o)	$C_6H_4 \cdot OH \cdot CH_3$		1·047
Dicyandiamide	$NH_2 \cdot C(:NH) \cdot NH \cdot CN$	orthorhombic	1·40
Fructose (D)	$C_6H_{12}O_6$	orthorhombic	1·60
Fumaric acid (trans)	$HOOC \cdot CH:CH \cdot COOH$	monoclinic	1·64
Glucose (dextrose)	$C_6H_{12}O_6 \cdot H_2O$	orthorhombic	1·56
Glutamic acid (D)	$(CH_2)_2 \cdot CH \cdot NH_2 \cdot (COOH)_2$		1·54
Glycine	$CH_2 \cdot NH_2 \cdot COOH$	monoclinic	0·83
Hydroquinone (p)	$C_6H_4(OH)_2$	monoclinic	1·33
o-Hydroxybenzoic acid	$C_6H_4 \cdot OH \cdot COOH$	monoclinic	1·44
m-Hydroxybenzoic acid	$C_6H_4 \cdot OH \cdot COOH$	orthorhombic	1·47
p-Hydroxybenzoic acid	$C_6H_4 \cdot OH \cdot COOH$	orthorhombic	1·47
Lactose	$C_{12}H_{22}O_{11} \cdot H_2O$	monoclinic	1·59
Maleic acid	$HOOC \cdot CH:CH \cdot COOH$	monoclinic	1·59
Malic acid	$HOOC \cdot CH_2 \cdot CH(OH) \cdot COOH$		1·60
Maltose	$C_{12}H_{22}O_{11} \cdot H_2O$		1·54
Melamine	$C_3N_3(NH_2)_3$	monoclinic	1·57
Mannitol (D)	$(CH_2OH)_2(CHOH)_4$	orthorhombic	1·49
Naphthalene	$C_{10}H_8$	monoclinic	1·15
Oxalic acid	$(COOH)_2 \cdot 2H_2O$	monoclinic	1·90
Pentaerythritol	$C(CH_2OH)_4$	tetragonal	1·40
Phenol	C_6H_5OH	orthorhombic	1·07
Phthalic acid (o)	$C_6H_4(COOH)_2$	monoclinic	1·59
Pyrocatechol (o)	$C_6H_4(OH)_2$	monoclinic	1·37
Raffinose	$C_{18}H_{32}O_{16} \cdot 5H_2O$	orthorhombic	1·47
Resorcinol (m)	$C_6H_4(OH)_2$	orthorhombic	1·27
Salicylic acid (o)	$C_6H_4 \cdot OH \cdot COOH$	monoclinic	1·44
Salol (o)	$C_6H_4 \cdot OH \cdot COOC_6H_5$	orthorhombic	1·26
Succinic acid	$(CH_2 \cdot COOH)_2$	monoclinic	1·57
Succinimide	$(CH_2CO)_2NH$	orthorhombic	1·42
Sucrose	$C_{12}H_{22}O_{11}$	monoclinic	1·59
Sulphanilic acid (p)	$C_6H_4 \cdot NH_2 \cdot SO_3H$		1·49
Tartaric acid (racemic)	$(CHOH \cdot COOH)_2 \cdot H_2O$	monoclinic	1·79
Thiourea	$NH_2 \cdot CS \cdot NH_2$	orthorhombic	1·41
Triglycine sulphate	$(NH_2CH_2COOH)_3H_2SO_4$	monoclinic	1·68
Urea	$NH_2 \cdot CO \cdot NH_2$	tetragonal	1·33
Uric acid	$C_5H_4O_3N_4$		1·89

Table A.8. DENSITIES OF SOME SATURATED AQUEOUS SOLUTIONS

Densities (g/cm³) at different temperatures (°C)

	Solute	0	10	20	30	40	50	60	70	80	90
Aluminium	potassium sulphate	1·03	1·04	1·05	1·07	1·10	1·14	1·21	1·29	1·44	
	sodium sulphate	1·26	1·30	1·31	1·33	1·31	1·32				
	thallium sulphate	1·03	1·04	1·06	1·08	1·12	1·18	1·28			
Ammonium	aluminium sulphate	1·03	1·04	1·05	1·06	1·09	1·12	1·17	1·21	1·29	
	benzoate			1·03	1·05						
	bromide		1·28	1·30							
	chloride	1·066	1·072	1·076	1·069	1·086	1·086	1·086	1·086	1·087	1·102
	dihydrogen phosphate	1·13	1·14	1·16	1·18	1·21	1·24	1·27	1·30		
	hydrogen phosphate		1·34								
	iodate		1·01	1·02							
	nitrate	1·26	1·29	1·31	1·33	1·35					
	oxalate		1·017	1·017	1·021	1·029	1·034				
	perchlorate	1·06	1·08	1·10	1·12	1·13	1·15	1·16	1·17	1·19	1·20
	salicylate			1·14	1·15						
	sulphate	1·243	1·245	1·246	1·248	1·248	1·252	1·252	1·255		
	sulphite	1·18	1·19	1·20	1·21	1·22	1·23	1·24	1·26	1·27	
Barium	bromide		1·70	1·73	1·75	1·76	1·80		1·90		2·10
	bromate		1·001	1·002	1·004						
	chlorate		1·23	1·27	1·31	1·32	1·33				
	chloride	1·20	1·27	1·29	1·32	1·36	1·40	1·43	1·47	1·51	1·55
	iodate			0·99	0·98						
	iodide	2·07	2·14	2·22	2·29	2·30	2·32	2·33	2·35		
	nitrate	1·04	1·05	1·06	1·09	1·10	1·12	1·14	1·15		
	perchlorate	1·78		1·91		2·01		2·07		2·11	2·14
Beryllium	sulphate	1·27	1·29	1·31		1·32		1·35			
Cadmium	bromide	1·44	1·57	1·73	1·87	1·97	1·96	1·95	1·94		1·93
	chloride	1·64	1·69	1·74	1·78	1·84	1·83	1·82	1·81	1·82	1·82
	iodide	1·57		1·59		1·62		1·65		1·68	1·71
	nitrate		1·78	1·78	1·76	1·77	1·78	1·79		1·83	1·85
	perchlorate			1·75							
	potassium bromide	1·72	1·74	1·84	1·92	1·97	2·03	2·08	2·13	2·17	2·21
	potassium chloride	1·20	1·23	1·26	1·30	1·33	1·39		1·44	1·45	1·47

continued

Table A.8. CONTINUED

Solute		Densities (g/cm³) at different temperatures (°C)									
		0	10	20	30	40	50	60	70	80	90
Caesium	bromate			1·02							
	bromide			1·69							
	chlorate			1·04							
	chloride	1·84	1·88	1·91	1·94	1·96	1·98	2·00	2·02	2·04	2·05
	iodate		1·01	1·02							
	iodide			1·54							
	nitrate	1·07	1·10	1·16	1·22	1·29	1·37	1·46	1·55	1·64	1·73
	sulphate	1·98	1·99	2·01	2·02	2·03	2·04	2·05	2·06	2·07	2·08
	aluminium sulphate	1·002	1·002	1·002	1·001	0·999	0·999	1·000	1·006	1·018	1·051
	perchlorate	1·00		1·01							
Calcium	chloride			1·74	1·80						
	chloride	1·36	1·38	1·43	1·49	1·51					
	ferrocyanide			1·35	1·37	1·38					
	nitrate			1·46	1·62						
	sulphate			0·999	0·998	0·996					
Cobalt	chlorate			1·86							
	nitrate	1·56	1·57	1·57	1·58	1·59					
	perchlorate			1·57							
Cupric	acetate	1·01	1·11	1·05	1·16						
	ammonium sulphate			1·13							
	bromide			1·84							
	chloride			1·70							
	nitrate			1·69							
	sulphate	1·15	1·18	1·20	1·23						
Ferric	chloride	1·61	1·63	1·65	1·66	1·68	1·70	1·71			
	perchlorate	1·14	1·16	1·19	1·22	1·25	1·29	1·32	1·35		
Ferrous	ammonium sulphate		1·47	1·44							
	chloride			1·50	1·51		1·57	1·58			
Lead	nitrate	1·54	1·55	1·56	1·57	1·58					
	perchlorate	1·14	1·20	1·26	1·41	1·60	1·88				
	acetate					1·26	1·30	1·34	1·40	1·43	1·50
	bromate			1·01	1·79		1·40	1·41	1·41		

		1·000	1·000	0·999	0·997	0·993	0·989	0·985	0·980	0·976	0·969
Lithium	iodide	1·000									
	nitrate	1·30	1·35	1·42	1·48	1·55	1·59	1·63			
	benzoate			1·10							
	bromate			1·83							
	chloride	1·27	1·28	1·29	1·30	1·30	1·31				
	citrate			1·21							
	fluoride			1·002	0·998	0·994					
	formate			1·14							
	iodate			1·56	1·55						
	iodide			1·32	1·83						
	nitrite			1·26	1·28	1·30					
	perchlorate	1·22	1·24	1·20	1·21						
	salicylate			1·61							
Magnesium	bromate	1·51	1·56	1·60	1·66	1·72	1·79	1·90	1·99	2·08	
	chlorate			1·34							
	chloride			1·42	1·50	1·54	1·57	1·61	1·61		
	chromate			1·07	1·08	1·09					
	iodate	1·04	1·06	1·15	1·16						
	molybdate	1·12	1·14	1·29	1·33	1·35	1·38	1·41	1·42	1·43	1·46
	sulphate	1·24	1·26	1·39		1·44	1·47	1·50			
Manganese	chloride		1·47	1·49	1·51	1·53					
	sulphate			1·05							
Mercuric	bromide			1·05							
	chloride		1·04	1·55	1·06	1·07					
Nickel	ammonium sulphate	1·04	1·03								
	chlorate		1·85								
	perchlorate	1·57	1·58	1·58	1·59	1·592					
	sulphate		1·31	1·36	1·40	1·42					
Potassium	acetate	1·398	1·400	1·403	1·41						
	aluminium sulphate	1·03	1·04	1·05	1·07	1·07	1·09	1·12	1·15	1·18	1·23
	bicarbonate			1·17							
	bromate	1·02	1·04	1·05	1·06						
	bromide	1·32	1·34	1·37	1·38	1·40					
	carbonate	1·546	1·549	1·559	1·557						
	chlorate	1·02	1·03	1·05	1·06						
	chloride	1·153	1·164	1·174	1·182	1·188	1·193	1·198	1·202	1·205	1·207
	citrate			1·51							
	chromate			1·38							

continued

Table A.8. CONTINUED

Solute		Densities (g/cm³) at different temperatures (°C)									
		0	10	20	30	40	50	60	70	80	90
Potassium	dichromate	1·11	1·06	1·07	1·17	1·20	1·23	1·26	1·31		
	dihydrogen phosphate	1·15	1·12	1·15	1·20	1·21	1·22	1·23			
	ferricyanide		1·16	1·18	1·19	1·21	1·24	1·26			
	ferrocyanide		1·13	1·15					1·27	1·29	1·30
	formate			1·58							
	iodate	1·04	1·05	1·06	1·08		1·11		1·46	1·48	1·50
	iodide	1·67	1·69	1·71	1·73	1·75	1·76	1·77	1·78	1·80	1·81
	nitrate	1·05	1·11	1·17	1·22	1·28	1·33	1·39	1·44	1·49	1·53
	oxalate			1·21	1·22		1·25				
	perchlorate	1·005	1·007	1·009	1·016	1·022	1·026	1·032	1·04	1·05	1·06
	permanganate			1·03	1·05						
	sodium carbonate		1·30	1·39							
	sodium tartrate			1·30	1·31						
	sulphate	1·06	1·07	1·08	1·09	1·10	1·105	1·110	1·114	1·117	1·119
Rubidium	aluminium sulphate	1·01	1·01	1·01	1·02	1·02	1·03	1·05	1·08		
	bromide			1·62							
	chlorate			1·04							
	chloride	1·44	1·47	1·49	1·51	1·53	1·54	1·55	1·57	1·58	1·59
	iodate			1·03							
	iodide			1·84							
	nitrate	1·08	1·20	1·32	1·43	1·55	1·67	1·77	1·86	1·94	2·01
	perchlorate	1·007	1·008	1·010	1·013	1·017	1·022	1·03	1·04	1·05	1·06
	sulphate	1·27	1·31	1·34	1·37	1·39	1·41	1·43	1·44	1·45	1·47
Silver	acetate			1·004	1·006						
	bromate			0·999		0·993					
	chlorate			1·12	1·13						
Sodium	acetate	1·158	1·162	1·17	1·19	1·20	1·23	1·26			
	ammonium sulphate		1·17	1·18							
	benzene sulphonate			1·07	1·08						
	benzoate			1·15	1·16						
	bicarbonate		1·018	1·025	1·035	1·040					

	1·209	1·204	1·200	1·196	1·191	1·187	1·183	1·178	1·175	1·170
chloride	1·06	1·09		1·12	1·14	1·19	1·22	1·26	1·28	1·29
citrate			1·26	1·28						
chromate			1·43							
dichromate			1·74							
dihydrogen phosphate			1·39							
ferrocyanide			1·11							
formate			1·32							
hydrogen phosphate			1·05							
iodate	1·02	1·04	1·07	1·09	1·10	1·12	1·14	1·16	1·190	1·192
iodide	1·86	1·88	1·91	1·94	1·97	2·01	2·16	2·15		2·14
molybdate			1·44							
nitrate	1·35	1·37	1·38	1·40	1·41	1·43	1·44	1·46	1·48	1·49
nitrite		1·34	1·36							
oxalate			1·02							
perchlorate		1·65	1·67	1·69	1·71	1·75				
periodate		1·04	1·05							
phosphate	1·068	1·071	1·09							
salicylate			1·23							
sulphate	1·04	1·08	1·15							
Strontium bromate	1·18	1·20	1·26							
chlorate	1·828	1·829	1·840	1·831	1·833	1·837	1·842	1·845	1·847	1·853
chloride	1·33	1·35	1·38	1·42						
nitrate	1·28	1·36	1·44	1·514	1·513	1·511	1·510	1·510	1·511	1·512
nitrite			1·45							
salicylate			1·01	1·02						
Thallium chloride	1·001	1·002	1·001	0·999	0·998	0·995	0·992	0·989	0·986	0·982
dihydrogen phosphate	2·54	2·69	2·83	2·95	3·15	3·20	2·79	4·17	4·69	5·71
hydroxide	1·23		1·32	1·34		2·19				
perchlorate	1·06	1·08	1·10	1·14	1·19	1·23	1·33	1·42	1·52	
nitrate	1·03	1·06	1·08	1·11	1·16		1·33	1·48	1·72	2·14
sulphate	1·02	1·03	1·04	1·05	1·06	1·07	1·08	1·09	1·10	1·11
Uranyl chloride			2·75							
Zinc acetate			1·16	1·17						
benzene sulphonate			1·18	1·19						
sulphate	1·38		1·46	1·52						
Sucrose					1·59		1·59	1·55	1·51	1·49
Urea			1·32		1·33	1·15	1·17		1·35	

Table A.9. DENSITIES OF SOME AQUEOUS INORGANIC SALT SOLUTIONS AT 20 °C

Solution concentration is expressed as a percentage by weight (g anhydrous solute/100g of solution)

% by wt.	NH₄Cl	NH₄NO₃	NH₄OH	(NH₄)₂SO₄	NH₄H₂PO₄	NH₄Al(SO₄)₂	(NH₄)₂Ni(SO₄)₂	BaCl₂
1	1·003	1·002	0·998	1·006	1·006	1·009	1·008	1·004
2	1·006	1·006	0·996	1·012	1·012	1·018	1·016	1·018
3	1·009	1·010	0·994	1·018	1·018	1·028	1·025	1·027
4	1·013	1·015	0·992	1·024	1·023	1·037	1·037	1·036
5	1·016	1·019	0·989	1·030	1·029	1·046	1·043	1·045
10	1·030	1·040	0·979	1·059	1·058			1·094
12	1·036	1·048	0·975	1·071	1·070			1·115
14	1·042	1·057	0·972	1·083	1·081			1·136
16	1·048	1·065	0·968	1·094	1·093			1·158
18	1·053	1·074	0·964	1·106	1·104			1·181
20	1·059	1·083	0·960	1·117	1·115			1·205
22	1·064	1·092	0·957	1·128	1·128			1·230
24	1·069	1·10	0·953	1·140	1·141			1·255
26		1·11	0·950	1·150				1·282
28		1·12	0·946	1·164				
30		1·13	0·943	1·176				
35			0·935					
40			0·926					

% by wt.	CaCl₂	CsCl	CoCl₂	CuSO₄	FeCl₃	Pb(NO₃)₂	LiCl	MgCl₂
1	1·008	1·008	1·009	1·010	1·009	1·009	1·006	1·008
2	1·017	1·015	1·018	1·021	1·017	1·018	1·012	1·016
3	1·025	1·023	1·027	1·031	1·026	1·027	1·018	1·025
4	1·033	1·031	1·032	1·042	1·034	1·038	1·023	1·033
5	1·042	1·040	1·046	1·053	1·043	1·045	1·029	1·041
10	1·085	1·082	1·096	1·109	1·087	1·093	1·058	1·084
12	1·103	1·100	1·117	1·132	1·106	1·114	1·069	1·101
14	1·122	1·118	1·138	1·157	1·125	1·135	1·081	1·118
16	1·141	1·138	1·161	1·182	1·144	1·157	1·093	1·136
18	1·160	1·157	1·184	1·208	1·164	1·180	1·105	1·154
20	1·180	1·178	1·207		1·184	1·204	1·118	1·173
22	1·200	1·199			1·204	1·229	1·129	1·192
24	1·220	1·221			1·225	1·254	1·142	1·211
26		1·243			1·247	1·280	1·155	1·230
28		1·267			1·270	1·305	1·168	1·250
30		1·291			1·293	1·330	1·181	1·271
35		1·35			1·33			

% by wt.	MgSO$_4$	MnSO$_4$	NiSO$_4$	KBr	K$_2$CO$_3$	KCl	K$_2$CrO$_4$	K$_2$Cr$_2$O$_7$
1	1·010	1·010	1·011	1·007	1·009	1·006	1·008	1·007
2	1·020	1·020	1·022	1·015	1·018	1·013	1·016	1·014
3	1·031	1·030	1·032	1·022	1·027	1·019	1·024	1·021
4	1·041	1·040	1·043	1·029	1·036	1·026	1·033	1·028
5	1·052	1·050	1·054	1·037	1·045	1·032	1·041	1·035
10	1·105	1·103		1·076	1·092	1·065	1·084	1·072
12	1·128	1·126		1·092	1·112	1·079	1·101	
14	1·150	1·149		1·109	1·131	1·092	1·120	
16	1·174	1·173		1·126	1·151	1·106	1·138	
18	1·198	1·197		1·144	1·171	1·120	1·157	
20	1·222	1·222		1·162	1·192	1·135	1·177	
22	1·247			1·181	1·213	1·149	1·196	
24	1·272			1·200	1·234	1·164	1·217	
26	1·298			1·220	1·256		1·238	
28				1·240	1·278		1·259	
30				1·262	1·300		1·281	
35				1·317	1·360		1·340	
40				1·377	1·417		1·399	

% by wt.	KI	KNO$_3$	K$_2$C$_2$O$_4$	KH$_2$PO$_4$	K$_2$HPO$_4$	K$_2$SO$_4$	KAl(SO$_4$)$_2$	AgNO$_3$
1	1·007	1·006	1·007	1·007	1·009	1·008	1·009	1·009
2	1·015	1·013	1·015	1·014	1·017	1·016	1·019	1·017
3	1·022	1·019	1·022	1·022	1·026	1·024	1·029	1·026
4	1·030	1·025	1·029	1·029	1·034	1·032	1·039	1·035
5	1·038	1·032	1·037	1·036	1·043	1·041	1·049	1·043
10	1·078	1·065	1·075	1·072	1·070	1·083		1·090
12	1·095	1·078						1·110
14	1·112	1·092						1·130
16	1·130	1·106						1·152
18	1·149	1·120						1·174
20	1·168	1·135						1·196
22	1·188	1·149						1·220
24	1·208	1·164						1·244
26	1·229							1·269
28	1·251							1·296
30	1·273							1·323
35	1·336							1·400
40	1·398							1·477

continued

Table A.9. CONTINUED

% by wt.	CH₃COONa	NaHCO₃	NaBr	Na₂CO₃	NaCl	NaOH	Na₂MoO₄	NaNO₃
1	1·005	1·007	1·008	1·010	1·007	1·011	1·009	1·007
2	1·010	1·015	1·016	1·021	1·014	1·022	1·017	1·014
3	1·015	1·022	1·024	1·031	1·021	1·033	1·026	1·020
4	1·021	1·029	1·032	1·042	1·029	1·045	1·035	1·027
5	1·026	1·036	1·040	1·052	1·036	1·056	1·044	1·034
10	1·051		1·082	1·105	1·073	1·111		1·069
12	1·062		1·100	1·126	1·088	1·133		1·084
14	1·072		1·118	1·148	1·103	1·155		1·090
16	1·083		1·137	1·168	1·118	1·177		1·114
18	1·093		1·157		1·134	1·199		1·129
20	1·104		1·177		1·150	1·221		1·145
22	1·115		1·197		1·166	1·243		1·161
24	1·126		1·218		1·183	1·265		1·177
26	1·137		1·240		1·199	1·287		1·194
28	1·148		1·263			1·309		1·211
30	1·160		1·286			1·330		1·228
35			1·350			1·380		1·270
40			1·416			1·432		1·320

% by wt.	Na₂HPO₄	Na₂SO₄	Na₂C₄H₄O₆	Na₂S₂O₃	Na₂WO₄	SrCl₂	ZnSO₄
1	1·010	1·009	1·007	1·008	1·009	1·009	
2	1·020	1·018	1·014	1·017	1·018	1·018	1·021
3	1·030	1·027	1·021	1·025	1·028	1·027	1·031
4	1·040	1·037	1·028	1·033	1·037	1·036	1·042
5	1·050	1·046	1·036	1·042	1·047	1·046	1·053
10		1·093	1·072	1·085		1·094	1·109
12		1·113	1·087	1·102		1·115	1·133
14		1·133	1·102	1·120		1·136	1·157
16		1·153	1·118	1·139		1·158	1·183
18		1·173	1·133	1·157		1·180	
20		1·194	1·149	1·176		1·203	
22		1·215	1·165	1·195		1·226	
24		1·236	1·182	1·215		1·250	
26			1·198	1·235		1·275	
28			1·216	1·255		1·301	
30				1·276		1·327	
35				1·325			

Table A.10. VISCOSITIES OF SOME AQUEOUS SOLUTIONS AT 25 °C

Solution concentrations are quoted in molality, m, i.e. gram-formula weight (anhydrous) per kg of water. Values in the final column marked with an asterisk (*) refer to saturated solutions at 25 °C.

$$(1 \text{ cP} = 0.01 \text{ g/cm s} = 10^{-3} \text{ kg/m s} = 10^{-3} \text{ N s/m}^2)$$

Solute		Viscosity (cP) at 25 °C			
		$0.5\,m$	$1\,m$	$5\,m$	$(x)\,m$
Ammonium	aluminium sulphate				1·26*
	acetate	1·0	1·1		
	bromide	0·88	0·86	0·84	(6) 0·85
	chloride	0·88	0·88	0·91	(6) 0·92
	chromate	0·95	1·03		
	dichromate	0·92	0·95		
	dihydrogen phosphate	1·05	1·16	1·59	(3) 2·0
	iodide	0·86	0·83	0·80	(6) 0·82
	nitrate	0·88	0·87	0·90	(6) 0·93
	sulphate	0·98	1·36	2·1	
Barium	acetate	1·3	1·8		
	chloride	1·0	1·1		1·3*
	nitrate		0·97		
Beryllium	sulphate	1·2			
Cadmium	chloride	1·00	1·02		
	nitrate	1·03	1·03		
	sulphate	1·20	1·19		
Calcium	acetate	1·3	1·9		
	chloride	1·03	1·2	4·1	(6) 7·6
	chromate	1·13			
	nitrate	1·00	1·15		(3) 2·2
Cobalt	chloride	1·07	1·27		
	nitrate	1·03			
	sulphate	1·2			
Copper (cupric)	chloride	1·07	1·27		(3) 2·3
	nitrate		1·26	4·6	
	sulphate	1·21	1·67		
Ferric	chloride	1·21	1·64	10·0	
Lead	acetate	1·15	1·39		
	nitrate	0·98	1·12		($1\frac{1}{2}$) 1·3
Lithium	acetate	1·1	1·3		
	bromate	0·90	1·04		
	chlorate		1·01	1·78	(35) 54
	chloride	0·95	1·01	1·69	(20) 17
	iodate		1·23		(3) 2·7
	hydroxide	1·0	1·1		(4) 2·6
	nitrate	0·92	0·99	1·5	(9) 2·4
	sulphate	1·15	1·48		

continued

Table A.10. CONTINUED

Solute		Viscosity (cP) at 25 °C			
		0·5 m	1 m	5 m	(x) m
Magnesium	chloride	1·07	1·3		(3) 3·0
	nitrate	1·04	1·22		(3) 2·5
	sulphate	1·22	1·72		(2) 3·4
Manganese	chloride	1·07	1·28		(3) 2·6
	nitrate	1·04	1·22		(3) 2·2
	sulphate	1·21	1·69		(3) 7·1
Nickel	ammonium sulphate				1·10*
	chloride	1·07			
	nitrate	1·05			
	sulphate	1·21			
Potassium	acetate	1·01	1·10		(4) 1·94
	aluminium sulphate				1·22*
	bromide	0·88	0·86	0·94	
	carbonate	1·04	1·19		(4) 2·9
	chlorate	0·89			
	chloride	0·89	0·89		(4) 0·94
	chromate	0·96	1·07		(2) 1·31
	bicarbonate	0·94	1·00		(2) 1·12
	bichromate				$(\frac{1}{2})$ 0·90
	bisulphate	0·95	1·01		(2) 1·17
	ferricyanide	0·98			
	ferrocyanide	1·10			
	fluoride		1·01	1·42	(6) 1·85
	hydroxide		1·01		(6) 1·95
	iodide	0·85	0·83	0·85	(9) 1·07
	nitrate	0·87	0·87		(2) 0·89
	oxalate	0·99	1·09		
Potassium	phosphate	1·13	1·53		
	hydrogen phosphate	1·07	1·34		(2) 1·94
	dihydrogen phosphate	1·11	1·15		
	sulphate	0·99			1·16*
	thiocyanate	0·87	0·85	0·93	
Silver	nitrate	0·98	0·94	1·33	(12) 2·4
Sodium	acetate	1·05	1·22		
	benzoate	1·12	1·41		
	bromate	0·94	0·99		
	bromide	1·19			
	carbonate	1·15	1·51		(2) 2·58
	bicarbonate	0·98	1·09		
	chlorate	0·93	0·96	1·52	
	perchlorate		0·93		
	chloride	0·97	0·97	1·52	
	chromate	1·11	1·46		
	dihydrogen phosphate	1·17	1·30		
	formate	0·97	1·07		
	lactate	1·09	1·31		

Table A.10. CONTINUED

| Solute | | Viscosity (cP) at 25 °C | | | |
		0·5 m	1 m	5 m	(x) m
	iodide	0·90	0·92	1·16	(8) 1·81
	nitrate	0·91	0·94	1·42	
	sulphate	1·10	1·38		1·25*
	bisulphate	0·99	1·11	2·55	
	propionate	1·09	1·34		
	salicylate	1·08	1·31		
	tartrate	1·17	1·61		
	thiosulphate				5·5*
Strontium	acetate	1·30	1·90		
	chloride	1·02	1·19		(3) 2·28
	nitrate	1·00	1·02		
Thallous	nitrate		0·85		
Zinc	chloride	1·06			
	nitrate	1·04			
	sulphate	1·22	1·69		(2) 3·36

Table A.11. DIFFUSION COEFFICIENTS OF CONCENTRATED AQUEOUS ELECTROLYTE SOLUTIONS AT 25 °C $(D \times 10^{-5}\ cm^2/s)$

mol/l	$BaCl_2$	$CaCl_2$	KCl	KBr	KI	LiCl	NaCl
0	1·385	1·335	1·993	2·016	1·996	1·366	1·610
0·05	1·18	1·12	1·86	1·89	1·89	1·28	1·51
0·1	1·16	1·11	1·84	1·87	1·86	1·26	1·48
0·2	1·15	1·11	1·84	1·87	1·85	1·26	1·48
0·3	1·15	1·12	1·84	1·87	1·88	1·26	1·48
0·5	1·16	1·14	1·85	1·88	1·95	1·27	1·47
0·7	1·17	1·17	1·87	1·91	2·00	1·28	1·47
1·0	1·18	1·20	1·89	1·97	2·06	1·30	1·48
1·5	1·18	1·26	1·94	2·06	2·16	1·33	1·50
2·0		1·31	2·00	2·13	2·25	1·36	1·52
2·5		1·31	2·06	2·19	2·34	1·39	1·54
3·0		1·27	2·11	2·28	2·44	1·43	1·57
3·5		1·20	2·16	2·35	2·53	1·46	1·58
4·0			2·20	2·43			1·59
4·5							1·59
5·0							1·59

mol/l	NaBr	NaI	NH_4Cl	NH_4NO_3	$(NH_4)_2SO_4$
0	1·625	1·614	1·994	1·929	1·530
0·05	1·53	1·52		1·79	0·80
0·1	1·51	1·52	1·84	1·77	0·83
0·2	1·50	1·53	1·84	1·75	0·87
0·3	1·51	1·54	1·84	1·74	0·90
0·5	1·54	1·58	1·86	1·72	0·94
0·7	1·56	1·61	1·88	1·71	0·97
1·0	1·59	1·66	1·92	1·69	1·01
1·5	1·62	1·75	1·99	1·66	1·05
2·0	1·66	1·84	2·05	1·63	1·07
2·5	1·70	1·92	2·11	1·61	1·09
3·0		1·99	2·16	1·58	1·11
3·5			2·20	1·55	1·12
4·0			2·24	1·52	1·14
4·5			2·26	1·50	
5·0			2·26	1·47	

The values at zero concentration are the Nernst limiting values (see p. 71).

Table A.12. ACTIVITY COEFFICIENTS OF ELECTROLYTES AT 25 °C

(molality = gram-formula weight of anhydrous solute per kg of water)

Molality	$AlCl_3$	$Al_2(SO_4)_3$	NH_4Cl	NH_4NO_3	$(NH_4)_2SO_4$	$BaCl_2$	$BaBr_2$	BaI_2
0·1	0·337	0·0350	0·770	0·740	0·423	0·508	0·517	0·536
0·2	0·305	0·0225	0·718	0·677	0·343	0·450	0·269	0·503
0·4	0·313	0·0153	0·665	0·606	0·270	0·411	0·440	0·504
0·6	0·356	0·0140	0·636	0·562	0·231	0·397	0·442	0·534
0·8	0·429	0·0149	0·617	0·530	0·206	0·397	0·452	0·581
1·0	0·539	0·0175	0·603	0·504	0·189	0·401	0·473	0·642
1·5	1·11		0·581	0·455	0·160	0·432	0·552	0·860
2·0			0·570	0·419	0·144		0·661	1·21
2·5			0·564	0·391	0·132			
3·0			0·561	0·368	0·125			
4·0			0·560	0·331	0·116			
5·0			0·562	0·302				
6·0			0·564	0·279				

Molality	$BeSO_4$	$CdCl_2$	$CdBr_2$	$CdSO_4$	$CaCl_2$	$CaBr_2$	CaI_2	$Ca(ClO_4)_2$
0·1	0·150	0·228	0·190	0·150	0·518	0·532	0·552	0·557
0·2	0·109	0·164	0·132	0·103	0·472	0·491	0·524	0·532
0·4	0·0769	0·114	0·0890	0·0699	0·448	0·482	0·535	0·544
0·6	0·0639	0·0905	0·0699	0·0553	0·453	0·504	0·576	0·589
0·8	0·0570	0·0765	0·0591	0·0468	0·470	0·542	0·641	0·654
1·0	0·0530	0·0669	0·0518	0·0415	0·500	0·596	0·731	0·743
1·5	0·0490	0·0525	0·0416	0·0346	0·615	0·796	1·063	1·08
2·0	0·0497	0·0441	0·0361	0·0321	0·792	1·12	1·617	1·63
2·5	0·0538	0·0389	0·0328	0·0317	1·06	1·65		2·62
3·0	0·0613	0·0352	0·0305	0·0329	1·48	2·53		4·21
4·0	0·0875	0·0306	0·0278		2·93	6·27		10·8
5·0		0·0279			5·89	18·4		26·7
6·0		0·0263			11·1	55·7		63·7

continued

Table A.12. CONTINUED

Molality	Ca(NO₃)₂	Cr₂(SO₄)₃	CoI₂	Co(NO₃)₂	CuSO₄	Pb(NO₃)₂	LiCl	MgCl₂
0·1	0·488	0·0458	0·56	0·521	0·150	0·405	0·790	0·528
0·2	0·429	0·0300	0·54	0·474	0·104	0·316	0·757	0·488
0·4	0·378	0·0207	0·57	0·448	0·0704	0·234	0·740	0·474
0·6	0·356	0·0182	0·64	0·451	0·0559	0·192	0·743	0·490
0·8	0·344	0·0185	0·74	0·468	0·0475	0·164	0·755	0·521
1·0	0·338	0·0208	0·88	0·493	0·0423	0·145	0·774	0·569
1·5	0·338		1·40	0·640		0·113	0·838	0·800
2·0	0·347		2·3	0·730		0·092	0·921	1·05
2·5	0·362		4·3	0·926			1·03	1·54
3·0	0·382		7·4	1·19			1·16	2·32
4·0	0·438		23	1·98			1·51	5·53
5·0	0·510		60	3·33			2·02	13·9
6·0	0·596		99				2·72	

Molality	MgBr₂	MgI₂	Mg(ClO₄)₂	Mg(NO₃)₂	MgSO₄	MnSO₄	NiCl₂	NiSO₄
0·1	0·542	0·571	0·577	0·522	0·150	0·150	0·523	0·150
0·2	0·512	0·550	0·565	0·480	0·107	0·105	0·479	0·105
0·4	0·520	0·575	0·599	0·465	0·0756	0·0725	0·460	0·0713
0·6	0·564	0·643	0·673	0·478	0·0616	0·0578	0·471	0·0562
0·8	0·627	0·742	0·780	0·501	0·0536	0·0493	0·496	0·0478
1·0	0·714	0·879	0·925	0·536	0·0485	0·0493	0·536	0·0425
1·5	1·05	1·41	1·51	0·661	0·0428	0·0372	0·683	0·0360
2·0	1·59	2·39	2·59	0·835	0·0417	0·0351	0·906	0·0343
2·5	2·56	4·27	4·78	1·09	0·0439	0·0349	1·24	0·0357
3·0	4·20	7·81	8·99	1·45	0·0492	0·0373	1·69	
4·0	12·0	28·6	33·3	2·59		0·0473	2·96	
5·0	36·1	113		4·74			4·69	
6·0								

Molality	KOH	KCl	KBr	KI	$KClO_3$	$KBrO_3$	KNO_3	K acetate
0·1	0·776	0·770	0·772	0·778	0·749	0·745	0·739	0·796
0·2	0·739	0·718	0·722	0·733	0·681	0·674	0·663	0·766
0·4	0·713	0·666	0·673	0·689	0·599	0·585	0·576	0·750
0·6	0·712	0·637	0·646	0·667	0·541		0·519	0·754
0·8	0·721	0·618	0·629	0·654			0·476	0·766
1·0	0·735	0·604	0·617	0·645			0·443	0·783
1·5	0·791	0·583	0·600	0·636			0·380	0·840
2·0	0·863	0·573	0·593	0·637			0·333	0·910
2·5	0·947	0·569	0·593	0·644			0·297	0·995
3·0	1·05	0·569	0·595	0·652			0·269	1·09
4·0	1·31	0·577	0·608	0·673				
5·0	1·67		0·626					
6·0	2·14							

Molality	KCNS	KH_2PO_4	K_2CrO_4	K_2SO_4	$K_3Fe(CN)_6$	$K_4Fe(CN)_6$	$AgNO_3$	NaOH
0·1	0·769	0·731	0·466	0·436	0·268	0·139	0·734	0·764
0·2	0·716	0·653	0·390	0·356	0·212	0·0993	0·657	0·725
0·4	0·663	0·561	0·320	0·283	0·167	0·0693	0·567	0·695
0·6	0·633	0·501	0·282	0·243	0·146	0·0556	0·509	0·683
0·8	0·614	0·456	0·259		0·135	0·0479	0·464	0·677
1·0	0·599	0·421	0·240		0·128		0·429	0·677
1·5	0·573	0·358	0·215				0·363	0·687
2·0	0·556		0·200				0·316	0·707
2·5	0·546		0·194				0·280	0·741
3·0	0·538		0·194				0·252	0·782
4·0	0·529						0·210	0·901
5·0	0·524						0·181	1·07
6·0							0·159	1·30

continued

Table A.12. CONTINUED

Molality	NaCl	NaBr	NaI	NaClO$_3$	NaClO$_4$	Na$_2$SO$_4$	Na$_2$CrO$_4$	Na$_2$S$_2$O$_3$
0·1	0·778	0·782	0·787	0·772	0·775	0·452	0·479	0·466
0·2	0·735	0·741	0·751	0·720	0·729	0·371	0·407	0·390
0·4	0·693	0·704	0·727	0·664	0·683	0·294	0·337	0·319
0·6	0·673	0·692	0·723	0·630	0·656	0·252	0·301	0·282
0·8	0·662	0·687	0·727	0·606	0·641	0·225	0·278	0·256
1·0	0·657	0·687	0·736	0·589	0·629	0·204	0·261	0·239
1·5	0·656	0·703	0·771	0·568	0·614	0·173	0·237	0·219
2·0	0·668	0·731	0·820	0·538	0·609	0·154	0·229	0·202
2·5	0·688	0·768	0·883	0·525	0·609	0·144	0·232	0·199
3·0	0·714	0·812	0·963	0·515	0·611	0·139	0·244	0·203
4·0	0·783	0·929			0·626	0·138	0·294	
5·0	0·874				0·649			
6·0	0·986				0·677			

Molality	NaBrO$_3$	NaNO$_3$	Na acetate	NaCNS	NaH$_2$PO$_4$	SrCl$_2$	SrBr$_2$	SrI$_2$
0·1	0·758	0·762	0·791	0·787	0·744	0·515	0·527	0·549
0·2	0·696	0·703	0·757	0·750	0·675	0·466	0·483	0·516
0·4	0·628	0·638	0·737	0·720	0·593	0·436	0·465	0·520
0·6	0·585	0·599	0·736	0·712	0·539	0·434	0·473	0·551
0·8	0·554	0·570	0·745	0·710	0·499	0·445	0·497	0·603
1·0	0·528	0·548	0·757	0·712	0·468	0·465	0·535	0·675
1·5	0·491	0·507	0·799	0·727	0·410	0·549	0·679	0·946
2·0	0·450	0·478	0·851	0·744	0·371	0·675	0·906	1·40
2·5	0·426	0·455	0·914	0·779	0·343	0·862		
3·0		0·437	0·982	0·814	0·320	1·14		
4·0		0·408		0·897	0·293	1·99		
5·0		0·386			0·276			
6·0		0·371			0·265			

Molality	Sr(NO₃)₂	UO₂(NO₃)₂	ZnCl₂	ZnBr₂	ZnI₂	Zn(NO₃)₂	ZnSO₄
0·1	0·478	0·543	0·518	0·547	0·572	0·530	0·150
0·2	0·410	0·512	0·465	0·510	0·550	0·487	0·104
0·4	0·348	0·518	0·413	0·504	0·573	0·467	0·0715
0·6	0·314	0·555	0·382	0·519	0·635	0·478	0·0569
0·8	0·292	0·608	0·359	0·537	0·713	0·499	0·0487
1·0	0·275	0·679	0·341	0·552	0·788	0·533	0·0435
1·5	0·251	0·899	0·306	0·568	0·936	0·650	0·0370
2·0	0·232	1·22	0·291	0·572	1·01	0·814	0·0357
2·5	0·223	1·60	0·287	0·582	1·05	1·05	0·0367
3·0	0·217	2·00	0·289	0·598	1·11	1·36	0·0408
4·0	0·212	2·64	0·309	0·664	1·24	2·30	
5·0		3·01	0·356	0·774	1·45	3·86	
6·0			0·420	0·930	1·75	6·38	

Table A.13. LIMITING EQUIVALENT IONIC CONDUCTIVITIES IN WATER, λ

Cations	0	25	50	100 °C	Anions	0	25	50	100 °C
H^+	230	350	465	640	OH^-	105	195	284	440
Li^+	19	39		110	F^-		55		
Na^+	26	50	82	150	Cl^-	41	76	116	210
K^+	41	74	115	200	Br^-	43	78		
Rb^+	44	78			I^-	41	77		
Cs^+	44	77			NO_3^-	40	72	104	200
Ag^+	33	62	101	180	ClO_3^-		65		
NH_4^+	40	74	115	190	BrO_3^-		56		
Be^{2+}		45			IO_3^-		41		
Mg^{2+}	29	53		170	ClO_4^-	37	67		190
Ca^{2+}	31	60	98	190	IO_4^-		55		
Sr^{2+}	31	60			HCO_3^-		46		
Ba^{2+}	33	65	104	200	Formate$^-$		55		
Cu^{2+}		54			Acetate$^-$	20	41	67	130
Zn^{2+}		53			Propionate$^-$		36		
Co^{2+}		55			Butyrate$^-$		33		
Pb^{2+}		70			Benzoate$^-$		32		
Cd^{2+}		52			Oxalate^{2-}	39	73	115	210
					SO_4^{2-}	41	80	125	260
					CO_3^{2-}		69		
					$P_3O_9^{3-}$		84		
					$P_4O_{12}^{3-}$		94		
					Citrate^{3-}	36	70	113	210
					$Fe(CN)_6^{3-}$		100		
					$Fe(CN)_6^{4-}$	58	110	173	320
					$P_2O_7^{4-}$		96		
					$P_3O_{10}^{5-}$		110		

$\lambda_0 = cm^2 \ \Omega^{-1} \ mol^{-1}$.

Table A.14. HEATS OF SOLUTION OF INORGANIC SALTS IN WATER AT APPROXIMATELY ROOM TEMPERATURE AND INFINITE DILUTION

A positive value indicates an exothermic, a negative value an endothermic heat of solution
[1 kcal/mol = 4·1868 kJ/mol]

	Substance	Formula	Heat of solution, kcal/mol
Aluminium	ammonium sulphate	$(NH_4)Al(SO_4)_2$	+30·5
	bromide	$AlBr_3$	+86·1
	chloride	$AlCl_3$	+77·9
	chloride	$AlCl_3 \cdot 6H_2O$	+13·1
	iodide	AlI_3	+90·6
	sulphate	$Al_2(SO_4)_3$	+120
	sulphate	$Al_2(SO_4)_3 \cdot 18H_2O$	+7·0
Ammonium	acetate	$NH_4C_2H_3O_2$	+1·7
	alum	$NH_4Al(SO_4)_2 \cdot 12H_2O$	−9·6
	bicarbonate	NH_4HCO_3	−6·7
	bichromate	$(NH_4)_2Cr_2O_7$	−12·9
	bisulphate	$(NH_4)_2HSO_4$	+0·56
	bromide	NH_4Br	−4·5
	chloride	NH_4Cl	−3·8
	chromate	$(NH_4)_2CrO_4$	−5·7
	cyanide	NH_4CN	−4·6
	dihydrogen phosphate	$NH_4H_2PO_4$	−3·86
	hydrogen phosphate	$(NH_4)_2HPO_4$	−3·13
	iodate	NH_4IO_3	−7·6
	iodide	NH_4I	−3·6
	nitrate	NH_4NO_3	−6·5
	nitrite	NH_4NO_2	−4·0
	oxalate	$(NH_4)_2C_2O_4$	−8·0
	oxalate	$(NH_4)_2C_2O_4 \cdot H_2O$	−11·5
	perchlorate	NH_4ClO_4	−6·3
	sulphate	$(NH_4)_2SO_4$	−1·5
	sulphite	$(NH_4)_2SO_3 \cdot H_2O$	−4·4
	thiocyanate	NH_4CNS	−5·5
	thiosulphate	$(NH_4)_2S_2O_5$	−6·2
Barium	acetate	$Ba(C_2H_3O_2)_2$	+6·0
	bromide	$BaBr_2$	+5·2
	bromide	$BaBr_2 \cdot 2H_2O$	−3·9
	carbonate	$BaCO_3$	−1·0
	chlorate	$Ba(ClO_3)_2$	−6·7
	chlorate	$Ba(ClO_3)_2 \cdot H_2O$	−11·0
	chloride	$BaCl_2$	+3·1
	chloride	$BaCl_2 \cdot 2H_2O$	−4·5
	cyanide	$Ba(CN)_2$	+2·7
	fluoride	BaF_2	−0·9
	hydroxide	$Ba(OH)_2$	+11·4
	hydroxide	$Ba(OH)_2 \cdot 8H_2O$	−14·5
	iodide	BaI_2	+10·3
	iodide	$BaI_2 \cdot 6H_2O$	−6·6
	nitrate	$Ba(NO_3)_2$	−9·6
	nitrite	$Ba(NO_2)_2$	−4·6
	phosphate	$Ba_3(PO_4)_2$	−16·6
	perchlorate	$Ba(ClO_4)_2$	−1·3
	sulphate	$BaSO_4$	−4·6

continued

Table A.14. CONTINUED

Substance		Formula	Heat of solution, kcal/mol
Beryllium	bromide	$BeBr_2$	$+62.6$
	chloride	$BeCl_2$	$+51.1$
	iodide	BeI_2	$+72.6$
	sulphate	$BeSO_4$	$+18.1$
	sulphate	$BeSO_4 \cdot 4H_2O$	$+1.1$
Boric acid		H_3BO_3	-5.4
Cadmium	bromide	$CdBr_2$	$+0.43$
	bromide	$CdBr_2 \cdot 4H_2O$	-7.3
	chloride	$CdCl_2$	$+3.11$
	chloride	$CdCl_2 \cdot 2\frac{1}{2}H_2O$	-3.0
	cyanide	$Cd(CN)_2$	$+8.7$
	fluoride	CdF_2	$+8.7$
	iodide	CdI_2	-3.3
	nitrate	$Cd(NO_3)_2$	$+8.1$
	nitrate	$Cd(NO_3)_2 \cdot 4H_2O$	-5.1
	sulphate	$CdSO_4$	$+12.8$
Caesium	bromate	$CsBrO_3$	-12.1
	bromide	$CsBr$	-6.2
	bicarbonate	$CsHCO_3$	-4.0
	chloride	$CsCl$	-4.6
	fluoride	CsF	$+9.0$
	fluoride	$CsF \cdot H_2O$	$+2.5$
	hydroxide	$CsOH$	$+17.0$
	hydroxide	$CsOH \cdot H_2O$	$+4.9$
	iodide	CsI	-7.9
	nitrate	$CsNO_3$	-9.6
	perchlorate	$CsClO_4$	-13.3
	sulphate	Cs_2SO_4	-4.9
	sulphide	Cs_2S	$+27.3$
Calcium	acetate	$Ca(C_2H_3O_2)_2$	$+7.0$
	acetate	$Ca(C_2H_3O_2)_2 \cdot H_2O$	$+5.9$
	bromide	$CaBr_2$	$+26.3$
	chloride	$CaCl_2$	$+19.8$
	chloride	$CaCl_2 \cdot 6H_2O$	-4.1
	chromate	$CaCrO_4$	$+6.4$
	fluoride	CaF_2	-2.2
	iodide	CaI_2	$+28.0$
	iodide	$CaI_2 \cdot 8H_2O$	$+1.7$
	nitrate	$Ca(NO_3)_2$	$+4.1$
	nitrate	$Ca(NO_3)_2 \cdot 4H_2O$	-8.0
	nitrite	$Ca(NO_2)_2$	$+2.2$
	sulphate	$CaSO_4$	$+5.2$
	sulphate	$CaSO_4 \cdot 2H_2O$	-0.18
	sulphide	CaS	$+4.5$
Cerium	chloride	$CeCl_3$	$+33.6$
	iodide	CeI_3	$+49.5$
Chromium	chloride	$CrCl_2$	$+18.6$
Cobalt	bromide	$CoBr_2$	$+18.4$
	bromide	$CoBr_2 \cdot 6H_2O$	-1.3
	chloride	$CoCl_2$	$+18.5$
	chloride	$CoCl_2 \cdot 6H_2O$	-2.9
	fluoride	CoF_2	$+14.6$
	iodate	$Co(IO_3)_2$	$+1.6$

Table A.14. CONTINUED

Substance		Formula	Heat of solution, kcal/mol
	iodide	CoI_2	$+18\cdot4$
	nitrate	$Co(NO_3)_2$	$+11\cdot9$
	nitrate	$Co(NO_3)_2 \cdot 6H_2O$	$-4\cdot9$
	sulphate	$CoSO_4$	$+25\cdot0$
	sulphate	$CoSO_4 \cdot 7H_2O$	$-3\cdot6$
Copper	acetate	$Cu(C_2H_3O_2)_2$	$+2\cdot5$
	bromide	$CuBr_2$	$+8\cdot8$
	chloride	$CuCl_2$	$+12\cdot2$
	iodide	CuI_2	$+6\cdot2$
	nitrate	$Cu(NO_3)_2$	$+10\cdot4$
	nitrate	$Cu(NO_3)_2 \cdot 6H_2O$	$-10\cdot7$
	sulphate	$CuSO_4$	$+15\cdot9$
	sulphate	$CuSO_4 \cdot 5H_2O$	$-2\cdot86$
Ferric	chloride	$FeCl_3$	$+31\cdot7$
	chloride	$FeCl_3 \cdot 6H_2O$	$+5\cdot6$
Ferrous	bromide	$FeBr_2$	$+19\cdot1$
	chloride	$FeCl_2$	$+17\cdot9$
	chloride	$FeCl_2 \cdot 4H_2O$	$+2\cdot7$
	iodide	FeI_2	$+19\cdot1$
	sulphate	$FeSO_4$	$+14\cdot9$
	sulphate	$FeSO_4 \cdot 7H_2O$	$-4\cdot4$
Lead	acetate	$Pb(C_2H_3O_2)_2$	$+2\cdot1$
	acetate	$Pb(C_2H_3O_2)_2 \cdot 3H_2O$	$-5\cdot9$
	bromide	$PbBr_2$	$-8\cdot8$
	chloride	$PbCl_2$	$-6\cdot5$
	iodide	PbI_2	$-15\cdot5$
	nitrate	$Pb(NO_3)_2$	$-9\cdot0$
Lithium	bromate	$LiBrO_3$	$-0\cdot34$
	bromide	$LiBr$	$+11\cdot5$
	bromide	$LiBr \cdot H_2O$	$+5\cdot6$
	bromide	$LiBr \cdot 2H_2O$	$+2\cdot1$
	carbonate	Li_2CO_3	$+3\cdot1$
	chloride	$LiCl$	$+8\cdot8$
	chloride	$LiCl \cdot H_2O$	$+4\cdot6$
	chloride	$LiCl \cdot 3H_2O$	$-2\cdot0$
	fluoride	LiF	$-1\cdot1$
	hydroxide	$LiOH$	$+5\cdot1$
	hydroxide	$LiOH \cdot H_2O$	$+1\cdot6$
	iodide	LiI	$+14\cdot9$
	iodide	$LiI \cdot H_2O$	$+7\cdot1$
	iodide	$LiI \cdot 2H_2O$	$+3\cdot5$
	iodide	$LiI \cdot 3H_2O$	$-0\cdot17$
	nitrate	$LiNO_3$	$+0\cdot47$
	nitrate	$LiNO_3 \cdot 3H_2O$	$-7\cdot9$
	nitrite	$LiNO_2$	$+2\cdot6$
	nitrite	$LiNO_2 \cdot H_2O$	$-1\cdot7$
	perchlorate	$LiClO_4$	$+6\cdot3$
	sulphate	Li_2SO_4	$+6\cdot7$
	sulphate	$Li_2SO_4 \cdot H_2O$	$+3\cdot8$
Magnesium	bromide	$MgBr_2$	$+43\cdot5$
	bromide	$MgBr_2 \cdot 6H_2O$	$+19\cdot8$
	chloride	$MgCl_2$	$+36\cdot0$
	chloride	$MgCl_2 \cdot 6H_2O$	$+3\cdot1$

continued

Table A.14. CONTINUED

Substance		Formula	Heat of solution, kcal/mol
Magnesium	chromate	$MgCrO_4$	$+2\cdot9$
	iodide	MgI_2	$+50\cdot2$
	nitrate	$Mg(NO_3)_2 \cdot 6H_2O$	$-3\cdot7$
	nitrate	$Mg(NO_3)_2$	$+20\cdot4$
	perchlorate	$Mg(ClO_4)_2$	$+31\cdot9$
	sulphate	$MgSO_4$	$+21\cdot1$
	sulphate	$MgSO_4 \cdot 7H_2O$	$-3\cdot18$
Manganese	acetate	$Mn(C_2H_3O_2)_2$	$+12\cdot2$
	bromide	$MnBr_2$	$+19\cdot4$
	chloride	$MnCl_2$	$+16\cdot0$
	chloride	$MnCl_2 \cdot 4H_2O$	$+1\cdot5$
	fluoride	MnF_2	$+20\cdot2$
	iodide	MnI_2	$+19\cdot7$
	nitrate	$Mn(NO_3)_2$	$+12\cdot7$
	nitrate	$Mn(NO_3)_2 \cdot 6H_2O$	$-6\cdot1$
	sulphate	$MnSO_4$	$+13\cdot8$
	sulphate	$MnSO_4 \cdot 7H_2O$	$-1\cdot7$
Mercuric	acetate	$Hg(C_2H_3O_3)_2$	$-4\cdot2$
	bromide	$HgBr_2$	$-2\cdot4$
	chloride	$HgCl_2$	$-3\cdot3$
Nickel	bromide	$NiBr_2$	$+19\cdot0$
	chloride	$NiCl_2$	$+19\cdot3$
	chloride	$NiCl_2 \cdot 6H_2O$	$-1\cdot15$
	iodate	$Ni(IO_3)_2$	$+0\cdot6$
	iodide	NiI_2	$+19\cdot4$
	nitrate	$Ni(NO_3)_2$	$+11\cdot7$
	nitrate	$Ni(NO_3)_2 \cdot 6H_2O$	$-7\cdot5$
	sulphate	$NiSO_4$	$+15\cdot1$
	sulphate	$NiSO_4 \cdot 7H_2O$	$-4\cdot2$
Potassium	acetate	$KC_2H_3O_2$	$+3\cdot5$
	aluminium sulphate	$K_2Al_2(SO_4)_4$	$+48\cdot5$
	aluminium sulphate	$K_2Al_2(SO_4)_4 \cdot 12H_2O$	$-10\cdot1$
	bicarbonate	$KHCO_3$	$-5\cdot1$
	bisulphate	$KHSO_4$	$-3\cdot1$
	bromate	$KBrO_3$	$-10\cdot1$
	bromide	KBr	$-5\cdot1$
	carbonate	K_2CO_3	$+6\cdot9$
	carbonate	$K_2CO_3 \cdot 1\frac{1}{2}H_2O$	$-0\cdot45$
	chlorate	$KClO_3$	$-10\cdot3$
	chloride	KCl	$-4\cdot4$
	chromate	K_2CrO_4	$-4\cdot8$
	cyanate	$KCNO$	$-5\cdot0$
	cyanide	KCN	$-2\cdot9$
	dichromate	$K_2Cr_2O_7$	$-17\cdot8$
	dihydrogen phosphate	KH_2PO_4	$-4\cdot5$
	ferricyanide	$K_3Fe(CN)_6$	$-14\cdot3$
	ferrocyanide	$K_4Fe(CN)_6 \cdot 3H_2O$	$-11\cdot0$
	fluoride	KF	$+4\cdot2$
	fluoride	$KF \cdot 2H_2O$	$-1\cdot7$
	hydroxide	KOH	$+13\cdot0$
	hydroxide	$KOH \cdot H_2O$	$+3\cdot5$
	hydroxide	$KOH \cdot 1\frac{1}{2}H_2O$	$+2\cdot5$
	iodate	KIO_3	$-6\cdot9$

Table A.14. CONTINUED

	Substance	Formula	Heat of solution, kcal/mol
	iodide	KI	$-5 \cdot 2$
	nitrate	KNO_3	$-8 \cdot 6$
	nitrite	KNO_2	$-3 \cdot 3$
	oxalate	$K_2C_2O_4$	$-4 \cdot 6$
	oxalate	$K_2C_2O_4 \cdot H_2O$	$-7 \cdot 5$
	perchlorate	$KClO_4$	$-13 \cdot 0$
	permanganate	$KMnO_4$	$-10 \cdot 5$
	sulphate	K_2SO_4	$-6 \cdot 3$
	sulphite	K_2SO_3	$+1 \cdot 8$
	sulphite	$K_2SO_3 \cdot H_2O$	$+1 \cdot 4$
	thiocyanate	KCNS	$-6 \cdot 1$
	thiosulphate	$K_2S_2O_3$	$-4 \cdot 5$
Rubidium	bicarbonate	$RbHCO_3$	$-4 \cdot 4$
	bisulphate	$RbHSO_4$	$-3 \cdot 3$
	bromate	$RbBrO_3$	$-11 \cdot 7$
	bromide	RbBr	$-6 \cdot 0$
	chlorate	$RbClO_3$	$-11 \cdot 4$
	chloride	RbCl	$-4 \cdot 2$
	fluoride	RbF	$+6 \cdot 3$
	fluoride	$RbF \cdot H_2O$	$+0 \cdot 10$
	fluoride	$RbF \cdot 1\frac{1}{2}H_2O$	$-0 \cdot 32$
	hydroxide	RbOH	$+15 \cdot 0$
	hydroxide	$RbOH \cdot H_2O$	$+4 \cdot 3$
	hydroxide	$RbOH \cdot 2H_2O$	$-0 \cdot 21$
	iodide	RbI	$-6 \cdot 2$
	nitrate	$RbNO_3$	$-8 \cdot 8$
	perchlorate	$RbClO_4$	$-13 \cdot 6$
	sulphate	Rb_2SO_4	$-6 \cdot 7$
Silver	acetate	$AgC_2H_3O_2$	$-5 \cdot 4$
	chlorate	$AgClO_3$	$-3 \cdot 92$
	fluoride	AgF	$+4 \cdot 85$
	nitrate	$AgNO_3$	$-5 \cdot 4$
	nitrite	$AgNO_2$	$-10 \cdot 5$
	perchlorate	$AgClO_4$	$-1 \cdot 65$
Sodium	acetate	$NaC_2H_3O_2$	$+4 \cdot 1$
	acetate	$NaC_2H_3O_2 \cdot 3H_2O$	$-4 \cdot 7$
	arsenate	Na_3AsO_4	$+31 \cdot 0$
	arsenate	$Na_3AsO_4 \cdot 12H_2O$	$-12 \cdot 6$
	bicarbonate	$NaHCO_3$	$-4 \cdot 2$
	borate (tetra)	$Na_2B_4O_7$	$+10 \cdot 3$
	borate (borax)	$Na_2B_4O_7 \cdot 10H_2O$	$-25 \cdot 8$
	bromate	$NaBrO_3$	$-6 \cdot 4$
	bromide	NaBr	$+0 \cdot 15$
	bromide	$NaBr \cdot 2H_2O$	$-4 \cdot 6$
	carbonate	Na_2CO_3	$+5 \cdot 6$
	carbonate	$Na_2CO_3 \cdot 10H_2O$	$-16 \cdot 2$
	chlorate	$NaClO_3$	$-5 \cdot 4$
	chloride	NaCl	$-0 \cdot 93$
	chlorite	$NaClO_2$	$+1 \cdot 15$
	chlorite	$NaClO_2 \cdot 3H_2O$	$-6 \cdot 8$
	chromate	Na_2CrO_4	$+2 \cdot 4$
	chromate	$Na_2CrO_4 \cdot 10H_2O$	$-15 \cdot 8$
	cyanate	NaCNO	$-4 \cdot 7$

continued

Table A.14. CONTINUED

	Substance	Formula	Heat of solution, kcal/mol
Sodium	cyanide	NaCN	$-0\cdot50$
	cyanide	$NaCN\cdot\frac{1}{2}H_2O$	$-0\cdot79$
	cyanide	$NaCN\cdot2H_2O$	$-4\cdot4$
	dihydrogen phosphate	NaH_2PO_4	$+5\cdot6$
	fluoride	NaF	$-0\cdot06$
	hydroxide	NaOH	$+10\cdot2$
	iodate	$NaIO_3$	$-4\cdot8$
	iodide	NaI	$+1\cdot5$
	iodide	$NaI\cdot2H_2O$	$-3\cdot9$
	nitrate	$NaNO_3$	$-5\cdot0$
	nitrite	$NaNO_2$	$-3\cdot6$
	oxalate	$Na_2C_2O_4$	$-5\cdot5$
	perchlorate	$NaClO_4$	$-3\cdot6$
	perchlorate	$NaClO_4\cdot H_2O$	$-5\cdot4$
	phosphate	Na_3PO_4	$+13\cdot0$
	phosphate	$Na_3PO_4\cdot12H_2O$	$-15\cdot0$
	pyrophosphate	$Na_4P_2O_7$	$+12\cdot0$
	pyrophosphate	$Na_4P_2O_7\cdot10H_2O$	$-12\cdot0$
	sulphate	Na_2SO_4	$+0\cdot28$
	sulphate	$Na_2SO_4\cdot10H_2O$	$-18\cdot7$
	sulphide	Na_2S	$+15\cdot2$
	sulphide	$Na_2S\cdot9H_2O$	$-16\cdot7$
	tartrate	$Na_2C_4H_4O_6$	$-1\cdot1$
	tartrate	$Na_2C_4H_4O_6\cdot2H_2O$	$-5\cdot9$
	thiocyanate	NaCNS	$-1\cdot8$
	thiosulphate	$Na_2S_2O_3$	$+1\cdot8$
	thiosulphate	$Na_2S_2O_3\cdot5H_2O$	$-11\cdot4$
Stannous	bromide	$SnBr_2$	$-1\cdot7$
	chloride	$SnCl_2$	$+0\cdot36$
	iodide	SnI_2	$-5\cdot8$
Strontium	acetate	$Sr(C_2H_3O_2)_2$	$+6\cdot2$
	acetate	$Sr(C_2H_2O_3)_2\cdot\frac{1}{2}H_2O$	$+5\cdot9$
	bromide	$SrBr_2$	$+16\cdot0$
	bromide	$SrBr_2\cdot6H_2O$	$-6\cdot4$
	chloride	$SrCl_2$	$-11\cdot5$
	chloride	$SrCl_2\cdot6H_2O$	$-7\cdot1$
	hydroxide	$Sr(OH)_2$	$+10\cdot3$
	hydroxide	$Sr(OH)_2\cdot8H_2O$	$-14\cdot3$
	iodide	SrI_2	$+20\cdot4$
	iodide	$SrI_2\cdot6H_2O$	$-4\cdot5$
	nitrate	$Sr(NO_3)_2$	$-4\cdot8$
	nitrate	$Sr(NO_3)_2\cdot4H_2O$	$-12\cdot4$
	nitrite	$Sr(NO_2)_2$	$-1\cdot1$
	sulphate	$SrSO_4$	$+2\cdot1$
Thallium	bromate	$TlBrO_3$	$-13\cdot8$
	bromide	TlBr	$-13\cdot7$
	chloride	TlCl	$-10\cdot0$
	hydroxide	TlOH	$-3\cdot2$
	iodide	TlI	$-17\cdot7$
	nitrate	$TlNO_3$	$-10\cdot0$
	sulphate	Tl_2SO_4	$-8\cdot3$
Uranyl	nitrate	$UO_2(NO_3)_2$	$+19\cdot0$
	nitrate	$UO_2(NO_3)_2\cdot6H_2O$	$-5\cdot5$

Table A.14. CONTINUED

Substance		Formula	Heat of solution, kcal/mol
	sulphate	$UO_2SO_4 \cdot 3H_2O$	+5·0
Zinc	acetate	$Zn(C_2H_3O_2)_2$	+9·8
	acetate	$Zn(C_2H_3O_2)_2 \cdot 2H_2O$	+4·0
	bromide	$ZnBr_2$	+15·0
	chloride	$ZnCl_2$	+15·7
	iodide	ZnI_2	+11·6
	nitrate	$Zn(NO_3)_2$	+20·1
	nitrate	$Zn(NO_3)_2 \cdot 6H_2O$	−5·9
	sulphate	$ZnSO_4$	+18·5
	sulphate	$ZnSO_4 \cdot 7H_2O$	−4·3

Table A.15. HEATS OF SOLUTION OF ORGANIC SOLIDS IN WATER AT APPROXIMATELY ROOM TEMPERATURE AND INFINITE DILUTION

A positive value indicates an exothermic, a negative value an endothermic heat of solution

$$[1 \text{ kcal/mol} = 4 \cdot 1868 \text{ kJ/mol}]$$

Substance	Formula	Heat of solution, kcal/mol
Acetamide	CH_3CONH_2	$-2 \cdot 0$
Aconitic acid	$C_3H_3(COOH)_3$	$-4 \cdot 2$
Benzoic acid	C_6H_5COOH	$-6 \cdot 5$
Chloral hydrate	$CCl_3CH(OH)_2$	$-0 \cdot 90$
Chloroacetic acid	$CH_2ClCOOH$	$-3 \cdot 4$
Citric acid	$C_3H_4OH(COOH)_3$	$-5 \cdot 4$
Dextrose (α-D-glucose)	$C_6H_{12}O_6$	$-2 \cdot 6$
Dextrose (hydrate)	$C_6H_{12}O_6 \cdot H_2O$	$-5 \cdot 0$
Fumaric acid	$HOOC \cdot CH : CH \cdot COOH$	$-5 \cdot 9$
Hydroquinone	$C_6H_4(OH)_2$	$-4 \cdot 4$
o-Hydroxybenzoic acid	$C_6H_4OHCOOH$	$-6 \cdot 3$
m-Hydroxybenzoic acid	$C_6H_4OHCOOH$	$-6 \cdot 2$
p-Hydroxybenzoic acid	$C_6H_4OHCOOH$	$-5 \cdot 8$
Itaconic acid	$HOOC \cdot C(:CH_2)CH_2COOH$	$-5 \cdot 9$
Lactose	$C_{12}H_{22}O_{11}$	$+2 \cdot 5$
Lactose (hydrate)	$C_{12}H_{22}O_{11} \cdot H_2O$	$-3 \cdot 7$
Maleic acid	$HOOC \cdot CH : CH \cdot COOH$	$-4 \cdot 4$
Malic acid (DL)	$CH \cdot OH \cdot CH_2(COOH)_2$	$-3 \cdot 3$
Malonic acid	$CH_2(COOH)_2$	$-4 \cdot 5$
Mannitol (D)	$(CH_2OH)_2(CHOH)_4$	$-5 \cdot 3$
p-Nitroaniline	$C_6H_4NH_2NO_2$	$-3 \cdot 7$
o-Nitrophenol	$C_6H_5OHNO_2$	$-6 \cdot 3$
m-Nitrophenol	$C_6H_5OHNO_2$	$-5 \cdot 2$
p-Nitrophenol	$C_6H_5OHNO_2$	$-4 \cdot 5$
Oxalic acid	$(COOH)_2$	$-2 \cdot 3$
Oxalic acid (hydrate)	$(COOH)_2 \cdot 2H_2O$	$-8 \cdot 5$
Oxamic acid	$NH_2CO \cdot COOH$	$-6 \cdot 9$
Phloroglucinol (1,3,5)	$C_6H_3(OH)_3$	$-1 \cdot 7$
Phloroglucinol (hydrate)	$C_6H_3(OH)_3 \cdot H_2O$	$-6 \cdot 7$
Phthalic acid	$C_6H_4(COOH)_2$	$-4 \cdot 8$
Picric acid (2,4,6)	$C_6H_2OH(NO_2)_3$	$-7 \cdot 1$
Pyrocatechol	$C_6H_4(OH)_2$	$-3 \cdot 5$
Pyrogallol (1,2,3)	$C_6H_3(OH)_3$	$-3 \cdot 7$
Raffinose	$C_{18}H_{32}O_{16}$	$+8 \cdot 4$
Raffinose (hydrate)	$C_{18}H_{32}O_{16} \cdot 5H_2O$	$-9 \cdot 7$
Resorcinol	$C_6H_4(OH)_2$	$-3 \cdot 9$
Salicylic acid	$C_6H_4OHCOOH$	$-6 \cdot 3$
Succinic acid	$(CH_2COOH)_2$	$-6 \cdot 4$
Sucrose	$C_{12}H_{22}O_{11}$	$-1 \cdot 3$
Tartaric acid (D)	$(CHOH \cdot COOH)_2$	$-3 \cdot 5$
Tartaric acid (DL)	$(CHOH \cdot COOH)_2$	$-5 \cdot 4$
Thiourea	NH_2CSNH_2	$-5 \cdot 3$
Urea	NH_2CONH_2	$-3 \cdot 6$
Urea nitrate	$NH_2CONH_2 \cdot HNO_3$	$-10 \cdot 8$

Table A.16. HEATS OF FUSION OF ORGANIC SUBSTANCES

$[1 \text{ cal/g} = 4 \cdot 1868 \text{ kJ/kg}]$

Substance	Formula	Melting point, °C	Heat of fusion, cal/g
o-Aminobenzoic acid	$C_6H_4NH_2COOH$	145	35·5
m-Aminobenzoic acid	$C_6H_4NH_2COOH$	180	38·0
p-Aminobenzoic acid	$C_6H_4NH_2COOH$	189	36·5
Anthracene	$C_{14}H_{10}$	217	38·7
Anthraquinone	$(C_6H_4)_2(CO)_2$	282	37·5
Benzene	C_6H_6	5·5	30·1
Benzoic acid	C_6H_5COOH	122	33·9
Benzophenone	$(C_6H_5)_2CO$	48	23·5
t-Butyl alcohol	C_4H_9OH	25	21·9
Cetyl alcohol	$CH_3(CH_2)_{14}CH_2OH$	49	33·8
Cinnamic acid	$C_6H_5CH:CH \cdot COOH$	133	36·5
p-Cresol	$CH_3 \cdot C_6H_4OH$	35	26·3
Cyanamide	NH_2CN	43	49·5
Cyclohexane	C_6H_{12}	6·7	7·57
p-Dibromobenzene	$C_6H_4Br_2$	86	20·6
p-Dichlorobenzene	$C_6H_4Cl_2$	53	29·7
p-Di-iodobenzene	$C_6H_4I_2$	129	16·2
o-Dinitrobenzene	$C_6H_4(NO_2)_2$	117	32·3
m-Dinitrobenzene	$C_6H_4(NO_2)_2$	90	24·7
p-Dinitrobenzene	$C_6H_4(NO_2)_2$	174	40·0
Diphenyl	$C_{12}H_{10}$	69	28·8
Ethyl alcohol	C_2H_5OH	−114	25·8
Hexamethylbenzene	$C_{12}H_{18}$	166	30·5
Hydroquinone	$C_6H_4(OH)_2$	172	58·8
Methyl alcohol	CH_3OH	−98	23·7
Naphthalene	$C_{10}H_8$	80	35·6
α-Naphthol	$C_{10}H_7OH$	95	38·9
β-Naphthol	$C_{10}H_7OH$	121	31·3
o-Nitroaniline	$C_6H_4NH_2NO_2$	70	27·9
m-Nitroaniline	$C_6H_4NH_2NO_2$	113	41·0
p-Nitroaniline	$C_6H_4NH_2NO_2$	147	36·5
Nitrobenzene	$C_6H_5NO_2$	6	22·5
Phenanthrene	$C_{14}H_{10}$	97	24·5
Phenol	C_6H_5OH	41	29·0
Pyrocatechol	$C_6H_4(OH)_2$	104	49·4
Resorcinol	$C_6H_4(OH)_2$	110	46·2
Stearic acid	$C_{17}H_{35}COOH$	68	47·6
Stilbene	$C_{14}H_{12}$	124	40·0
Toluene	$C_6H_5CH_3$	−95	17·2
o-Toluic acid	$CH_3C_6H_4COOH$	104	35·4
m-Toluic acid	$CH_3C_6H_4COOH$	109	27·6
p-Toluic acid	$CH_3C_6H_4COOH$	180	40·0
p-Toluidine	$CH_3C_6H_4NH_2$	43	39·9
Thymol	$C_{10}H_{13}OH$	50	27·5
o-Xylene	$C_6H_4(CH_3)_2$	−25	30·6
m-Xylene	$C_6H_4(CH_3)_2$	−48	26·1
p-Xylene	$C_6H_4(CH_3)_2$	13	38·5

Table A.17. COMMON NAMES FOR SOME CRYSTALLINE SUBSTANCES

Common name	Chemical name	Formula
ADP	ammonium dihydrogen phosphate	$NH_4H_2PO_4$
Alum	$M_2^IM_2^{III}(SO_4)_4 \cdot 24H_2O$	e.g., $K_2SO_4 \cdot Al_2(SO_4)_3 \cdot 24H_2O$
Alumina	aluminium oxide	Al_2O_3
Anhydrite	anhydrous calcium sulphate	$CaSO_4$
Antifebrin	acetanilide	$C_6H_5 \cdot NH \cdot COCH_3$
Aspirin	acetylsalicylic acid	$C_6H_4 \cdot OCOCH_3 \cdot COOH$
Baryta	barium oxide	BaO
Bayer hydrate	alumina trihydrate	$Al_2O_3 \cdot 3H_2O$
Beet sugar	sucrose	$C_{12}H_{22}O_{11}$
Betol	β-naphthylsalicylate	$C_6H_4 \cdot OH \cdot COOC_{10}H_7$
Biuret	ureidoformamide	$NH(CONH_2)_2$
Blance fixe	precipitated barium sulphate	$BaSO_4$
Blood sugar	dextrose (α-D-glucose)	$C_6H_{12}O_6 \cdot H_2O$
Blue vitriol	copper sulphate	$CuSO_4 \cdot 5H_2O$
Boracic acid	boric acid	H_3BO_3
Borax	sodium tetraborate	$Na_2B_4O_7 \cdot 10H_2O$
Calamine	zinc carbonate	$ZnCO_3$
Calomel	mercuric chloride	$HgCl$
Cane sugar	sucrose	$C_{12}H_{22}O_{11}$
Carbolic acid	phenol	C_6H_5OH
Carnallite	magnesium potassium chloride	$MgCl_2 \cdot KCl \cdot 6H_2O$
Caustic potash	potassium hydroxide	KOH
Caustic soda	sodium hydroxide	$NaOH$
Caustic, lunar	silver nitrate	$AgNO_3$
Ceruse	basic lead carbonate	$2PbCO_3 \cdot Pb(OH)_2$
Chile saltpetre	sodium nitrate	$NaNO_3$
Chrome alum	$M_2^ICr_2(SO_4)_4 \cdot 24H_2O$	e.g., $K_2SO_4 \cdot Cr_2(SO_4)_3 \cdot 24H_2O$
Chrome green	chromium oxide	Cr_2O_3
Chrome red	basic lead chromate	$PbCrO_4 \cdot PbO$
Chrome yellow	lead chromate	$PbCrO_4$
Common salt	sodium chloride	$NaCl$
Copperas	ferrous sulphate	$FeSO_4 \cdot 7H_2O$
Corn sugar	dextrose (α-D-glucose)	$C_6H_{12}O_6 \cdot H_2O$
Corrosive sublimate	mercuric chloride	$HgCl_2$
Cream of tartar	potassium hydrogen tartrate	$KHC_4H_4O_6$
DAP	diammonium phosphate	$(NH_4)_2HPO_4$
Dextrose	α-D-glucose	$C_6H_{12}O_6 \cdot H_2O$
EDT	ethylenediamine tartrate	$(CH_2 \cdot NH_2)_2(C_4H_6O_6) \cdot H_2O$
Epsom salts	magnesium sulphate	$MgSO_4 \cdot 7H_2O$
Flowers of sulphur	sulphur	S
Fruit sugar	fructose	$C_6H_{12}O_6$
Glaserite	potassium sodium sulphate	$3K_2SO_4 \cdot Na_2SO_4$
Glauber's salt	sodium sulphate	$Na_2SO_4 \cdot 10H_2O$
Glycine	amino-acetic acid	$NH_2 \cdot CH_2 \cdot COOH$
Grape sugar	dextrose (α-D-glucose)	$C_6H_{12}O_6 \cdot H_2O$

Table A.17 CONTINUED

Common name	Chemical name	Formula
Green vitriol	ferrous sulphate	$FeSO_4 \cdot 7H_2O$
Gypsum	calcium sulphate	$CaSO_4 \cdot 2H_2O$
Halite	sodium chloride	$NaCl$
Hexamine	hexamethylenetetramine	$(CH_2)_6N_4$
Horn silver	silver chloride	$AgCl$
Hypo	sodium thiosulphate	$Na_2S_2O_3 \cdot 5H_2O$
Iron alum	$M_2Fe_2(SO_4)_4 \cdot 24H_2O$	e.g., $K_2SO_4 \cdot Fe_2(SO_4)_3 \cdot 24H_2O$
Iron black	precipitated antimony	Sb
KDP	potassium dihydrogen phosphate	KH_2PO_4
Lemon chrome	barium chromate	$BaCrO_4$
Litharge	red-brown lead monoxide	PbO
Liver of sulphur	potassium bisulphide	KHS
Lunar caustic	silver nitrate	$AgNO_3$
MAP	monoammonium phosphate	$NH_4H_2PO_4$
Massicot	yellow lead monoxide	PbO
Microcosmic salt	sodium ammonium hydrogen phosphate	$NaNH_4HPO_4 \cdot 4H_2O$
Milk sugar	lactose	$C_{12}H_{22}O_{11}$
Mohr's salt	ferrous ammonium sulphate	$FeSO_4 \cdot (NH_4)_2SO_4 \cdot 6H_2O$
Mosaic gold	stannic sulphide	SnS_2
Muriate of potash	potassium chloride	KCl
Nickel vitriol	nickel sulphate	$NiSO_4 \cdot 7H_2O$
Nitre	potassium nitrate	KNO_3
Paris yellow	lead chromate	$PbCrO_4$
Pearl ash	potassium carbonate	K_2CO_3
Plaster of Paris	calcium sulphate	$CaSO_4 \cdot \frac{1}{2}H_2O$
Potash alum	potassium aluminium sulphate	$K_2SO_4 \cdot Al_2(SO_4)_3 \cdot 24H_2O$
Prussiate of potash (yellow)	potassium ferrocyanide	$K_4Fe(CN)_6 \cdot 3H_2O$
Prussiate of potash (red)	potassium ferricyanide	$K_3Fe(CN)_6$
Quinol	hydroquinone (p)	$C_6H_4(OH)_2$
Racemic acid	DL-tartaric acid	$(CHOH \cdot COOH)_2$
Red precipitate	mercuric oxide	HgO
Rochelle salt	sodium potassium tartrate	$NaKC_4H_4O_6 \cdot 4H_2O$
Rock crystal	quartz	SiO_2
Rock salt	sodium chloride	$NaCl$
Sal-ammoniac	ammonium chloride	NH_4Cl
Salol	phenyl salicylate	$C_6H_4 \cdot OH \cdot COOC_6H_5$
Salt cake	sodium sulphate	Na_2SO_4
Salt of tartar	potassium carbonate	K_2CO_3
Saltpetre	potassium nitrate	KNO_3
Salts of lemon	potassium hydrogen oxalate or potassium tetroxalate	$KHC_2O_4 \cdot H_2O$ or $KHC_2O_4 \cdot H_2C_2O_4 \cdot 2H_2O$
Salts of sorrel	potassium hydrogen oxalate	$KHC_2O_4 \cdot H_2O$

continued

Table A.17 CONTINUED

Common name	Chemical name	Formula
Scheele's green	hydrogen cupric arsenite	$HCuAsO_3$
Schlippe's salt	sodium thioantimonate	$Na_3SbS_4 \cdot 9H_2O$
Schoenite	magnesium potassium sulphate	$MgSO_4 \cdot K_2SO_4 \cdot 6H_2O$
Seignette salt	sodium potassium tartrate	$NaKC_4H_4O_6 \cdot 4H_2O$
Silver salt	sodium anthraquinone β-sulphonate	$C_{14}H_7O_2 \cdot SO_3Na$
Soda ash	anhydrous sodium carbonate	Na_2CO_3
Soluble tartar	potassium tartrate	$K_2C_4H_4O_6 \cdot \frac{1}{2}H_2O$
Spinel	$M^{II}O \cdot M_2^{III}O_3$	e.g., $MgO \cdot Al_2O_3$
Sugar	sucrose	$C_{12}H_{22}O_{11}$
Sugar of lead	lead acetate	$Pb(CH_3COO)_2 \cdot 3H_2O$
Tartar emetic	potassium antimonyl tartrate	$KSbO \cdot C_4H_4O_6 \cdot \frac{1}{2}H_2O$
Tetralin	tetrahydronaphthalene	$C_{10}H_{12}$
TGS	triglycine sulphate	$(NH_2CH_2COOH)_3H_2SO_4$
Tin crystals	stannous chloride	$SnCl_2$
Trona	natural sodium sesquicarbonate	$Na_2CO_3 \cdot NaHCO_3 \cdot 2H_2O$
Tutton salts	$M_2^I M^{II}(SO_4)_2 \cdot 6H_2O$	e.g., $NiSO_4 \cdot (NH_4)_2SO_4 \cdot 6H_2O$
Washing soda	sodium carbonate	$Na_2CO_3 \cdot 10H_2O$
White arsenic	arsenious oxide	As_2O_3
White lead	basic lead carbonate	$2PbCO_3 \cdot Pb(OH)_2$
White precipitate	mercuric ammonium chloride	$HgNH_2Cl$
White vitriol	zinc sulphate	$ZnSO_4 \cdot 7H_2O$
Whiting	calcium carbonate	$CaCO_3$

Author index

Abegg, C. F., 50, 346, 370
Acrivos, A., 98
Adams, J. R., 299
Adamski, T., 178, 229
Albertins, R., 298
Alexander, A. E., 98
Allen, T., 396, 415
Albon, N., 158, 227
Amatavivadhana, A., 197, 229, 230
Amin, A. B., 346, 370
Anikin, A. G., 298
Armstrong, A. J., 333
Armstrong, R. M., 333
Arnold, P. M., 275
Arrhenius, S., 142, 155, 195
Aston, J. G., 242, 257
Ayerst, R. P., 229, 230

Badger, W. L., 335, 369
Bailey, A. E., 135, 241, 257
Bain, J., 196, 230
Balakrishnan, N. S., 350, 370
Bamforth, A. W., 266, 267, 297, 302, 309, 319, 326, 330, 331, 333
Barduhn, A. J., 298
Barrer, R. M., 252, 257
Bass, S. J., 257
Bates, O. K., 82, 99
Baum, S. J., 231
Becker, G. W., 346, 370
Becker, R. von, 138, 145, 226
Beezer, A. E., 135
Behnken, D. W., 339, 370
Behre, H. A., 373, 415
Belyustin, A. V., 181, 214, 229, 232
Bennema, P., 158, 173, 227, 228, 230
Bennett, J. A. R., 227
Bennett, M. C., 230
Bennett, R. C., 325, 333, 350, 365, 370, 371
Berg, W. F., 164, 215, 228
Berkeley, Earl of, 228
Berthoud, A., 161, 227
Bernal, J. D., 50
Berry, C. R., 224, 232

Betts, W. D., 298
Bhatia, V. K., 416
Bienfait, M., 173, 228, 230
Bigg, E. K., 228
Bingham, E. C., 62
Blackadder, D. A., 225
Blasdale, W. C., 135
Blatchly, J. M., 212, 231
Blecharczyk, S. S., 265, 297
Blitznakov, G., 198, 230
Boistelle, R., 230
Bolsaitis, P., 298
Bose, E., 2
Bosňjakovič, F., 91
Botsaris, G. D., 198, 229, 230
Bovington, C. H., 228
Bowden, F. P., 99
Bowden, S. T., 135
Bowen, H. J. M., 27
Brandes, H., 154, 227
Bransom, S. H., 192, 229, 230, 370, 338–339, 349, 369
Bragg, W. L., 27
Bravais, A., 14, 15, 170
Brice, J. C., 165, 170, 225, 228
Bricknell, W. S., 257
Brodie, J. A., 280, 298
Brooks, R., 214, 232
Brown, D. E., 229
Bruins, P. F., 186, 229
Bruno, A. J., 232
Bryant, H. S., 298
Buckley, H. E., 27, 152, 209
Bujac, P. D. B., 230, 231
Bunn, C. W., 27, 197, 215, 228, 230
Burac, N., 299
Burton, W. K., 157, 227

Cabrera, N., 157, 158, 159, 199, 227
Cahn, J. W., 151, 227
Calderbank, P., 231
Caldwell, H. B., 331, 333
Callis, C. F., 299
Campbell, A. N., 135

465

Wilcox, W. R., 215, 226, 232, 257, 298
Wilke, C. R., 68, 98
Willard, H. H., 216, 232
Wise, W. S., 43, 56
Wolff, P. R., 229, 345, 370
Wollaston, W. H., 3
Wood, J. T., 242, 257
Wulff, G., 152, 227
Wyckoff, R. W. G., 27
Wynne, E. A., 247, 257

Young, S. W., 228

Zaki, W. N., 412, 416
Zdansky, A. E., 310
Zenz, F. A., 416
Zettlemoyer, A. C., 226
Zief, M., 226, 257, 298
Zimmerman, H. K., 42, 56
Zweitering, Th. N., 407, 416

Subject index